The Construction Industry in the Fourth Industrial Revolution

Clinton Aigbavboa · Wellington Thwala
Editors

The Construction Industry in the Fourth Industrial Revolution

Proceedings of 11th Construction Industry Development Board (CIDB) Postgraduate Research Conference

Springer

UNIVERSITY OF JOHANNESBURG

cidb
Construction Industry Development Board
DEVELOPMENT THROUGH PARTNERSHIP

Editors
Clinton Aigbavboa
Construction management
and Quantity Surveying
University of Johannesburg
Johannesburg, Gauteng
South Africa

Wellington Thwala
Construction management
and Quantity Surveying
University of Johannesburg
Johannesburg, Gauteng
South Africa

ISBN 978-3-030-26527-4 ISBN 978-3-030-26528-1 (eBook)
https://doi.org/10.1007/978-3-030-26528-1

This Springer imprint is published by the registered company Springer Nature Switzerland AG
The registered company address is: Gewerbestrasse 11, 6330 Cham, Switzerland

Contents

A Systems Thinking Approach to Construction Project Management

Joe Frederick Cobbinah[1(✉)], Clinton Ohis Aigbavboa[1], Wellington Didibhuku Thwala[1], and Kwamina Ewur Banson[2]

[1] Department of Engineering Management, University of Johannesburg, Johannesburg, South Africa
petsdf@yahoo.com

[2] Department of Technology Transfer Centre (TTC), Biotechnology and Nuclear Agriculture Research Institute (BNARI), Ghana Atomic Energy Commission (GAEC), Legon-Accra, Ghana

Abstract. Construction projects plays an important role in nation's development. The complexity and dynamic environment of the construction industry characterized it frequent cost, time overruns, low quality, etc. for construction project management. This study therefore uses a systems thinking approach to identify the key variables that are affecting this trend. Casual loop and Systems Archetypes were used to develop systems models to determine the components and interactions between the policy and the social, environmental and economic dimensions of the industry. As demonstrated in the aforementioned CLD and archetypes, construction projects will remain to be the engine that develops and empowers the emerging and existing commercial business, housing sectors and nation building across Africa. This systems thinking approach will also provide more clarity in dealing with the complex management challenges and gradually replace the traditional theoretical approach of dealing with construction project management.

Keywords: Construction projects · Project management · Systems thinking · Technology · Contractors

1 Introduction

The growing complexity in the field of construction projects poses substantial management challenges [1, 2]. Each construction project is unique in nature and operates in complex environment which requires its own sets of managerial techniques [3]. These managerial techniques span long time intervals, in some cases decades or more [4]. Large investments, at times billions of dollars, are involved. Debrah and Ofori [5] and Yip, Rowlinson [6] assert that construction project management are complex because of the extensive use of sophisticated plant, equipment, modern methods of construction, and multi-disciplinary and multi-tasked aspects of its team and workforce. It was further stated that these complexities are very high and could lead to cost escalation, delays, and technical problems which can undermine the financial feasibility of the project, jeopardize it completion, threatens the solvency of stakeholders, and sometimes provoke legal disputes.

C. Aigbavboa and W. Thwala (Eds.): CIDB 2019, *The Construction Industry in the Fourth Industrial Revolution*, pp. 1–11, 2020.
https://doi.org/10.1007/978-3-030-26528-1_1

Despite advances in project management methodologies many projects continue to fail. Modern project management methods have evolved tremendously over the years to correct this undesirable situation but with limited success [7]. Introduction of production management tools along with project management tool: critical path method (CPM) and a number of ingenious cost management systems have and in the best cases only resulted in marginal improvements [8]. These hitches and challenges cannot be solved with the same level of thinking that created them and tools of the past [9]. The solution requires moving away from tradition linear approached to systems thinking holistic approach which integrates social, environment, economic, and politics for the control of construction project to achieve the sustainable outcomes desired. The key premise of systems design thinking is that people are an integral part of the system alongside technology, social, environment, economic and politics and should be regarded as such through the entire design process [10, 11]. Concerns have been raised as to the relevance of the traditional construction project management tools and techniques to the management of modern construction projects, hence a proposal has been put forward for an alternative theoretical approach by academicians and practitioners from both developed and developing countries [12]. This study will fill this gap by applying a systems thinking approach as an alternative theoretical tool for construction project management. Systems thinking application which is a new approaches to construction project management allows for more effective selection of key measures, a less arbitrary process for setting objectives, and most importantly, creating a true model of the business that can be tested and used to support improved decision making. The research used casual loop diagrams and systems archetypes as key tools. Finding of this research will add to the knowledge and understanding of the subject of construction project complexity management and its application by project managers to enhance performance in the construction industry.

1.1 Systems Thinking

The systems thinking framework view 'problems' as fragments of a complete system and addresses the root causes of challenges thus, transdisciplinary approach [13]. Currently, the words "sustainability" or "green" is the key word of many researchers and is applied to anything from ground preparation to consumption [14]. Overcoming today's problems requires a shift from a "traditional" way of thinking to a "systems" viewpoint that organizes thought and behaviour in line with the natural acts of sustainability [13, 15]. Systems thinking is a "framework for seeing inter-connectedness relationships rather than events, for seeing patterns of alteration rather than static portraits" [16, 17]. Systems thinking fosters new ways and levels of rational and enables people to sustain and manage states of complexity and uncertainty which have no simple answers. The construction industry and its sustainable management might profit from a systems thinking approach to situational interventions and capacity development, to address problematic issues holistically and recommend the desirable sustainable results.

1.2 Systems Archetypes

A well-defined structure, that unveils distinct behavior over time chart, and has a defined strategies for dealing with the underlying organization of a system being studied is called system archetype [18]. System archetypes do not specifically describe any one problem at a time but describe a spiral of problems generally [18]. Their usefulness comes from the discernments they offer from the vigorous relations of complex systems components. This framework reveals one's understanding of a specific system's organization and behaviour, which nurture communication and identification of high leverage points for problematic complex structure.

1.3 Causal Loop Diagrams

A causal loop diagram (CLD) is a causal diagram that aids in visualising how different component in a system interacts and it is the first-step to modelling in systems thinking to convert complex structure into an easy understandable structure [19]. They are connected by key causal relationships to represent realism used to exhibit the behavior of cause and effect from a system position [20]. Arrows connect variables in Causal loop diagrams (CLDs) with polarities such as 'negative' '−' and 'positive' '+' signs and delays '||' to describe the causal relations [13, 17, 21]. The cause and effect that take on a life of their own are represented by a feedback loops.

2 Research Methods

National data sources includes qualitative and quantitative data from participants' project records, documents (media reports, publications, policy documents, previous studies and statistical records). Stakeholder's workshop and focus group discussions were also conducted to collect data. Structured and open ended questionnaires were used to direct discussions held with stakeholders individually and in groups. All the data obtained from the study areas made used of the four levels of thinking model (Fig. 1) to provide an overview of the current state and impact of management strategies in the construction industry ripple by multiple feedback loops. Data collection began by assembling the mental models of all stakeholders involved in the construction industry of Ghana in the workshop. They also analyse the systems' barriers and drivers to construction management.

The model which starts at the 'first level of thinking', involves a series of literature reviews and interviews with experts in the field. Data on stakeholder opinions and perceptions of how the construction project system works, the barriers to success, the system drivers and the possible strategies (solutions) to overcome these problems were obtained through focus group discussions with a group of experts (e.g., construction managers and site managers or foremans and labourers) to obtain their mental models of the construction sector under deliberation. Using the CLD modelling approach with the VENSIM Software program (Ventana Systems UK), variables were processed into CLD loops concerning the issues under consideration. When unintended consequences and potential barriers have been identified, results are interpreted and patterns explored

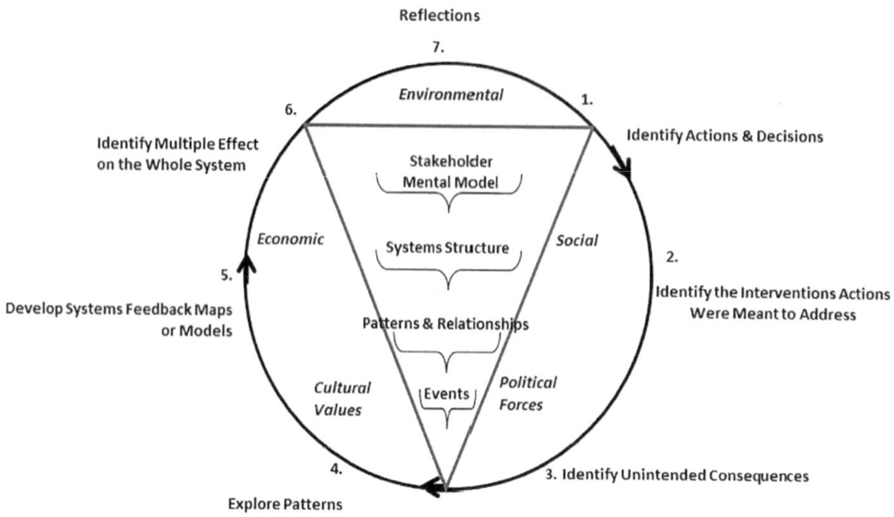

Fig. 1. Feedback learning laboratory [19].

using the 'second level of thinking'. Also the kind of feedback loops, reinforcing loops and balancing loops that were generated are analsed.

Step three precedes, 'third level of thinking', in which pre-existing system archetypes influencing the construction industry are identified. Finally, the 'fourth level of thinking' emphasises the strategies needed to overcome the problems and challenges fronting the entire complex construction system. The research process used concentrated on the use of the Evolutionary Learning Laboratory (ELLab) as knowledge integration tools in sustainable management and policymaking [22]. The ELLab demonstrations a new way of 'thinking' beyond the traditional 'linear' approach to help equip policy makers, researchers and all relevant stakeholders of solving problems. This framework is useful for prioritisation of management interventions and policy support.

3 Results and Discussion

3.1 The Mental Model of the Construction Industry

The relationships of their views or mental model are drawn out in a causal loop diagram (CLD) in Fig. 2. In this diagram, successful project delivery is in direct conflict with client payment delay which is in opposite direction to project funding and "Late payment of salaries or wages" which in turn demotivate workers leading to delay of contract and theft as shown in the Fig. 2 (B2, R4, B3). Delay of contract is in the same direction with inflationary effect which leads to price increases leading to high project cost and this in turn affect project design to suit current budget. This results in project variation which can affect construction project R&D.

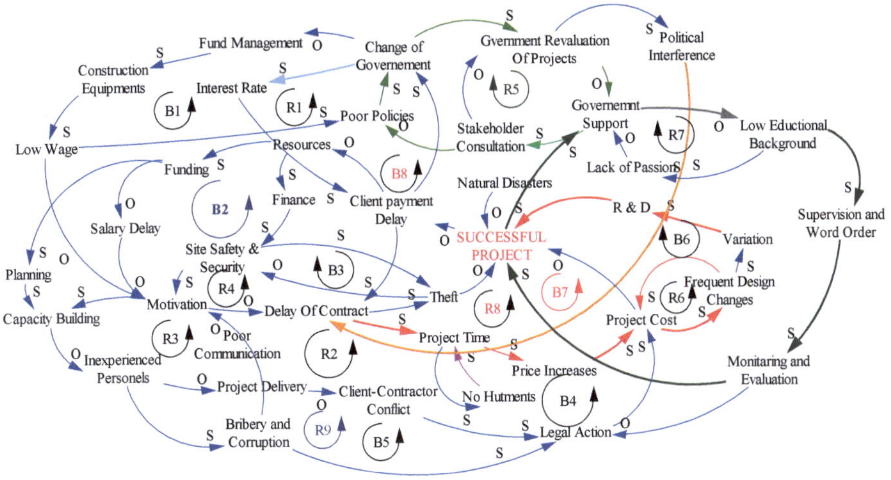

Fig. 2. CLD of the construction industry

The Casual Loop Diagram (CLD) of Fig. 2 gave rise the preceding underlining systems archetypes which disentangle the complexity of the construction industry. Systems archetypes have proven to be effective in unraveling complexity and dynamic forces within a complex system [22–24].

3.2 Escalating Archetype

At the heart of an escalation dynamic are two (or more) parties, each of whom feels threatened by the actions of the other as shown in Fig. 3. Each side attempts to keep things under control by managing its own balancing process. Actions taken by contractor A, for example, improve contractor A's result by submitting lower bid relative to contractor B. This decreases contractor A's feeling of threat, so contractor A eases off its activities (B 1) as depicted by the Escalating Archetype.

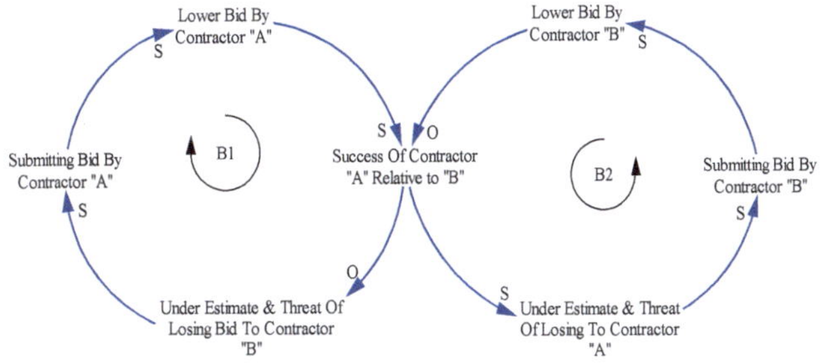

Fig. 3. Escalating archetype

Contractor B, on the other hand, now feels threatened by contractor A's relative advantage and increases its activities by submitting lower bid in order to improve its result over contractor A (B2). The interaction of the two contractors trying to unilaterally maintain control produces a reinforcing spiral in which nobody feels in control.

One of the reasons we get caught in escalation dynamics may stem from our view of competition. This archetype suggests that cutthroat competition serves no one well in the long run. The archetype provides a way to identify escalation structures at work and shows how to break out of them or avoid them altogether. The purpose of bidding is of course to enhance competition and transparency. The drawback is that it leads to under estimate of project cost and hence results in shoddy work and conflict.

Effective Strategies. Nothing comes without a cost and there is a limit to the results that one can generate. In this archetype, "contractor A" and "contractor B" are on a mutually destructive course. A point must be reached where the structure can grow no more because nothing grows forever. There are two effective strategies for dealing with this structure. One approach is to disconnect the two loops so "contractor A" and "contractor B" are no longer competing with each other but competing with themselves. This produces two reinforcing loops. The second approach is to begin evaluating the composite of "contractor A" and "contractor B's" actions rather than their individual actions. In this way, they begin to see the value of cooperation rather than competition and the structure turns into two synergistic reinforcing loops.

3.3 Fixes that Fails Archetype

The Fixes That Fail structure consists of a balancing loop and a reinforcing loop as shown in Figs. 4 and 5. These two loops interact in such a way that the desired result initially produced by the balancing loop is, after some delay, offset by the actions of the reinforcing loop. A paradox and I quote; "why don't people have time to do things right in the first place, but have time to fix them over and over again? is posed.

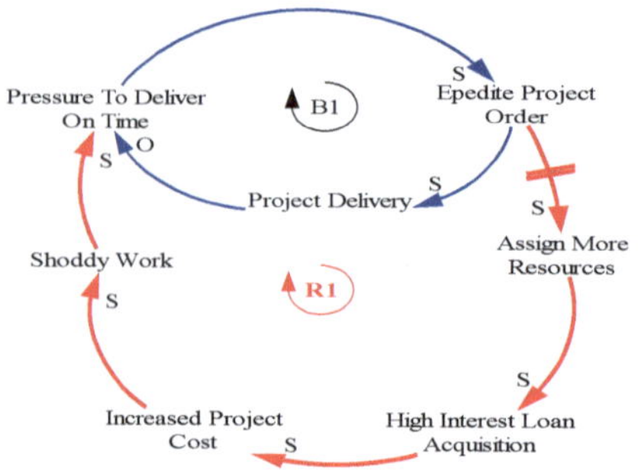

Fig. 4. Fixes that fails archetype A

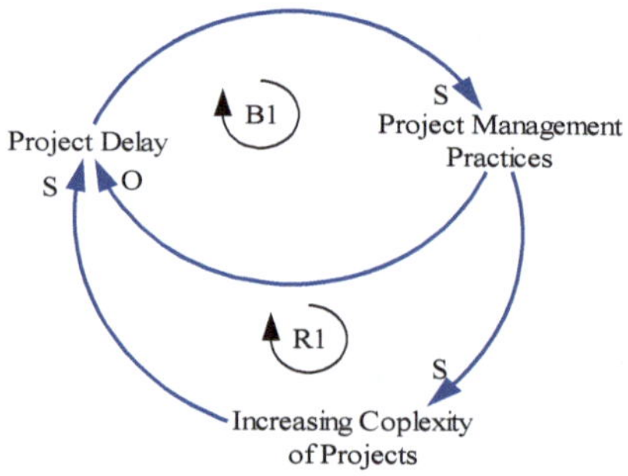

Fig. 5. Fixes that fails archetype B

The internal balancing loop operates in the standard balancing loop fashion. The Action to expedite project order in Fig. 4 increase pressure to deliver project on time which leads to assigning more resources which may be obtained at high interest rate increasing project cost and may lead to some Unintended Consequences such as shoddy works. These Unintended Consequences subsequently impede the objective of expediting project order in the intended direction. The fixes that fail archetype highlights how one can get caught up in a dynamic that reinforces the need to continually implement quick fixes to address the problem. In Fig. 5, a problem symptom gets bad enough that it captures attention: for example, project management practices, leads to increasing complexity of projects (quick fix) that makes the symptoms go away (project delay). However, that action initiates unintended consequences which make the initial symptoms resurface with time which is normally worse than previous. Consequently, restraining business enlargement and progress.

Effective Strategies. A "fixes that fail" structure is often part of a more elaborate structure in which the fixes that fail structure simply represents dealing with the symptoms rather than the root cause underlying the real problem. A "fixes that fail" structure often results in becoming dependent on the fix, thus applying it over and over. The most effective strategy for dealing with this structure is advance planning. Since one can never do just one thing, as everything affects everything else, before taking action to change the current state, one has to think about what else that action is apt to affect. And, what the impact will be. Sometimes the unexpected consequences may be several affects away, so one be able to forecast. Fundamentally, identify the unanticipated consequently, which makes it no longer unexpected than before. An alternative strategy is to disengage the unexpected consequence from affecting the current state in time, it wouldn't be a consequence any longer.

3.4 Limit to Growth or Success

Figure 6, reveals the limit to growth archetype. In this, certain actions such as fund management initially lead to successful project delivery and progression, inspiring the contractor to continue in more of those similar efforts. Later, nonetheless, the system bump into limit that slow progressions. As progression deteriorations, the contractor tends to apply the actions that originally led to progression. Marketing is a common engine of growth in the Limit to Growth Archetype for the construction industry. Contractors assign a convinced fraction of the budget to marketing and participate in numerous marketing deeds that can attract and awards contracts. More contracts lead to increased revenue and a corresponding rise in fund management (R1). However, as contract awards grow, the client base naturally expands, along with following up calls to build customer relationship marketing and minimize contractor client conflict (B1). If the company fund management support capacity does not increase fast enough to meet the growing demand, the adequacy of that capacity begins to drop. This trend leads to lower customer satisfaction and downwards pressure on fund management. Contractors begin to seek for substitute occupation or decrease capital in the current contract for other segments.

Effective Strategies. The best defence is a good offence. As defined in the effective strategies for the Reinforcing Loop, if there is a Reinforcing Loop operating start looking for what is going to become a **limiting factor**, and remove it before it even has a chance to create a substantial impact on **results**.

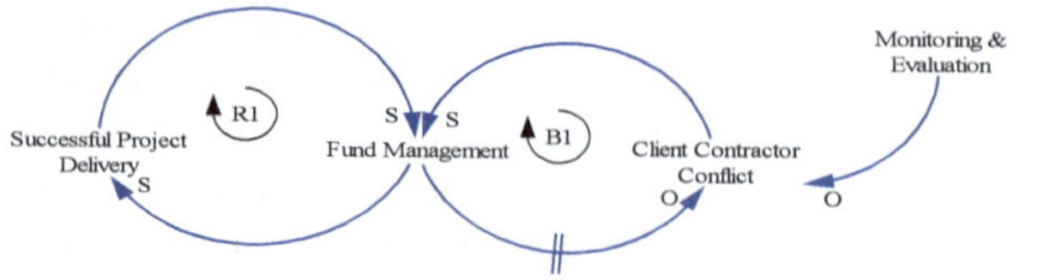

Fig. 6. Limit to growth archetype

If the structure is already at a stage where the **limiting factor** is interacting with **results** to limit them the options are the following: 1. Modify the **limiting factor** to the extent that it no longer intermingles with the **results** which creates a reducing action, 2. Find a way to separate the **results** from the **reducing action** so that, it no longer occurs. 3. Disengage the **reducing action** from the **consequences** so it can have no effect on **results**.

3.5 Shifting the Burden Archetype

In the 'shifting the burden' shown in Fig. 7, most self-depended contractors opt to depend upon private consultations rather than stakeholder consultation to mobilize adequate information during crisis, which in long run, was unable to solve the problem of delay crisis (Fig. 7). Very often, leading to shoddy works which attracts government intervention for project revaluation. The construction industry needs a radical policy reform to support government's agenda of "trade beyond aid". Currently, the industry is faced with a lot of challenges including; Delayed payment leading to cash flow challenges that threaten the sustainability of construction firms and banks; Lack of capacity of local contractors to deliver big projects and quality; Unfavourable competition from foreign players; and Unwillingness of banks to advance credits due to high levels of unserviced debt [25].

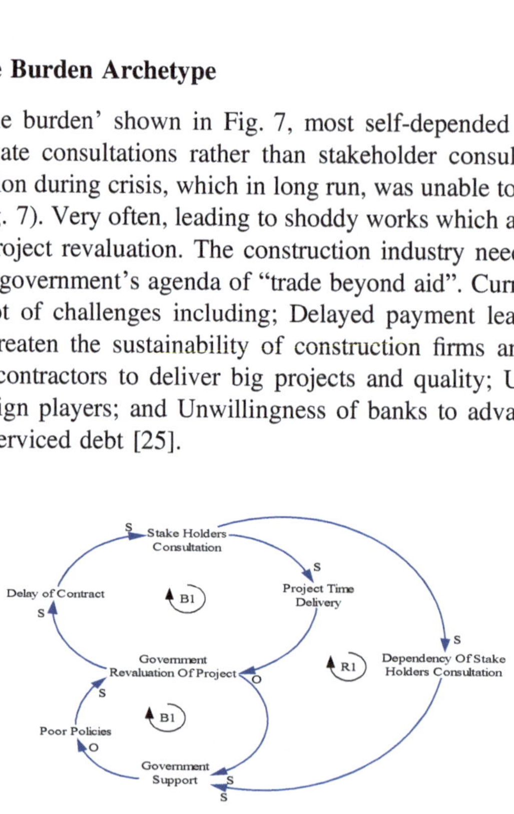

Fig. 7. Shifting the burden archetype

Effective Strategies. When dealing with the delay of contract problems, a manager must ask himself whether he is avoiding future revaluation of projects by the Government or addressing the root cause of the problem. Regularly, out of convenience, self-depended or private consultation as a answer is vital. Dealing with a Shifting the Burden structure, the most effective strategy is by obtaining sufficient information through stakeholder discussion of that business. Hence, one resolves the pertaining problem and make sure that it do not reoccur.

3.6 Success to Successful Archetype

The demand made by domestic contractors in comparison to foreign contractors for a common resource such as funds, labour, etc. are linked by two reinforcing loops as shown in the Success to successful Archetype in Fig. 8. As the resources dedicated to 'local contractors' and 'local contractor's success both increase, the resources invested in 'foreign contractors' and in turn 'foreign contractors' success decline. Thus, 'foreign contractor's net returns stay low or negative longer than 'local contractors', and 'foreign contractors' begins to look less attractive as an alternative. Thus, a lesser amount is invested in 'foreign contractors', which delays 'foreign contractors' attainment of success even further. Eventually, resources, such as people, equipment, etc. is even taken away from 'foreign contractors' because government don't want to waste resource on "lost cause". In turn, 'foreign contractors' performance only drops further. Finally, 'foreign contractors' will be considered a undependable contractors and abandoned.

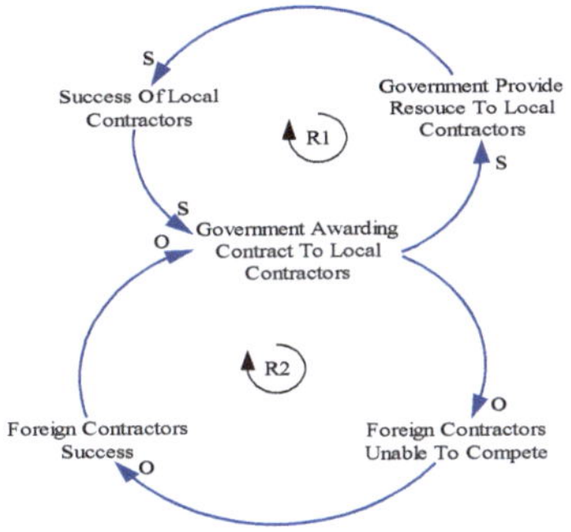

Fig. 8. Success to successful archetype

Effective Strategies. There are two strategies for dealing with a Success to the Successful situation; 1. Identify the resource(s) being unequally distributed and balance the distribution. 2. Disconnect the two reinforcing structures so they are not dependent on the allocation of shared resource(s).

4 Conclusion

To many construction project managers, a complex project seems to be a labyrinth with many hidden dangers. This research presents a systems-based approach to construction project management that can facilitate a greater understanding of the complexity inherent in construction projects and how that complexity can be effectively managed. The systems approach permits the onsite construction project manager to take a complex construction project, holistically manage it and ensure that all systems are in alignment with the original goal of the project. This approach combines industrial engineering, project management, and finance into a unified approach for effective management of complex construction projects, ranging from a power plant to a highway project. The system archetypes revealed insights into the complexity in construction projects that previously exist to antedate potential problems and its symptoms. With the archetypes, it has been recognized that superior traditional or linear answers do not solve the problems, but systemic approach to the problems could lead to the delivery of the accurate management policies.

Acknowledgements. The authors would like to express a great gratitude to all the respondents and construction stakeholders of Ghana and Tarkoradi Technical University students of the faculty of Built for their time, willingness and contributions to this study.

References

1. Hartlieb, S., Silvius, G.: Dealing with uncertainty in projects: what project management can learn from business development. Proj. Manag. Dev.–Pract. Perspect. 141 (2016)
2. Mir, F.A., Pinnington, A.H.: Exploring the value of project management: linking project management performance and project success. Int. J. Proj. Manage. **32**(2), 202–217 (2014)
3. de Blois, M., Lizarralde, G., De Coninck, P.: Iterative project processes within temporary multi-organizations in construction: the self-, eco-, re-organizing projects. Proj. Manage. J. **47**(1), 27–44 (2016)
4. Weil, H.B., Dalton, W.J.: Risk management in complex projects. Syst. Dyn. (1992)
5. Debrah, Y.A., Ofori, G.: The State, skill formation and productivity enhancement in the construction industry: the case of Singapore. Int. J. Hum. Resour. Manage. **12**(2), 184–202 (2001)
6. Yip, B., et al.: Job burnout within the Hong Kong construction industry: a cultural perspective. In: Proceedings of the CIB W92/T23/W107 International Symposium on Procurement Systems (2005)
7. Nicholas, J.M., Steyn, H.: Project Management for Engineering, Business and Technology. Routledge, Abingdon (2017)
8. Aziz, Z., et al.: Intelligent wireless web services for construction—a review of the enabling technologies. Autom. Constr. **15**(2), 113–123 (2006)
9. Banson, K.E., Nguyen, N.C., Bosch, O.J.: Systemic management to address the challenges facing the performance of agriculture in Africa: case study in Ghana. Syst. Res. Behav. Sci. **33**(4), 544–574 (2016)
10. Akadiri, P.O., Chinyio, E.A., Olomolaiye, P.O.: Design of a sustainable building: a conceptual framework for implementing sustainability in the building sector. Buildings **2**(2), 126–152 (2012)
11. Bosch, O., Nguyen, N.: A new approach for cross sectoral and organizational collaboration and communication to deal with increasing complexity and promote effective change. In: IUAES/AAS/ASAANZ Conference (2011)
12. Kerzner, H.: Project Management: A Systems Approach to Planning, Scheduling, and Controlling. Wiley, Hoboken (2017)

13. Banson, K.E., et al.: A systems thinking approach to address the complexity of agribusiness for sustainable development in Africa: a case study in Ghana. Syst. Res. Behav. Sci. **32**(6), 672–688 (2015)
14. Kim, S.: Can systems thinking actually solve sustainability challenges? Part 1, the diagnosis (2012). http://www.fusbp.com/wp-content/uploads/2010/07/Can-Systems-Thinking-Actually-Solve-Sustainability-Challenges.pdf. Accessed 24 Apr 2018
15. Doppelt, B.: Business leaders need systemic thinking for sustainability (2012). https://www.theguardian.com/sustainable-business/blog/business-leaders-systems-thinking-sustainability. Accessed 24 Apr 2018
16. Nguyen, N.C., Bosch, O.J.: A systems thinking approach to identify leverage points for sustainability: a case study in the Cat Ba Biosphere Reserve, Vietnam. Syst. Res. Behav. Sci. **30**(2), 104–115 (2012)
17. Sherwood, D.: Seeing the Forest for the Trees: A Manager's Guide to Applying Systems Thinking. Nicholas Brealey Publishing, Boston (2002)
18. Braun, W.: The system archetypes. System **2002**, 27 (2002)
19. Banson, K.E., Nguyen, N.C., Bosch, O.J.: Using system archetypes to identify drivers and barriers for sustainable agriculture in Africa: a case study in Ghana. Syst. Res. Behav. Sci. **33**(1), 79–99 (2016)
20. Toole, T.M.: A project management causal loop diagram. In: ARCOM Twenty First Annual Conference (2005)
21. Senge, P.M.: The Fifth Discipline: The Art and Practice of the Learning Organization. Crown Business, New York (2006)
22. Bosch, O.J., et al.: Managing complex issues through evolutionary learning laboratories. Syst. Res. Behav. Sci. **30**(2), 116–135 (2013)
23. Sherwood, D.: Seeing the Forest for the Trees: A Manager's Guide to Applying Systems Thinking. Nicholas Brealey Publishing, Boston (2011)
24. Banson, K.E., Nguyen, N.C., Bosch, O.J.: A systems thinking approach to the structure, conduct and performance of the agricultural sector in Ghana. Syst. Res. Behav. Sci. (2016)
25. AGI Construction Sector: What is gov't Strategic Agenda for the construction sector? (2017). https://www.myjoyonline.com/opinion/2017/december-13th/what-is-govt-strategic-agenda-for-the-construction-sector.php. Accessed 21 Mar 2019

Barriers of Project Bond Initiatives in Infrastructure Financing in Ghana

De-Graft Owusu-Manu[1], Caleb Debrah[1](✉), Prince Antwi-Afari[1], and David John Edwards[2]

[1] Department of Construction Technology and Management, College of Art and Built Environment, Kwame Nkrumah University of Science and Technology, PMB, Kumasi, Ghana
d.owusumanu.cap@knust.edu.gh, cal.debrah@gmail.com, pantwiafari.carp@st.knust.edu.gh

[2] Faculty of Technology Environment and Engineering, Birmingham City University, Birmingham, UK
david.edwards@bcu.ac.uk

Abstract. This research was centred on examining the barriers of project bond initiatives in infrastructure financing in Ghana. The use of project bond; an alternative form of financing infrastructural projects throughout the world became intensified after the global financial crisis which led to stricter banking regulations. The introduction of project bonds in Ghana has not seen much growth due to some inherent challenges identified. The analysis depicted predominant challenges like lack of industrial expertise, currency issues, being the cedi-dollar exchange rate due to the constant depreciation of the Ghanaian Cedi and unavailability of project bond investors. The gaps in literature identified in this study set the course for future studies in this field in Ghana. The paper provides information for policy makers and investors on the major barriers affecting project bond initiatives in infrastructure in Ghana. The evidence suggests that these challenges have curtailed project bond initiatives in the country.

Keywords: Bonds · Construction · Ghana · Investors · Project financing

1 Introduction

Traditional means of financing infrastructural development remain important, but private investment is very crucial to solving the infrastructural gap and meeting demand in Africa [1]. Briceño-Garmendia et al. agree that governments are the main financiers of infrastructure investment in Sub- Saharan Africa [2]. Badu et al. opined that infrastructure deficit is directly linked to funding availability especially in developing countries [3]. Private Participation in Infrastructure (PPI) have over the years focused on investment of the private sector through the public sector of multifaceted funding [4]. The traditional methods of financing infrastructure have failed to solve Ghana's infrastructure deficit [3], nonetheless, participation in infrastructural projects through the private sector as posited by the World Bank has augmented in most economies [4].

C. Aigbavboa and W. Thwala (Eds.): CIDB 2019, *The Construction Industry in the Fourth Industrial Revolution*, pp. 12–21, 2020.
https://doi.org/10.1007/978-3-030-26528-1_2

The stringent guidelines and capital requirements imposed on the banking industry after the global financial distress have restricted their urge to invest in long-term infrastructural projects [5]. Sponsoring Public Private Partnerships (PPP) through more innovative means has been the quest for many developers recently hence the advancement of project bonds [6].

Project bonds were first issued in the 1990s and were an important source of capital for infrastructure developments in Western Europe and North America [7]. Deloitte states emphatically that project bonds offer long-term investors attractive yields and significant credit spreads with the comfort of an underlying asset, often in a quasi-monopolistic position and often with indirect access through contractual payments from governments to the issuer, to sovereign credit rating [6]. Culp et al. also put it across that pension plans, insurance companies, hedge funds, finance companies, and other asset managers are all examples of investors in project bonds [8].

Project bonds have been utilized in Europe and America to fund infrastructure projects [6]. Also, tax-exempt municipal bonds, corporate bonds of utility firms, as well as Project Financing Initiatives (PFI) bonds in the UK are examples of institutional investor portfolios of project bonds [9]. In countries like Kenya, Senegal and Ghana, there are some examples of government bonds earmarked for infrastructure investments although it is not clear the extent the 'infrastructure label' leads to actual investment [9]. Ghana acquired a 15-year $1 Billion Bond guaranteed by World Bank. The Sankofa Integrated Oil and Gas Project, being US $700 million, was a program under the World Bank guarantee [10]. In Ghana, bonds are usually used mostly to service debts and not primarily for financing project development [11]. World Bank makes it known that Ghana in 2007 issued its first Eurobond and successfully raised $750 million and in 2014 and 2015 $2 billion was raised. An enonomy's investor base is broadened by bond issuance [12].

Project bond is an initiative being taken by various governments to ensure the country's development especially in the infrastructural sector [11]. This initiative has gained grounds in continents like Europe and America with a lot of researches. Africa, including Ghana, is also developing much interest in project bonds. Green bonds in Africa and Ghana is a developing variance in project bond financing that needs much research to ascertain the challenges affecting such an alternative in project financing in Ghana. Green bonds gained grounds in 2007 as a tool to finance a more sustainable development and green bonds have a market of unexploited potential [13] in Africa and the world at large. Academic literature has not yet covered much on project bonds in Ghana. Less is also said about the accompanying challenges of project bonds although the potentials in its exploitation are numerous for Ghana. With this gap in literature, this research seeks to explore the challenges facing the alternative means of financing infrastructure in Ghana through project bonds.

2 Literature Review

With the current scarcity of bank loans, project sponsors now agree that they must now exploit the project bond market to finance any project that needs US $1 billion or more in financing even in the early stages of the project [14]. Arca then alludes that early project bonds throughout the world over benefitted from guarantees given by monolines [15]. Monoline insurers offer investors a guarantee on returns from Structured Mortgage Products (SMP) bonds [16]. Although monoline bond insurers gave out payment guarantees against risk premiums for project bonds, this halted in the event of the 2008 financial crisis [17]. As technical as project bonds can be, it has its own drawbacks like all other developments and Arca could not agree much by saying that every peculiar project bond has its characteristic challenges [15]. Discussed below are some examples of the setbacks facing the project bond industry in Ghana, Africa and the world at large.

2.1 Construction Risk

A construction project is the riskiest adventure any firm does in their normal day to day business [18]. The construction industry has touted as highly risky; as a result, bond investors who are risk averse avoid funding infrastructure developments through project bonds [6]. Nonetheless, not all projects have a construction risk. Project bonds devoid of construction risk are known as operating asset securitizations or the take-out phase of financing project [15]. He further points out the two elements to construction risk being credit risk and timing risk.

Credit Risk. The fear of defaulting bond is referred to as credit risk. It means in the event project is not completed, the unfinished surety must be sold. This means no revenue but a loss [15]. Babatunde and Perera add credit or default risk to their list of risk linked with corporate bonds [5]. Municipal bonds are classified as one of the safest types of investment but there remain some risks associated with it which includes credit risk, call risk, interest risks [19].

Timing Risk. According to Arca, delay risk like a delay in construction is the main cause of timing risk leading to delayed payments or revenue generation which are mostly addressed with reserve accounts [15].

2.2 Currency Issues

Arca mentions some of the risks associated with the currency market bonds include; a dollar denominated mode of payments, an overly dollarized economy, overcollaterization and the issue of currency hedges from a credit-worthy counterparty typically banks. The major international currencies like the US dollars deepen the demand in the international bond market although there have been bits of interest on the local currency bond issues [15].

2.3 Rating Ceilings

Verdouw et al. state that the basics for acquiring funding from the capital market is to get a credit rating at or above the investment grade from rating agencies. These are internationally recognised which are done after a thorough and an in-depth review of the project's ability to meet all its financial commitments [20]. Arca then emphasizes that few countries over the years in the new markets have a rated investment grade on a foreign currency based on issues like convertibility and transfer of foreign currency payments, economic and political stability within the issuing country [15]. Ratings are very important, and they sometimes breach the sovereign ceiling [20].

2.4 Industrial Expertise

Typical financial advisers have a direct experience on bond financing as they mostly have the technical know-how. Arca agrees that a team knowledgeable in bonds issues explicitly country of issue, asset type etc. are always present in banks involved in project lending. However, the financial institutions devote more time to learn about it than it is normally required in bond execution anytime they lack experts in countersigning a bank loan and documenting it [15]. Banks consistently involved in project lending are expertise in extended variety of investments as fixed income investment likened to institutional investors with some level of knowledge which is incomparable to the bank connoisseurs in bonds.

2.5 Political Risk

Gardner et al. explain political risk to mean the risk that an investment suffers due to political instability or changes in government in a country. This could be wars, riots, breaches in contract by government [21]. Chan and Kumaraswamy put it across that political "force majeure" is contractually borne by the government in most PPP programs, the efficacy of that solution to investors would be annulled by a strategic sovereign default. Political risk can hardly be checked in the insurance markets. Varying degrees of political risk insurance can be obtained through the use of financing products available from multilateral and export credit agencies [22].

3 Methodology

The philosophical concepts guiding this study is that of the positivist and objectivist paradigm [3, 5]. In the data collection, a survey questionnaire was employed. According to Babatunde and Perera, one of the most accepted social research methods is survey questionnaire [5]. Prior to the questionnaire development and distribution, an extensive literature on the challenges of project bonds were perused [cf. 3, 4, 15]. Fifteen barriers to project bond initiatives were identified after an extensive review of extant literature which aided the design of the questionnaire. The Likert scale was used in the questionnaire of the study and was administered to professionals in the banking sector, insurance firms and managers of pension funds involved in project bonds. Convenient and purposive sampling were used to limit the population to just financial institutions and subsequently census sampling was used using the Kish formula to obtain the sample size. The study was conducted in Kumasi, the capital city of the Ashanti Region of Ghana where most of the financial institutions are concentrated. There are 34 Banks in Ghana [23], 33 registered pension fund managers [24] and 40 insurers and reinsurers in Ghana [25]. A total population of 107 commercial banks, pension fund managers and insurance firms were present during the time of the study.

A total of 63 questionnaires were distributed among sampled branches in Kumasi representing 58.88% of the total populations. The anonymity and confidentiality of respondents were assured. The response rate used for the data analysis was 57% (36 out of the 63 questionnaires administered). Relative Important Index, standard deviation, mean statistics and standard error were employed as the data analysis tool. The Statistical Package for Social Science (SPSS) version 21 was used in coding and analysing data collated.

4 Analysis of Results

Several preliminary procedures were employed to ensure the consistency and reliability of the quantitative data to guarantee a meaningful interpretation of data. Prevalent among them were categorising, coding, editing and generating data files. The demographics analysed encompassed position of respondents, work experience and roles of respondents. The respondents were asked to indicate their positions in the banks, insurance companies and pension funds. Out of the total of 36 responses received, 41.7% were managers, 30.6% were customer service personnel, 11.1% were Executive Directors, and with Chief Executive Officers (CEO) and Senior Executives, a response rate of 8.3% each were received. According to EIB, Project Bond Initiatives are structured to attract investors such as insurance companies and pension funds [26]. It is in this vein that the researchers targeted pension fund managers, top executives of banks and insurance as the respondents of the study.

The Relative Important Index (RII) was used to analyse the data gathered. RII is calculated by means of

$$\mathrm{RII} = \frac{\Sigma \mathrm{W}}{\mathrm{A} * \mathrm{N}}$$

N being the overall respondents; A - Higher response integer (5); W - weighting ranging from 1 to 5 as factors given by respondents [21].

Gardner and Wright put it this way by saying that the RII has been preferred over the mean and standard deviation as they are not credible and reliable tools for examining the overall ranking of key attributes [22]. The mean, standard deviation, RII, standard mean error was calculated for each variable as shown in Table 1. To ensure that the results obtained were reliable, the Cronbach's Alpha was used to check the internal consistency. Validating the reliability of the measuring instrument is one of the basic elements in evaluation [27].

The Cronbach's Alpha for the data was 0.904 for 15 items or variables of study which was a most significant thing to consider in this test. For a scale to be reliable, ideally the Cronbach's Alpha coefficient should exceed 0.70 [28]. From the results obtained, it proved that the scale that was used was very reliable. Before the data was analysed, the significance level was set at 95% confidence interval, thus, based on the five-point Likert scale rating, a success criterion deemed significant if it had a mean of 3.5 or more. Where two or more criteria have the same mean, the one with the lowest standard deviation assigned the highest significance ranking [29]. From the result from

Table 1. RII of the challenges or barriers of project bond initiative

Challenges of project bond initiatives	N	Sum	Mean		Std. deviation	RII	Ranking
			Statistic	Std. error			
Lack of industrial expertise on project bonds	36	164	4.556	0.084	0.504	0.911	1st
Currency issues (cedi-dollar exchange rate)	36	160	4.444	0.109	0.652	0.889	2nd
Availability of project bond investors	36	153	4.25	0.101	0.604	0.85	3rd
Delayed revenue generations or repayments	36	151	4.194	0.096	0.577	0.839	4th
Limiting government policies	36	144	4	0.138	0.828	0.8	5th
Risk to investors in new markets	36	140	3.889	0.087	0.523	0.778	6th
Poor structuring of bond issued	36	140	3.889	0.142	0.854	0.778	7th
Constructions risk	36	139	3.861	0.139	0.833	0.772	8th
Economic instability of issuing nation	36	138	3.833	0.129	0.775	0.767	9th
Fluctuations and instability in the bond market	36	134	3.722	0.13	0.779	0.744	10th
Rating agencies of project bonds in Ghana	36	131	3.639	0.144	0.867	0.728	11th
Sustainability of project bonds in the long run	36	127	3.528	0.084	0.506	0.706	12th
Political instability of issuing nation	36	123	3.417	0.115	0.692	0.683	13th
Over reliance on traditional means of financing projects	36	110	3.056	0.159	0.955	0.611	14th
Emergence of innovative and new ways of financing	36	100	2.778	0.127	0.76	0.556	15th

Field Survey (2018)

Table 1, all the mean statistics range between 4.55 and 3.52. This meant that all the barriers were critical to the success of project bonds in Ghana except political instability, over reliance on traditional means of financing projects, and emergence of innovative and new ways of financing which had their means below the accepted mean to be important for the study. Also, standard deviation shows how close the individual data values are from the mean value [30]. This helps us to ascertain how the sample mean is from the true mean population. A standard deviation over 1 means the responses were away from the mean. All the standard deviations were below one proving that our respondents agreed to most of the barriers of project bonds in Ghana. A small standard error is an indication of the reliability of the mean. The smaller the standard error, the more accurate the actual sample mean [30]. It could be deduced from Table 1 that all the standard errors were closer to zero, proving that the responses were accurate and critical barriers to project bonds in Ghana.

Project bond initiatives in infrastructure has come as one of the greatest innovations in infrastructural financing. Nonetheless, every innovation or any new development has its own inherent challenges. One of the extreme challenge impeding project bond initiative in infrastructural development globally is the lack of industrial expertise in project bond structuring and administration. This is the first ranked variable in this section. Several authors and scholars over the year have written on this issue and have suggested several remedies, but it is still the reason why potential investors in so many countries hold little infrastructure bond [15]. Critical among the challenges is the currency in which project bonds is administered. Gardner and Wright [22] put this across that hard currency loans creates currency risk if revenues are in the local currency. Liabilities grow by 15% if the Ghanaian cedi denigrates to dollar by 15% granted revenue does not change. This can lead to an asset-liability current disparity if Ghana procures a power plant through bonds financed in dollars, causing an upsurge in tariffs charged in Ghanaian cedis. It is for this reasons that currency issues were ranked as 2nd variable among the challenges of project bonds initiatives. The need to build a great pool of indigenous long-term investments and savings through the issuance of domestic insurance and pension funds. There is also the element that makes the bond attractive to foreign investors as well [31]. This introduces us to the third ranked variable being availability of project bond investors in the industry.

Ostendorf [32] could not agree more that project bonds have become an established primary source of funds for large-scale non-resource undertakings for over 20 years now. This has led to a diversified investor base to support the product syndication with each class of investor approaching the credit process in their varying dimensions.

5 Discussion of Findings

After the data analysis, the emergence of innovative and new ways of financing infrastructure other than project bonds were the least of the barriers. Nonetheless, the lack of industrial expertise has had a toll on the development of project bonds in infrastructure in Ghana. Africa as a continent, with Ghana inclusive, has not exploited project bond to its full potential. Research on project bonds in Ghana is not sufficient if not lacking. Mengel [33] agreed to the above assertion that projects being financed with bonds is a common phenomenon in the US, Canada and Australia with project bonds being relatively new to the market. In Africa; Nigeria, South Africa and Kenya have more examples with project bonds and a few in Ghana like the 10-year development capital bond by their government [19]. It is with this development that this research was conducted to verify the reason for this slow progress in adopting project bonds to finance infrastructure.

The analysis of the results revealed that the lack of industrial expertise has brought about a major setback in the project bond industry in Ghana. As technical as the project bond market is, a lot of expertise is needed for the structuring and administration of project to be effective for yielded returns. The industry has over the years been crippled with this major setback of lack of industrial expertise as agreed by [34]. The results further went to identify currency risk as another challenge of project bond initiatives in infrastructure. The issues surrounding the currency in which a bond is issued especially with project bonds can never be overlooked as the research pointed out from the analysis of the results. This is due to recurring depreciation of the local currency in the events of the issuance of bonds in the US dollar which is the most preferred international currency for issuing bonds. The constant depreciation of the Ghanaian Cedi to the dollar has been one the greatest barriers the project bond industry has been facing over the years. It makes servicing bonds costlier to the issuers and sponsors. This tend to discourage most investors due to the uncertainty surrounding the local currency, the Ghanaian Cedi. The above challenges discussed are among the many reasons leading to the next major barrier of project bond development in Ghana, availability of project bond investors. Results from the data analysis revealed that there is deficient pool of project bond investors in Ghana. This has not really helped the project bond industry and its development in Ghana.

6 Conclusion

This study delved into understanding the barriers of project bond initiatives in infrastructure financing in Ghana. Project bonds throughout extant literature were seen to be the new mode of financing infrastructure in most developed worlds. Although the infrastructure deficit in Africa is still lingering and worsening, literature revealed that Africa and Ghana in particular have not assimilated the idea of project bonds to finance infrastructure. The study revealed some of the barriers of project bonds in Ghana. Prevalent among them were lack of industrial expertise on project bonds, weak currency issues, non-availability of project bond investors, delayed revenue generations or repayments, limiting government policies. Also, risk to investors in a new market like Ghana, poor structuring of bonds issued, construction risk, economic instability, fluctuations and instability in the bond market as well as ratings of project bonds in Ghana by rating agencies were some of the barriers identified to be impeding the use of project bonds in Ghana.

These findings on the barriers of project bond initiatives in infrastructure in Ghana will help us comprehend how far project bonds have travelled over the years globally with the factors supporting such steady growth. This will help both investors and bond issuers to understand the factors surrounding project bonds in infrastructure better for an informed decision in both the structuring and administration. Moreover, the results obtained from this revealed some of the challenges project bonds have been facing over the years which would help both investors and sponsors to know forehand what is ahead of them so they could develop strategic solutions in meeting such barriers. The Ghanaian government and Africa in general must do well to promote innovation in the bond markets to promote development. In discovering an alternate to bank financing and diversifying sources of funding, local powers that be can tap into the bond market.

The research was non-empirical but purely exploratory and can therefore not be generalized. It was discovered throughout the research that studies had not covered some areas under project bond. A further researcher to delve more into innovations in project bonds like green bonds in Ghana, how to manage expectations in project bond and the Ghanaian project bond market is encouraged as there is little information in this area particularly. This would help improve the understanding of the adoption of project bonds in Ghana and largely Africa.

References

1. Mafusire, A., Anyanwu, J., Brixiova, Z., Mubila, M.: Infrastructure deficit and opportunities in Africa. Afr. Dev. Bank Econ. Brief **1**, 1–15 (2010)
2. Briceño-Garmendia, C., Smits, K., Foster, V.: Financing public infrastructure in sub-Saharan Africa: patterns and emerging issues. Background Paper, vol. 15 (2008)
3. Badu, E., Owusu-Manu, D.G., Edwards, D.J., Holt, G.D.: Innovative financing (IF) of infrastructure projects in Ghana: conceptual and empirical observations. Eng. Proj. Organ. J. **1**(4), 255–268 (2011)
4. World Bank: Billion-Dollar Projects Fuel Spike in Private Sector Infrastructure Investment (2017). http://www.worldbank.org/en/news/press-release/2017/10/20/billion-dollar-projects-fuel-spike-in-private-sector-infrastructure-investment. Accessed 27 Jan 2018
5. Babatunde, S.O., Perera, S.: Barriers to bond financing for public-private partnership infrastructure projects in emerging markets: a case of Nigeria. J. Financ. Manage. Property Constr. **22**(1), 2–19 (2017)
6. Deloitte Touche Tohmatsu, Project Bonds.: An alternative source of financing infrastructure projects (2017). https://www2.deloitte.com/za/en/pages/finance/articles/project-bonds-an-alternative-to-financing-infrastructure-projects.html. Accessed 12 Oct 2017
7. Kamin, S.B., Von Kleist, K.: The evolution and determinants of emerging markets credit spreads in the 1990s (1999)
8. Culp, C.L., Forrester, J.P.: Structured financing techniques in oil & gas project finance: future flow securitizations, prepaids, volumetric production payments, and project finance collateralized debt obligations. Soc. Sci. Res., 1–5 (2009)
9. Inderst, G., Stewart, F.: Institutional investment in infrastructure in developing countries: introduction to potential models. The World Bank (2014)
10. World Bank: New World Bank Guarantee Helps Ghana Secure $1 Billion, 15- Year Bond (2015). http://www.worldbank.org/en/news/feature/2015/11/18/new-world-bank-guarantee-helps-ghana-secure-us1billion-15-year-bond. Accessed 19 Jan 2018
11. Quist, E.: Ghana's 2015 Eurobond; Overview, prospects and challenges. Economists' Corner (2015). http://www.pulse.com.gh/news/business/economist-corner-ghana-s-2015-eurobond-overview-prospects-and-challenges-id4233502.html. Accessed 19 June 2018
12. Inderst, G.: Infrastructure as an asset class (2011)

13. Mezui, C.A.M., Hundal, B.: Structured finance: conditions for infrastructure project bonds in African markets. African Development Bank (2013)
14. Reed, P., Cort, T., Yonavjak, L.: Data-driven green bond ratings: a market catalyst (2017)
15. Arca, E.: The future of project bonds in Latin America. Harv. Bus. Law Rev. Online **3**, 187–195 (2013)
16. Petersen, M.A., Mukuddem-Petersen, J., Mulaudzi, M., De Waal, B., Schoeman, I.M.: Subprime risk and insurance with regret. Discrete Dyn. Nat. Soc., 1–53 (2010)
17. Akintoye, A., Beck, M., Hardcastle, C., Chinyio, E., Asenova, D.: The financial structure of private finance initiative projects. In: Proceedings of the 17th ARCOM Annual Conference, Salford University, Manchester, vol. 1, pp. 361–369, September 2001
18. Langer, B., Schwenke, M.: EU support for project bonds. CEP **18**, 1–4 (2011)
19. Oji, C.K.: Bonds: a viable alternative for financing Africa's development (2015)
20. Verdouw, W., Uzsoki, D., Ordoñez, C.D.: Currency risk in project finance. International Institute for Sustainable Development Discussion Paper, Winnipeg (2015). https://www.iisd.org/sites/default/files/publications/currency-risk-project-finance-discussion-paper.pdf
21. Somiah, M.K., Osei-Poku, G., Aidoo, I.: Relative importance analysis of factors influencing unauthorized siting of residential buildings in the Sekondi-Takoradi Metropolis of Ghana. J. Build. Constr. Plann. Res. **3**(3), 117 (2015)
22. Gardner, D., Wright, J.: Project finance. Encyclopedia of debt finance (2012)
23. BoG.: Bank of Ghana, Annual Report 2017 (2017). https://www.bog.gov.gh/privatecontent/Publications/Annual_Reports/Annual%20Report%202017%2012th%20July.pdf. Accessed 18 Apr 2018
24. NPRA.: National Pensions Regulatory Authority (NPRA). Public Notice, List of Registered Pension Fund Managers that have fulfilled the prescribed requirements of the Authority for the 2017 Renewal Process (2017). http://www.npra.gov.gh/pdf/Approved-Pension-Fund-Manager-for-Website-31-07-2017-1.pdf. Accessed 18 Apr 2018
25. NIC.: National Insurance Commission, Insurers and Reinsurers (2018). http://nicgh.org/insurers-reinsurers/non-life-companies/. Accessed 18 Apr 2018
26. EIB: European Investment Bank, The Europe 2020 Project Bond Initiative – Innovative infrastructure financing (2012). https://www.eib.org/en/products/blending/project-bonds/index.htm. Accessed 18 Apr 2018
27. Chan, D.W.M., Kumaraswamy, M.M.: A comparative study of causes of time overruns in Hong Kong construction projects. Int. J. Proj. Manage. **15**(1), 55–63 (1997)
28. Tavakol, M., Dennick, R.: Making sense of cronbach's alpha. Int. J. Med. Educ. **2**, 53 (2011)
29. Howland, J., Wedman, J.: A process model for faculty development: individualizing technology learning. J. Technol. Teach. Educ. **12**(2), 239–264 (2004)
30. Lee, D.K., In, J., Lee, S.: Standard deviation and standard error of the mean. Korean J. Anesthesiol. **68**(3), 220 (2015)
31. Eichengreen, B., Hausmann, R.: Exchange rates and financial fragility (no. w7418). National bureau of economic research (1999)
32. Ostendorf, C.: Effects of Basel III on the business segment of international project finance (2017)
33. Mengel, K.: African project finance: bonds and alternative financing (2017). https://www.insideafricalaw.com/blog/african-project-finance-bonds-and-alternative-financing. Accessed 8 May 2018
34. Garcia-Kilroy, C., Rudolph, H.P.: Private financing of public infrastructure through PPPs in Latin America and the Caribbean (2017)

Factors Influencing the Cost Performance of Traditional and Public-Private Partnership Procured Housing Projects in Nigeria

Olusola Festus Akinradewo(✉) and Alexander Ejiro Omoraka

Department of Quantity Surveying,
Federal University of Technology, Akure, Ondo, Nigeria
{foakinradewo,aeeomoraka}@futa.edu.ng

Abstract. Successive administrations in Nigeria have expended several efforts on housing delivery through either the traditional (TRD), or public-private partnership (PPP) procurement route among others. Considering the relationship between housing demand and supply, there is no record of meaningful achievements due to several factors. This study explored the factors affecting the cost performance of TRD and PPP with a view to enhance their performance. Eighty questionnaires retrieved from two sets of respondents were analysed on housing project at Suleja, Niger State, Nigeria. Mean score and Mann-Whitney test were used to assess the influence of these factors, and test the significance of the difference between the views of both set of respondents. The study revealed that economic conditions, unforeseen ground conditions and cash flow/finance as the key factors influencing TRD, while fluctuation in prices of materials, late delivery of materials, and economic conditions are the key factors influencing PPP.

Keywords: Cost performance · Housing project · Nigeria · Public-private partnership · Traditional Procurement Route

1 Introduction

The need to improve construction project performance in relation to time, budget and project completion has given birth to various procurement routes [1]. Ogunsanmi [2] defined procurement in construction as a process through which a facility is acquire through leasing, design and construction; or outright buying to fulfil a peculiar need. While, procurement route is posited as a system that involves the management of the entire process required to fulfil the delivery of a construction project [3, 4].

Ojo, Adeyemi, and Fagbenle [5] adjudged TRD as a prominent procurement method in the world and according to Sanni [6], the use of PPP to deliver infrastructure and services are on the increase in both the developed and developing countries; and Nigeria is not an exception. Irrespective of the procurement method adopted, the construction industry as a key part in determining the fate of the economy has been attributed with under-performance in comparison to other industries [7]. The low performance has been associated with the influence of certain factors on the procurement methods. Therefore, this study examined the factors affecting the cost performance of TRD and PPP with a

C. Aigbavboa and W. Thwala (Eds.): CIDB 2019, *The Construction Industry in the Fourth Industrial Revolution*, pp. 22–33, 2020.
https://doi.org/10.1007/978-3-030-26528-1_3

view to enhance the performance of these two procurement routes in Nigeria. The study provided an insight into the factors having the most influence on the cost performance of TRD and PPP projects. It also equips stakeholders with requisite knowledge on the key factors to focus their efforts on with a view to enhance the cost performance of both TRD and PPP procured projects. The study also contributes to the cost performance literature and will benefit the society; because its findings provide a reasonable basis for action to ameliorate the cost performance of TRD and PPP procured projects.

2 The Traditional Procurement Route (TRD)

The Traditional Procurement Route (TRD) according to Ojo, Adeyemi, and Fagbenle [5] is a prominent method of procurement employed in construction. This is in consonant with Ogunsanmi [2], which posited that most projects in Great Britain, Nigeria and other Commonwealth countries were procured via the TRD. This position is strengthen by the literature [1, 8], which opined TRD as the most commonly employed method of procurement in Nigeria. Turner [9] asserted that TRD has been in use for centuries with the function of design been provided directly to the client. The study explained further that TRD is a procurement method in which the designer doubles as the one to manage the construction process. The study concluded that, TRD involves the client appointing consultants for design and cost control; and when the design has attained a certain milestone, the main contractor is employed to carry out the construction work.

One key shortcoming of the TRD identified by Dada [10] is that it creates a contractual and organisational dichotomy between design and construction. Ojo et al. [5] opined that there is need to eliminate this dichotomy and integrate the aspects of design and construction as construction projects are becoming more complex. This is because separation of construction projects into two different entities created a gap for effective coordination and communication [5]. The need to integrate design and construction; and the low performance of the TRD in the timely delivery of projects in the United Kingdom, led to the emergence of several alternative procurement routes, and one of these routes is the PPP.

3 Public Private Partnership (PPP)

PPP is an arrangement between the government and the private sector, with the aim to provide community facilities, public infrastructure, and relevant services [11, 12]. The basic aim of involving the private sector is to mitigate infrastructural deficit without the financial burden on the government through the provision of public facilities through partnerships [13]. Globally, Government have increasingly turned to the private sector to provide infrastructure services in area of energy and power; communication, transport and water sectors which were once delivered by the public sector. One of the main reasons behind such growing involvement of the private sector is hinged on the need to provide sustainable development in infrastructural facilities and services; and to increase efficiency in project delivery and operation. This however, is due to the availability of additional resources to meet the increasing needs of investment in infrastructure services and access to advanced technology. In Nigeria, the government awards public projects through various methods and one of these methods is the PPP. Oyewobi, Ibrahim, and Ibrahim [14] indicated that PPP is the panacea for inability of governments to finance the construction of major infrastructures. According to Ogunsanmi [15], PPP arrangements have been made by several countries for their infrastructure provisions; and Nigeria is trailing this path for her infrastructural development. Thus, there is a need to investigate the factors affecting the cost accomplishment of PPP projects in Nigeria.

4 Cost Performance

Cost performance is evaluated through the comparison of final cost of a completed project and the initial or budgeted cost; it has been adjudged as a major criterion of project success over the years [20]. This conforms to Memon *et al.* [22], which asserted construction cost as the most crucial measure of project success. From a study conducted in Malaysia on 51 private projects and 308 public projects, Endut, Akintoye and Kelly [23] discovered that only 37.2% and 46.8% of private and public sector projects were completed within budgeted cost respectively. In Nigeria, Omoregie and Radford [21] reported that the minimum average project cost overrun was 14%. Therefore, it is not oblivious to assert that the occurrence of cost overruns exceeds the completion of projects within budgeted cost, this thus necessitates the identification of factors influencing project cost performance from the inception; because proper handling of these will mitigate the occurrence of cost overruns and improve the cost performance of projects [20].

5 Methodology

The study explored the factors affecting the cost performance of TRD and PPP with a view to enhance the performance of these two procurement routes. In other to obtain the opinion of respondents using a uniform basis, a well-structured questionnaire was the instrument used for data collection from two sets of respondents. These included the in-house construction professionals within the Federal Ministry of Power, Works and Housing (FMPW&H), and the Contractors at Suleja, Niger State. The choice of restricting the target professionals in FMPW&H is because these are involved in the key projects constructed in the state and work with elite contractors. This is because most of the major projects in the State are done by the government and are handled by the ministry. The questionnaire has two sections. Section A dwelt on the background information of respondents, while, section B assessed the factors affecting the cost performance of TRD and PPP procured housing projects in Suleja, Niger State. The 5-point Likert scale was used to rate the factors, with five being the highest and one being the lowest. Due to the manageable population size, census, a variant of the non-probability sampling technique was used for the distribution of questionnaire. The respondents included Architects, Builders, Quantity Surveyors, Urban and Regional

Planners, Land Surveyors, Estate Surveyors, and Engineers working as in-house construction professionals within the Federal Ministry of Power, Works and Housing (FMPW&H), or as Contractors at Suleja, Niger State, Nigeria. A total number of 82 questionnaires were distributed but only 80 questionnaires were completed, returned and good for analysis. In analysing the data, mean item score (MIS) was computed to identify the factors that had the highest influence on the cost performance of TRD and PPP procured housing projects respectively. While Mann-Whitney U Test (MWT) was done to test the significance of the difference between the views of both set of respondents.

6 Results

The background information indicated in Table 1 shows that the in-house professionals of the FMPW&H constituted a greater percentage of the respondents in comparison to the contractors. The Table also indicates that most of the respondents are Builders, Engineers and Architects, which constituted 25%, 22.5% and 21.3% of the population respectively. The least represented professionals are the Urban and Regional Planners and Land Surveyors with 3.8% each. Furthermore, about 79% of the respondents had a Higher National Diploma and Bachelor degree, while the remaining 21% had Master's degree. This implies that they are educated enough to provide reasonable input to the study. In addition, the average year of experience is 9 years. Thus implies that they possess sufficient experience to attend to the questionnaire. In addition, the average number of projects executed using the TRD and the PPP procurement route are 12 and 7 projects respectively. This connotes that they have sufficient exposure to rate the factors affecting the cost performance of TRD and PPP procured housing projects.

6.1 Factors Influencing the Cost Performance of TRD Procured Housing Projects (PHP)

The 45 identified factors affecting cost performance of both procurement methods were grouped into seven (7) groupings [16]. Result in Table 2, shows that under the client related factors, the overall MIS revealed cash flow/Finance and payment for work; and financial difficulties of client, as the two key elements responsible for the cost accomplishment of traditionally procured housing projects with an overall MIS of 4.28 and 4.08 respectively. Mann-Whitney U Test conducted shows that at 95% confidence level, a significant p-value of above 0.05 was generated for both factors. This implies that there is no significant difference in the opinion of the two categories of respondents on the level of influence of these elements have on the cost accomplishment of traditionally procured housing projects.

From Table 2, the overall mean item score under the contractor related elements, consultants related factors and materials related factors as the key factors influencing the cost performance of traditionally procured housing projects with a mean score of 3.71. The overall assessment also indicate the incomplete design at the time of tender with a mean score of 3.86 and fluctuation of prices of materials with a mean score of

Table 1. Background of respondents

Category	Classification	Frequency	Percentage
Category of respondents	FMPW&H	57	71.30
	Contractor	23	28.80
	Total	**80**	**100.00**
Professions	Architect	17	21.25
	Builder	20	25.00
	Quantity surveying	14	17.50
	Urban and regional planner	3	3.75
	Land surveyor	3	3.75
	Estate surveyor	5	6.25
	Engineers	18	22.50
	Total	**80**	**100.00**
Level of qualification	Higher national diploma	26	32.50
	B.Sc./B.Tech.	37	46.25
	M.Sc./M.Tech.	17	21.25
	Total	**80**	**100.00**
Years of experience	0–5 years	18	22.50
	6–10 years	38	47.50
	11–15 years	13	16.25
	16–20 years	7	8.75
	21 years and above	4	5.00
	Total **Average**	**80** **9**	**100.00**
Number of projects executed using TRD	1–10	42	52.50
	11–20	21	26.25
	21–30	9	11.25
	31–40	5	6.25
	41 and above	3	3.75
	Total **Average**	**80** **13**	**100.00**
Number of projects executed using PPP	1–10	62	77.50
	11–20	18	22.50
	Total **Average**	**80** **7**	**100.00**

3.79 respectively. The Mann-Whitney U Test also for these three groups of factors revealed that there is no significant difference in the opinion of both categories of respondents as a significant p-value of above 0.05 was established for all the assessed factors.

Table 2. Factors Influencing the Performance of TRD PHP

Factors	In-house		Contractor		Overall		MWT
	MIS	Rk	MIS	Rk	MIS	Rk	P-Value
Client factors							
Cash flow/Finance and payment for work	4.32	1	4.17	1	4.28	1	0.664
Financial difficulties of client	4.05	2	4.13	2	4.08	2	0.635
Slow decision-making by clients	3.79	3	3.61	4	3.74	3	0.330
Client interference	3.74	4	3.65	3	3.71	4	0.890
Unrealistic imposed contract duration	3.60	5	3.57	5	3.59	5	0.921
					3.88		
Contractor factors							
Construction methods	3.68	4	3.78	1	3.71	1	0.842
Poor contractor's estimation	3.70	1	3.70	3	3.70	2	0.920
Mistakes during construction	3.68	4	3.74	2	3.70	2	0.752
Schedule delay	3.70	1	3.65	4	3.69	4	0.916
Inadequate monitoring and control	3.70	1	3.61	6	3.68	5	0.956
Inadequate contractor experience	3.68	4	3.65	4	3.68	5	0.934
Poor site management and supervision	3.63	7	3.57	8	3.61	7	0.795
Poor financial control on site	3.61	8	3.61	6	3.61	7	0.869
Incompetence of Subcontractors	3.51	9	3.39	9	3.48	9	0.700
					3.65		
Consultant factors							
Incomplete design at the time of tender	3.84	2	3.91	1	3.86	1	0.619
Contractual claims, i.e. cost claims	3.86	1	3.82	5	3.85	2	0.804
Inaccurate cost estimate	3.81	3	3.91	1	3.84	3	0.570
Change in the scope of the project	3.74	4	3.91	1	3.79	4	0.374
Quality assurance/control	3.74	4	3.77	6	3.75	5	0.964
Delays in decisions making	3.67	6	3.91	1	3.74	6	0.284
Frequent design changes	3.63	7	3.61	9	3.63	7	0.925
Approval of test and inspections duration	3.60	8	3.64	8	3.61	8	0.852
Poor contract management	3.54	9	3.70	7	3.59	9	0.648
Delay preparation and approval of drawings	3.39	10	3.39	10	3.39	10	0.965
					3.70		
Material factors							
Fluctuation of prices of materials	3.74	2	3.91	1	3.79	1	0.475
Shortage in material	3.75	1	3.83	4	3.78	2	0.820
Late delivery of material	3.67	3	3.91	1	3.74	3	0.356
Quality of material	3.63	4	3.87	3	3.70	4	0.463
					3.75		

(*continued*)

Table 2. (*continued*)

Factors	In-house		Contractor		Overall		MWT
	MIS	Rk	MIS	Rk	MIS	Rk	P-Value
Labour and equipment factors							
Shortage of technical personnel (skilled labour)	4.00	1	4.26	1	4.08	1	0.459
High cost of labour	3.96	2	4.13	2	4.01	2	0.479
Equipment availability and failure	3.93	3	4.04	4	3.96	3	0.545
Labour productivity	3.91	4	4.04	4	3.95	4	0.585
Shortage of site workers	3.88	5	4.09	3	3.94	5	0.459
Late delivery of equipment	3.70	6	4.00	6	3.79	6	0.258
Labour absenteeism	3.11	7	3.30	7	3.16	7	0.371
					3.84		
Contractual relationships factors							
Lack of coordination between parties	3.47	1	3.35	1	3.44	1	0.523
Lack of communication between parties	3.44	2	3.22	3	3.38	2	0.452
Inappropriate organizational structure linking parties	3.19	3	3.30	2	3.23	3	0.530
Slow information flow between parties	2.96	5	3.22	3	3.04	4	0.148
Major disputes and negotiations	3.00	4	3.13	5	3.04	4	0.467
					3.22		
External factors							
Economic conditions	3.93	2	4.17	1	4.00	1	0.155
Unforeseen ground conditions	3.98	1	3.96	3	3.98	2	0.919
Regulatory changes and building code	3.87	4	3.98	2	3.93	3	0.622
Political conditions	3.93	2	3.83	4	3.90	4	0.933
Weather condition	3.63	5	3.65	5	3.64	5	0.939
					3.89		

6.2 Factors Influencing the Cost Performance of PPP PHP

Assessing the influence of the factors assessed in Table 2 on the cost accomplishment of PPP procured housing projects, Table 3 shows that under the client related factors, the overall mean score revealed that client's interference and slow decision-making by clients are the major factors influencing the cost performance of PPP procured housing projects. These two factors had a MIS of 3.54 and 3.41 with a significant p-value of 0.166 and 0.262 respectively. This implies that there is no significant disagreement in the view of both set of respondents as regards the significance of these factors.

Table 3. Factors influencing the cost performance of PPP PHP

Factors	Client		Contractor		Overall		MWT
	MIS	Rk	MIS	Rk	MIS	Rk	P-Value
Client factors							
Client interference	3.63	1	3.30	2	3.54	1	0.166
Slow decision-making by clients	3.33	2	3.61	1	3.41	2	0.262
Unrealistic imposed contract duration	3.02	3	3.26	3	3.09	3	0.547
Cash flow/Finance and payment for work	2.86	4	2.91	4	2.88	4	0.890
Financial difficulties of client	2.82	5	2.87	5	2.84	5	0.922
					3.15		
Contractor factors							
Poor contractor's estimation	3.40	1	3.61	1	3.46	1	0.517
Inadequate monitoring and control	3.21	2	3.35	3	3.25	2	0.715
Poor site management and supervision	3.19	3	3.30	6	3.23	3	0.746
Incompetence of Subcontractors	3.07	7	3.57	2	3.21	4	0.064
Inadequate contractor experience	3.11	5	3.35	3	3.18	5	0.291
Poor financial control on site	3.19	3	3.30	6	3.15	6	0.460
Construction methods	3.09	6	3.26	9	3.14	7	0.537
MIS value takes during construction	3.04	8	3.30	6	3.11	8	0.285
Schedule delay	2.98	9	3.35	3	3.09	9	0.200
					3.20		
Consultant factors							
Incomplete design at the time of tender	3.84	1	4.09	1	3.91	1	0.419
Poor contract management	3.74	3	3.96	2	3.80	2	0.488
Frequent design changes	3.77	2	3.52	6	3.70	3	0.346
Inaccurate quantity take-off	3.54	4	3.91	4	3.65	4	0.251
Delays in decisions making	3.19	6	3.91	4	3.40	5	0.016**
Change in the scope of the project	3.14	9	3.96	2	3.38	6	0.010**
Quality assurance/control	3.19	6	3.48	7	3.28	7	0.305
Delay in preparation and approval of drawings	3.21	5	3.39	8	3.26	8	0.367
Waiting time for approval of test and inspections	3.18	8	3.26	9	3.20	9	0.692
Contractual claims, i.e. cost claims	2.96	10	3.17	10	3.03	10	0.550
					3.46		
Material factors							
Fluctuation of prices of materials	3.93	1	3.61	2	3.84	1	0.194
Late delivery of material	3.56	2	3.87	1	3.65	2	0.351
Quality of material	3.54	3	3.52	3	3.54	3	0.877
Shortage in material	3.44	4	3.43	4	3.44	4	0.877
					3.62		

(*continued*)

Table 3. (*continued*)

Factors	Client		Contractor		Overall		MWT
	MIS	Rk	MIS	Rk	MIS	Rk	P-Value
Labour and equipment factors							
High cost of labour	3.91	1	3.61	2	3.83	1	0.195
Shortage of technical personnel (skilled labour)	3.70	2	3.78	1	3.73	2	0.754
Shortage of site workers	3.49	3	3.48	3	3.49	3	0.938
Equipment availability and failure	3.44	4	3.22	6	3.38	4	0.452
Late delivery of equipment	3.19	5	3.30	4	3.23	5	0.530
Labour productivity	3.05	6	3.26	5	3.11	6	0.463
Labour absenteeism	3.00	7	3.13	7	3.04	7	0.467
					3.40		
Contractual relationships factors							
Slow information flow between parties	3.86	1	3.48	1	3.75	1	0.069
Lack of communication between parties	3.54	2	3.39	3	3.50	2	0.494
Major disputes and negotiations	3.47	4	3.35	4	3.44	3	0.523
Lack of coordination between parties	3.51	3	3.17	5	3.41	4	0.307
Inappropriate organizational structure linking parties	3.25	5	3.43	2	3.30	5	0.391
					3.48		
External factors							
Economic conditions	3.93	1	3.87	1	3.91	1	0.790
Unforeseen ground conditions	3.75	2	3.52	2	3.69	2	0.261
Political conditions	3.79	3	3.39	4	3.68	3	0.174
Regulatory changes and building code	3.40	4	3.48	3	3.43	4	0.939
Weather condition	3.42	5	3.17	5	3.35	5	0.194
					3.61		

**Significant at 0.05

Table 3 further revealed poor contractor's estimation with a MIS of 3.46, incomplete design at the time of tender with a MIS of 3.91; and fluctuation of prices of materials with a MIS of 3.84 as the top factors affecting the cost performance of PPP procured housing projects under the factors relating to contractors, consultants and materials respectively. The Mann-Whitney U Test shows that there is no significant disagreement in opinion of both set of respondents on most of the factors under these groups except for "delays in decisions making" and "change in the scope of the project" which had a p-value of 0.016 and 0.010 respectively.

In addition, under the labour and equipment related elements, contractual relationship factors and external related factors; the key factors influencing the cost performance of PPP procured housing projects included high cost of labour (overall MIS of 3.83), slow information flow between parties (overall MIS of 3.75), and economic conditions (overall MIS of 3.91) respectively. The Mann-Whitney U Test posited no significant difference in the view of both set of respondents, as a significant p-value of above 0.05 was derived for all assessed factors.

7 Discussion of Findings

The Findings revealed that the main element influencing the cost performance of TRD procured housing projects could be categorised under the external factors, client factors, and labour and equipment factors. Key factors under these groups include; economic conditions, unforeseen ground conditions, cash flow/finance and payment for work, financial difficulties of client, shortage of skilled labour, and high cost of labour. For housing projects procured through PPP, the key factors are under the material factors, and external factors. Key factors under these groups include; unstable prices of materials, late delivery of materials, economic conditions, unforeseen ground conditions and political conditions.

Thus from the findings, it can be deduced that issues relating to external factors such as economic conditions, political conditions and unforeseen ground conditions are key factors that affects the cost performance of housing projects delivered through the traditional and PPP procurement systems. According to Ubani, Okorocha, and Emeribe [16], external factors are frequently perceived as reflex factors, which make their control difficult. Furthermore, the economic situation such as instability within the country has crucial effect on the cost delivery of projects. The political condition is also an important circumstance that could influence the cost performance of housing projects. This is in agreement with Sanni [6], which posited political support as one of the two key elements influencing the performance of projects delivered via the PPP. This is not alien because, if the political leader do not formulate policies that are mutually beneficial for the private sector to collaborate with the government, PPP procured projects will be difficult to attain. This is because no private entity is willing to make an investment that is not profitable. These findings are in line with Iyer and Jha [17] and Oyedele [18], which states that the economic and political condition within an environment tends to influence the cost of projects within such areas.

Another crucial factor identified is cash flow, which is in resonance with Memon et al. [19], as cash flow and payment for finished jobs were the major factors that influence the cost performance of projects. This is because; any delay in payment to contractors can lead to suspension of activities on site, which might give rise to time overrun and ultimately cost overrun. The issue of price fluctuation, which influences the cost performance of PPP delivered through housing projects, has become a common phenomenon in the Nigeria construction industry. Aje, Omoraka and Ariyo [20] posited price fluctuation as the factor with the greatest effect on cost overrun. Omoregie and Radford [21]; Memon, Rahman, Abdullah and Azis [22] strengthened this position; they identified fluctuating prices as the major cause of project cost increase in construction projects. This was linked to the limitation in exchange rate, which in turn influences the prices of construction materials and the general level of price.

8 Conclusion and Recommendations

The study revealed that economic conditions, unforeseen ground conditions, cash flow/finance and payment for work, financial difficulties of client, shortage of skilled labour, and costly labour as the key factors influencing TRD. In addition, fluctuation in prices of materials, delay in materials delivery, economic conditions, unforeseen ground conditions and political conditions are the key factors influencing the cost performance of PPP. The MWT on the factors influencing TRD revealed that there is no significant difference in the views of the respondents on cost performance. The test was repeated for the PPP, the same results were obtained except for these factors: delays in decisions making, and change in the scope of the project. The study thus recommends that adequate pre-cautionary measures such as proper site investigation before the commencement of projects, proper feasibility and viability studies, should be put in place to check these external factors and cushion their effect on the cost performance of housing projects. In addition, in the case of the traditional procurement option, clients should ensure that finance to cover a proposed housing project is in available before the award of such contract. In the same vein, developers seeking to execute housing projects via the PPP route should ensure that they have the financial capability to handle such projects. Furthermore, fluctuation in the price of materials should be adequately predicted, and included in the initial estimate to avoid cost overrun or even shortage of funds. Although, the study focused on Niger State, it also provides useful insights on the factors influencing the cost performance of TRD and PPP in other states of the country. This is based on the premise that most professionals and contractors are not restricted by state boundary; and the wealth of experience gathered from different states influenced their responses to the questionnaire. However, this research can be replicated in other states and comparison made. Future studies can also assess the factors influencing the time performance of TRD and PPP.

References

1. Babatunde, S.O., Opawole, A., Ujaddugbe, I.C.: An appraisal of project pro- curement methods in the Nigerian construction industry. Civil Eng. Dimens. **12**(1), 1–7 (2010)
2. Ogunsanmi, O.E.: Correlates of procurement performance of traditional and labour-only methods in Nigeria. Ethiop. J. Environ. Stud. Manage. **6**(2), 182–190 (2013)
3. Ogunsanmi, O.E., Bamisile, A.: Factors affecting the selection of project procurement methods. Build. Mag. **12**(1), 11–16 (1997)
4. Ashworth, A., Hogg, K.: Willis's Practice and Procedure for Quantity Surveyor. Blackwell Publishing Ltd., Oxford (2007)
5. Ojo, S.O., Adeyemi, A.Y., Fagbenle, O.I.: The performance of traditional contract procurement on housing projects in Nigeria. Civil Eng. Di-mens. **8**(2), 81–86 (2006)
6. Sanni, A.O.: Factors determining the success of public private partnership projects in Nigeria. Constr. Econ. Build. **16**(2), 42–55 (2016)

7. Idiake, J.E., Shittu, A.A., Anunobi, A.I., Akanmu, W.P.: A comparative analysis of traditional and design & build methods of procurement in the Nigerian construction industry. Int. J. Constr. Eng. Manage. **4**(1), 1–12 (2015)
8. Ojo, S.O.: Benchmarking the performance of construction procurement methods against selection criteria in Nigeria. Civil Eng. Dimens. **11**(2), 106–112 (2009)
9. Turner, A.: Building Procurement. Macmillan Education Ltd., Basingstoke (1990)
10. Dada, M.O.: Predictors of procurement selection: an investigation of traditional and integrated methods in Nigeria. J. Constr. Dev. Ctries. **17**(1), 69–83 (2012)
11. Olaniyan, O.: Appraisal of critical success factors (CSF) for implementation of public-private partnership in Lagos State. M.Sc. Thesis, University of Lagos, Lagos (2013)
12. Egbewole, Q.A.: Examining public-private partnership in Nigeria: potentials and challenges. LLB Long Essay of University of Ilorin, Ilorin (2011)
13. Syuhaida, I., Aminah, M.Y.: The provision of infrastructure via private finance initiative. Theor. Empir. Res. Urban Manage. **4**, 76–86 (2009). Special Number 15/April 2009: Urban Issues in Asia
14. Oyewobi, O.L., Ibrahim, A.D., Ibrahim, Y.M.: Investigating optimum conditions for public-private partnership in health, education and housing sectors in Nigeria. In: Laryea, S., Agyepong, S.A., Leiringer, R., Hughes, W. (eds.) Proceedings of 4th West Africa Built Environment Research (WABER) Conference, 24–26 July 2012, Abuja, Nigeria, pp. 1261–1274 (2012)
15. Ogunsanmi, O.E.: Critical success factors (CSF) determining the implementation of public-private partnership projects. Covenant J. Res. Built Environ. (CJRBE). **1**(2), 41–66 (2013)
16. Ubani, E.C., Okorocha, K.A., Emeribe, S.C.: Analysis of Factors influencing time and cost overrun on construction projects in South Eastern Nigeria. Int. J. Manage. Sci. Bus. Res. **2** (2), 73–84 (2013)
17. Iyer, K.C., Jha, K.N.: Factors affecting cost performance: evidence from Indian construction projects. Int. J. Proj. Manage. **23**(4), 283–295 (2005)
18. Oyedele, E.: The British and the making of a Capital City. In: Ashafa, A.M. (ed.) 1913–1960 Urbanization and Infrastructure in Nigeria since the 20th Century. Kaduna State University, Kaduna, Nigeria, pp. 20–60 (2011)
19. Memon, A.H., Rahman, I.A., Abdullah, M.A., Aziz, A.A.A.: Time overrun in construction projects from the perspective of project management consultant (PMC). J. Surv. Constr. Property **2**(1), (2011)
20. Aje, I.O., Omoraka, A.E., Ariyo, I.E.: Assessment of factors inhibiting cost performance of construction projects in Nigeria. Quant. Surv. **60**, 21–31 (2015)
21. Omoregie, A., Radford, D.: Infrastructure delay and cost escalations: causes and effects in Nigeria. School of Architecture, De Montford University, Leicester, LE 1 9BH, England (2006)
22. Memon, A.H., Rahman, I.A., Abdullah, M.R., Azis, A.A.: Factors affecting construction cost performance in project management projects: case of Mara large projects. In: Proceedings of Post Graduate Seminar on Engineering, Technology and Social Science. Universiti Tun Hussein Onn Malaysia (2010)
23. Endut, I.R., Akintoye, A., Kelly, J.: Cost and time overruns of projects in Malaysia, pp. 243–252 (2019). http://www.irbnet.de/daten/iconda/CIB10633.pdf. Accessed 26 Apr 2019

Assessment on the Influence of Stakeholders on Sustainable Building Construction in Ondo State, Nigeria

Ayodeji Emmanuel Oke, Diekolola Abiola-Ogedengbe(✉), and Ainlolu Akinseli

Department of Quantity Surveying, Federal University of Technology Akure, Akure, Ondo State, Nigeria
emayok@gmail.com, abyolardieko@gmail.com

Abstract. Stakeholders' influence on construction project is of great importance in any sustainable construction development. This study presents the assessment of the influence of stakeholder's contribution on sustainable building construction, with a view to ascertaining the respective role of stakeholders on construction projects. Data obtained for the study were, through well-structured questionnaires distributed to professionals in construction industry. Out of the eighty-six (86) questionnaires that was administered, seventy-three (73) questionnaires were collected and analysed using Mean Item Score (MIS). The data were analysed and presented with tables, considering the influences of different stakeholders at different stages of construction stages, Mean Item Scores (MIS) was used to rank the divers' influences that stakeholders can have on a project. The results show that Government authorities were highly ranked for the pre-construction stage. For the construction stage, stakeholders have the highest dominant role in the construction stage. It was therefore concluded that the general public ranked the highest influence in the post construction stage.

Keywords: Construction professionals · Project success · Project stakeholders · Sustainable construction

1 Introduction

Stakeholder's influence on construction project is of great importance in any construction development. Traditionally, the bedrock of any construction project coalition are the architect, the client and the contractor [1]. To decipher and comprehend the perception of stakeholders in construction, this study helps to establish the theoretical backgrounds of contribution of stakeholder to sustainable construction. However, the secondary business goal is utilizing values/profits because of three reasons: Firstly, the dominant factor is the good accomplishment for society rather than value added to stakeholders only. Secondly, existence of other sustainability construction objectives is important. Thirdly, a new managerial way to handle and manage stakeholders is needed due to the construction dynamics [2]. Similarly in developing economy, construction projects are the major investment that create avenue for profit or loss. The complexities

C. Aigbavboa and W. Thwala (Eds.): CIDB 2019, *The Construction Industry in the Fourth Industrial Revolution*, pp. 34–42, 2020.
https://doi.org/10.1007/978-3-030-26528-1_4

of projects among several parties are hugely committed to closely interact with others endeavours to achieve its own comforts [3].

The increasing form of fragmentation in construction projects, have also contributed to its intricacies; annexing the importance of firm coherence, that guarantees high form of incorporation among stakeholders (demand and supply sides), whether the project is to be concluded properly and effectively. Also, when there is a flow in communication among suppliers, co-operation, openness and transparency will be upheld. However, in developing economy, sustainable construction needs certain level of improvement as the entire construction industry requires a cognisant and managed process to advance the efficiency and efficacy of the construction industry to reach same standard with the national economic requirement for building and civil engineering products, to upkeep social development objectives and sustained national economic [4]. This improvement of the industry can be achieved by refining on products that are used in buildings as well as processes of construction of a sustainable development and buildings design [5]. However, the traditional project's success measures, which are cost, time, and quality are not adequate to measure project success, as powerful stakeholders influence policymaking at various facet of the entire project life span for well-defined sustainable construction.

2 Types of Stakeholders

Generally, in sustainable construction stakeholders are divided into internal stakeholders and external stakeholders. The internal stakeholders are directly involved in making decision in an organization (e.g. customers, supplies, employees, and owners) while the external stakeholders affect the organisation in a crucial way (e.g. General public, local authorities and neighbours). There has been a resilient emphasis in the relationship of internal stakeholder such as site management in construction and procurement, while external stakeholder relationships at certain level are emphasised as goal for public officials through the legislation and rules that regard facility development [6]. Newcombe [7] has a similar classification as inside stakeholders and outside stakeholder while direct and indirect stakeholders was used by Smith and Love [8]. Primary stakeholders' vs secondary stakeholders were considered by delineation according to Carroll and Buchholtz [9]. To demarcate between the various types of stakeholders, we can say that primary stakeholder group, is one whose participation in the corporation are temporal whereas, secondary stakeholders are characterized by the firm. Stakeholders could also be challenged between those contracted to provide services (e.g. Consultant, subcontractors, contractors and engineers) that is, in a direct or primary relationship with an organization; in opposition to those that have no contracted duties or formal redress, but are in a secondary or indirect relationship with an organisation [10]. Based on the above explanations, examples of construction project stakeholders are listed in Table 1.

Table 1. Example of project stakeholders in construction

Stakeholders	Objectives and role
Client	Depending on the role, the client can act in public construction project in private construction project. The main difference between the two is that the client and the benefactor act in like manner in private project and in the public project the ideal starter is the government and benefit is geared towards the affected community
Contractor/subcontractors	They act in construction based on specifications, designs, contract documents verbalized by the applicable parties
Funding body/donor	E.g. IDB, UN, ICRC. Address humanitarian issues, they provide liquid to the community project. They also ensure liquids are utilized for the purpose to which it was assigned. E.g. if a precondition is issued to spend the liquid on certain activity, the donor must ensure that the liquids are exhausted for same activity
International non-governmental organization/non-governmental organization	They act as the intermediaries of the liquidity body and the government. They help assist in molding numerous temporary shelters and permanent homes
Government	The government helps in the formulation and maintenance of regulations, rules, policies and also monitors the adherence. Creating the standards relating to the delivery of housing projects
General public	They are involve in the construction phase for the provision of labour and clearing the debris.
Land owners/neighbourhood	They ensure that their interests will not be jeopardized by the project

Source; Salah [10]

3 Methodology

This study adopted a well-structured closed-ended research questionnaire approach as the data collection instrument. This approach is suitable for the study because it portrayed the objective and precise collection of each respondent's opinion and knowledge on the influence of stakeholders on sustainable construction project, out of the eighty-six (86) questionnaire administered seventy-three (73) was retrieved. Nevertheless, to facilitate the objectives of the study, 5-point Liker scales was used (1 = very low; 2 = low; 3 = medium; 4 = High; and 5 = Very High). A purposive sampling techniques was adopted, to focus more on those with professional membership qualification. As a result, mean item score MIS, descriptive statistics of percentage), significant important index (SII) and ranking system were used to analyse the findings.

The study was conducted in Ondo State of Nigeria. Furthermore, data were collected from engineers in telecommunication, quantity surveyors, builders, and architects currently involved in construction practices.

4 Data Analysis

The category of Firms/Organisation visited includes 39.70% contracting firms and 60.30% consulting firms. Their years of involvement in a project reveals that 17.80% of the respondent have handled less than 5 project, while 19.20% have handled between 6–11 projects, 26% have handled between 11–15 project, respondent that have handled between 16–20 are of 6.80%, 2.70% have handled between 21–25 projects project. 16.40% have been involved in project between 26–30 while, 11.00% have been involved in more than 31 projects.

Years of experience reveals that, 1–10 and 11–20 years of experience falls on the same percentage of 21.90%, which holds the highest percentage, 13.70% falls between 21–30 years of experience, 11.00% falls between 31–40 years of experience. 8.20% are between 51–60 years of experience. 6.80% are between 41–50 which are the lesser of all the percentage (6.80%) in the years of experience.

Also, statistics for professionals' reveals that 16% of the respondent were quantity surveyors while 18% are Architect, 17% are builder, 22% are Engineer while, they also belong to their professional membership.

Table 2. Influence of stakeholders on sustainable construction

Influence on sustainability	SII	Rank
Life-cycle costs such as construction, operations, and maintenance to be used more efficiently into project design	90.96	1
Innovative methods of project delivery	88.77	2
Strategic, concept planning, project programming and needs of each construction stages	88.49	3
Control of the total project cost	86.3	4
Specifications that go beyond best practice, and specify construction	85.48	5
Quality-based selection in design-build	84.93	6
Contracting methods best suited for the physical construction	84.93	7
Targeted support for selected technologies and for taking risks on innovative	81.1	8
Financing alternatives, leading to say certain levels of cash flow	81.1	8

Table 2 above shows that life-cycle cost such as construction, operations, and maintenance to be used more efficiently into project design is ranked the highest with the significant importance index 90.96, similarly Innovative method of project delivery and Strategic, concept planning, project programming and needs of each construction stages ranked directly below as 2nd and 3rd with significant importance index 88.77 and 88.49 respectively.

Table 3. Influences of stake holders on sustainable building construction during pre-construction, construction and post-construction

Professionals	Stages	N	Mean rank
Project manager	Pre-construction stage	73	4.11
	Construction stage	73	4.57
	Post construction stage	73	3.67
	Total	219	
Architect	Pre-construction stage	73	3.98
	Construction stage	73	4.50
	Post construction stage	73	4.27
	Total	219	
Client	Pre-construction stage	73	4.32
	Construction stage	73	4.47
	Post construction stage	73	4.51
	Total	219	
Contractor	Pre-construction stage	73	3.57
	Construction stage	73	4.06
	Post construction stage	73	2.56
	Total	219	
Quantity surveyor	Pre-construction stage	73	3.98
	Construction stage	73	3.67
	Post construction stage	73	2.45
	Total	219	
Engineers	Pre-construction stage	73	
	Construction stage	73	4.68
	Post construction stage	73	3.21
	Total	219	
Builder	Pre-construction stage	73	4.21
	Construction stage	73	4.62
	Post construction stage	73	3.24
	Total	219	
Funders	Pre-construction stage	73	2.78
	Construction stage	73	3.15
	Post construction stage	73	2.10
	Total	219	
Property advisors	Pre-construction stage	73	4.29
	Construction stage	70	3.75
	Post construction stage	73	4.64
	Total	216	
Consultants planner	Pre-construction stage	73	4.51
	Construction stage	73	2.38
	Total	219	

(*continued*)

Table 3. (*continued*)

Professionals	Stages	N	Mean rank
Local community	Pre-construction stage	73	3.47
	Construction stage	73	3.65
	Post construction stage	73	4.63
	Total	219	
Sub-contractors	Pre-construction stage	73	4.53
	Construction stage	73	2.40
	post construction stage	73	4.58
	Total	219	
Supplier	Pre-construction stage	73	2.12
	Construction stage	73	3.62
	Post construction stage	73	4.54
	Total	219	
Financial firms (banks)	Pre-construction stage	73	3.41
	Construction stage	73	3.45
	Post construction stage	73	3.51
	Total	219	
Trade associations	Pre-construction stage	73	3.12
	Construction stage	73	3.62
	Post construction stage	73	3.53
	Total	219	
Government authorities	Pre-construction stage	73	4.67
	Construction stage	73	2.10
	Post construction stage	73	1.97
	Total	219	
Investors	Pre-construction stage	73	4.49
	Construction stage	73	3.69
	Post construction stage	73	2.58
	Total	219	
Labour	Pre-construction stage	73	3.29
	Construction stage	73	4.71
	Post construction stage	73	3.65
	Total	219	
General public	Pre-construction stage	73	4.50
	Construction stage	73	3.68
	Post construction stage	73	4.69
	Total	219	

Table 3 Shows the influences of different stake holders, at different stages of pre-construction stages, Government authorities have the highest ranking, having the mean rank of 4.67, also mean rank of 4.51 for consultant planner, while the 3rd place with the mean ranking 4.50 for general public, investors also have the 4th mean ranking of 4.49, followed by the client and the property advisers, mean rank of 4.32 and 4.29 respectively.

While for the construction stage, the stakeholder that has the highest dominant role in the construction stage is the Labourers, mean ranking of 4.71, followed by the Engineers with mean rank of 4.68, also the 3rd place with the mean ranking 4.62 for Builder. The project manager was ranked 4th having the mean rank of 4.57, the 5th and 6th place according to their level of influence on construction stage were selected for the sub-contractor and the architect mean rank of 4.53 and 4.50 respectively. The 7th place goes for the client mean rank of 4.47.

The post construction stage, General public was selected to hold the highest influence in the post construction stage with the mean ranking of 4.69, while property advisors with mean rank of 4.64 rated 2nd highest influence on post building construction. The local community was ranked 3rd with mean rank of 4.63, also mean rank of 4.58 for sub-contractor. The 5th and 6th rank goes for the suppler and the client with mean rank 4.54 and 4.51 respectively.

Table 4. Test statistic on the level of influence of stake holder's on the sustainable building construction project

Test statatistic A, B (A) Kruskal wallis test (B) Grouping variable; stages	Chi-square	DF	Asymp sig
Project manager	31.95	2	.000
Architect	38.83	2	.000
Client	8.228	2	.016
Contractor	10.500	2	.005
Quantity surveyor	5.861	2	.053
Engineers	57.930	2	.000
Builder	58.030	2	.000
Funders	3.78	2	.151
Property advisors	6.892	2	0.32
Consultants planner	10.846	2	0.004
Local community	1.933	2	0.380
Sub-contractors	50.605	2	0.00
Supplier	54.311	2	0.00
Financial firms (banks)	5.945	2	.051
Trade associations	4.886	2	.081
Government authorities	13.753	2	.001
Investors	11.753	2	.003
Labour	76.529	2	.000
General public	22.066	2	.000

Table 4 according to the decision rule any Asymptotic significant (Asymp). Sig ≤ 0.05 has a significant influence. The table shows that the project manager, Architect, Contractor, Engineer, builder, sub-contractors, Supplier, Labour, General public, Government authorities, Quantity Surveyor, Financial bank, Investors, consultant planner, Client, all falls below to the Asymp sig of Sig $\leq$ 0.05 has more significant influence of sustainable construction while those that falls on >0.05 has lesser significant influence, which are Trade associate, Local community, property advisers, Funders.

5 Discussion of Findings

Marthur, Price and Austin [5] determined the essence of allotting different stakeholders in the quest for the assessment of sustainability. The study highlighted the benefits of stakeholder allotment for showcasing social learning and by implication, increasing the sustainability dispositions of professional in the construction industry. This paper further the study on stakeholder's influence on sustainable construction projects, by assessing their influences on each construction phase. Korkmaz, Riley and Horman [11] also conducted a survey to develop a kind of metrics that can basically be employed to assess high-performance sustainable building projects, commencing from contract structure to owner's attitude about sustainable building projects. The survey took a slight view, asking indeptly about the role of the general contractor in delivering sustainable projects. Korkmaz et al. [11] gave a clear assertion that when a contractor becomes involved in the post contract stage of construction work is panacea to achieving sustainable building goals, and these work supports this finding. Tan, Shen and Yao [12] studied further on evaluating sustainable project delivery and discuss the role of sustainable construction practices, by improving the competitiveness of the contractor in the marketplace. That work focuses on those practices entirely within the contractor's purview, e.g., waste management on site, while this research seeks the influences of stakeholders like the government authorities, general public, contractors and the likes has on each phase of construction work.

6 Conclusions and Recommendation

Results of the questionnaire survey of construction professionals as well as; other stakeholders involved in construction process and data analysis indicate that, most of the respondent have long years of experience, which did influence the nature of the answers obtained. From the analysis, government authorities have the highest authorities and influence on sustainable construction at the pre-construction stage of the project. For the construction stage, the study indicated that the labour(s) has the dominant role among other stakeholder at this stage. The post construction stage, general public was selected to hold the highest influence in the post construction stage. This study also indicate that life-cycle cost such as construction, operations, and maintenance to be used more efficiently into project design as the most important area of influence of the stakeholders. Therefore, government authorities, labour, general public as stakeholders, should work manageably for the sustainable project actualization.

References

1. Boesso, G., Kumar, K.: An investigation of stakeholder prioritization and engagement: who or what really counts. J. Acc. Organ. Change **5**(1), 62–80 (2009)
2. Jergeas, G.F., Williamson, E., Skulmoski, G.J., Thomas, J.L.: Stakeholder management on construction projects. AACE Int. Trans. **P12A**, 12.1–12.6 (2000)
3. Karlsen, J.T.: Project stakeholder management. Eng. Manage. J. **14**(4), 19–24 (2002)
4. Ofori, G.: Globalization and construction industry development: research opportunities. Constr. Manage. Econ. **18**(3), 257–262 (2000)
5. Mathur, V.N., Price, A.D.F., Austin, S., Moobela, C.: Defining, identifying and mapping (2007)
6. Atkin, B., Skitmore, M.: Editorial: stakeholder management in construction. Constr. Manage. Econ. **26**(6), 549–552 (2008)
7. Newcombe, R.: From client to project stakeholders: a stakeholder mapping approach. Constr. Manage. Econ. **21**(8), 841–848 (2003)
8. Smith, J., Love, P.E.D.: Stakeholder management during project inception: strategic needs analysis. J. Archit. Eng. **10**(1), 22–33 (2004)
9. Carroll, A.B., Buchholtz, A.K.: Business and Society: Ethics and Stakeholder Management, 6th edn. Thomson South-Western, Mason (2006)
10. Salah, H.: Investigating the stakeholder management in construction projects in the Gaza Strip. Thesis submitted to the department of Civil Engineering – Construction Management, The Islamic University of Gaza (2013)
11. Korkmaz, S., Riley, D., Horman, M.: Piloting evaluation metrics for sustainable high-performance building project delivery. J. Constr. Eng. Manage. **136**(8), 877–885 (2010)
12. Tan, Y., Shen, L., Yao, H.: Sustainable construction practice and contractors' competitiveness: a preliminary study. Habit. Int. **35**(2), 225–230 (2011)

Investigating a Male-Dominated Space: Female Students' Perceptions of Gendered Cultures in Construction Workplaces

Mariam Akinlolu[1(✉)] and Theo C. Haupt[2]

[1] School of Engineering, University of KwaZulu-Natal, Durban, South Africa
akinlolumariam@gmail.com
[2] Faculty of Engineering, Mangosuthu University of Technology, Durban, South Africa
theo.haupt@mut.ac.za

Abstract. Despite its significant contribution regarding employment in South Africa, women remain severely underrepresented in the construction industry. Men have become the primary beneficiaries of these job opportunities and the determinants of the construction industry's culture. This extreme gender stratification and macho nature of construction work has been identified as one of the main obstacles influencing women's career decisions and forms the basis for their significantly low participation in the construction industry. This study explores female students' perceptions and experiences of gendered cultures in male-dominated spaces, such as the construction workplace. The study adopted a quantitative descriptive survey design. Thirty-five purposively sampled female students, across different levels of study, completed a questionnaire with closed-ended questions. The samples were drawn from student cohorts enrolled in the construction studies; civil engineering, property development, land surveying and civil engineering program. Responses were that a majority of the students experienced consistent masculine gendered cultures or practices at their workplaces. The experiences presented in the study could be adapted to promote and enhance the inclusiveness of all genders. The study is a reflection of female students' memoirs of experiences at the workplace. The sample is drawn from only one university in South Africa and is not confirmed as a representation of women employed in the construction profession in South Africa.

Keywords: Construction workplace cultures · Gender · Male-dominated work · Retention · Women in construction

1 Introduction

Numerous studies have reported that despite an extensive range of global legislation developed to promote women's growth in the economy, women are still underrepresented in the construction industry and more so among construction students [1–4]. Although both first and third world countries share the problem of inadequate representation of women, women are present in the construction workforce of some developing countries compared to European countries [5]. Findings from studies

C. Aigbavboa and W. Thwala (Eds.): CIDB 2019, *The Construction Industry in the Fourth Industrial Revolution*, pp. 43–55, 2020.
https://doi.org/10.1007/978-3-030-26528-1_5

conducted in the European Union and America revealed a low representation of women in the professions with a review of women in construction in these countries [6–8]. Studies revealed percentages as low as 3.3% per annum, somewhat below the overall average employment growth rate of 3.4% [6]. While women in some developing countries outnumber men, their involvement in skilled labour is particularly low [8–10]. The Engineering Council of South Africa reported that out of their 11% female registered members, only 4% were professionally active [11].

Of the total employed South African population in 2017, 44% were women, and only 13% of these women were employed in the construction industry [12]. No changes have occurred over the years, as it has been 44% since September 2002. Furthermore, sectors of the economy such as mining and transportation had low concentrations of female employees [12]. Out of the total female workforce which is estimated at ten million and two hundred thousand, the construction industry has only one hundred and seventy-four thousand [12].

Although women are more successful than their male counterparts as students [6], women leave the construction profession at higher rates than men [1]. The masculine culture of the workplace, where interactions marginalizes the identities of female engineers has been identified as a prevalent contributor towards women's refusal to take up careers in the construction industry [13]. Gender divisions in the workplace are established by vertical segregation and gender stereotyping [2]. More specifically in the construction industry with a low number of women, there is an indication of gender segregation and a shortage of skilled females in the workforce [2].

Although significant for both genders, the culture of the industry was found to be principally a significant barrier for women. Women have to circumnavigate common issues such as discriminatory attitudes perpetrated by the dominant white male management and adversarial business relationships [14].

Numerous studies have shown that women who look to pursue careers in the construction industry either have to behave like men to be successful, leave if they cannot adapt to the masculine culture or maintain their feminine attitudes to occupy minor positions. [10, 15, 16].

Worldwide, the construction industry is currently facing skill shortages, and diversity-based levels are inadequate in meeting the skill gap [17]. Encouraging the participation of women would enable the untapped resources, promote inclusivity and improve the skill gap in the industry [1, 10, 17].

Numerous initiatives targeted to increase the representation of women within the construction industry have been established [17, 18]. Initially, these programmes were designed to create awareness and familiarise female engineering students with engineering courses, and eventually incorporated initiatives focused on improving curriculum, in light of recommendation from studies that concentrate on gender inclusivity in engineering curricula [6, 19, 20].

Emphasis has been placed on the applications of engineering science and project-based learning [21]. The role of the curriculum is to educate engineering students on how to manage and respond to social challenges at the workplace and should be part of the engineering curriculum [6]. Comprehensive and inclusive curricula should prepare students for the engineering workplace culture [22]. However, suitability of the curriculum as an instrument to engage students on social issues in the industry has not received much attention, which is disturbing, as engineering students experience workplace cultures even before they are out of school [4]. At most universities and higher learning institutions offering engineering degree programmes in South Africa, students are required to undergo at least 12 months of relevant work experience.

A few studies have investigated the experience of gendered cultures, and gender inclusivity of students in male-dominated workplaces [2, 4, 6, 15, 23]. In Australia, nine female and four male students from three universities were interviewed about their experiences and responses to gendered cultures during their workplace learning placements [4]. All students reported experiences consistent with gendered cultures. The study identified marginalization of women or stereotypically feminine practices or privilege of stereotypically masculine practices or traits.

In the United Kingdom, an investigation of the perception and responses of female students to the construction industry's masculine culture during their industry placements through interview sessions for 26 female engineering students [23]. The study highlighted the experiences of female students as being assigned to fill in co-water position and execute supporting activities. The students further reported that regularly work twice as hard as their male colleagues to prove themselves.

In South Africa, 17 female students were interviewed about their perceptions and preparedness for male-dominated workplaces, such as the construction industry [6]. Findings showed that the students perceived that they need to act their male counterparts to be able to prove their worth and continually accept discrimination from male colleagues. Responses from the study are similar to findings from a study investigating how women in engineering professions in the United States of America created their professional identities and how their interpersonal interactions at the workplace influenced their sense of belonging at the workplace [13]. Using data from interviews with 52 female engineers, the study identified women proving themselves and continuously striving to achieve a reputation. Also, the study found that women responded to these masculine cultures by rejecting gendered expectations.

A survey of 1435 industry practitioners, 141 first and final year construction students and 17 professional women revealed experiences of gender-based discrimination and sexual harassment [2]. In summary, numerous studies in Europe, America, Australia and Africa have revealed attitudes and experiences consistent with gendered cultures in male-dominated workspaces.

In comparison of existing literature concerning women's employment, retention and career progression in the construction industry, similar challenges are encountered. These are antagonistic business relationships, poor working practices and working conditions [21], Lack of training and education, lack of role models, lack of promotion or the glass ceiling [10, 16].

1.1 Theoretical Framework

In the feminist theory, the issue of gender is emphasised, and the differences in the experiences of men and women are acknowledged [24]. Feminism recognises that the oppressions and injustices women encounter have developed over time and has emanated from structural problems embedded in the society [25]. The study is conceptualized from an understanding of masculine cultures in male-dominated workplaces that relegate women and stereotypical feminine traits while promoting men and appreciating stereotypical masculine traits.

Specifically, the study relies on the feminist theory for an in-depth understanding of the influence of social and structural issues on gender and how it relates to the academic and workplace experiences of female students in construction. The feminist theory creates knowledge by reflecting critically on the experiences and perceptions of women [26].

Construction has a deficient female representation at all levels including management. Most times women find themselves in situations where they are the only female engineers in a workplace, and their needs are often neglected. Physical strength, technical skills and knowledge of construction support stereotypically masculine practices, which are considered as significant in the industry. Unlike professions like medicine, dentistry and law, in the construction industry, the majority of the engineers engage with people in stereotypically masculine domains such as technicians and tradespeople rather than those occupying administrative positions [27, 28].

For a better understanding of the workplace experiences and perceptions of women in male-dominated spaces, the study sought to identify exhibitions of gendered cultures as experienced by female students and the coping mechanisms adopted by the students. A study by [13] identified imposed gendered expectations, demeaning women and making requests based on gender as types of personal interactions reported by female students that diminish the professional identity of women in male-dominated workspaces.

1.2 Aim

The overarching aim of the study is to identify the perceptions, experiences and responses and coping mechanisms to gendered cultures in the workplace by female students in construction-related disciplines. The objectives were to make recommendations for the improvement and promotion of immediate and long-term inclusivity in the workplace interactions and experiences of female students.

2 Methodology

The study implemented a descriptive survey design adopting a quantitative research approach. Quantitative research is the systematic and objective process of using numerical data from a selected subject of the population to carry out a study [29]. Data was obtained through self - administration of a questionnaire with close-ended questions. SPSS version 25.0 was used to capture and compute the data. The descriptive analysis consisting of means, percentage and the standard deviation were used to analyse the data.

Female students were purposively sampled as the criteria for inclusion in the study was gender, which was female and were majoring in Property development, land surveying, civil engineering and construction studies. Students at this stage were first to third-year students enrolled at a university in the KwaZulu-Natal province of South Africa. Demographic information of respondents and details about their most significant work placements are presented in Tables 1 and 2.

Table 1. Participant demographics

Characteristics	Description	No of participants	%
Discipline	Property development	18	51.4
	Construction studies	6	17.1
	Land surveying	7	20.0
	Civil engineering	4	11.4
Level of study	1st year	2	5.7
	2nd year	16	45.7
	3rd year	14	48.6
Degree completed at time placement	1st year	13	37.1
	2nd year	13	37.1
	3rd year	9	25.7
Ethnicity	Black	21	60.0
	White	4	11.4
	Colored	2	5.7
	Indian	8	22.9

2.1 Material

The questionnaire content used in this study was adapted from themes identified from student's responses to gendered cultures in studies by [13, 23]. A Five-point Likert-type scale was used in the quantitative survey. The questionnaire included questions in the following relevant areas:

Section A: Demographic information
Section B: Details about respondents' placements
Section C: Experiences of interactions consistent with gendered workplaces

In Section A, participants were expected to provide details on their personal information. In Section B and C, respondents were required to respond according to the Likert scale provided. The first two set of questions included response options; Never, rarely, sometimes, often, always. The second set required participants to respond to the dichotomous scale; yes/no.

2.2 Participants

Participants in the study were confirmed to represent students in male-dominated disciplines. 35 female undergraduate students at a University in the KwaZulu-Natal province of South Africa were the participants in this study. Table 1 presents the demographic information of students who participated in the study.

51.4% (18) of the respondents were enrolled in the property development program, 7.1% (6) were enrolled in construction studies, 20% (7) were enrolled in land surveying, and 11.4% (4) were studying civil engineering. 5.7% (2) were 1st-year

Table 2. Details about participant's most influential placement

Characteristics	Description	No of participants	%
Significant work placement	Part-time work	12	34.3
	Vacation employment	11	31.4
	Internship	12	34.3
Number of professional female engineers at the workplace	1	4	11.4
	2	10	28.6
	3	11	31.4
	4	5	14.3
	5	3	8.6
	6	1	2.9
	7	1	2.9

students, 45.7% (16) were in their 2nd year, and 48.6% (14) were 3rd-year students. 37.1% (13) indicated that they were in their 1st year of study at the time of placement, 37.1% (13) were also in their 2nd year at the time of work placement, and 25.7%(9) were in their 3rd year. 60% (21) were Blacks, 11.4% (4) were White, 5.7% (2) were Coloreds, and 22.9% (8) were Indians.

Details on respondents nominated most prominent placements are represented in Table 2 34.3% (12) respondents indicated that undertaken part-time work was their most influential work placement. 31.4% (11) indicated vacation employment and 34.3% (12) confirmed internship as their most significant work placement.

Respondents were required to specify the number of professional female engineers employed at their workplace. It is evident from Table 2 that the maximum number of female engineers was 7 and was indicated by 1 student (2.9%).

3 Analysis

3.1 Frequency of Interactions with Professional Engineers

Relative to being asked about how often the respondents interacted with professional engineers at their workplace; almost all respondents reported that they had little or no interactions with the engineers at their place of work. 17.1% (6) indicated that they had "Never" interacted while 8.6% (3) students reported that they "Always" had interactions with professional engineers at their workplaces (Table 3).

Table 3. Frequency of interaction with professional engineers

Response	N	%
Never	6	17.1
Rarely	6	17.1
Sometimes	9	25.7
Often	11	31.4
Always	3	8.6
Total	35	100

3.2 Experience of Gendered Cultures

This section sought to investigate the experience and perceptions of respondents on the occurrence of gendered cultures at their work placements. Students were required to rate the level to which they experienced and observed a set of gendered cultures using a Five-point Likert scale, where 1 = Never, 2 = Rarely, 3 = Sometimes, four = Often, and 5 = Always. Table 4 shows that belittling women or drawing attention to their gender, making requests based on gender, marginalizing stereotypically feminine interests and lack of respect from tradespeople and technicians were rated as the most frequently experienced gendered interactions by the respondent. With a mean score of (2.0), students indicated that they least experienced negative attitudes from their superiors.

Table 4. Experience of gendered cultures

Experiences	1	2	3	4	5	T	M	SD	R
Belittling women or drawing attention to their gender	5.7	17.1	8.6	17.1	51.4	35	3.91	1.36	1
Making requests based on gender e.g. asking female engineers to do secretarial work	5.7	11.4	17.1	51.4	14.3	35	3.57	1.07	2
Marginalizing stereotypically feminine interests	11.4	8.6	40.0	17.1	22.9	35	3.31	1.25	3
Rough culture on site e.g. swearing, sexist comments, unwelcome body contact, unwelcome sexual connotations	14.3	20.0	22.9	25.7	17.2	35	3.29	1.86	4
Lack of respect from tradespeople or technicians e.g. being ignored by male team members	2.9	25.7	34.3	31.4	5.7	35	3.11	0.96	5
Imposing gendered expectations e.g. keeping female engineers from going site	17.0	20.0	31.4	22.9	8.6	35	2.86	1.21	6
Difficulty asking for support e.g. difficulty seeking help due to macho expectations	14.3	28.6	40.0	5.7	11.4	35	2.71	1.15	7
Unfair judgement of women's work	22.9	31.4	20.0	17.1	8.6	35	2.57	1.27	8
Negative attitude from superiors	45.7	25.7	14.3	11.4	2.9	35	2.00	1.16	9

3.3 Responses to Gendered Cultures

Respondents were required to indicate how they responded or coped with the gendered cultures they experienced at their work placements. 25.7% (9) reported that they left the job to avoid the culture while 71.4% (25) indicated that they stayed in the workplace. 68.6% of the students reported that they tolerated and adapted to the culture. In terms of justifying the gendered interactions experienced, 34.3% (12) indicated that they made excuses for the culture while 65.8% (23) reported otherwise. Notably, a large proportion of 74.3% (25) of the respondents indicated that reported the issues of gendered interaction they encountered (Table 5).

Table 5. Responses/coping mechanisms to gendered cultures

Responses	Yes		No	
	N	%	N	%
Leaving the workplace	9	25.7	25	71.4
Tolerating and adapting	24	68.6	11	31.4
Justifying interactions experiences	12	34.3	23	65.8
Denying the gendered culture	15	42.9	20	57.1
Reporting	26	74.3	9	25.8

4 Findings and Discussion

4.1 Interactions Consistent with Gendered Cultures

Belittling Women or Drawing Attention to Their Gender. Findings from the data analysis showed that students experienced interactions that were demotivating and consistent with gendered cultures. Respondents reported that they mostly experienced interactions that belittled women and drew attention to their gender. Interactions that demeaned women as marginalizing the interests and identities of professional women engineers were found in Hatmaker's study [13]. A similar study reported discomfort experienced by female students as a result of comments made about their gender [4, 15]. Examples were comments made by male contractors about prostitutes and comments on avoiding the use swear words because a woman was present.

Making Requests Based on Gender. The survey showed that students confirmed that during their placements, tasks were assigned to them based on their gender. Findings from [23, 30] confirmed that students undertook feminine stereotypical work employments. Women were assigned to undertake supporting roles such as secretarial duties which limits their opportunities and career progression [4, 6]. Although none of the women in the study intended to leave their male-dominated occupations, they demonstrated a reluctance to progress into the more intensely competitive male roles because of these negative self-perceptions. Rather, they opted for those 'softer' roles while remaining in male-dominated environments.

Marginalizing Stereotypically Feminine Interests. It is apparent from the study that students experienced a gendered workplace culture where stereotypically masculine interest where prioritised over feminine activities. Previous studies identified the presence of multiple masculine traits and practices in male-dominated workplaces which were mostly desired and given preference over stereotypically feminine traits [4, 31]. A past study found that predominant gender stereotypes exist in the workplace and form the basis for discriminatory employment regulations and management strategies exist in the workplace and form the basis for discriminatory employment regulations and management strategies that hinder women from making progress in male-dominated professions [32].

Rough Culture on Site. Gendered interactions consistent with a rough culture on site were identified by [4]. The study that although the rough culture on construction site cannot be avoided both men and women, in most cases the cultures on site are usually gendered. In this study, respondents reported experiences of swearing, sexist comments, unwelcome body contact and unwelcome sexual connotations.

Lack of Respect from Tradespeople and Technicians. In this study, students reported that they experienced disrespect or lack of support from tradespeople and technicians at their workplaces. There is a need to receive support and respect from tradespeople as reported by female students [15]. [23] described a situation where the male team members were ignoring the women as an example of such gendered interactions. Previous studies acknowledged the fact that women had to work twice as hard as men to build a reputation as a barrier for women in male-dominated professions [4, 33]. It was further noted that despite efforts to recruit and retain women, they felt unaccommodated by male colleagues at the workplace and experienced vindictive behaviour [32]. In [6] women experienced inherent forms of discrimination in interactions with their male team members in projects where their contributions were disregarded.

4.2 Students Coping Strategies for Gendered Cultures

Findings from the study showed that students' response to the gendered interactions they experienced are as follows;

- Reporting
- Tolerating and Adapting
- Leaving the workplace
- Justifying gendered culture
- Denying the gendered culture

Reporting. In this study, respondents reported that they reported the gendered interactions that they experienced. Reporting discriminatory incidents and harassments to the appropriate authorities is one of the few responses that could initiate change and provide support [4].

Tolerating and Adapting. Tolerating and adapting the interaction was one of the coping mechanisms reported by students as a coping mechanism for gendered cultures.

Many women choose to accept the masculine culture of the industry by trying to fit in because of the financial benefits they get from their jobs [33]. They are less concerned with the vulgar language and discrimination they experience. Women are either forced to 'undo' their gender to be able to cope with the disadvantages they encounter at their workplaces [34]. In a previous study, it was observed that women made efforts to adopt male practices and moderated their feminine traits and appearance to fit into the environment [32].

Leaving the Workplace. Majority of students in this study reported that they did not resign from their jobs because of the gendered cultures they experienced at their work placements. Although, [32] found that despite the negative work-identity interactions experienced, none of the women reported intentions to leave the workplace. However, the women displayed an attitude of low self-esteem and lack of confidence.

Justifying the Gendered Culture. Findings from the study indicate that gendered interactions experienced by the students were not justified. Contrary to findings in this study, female engineers were found to justify and make excuses for the gendered interactions at their workplace which discriminated against stereotypically feminine practices and marginalised their professional identities [13]. An example of excuses respondents made was making arguments that it was challenging for an employer to accommodate the needs of different genders.

Denying the Gendered Culture. The study findings suggest that students were indifferent in their response to denying gendered interactions at their workplace. A similar study found that students did not recognise the interactions they experienced as gendered. The studies concluded that there was uncertainty about whether the students did not recognize these cultures or chose to deny it [23, 30]. However, other similar studies reported responses consistent with denying the gendered culture [23, 27]. Female students were found not to notice any differences in their treatment and that of their male counterparts. Furthermore, these that found differences were more likely to deny or trivialize these gendered cultures [35].

5 Conclusion and Recommendations

The study concludes that the unique needs of women employed in male-dominated professions are unaccommodated at the workplace as a result of rooted stereotypical gendered cultures. The working practices in the environment are considered undesirable, and women are left to circumnavigate the problems they encounter in their respective professions. The study described gendered workplace interactions that female students experienced in engineering workplaces. Students experienced a rough culture on site, interactions that demeaned and drew attention to their gender, assignments based on their gender and lack of support and respect from team members. Students also perceived discriminations against their femininity and were obligated to conform to gendered expectations. Findings from the study highlighted prevalent structural and sociocultural influences that hinder the progress of women at the workplace.

Concerning the coping strategies adopted by students in response to the gendered cultures, the responses reported in this study were consistent with findings from previous studies [4, 13, 23, 32].

Acknowledging the gendered interactions of female students in workplaces in programmes to improve gender parity in male-dominated environments has been long overdue. Redressing the discriminations experienced by women in terms of recruitment, training and retention requires affirmative action. The gendered workplace cultures and coping strategies reported shows that employers need to make efforts to recruit women by proactively providing support and opportunities to attract them into male-dominated professions. To promote participation and career progression of women, employers and educators must improve workplace cultures and prepare students for the reality of the masculine engineering culture. Employers need to review their workplace policies and introduce solid initiatives geared towards creating a conducive environment which provides mentorship and accommodates the needs of women who are considering taking up careers in male-dominated fields.

The realities of gender discrimination should be addressed in the development of curricula. Female students need to understand the layers of discrimination and the challenges in male-dominated industries from their first contact with the occupations. This knowledge will assist students achieve their goals effectively without experiencing sexism at the workplace.

Acknowledgments. The authors greatly acknowledge the support of the National Research Fund.

References

1. Gurjao, S.: Inclusivity: the changing role of women in the construction workforce. The Chatered Institute of Building (CIOB), London (2006)
2. Madikizela, K., Haupt, T.: Influences on women's career choices of careers in construction: a South African study. Aust. J. Constr. Econ. Build. **10**, 1–15 (2010)
3. English, J., Hay, P.: Black South African women in construction: cues for success. Eng. Design Technol. **13**, 144–164 (2015)
4. Male, S.A., Gardner, A., Figueroa, E., Bennett, D.: Investigation of students' experiences of gendered cultures in the engineering workplaces. Eur. J. Eng. Educ. **3**, 360–377 (2017)
5. English, J.: The development of women in the South African construction workforce in terms of employment, training and skills acquisition. Unpublished Ph.D. thesis, School of the Built and Natural Environment, Glasgow Caledonian University (2007)
6. English, J., Alves, S.: Female students' preparedness dominated disciplines and careers. South Afr. J. Ind. Eng. (SAJIE) **3**, 112–125 (2018)
7. Lourens, A.: The development of co-curricular interventions to strengthen female engineering students' sense of self-efficacy and to improve the retention of women in traditionally male- dominated disciplines and careers. South Afr. J. Ind. Eng. (SAJIE) **3**, 112–125 (2014)
8. Navarro-Astor, E., Román-Onsalo, M., Infante-Perea, M.: Women's career development in the construction industry across 15 years: main barriers. J. Eng. Design Technol. **2**, 99–221 (2017)

9. Ahuja, V., Kumari, S.: Issues and challenges for women in construction industry: global as well as Indian perspective. In: Proceedings of the 18th Annual Convention and Seminar on Training Skill Upgradation and Competence Development in Building Industry, New Delhi, pp. 55–60 (2012)
10. English, J., Le Jeune, K.: Do women and tradeswomen in construction share common employment barriers? J. Prof. Issues Eng. Educ. Pract. (2012)
11. Thompson, M.: Women engineers quit over men's attitudes. Mail Guardian (2015)
12. Statistics South Africa.: Quartery Labour Force Survey, Quarter 2. Pretoria (2017)
13. Hatmaker, D.M.: Engineering identity: gender and professional identity negotiation among women engineers. Gend. Work Organ. **20**, 382–396 (2013)
14. Haupt, T.C.: Women in construction development, empowerment and sustainability (CODES), Report to DBSA, Halfway House, Midrand (2010)
15. Male, S., MacNish, C.: Pilot exploration of gender inclusivity of engineering students' exposure to engineering practice in an Australian university. Aust. J. Eng. Educ. **20**, 135–144 (2015)
16. Haupt, T., Fester, F.: Women-owned construction enterprises: a South African assessment. J. Eng. Design Technol. **1**, 52–71 (2012)
17. Worrall, L., Harris, K., Stewart, R., Thomas, A., Mcdermott, P.: Barriers to women in the UK construction industry. Eng. Constr. Archit. Manage. **17**, 268–281 (2010)
18. Fox, M., Sonnert, G., Nikiforova, I.: Programs for undergraduate women in science and engineering: issues, problems, and solutions. Gend. Soc. **25**, 589–615 (2011)
19. Godfrey, J.: The culture of engineering education and its interaction with gender: a case study of a New Zealand University, Ph.D., Curtin University of Technology (2003)
20. Godfrey, J., King, R.: Curriculum specification and support for engineering education: understanding attrition, academic support, revised competencies, pathways and access. Australian Learning and Teaching Council, Strawberry Hills (2011)
21. Louw-Harmse, Y.S.: Gender imbalance: a focus on senior management positions in the IT sector in South Africa. Mtech Dissertation, Cape Peninsula University of Technology, South Africa (2015)
22. Gill, J., Julie, E., Mills, S., Rhonda, S.: Oh you must be very clever! High-achieving women, professional power and the ongoing negotiation of workplace identity. Gend. Educ. **20**, 223–236 (2008)
23. Powell, A., Bagilhole, B., Dainty, A.: 2009: how women engineers do and undo gender: consequences for gender equality. Gend. Work Organ. **16**, 411–428 (2009)
24. Ropers-Huilman, R., Kelly, W.: Feminist methodology in higher education. J. High. Educ. **82**, 667–690 (2011)
25. Smith, K.N., Joy, G.G.: Setting up for the next big thing: undergraduate women engineering students' postbaccalaureate career decisions. J. Coll. Stud. Dev. **58**, 1201–1217 (2017)
26. Intemann, K.: 25 years of feminist empiricism and standpoint theory: where are we now? Hypatia **25**, 778–796 (2010)
27. Faulkner, W.: Doing gender in engineering workplace cultures. I. Observations from the Field. Eng. Stud. **1**, 3–18 (2009)
28. Faulkner, W.: Nuts and bolts and people: gender-troubled engineering identities. Soc. Stud. Sci. **3**, 331–356 (2007)
29. Maree, K., Pietersen, J.: Surveys and the use of questionnaires. Van Schaik, Pretoria, pp. 155–171 (2007)
30. Seron, C., Silbey, S., Cech, E., Rubineau, B.: Persistence is cultural: professional socialization and the reproduction of sex segregation. Work Occup. **43**, 178–214 (2016)
31. Connell, R.W.: The Men and the Boys, 1st edn. Wiley, Oxford (2013)

32. Martin, P., Barnard, A.: The experience of women in male-dominated occupations: a constructivist grounded theory inquiry. SA J. ind. Psychol. **39**, 2–12 (2013)
33. Franzway, S., Rhonda, S., Julie, E., Judith, G.: Engineering ignorance: the problem of gender equity in engineering. Front.: J. Women Stud. **30**, 89–106 (2009)
34. Hartman, H., Hartman, M.: Do gender differences in undergraduate engineering orientations persist when major is controlled. Int. J. Gend. Sci. Technol. **1**, 62–82 (2009)
35. Tonso, K.: Engineering gender—gendering engineering: a cultural model for belonging. J. Women Minor. Sci. Eng. **5**, 365–404 (1999)

Construction Transaction Costs in a Developing Economy: The Nigerian Construction Perspective

Matthew Osivue Ikuabe[1(✉)] and Ayodeji Emmanuel Oke[2]

[1] Department of Quantity Surveying, University of Benin, Benin City, Nigeria
matthew.ikuabe@uniben.edu
[2] Department of Construction Management and Quantity Surveying, University of Johannesburg, Johannesburg, South Africa
emayok@gmail.com

Abstract. The execution of construction projects involves cost other than that of production, erecting and putting in place structures; usually this is as a result of economic exchange. Optimization of construction project transaction cost is a much desired eventuality mostly on the part of owners of construction projects. Hence, construction project transaction cost is a vital anchorage in the deliberations of project success. This study sets out to examine the influencing factors of construction project transaction costs with Lagos state, Nigeria as the research area. Responses were gotten from construction professionals through the use of questionnaire survey with three hundred and thirty-seven distributed and two hundred and sixty-four retrieved and was fit for analysis. Mean Item Score, ANOVA, Kruskal-Wallis and Student Newman Kaul (SNK) Post-Hoc Test were adopted for analysis of obtained data. Findings from the study revealed that the major influencing factors of construction project transaction costs are Leadership, Quality of decision making and Competition between bidders. Equally, factor analysis revealed four underlying constructs which are Client's behavioral traits, Contractor's behavioral traits, Managerial skills and Project characteristics. The hypothesis of the study was subjected to a test; the result revealed that there is a significant difference in professionals' perceived influencing factors of construction project transaction costs. The study concluded by asserting the need for collaborators of construction project execution to as a matter of necessity ensure that actions and inactions depicted through the course of administering construction projects should be geared towards the optimization of transaction costs.

Keywords: Construction · Contractors · Project · Transaction cost · Professionals

1 Introduction

Construction works involves huge financial outlay before and during the execution of the project. Project owners and Investors would adopt all measures to bring about optimization of committed funds to any given project. Contractors expend substantial amount of time and resources to ascertain construction costs and then assessing the

C. Aigbavboa and W. Thwala (Eds.): CIDB 2019, *The Construction Industry in the Fourth Industrial Revolution*, pp. 56–61, 2020.
https://doi.org/10.1007/978-3-030-26528-1_6

price to be put to the owner. The rate at which exchange may take is then the price. The contractor's view of price translates to cost from the owner's perspective [1].

There are costs incurred other than production costs, they are referred to as transaction costs [2]. Production costs can be referred to the costs of transmuting inputs into outputs, while transaction costs is as a result of economic exchange. Transactions costs include costs incurred during construction activities such as preparing documents for bidding, the act of estimating, putting in place a contract, contract administration, and dealing with any deviation from contract conditions. Transaction costs also include costs connected with infringement of contractual promises; the costs of acquiring and processing information, organizational costs, legal costs, likewise costs in relation with inefficient pricing and production behaviour makes up transaction costs [3, 4].

Transaction cost is a vital construct in the discussion of project success delivery. Therefore this study sets out to identify and examine the factors affecting construction projects transaction costs in a developing country using Lagos state, Nigeria as its study area.

Factors Affecting Construction Transaction Costs

It has been established that transaction cost should be given much needed attention as the cost of production, as it is termed to be connected with cost of economic exchange. Construction sector is distinguishingly unique from other economic sectors. Its *modus operandi* in delivering the final end product is distinct and differs in relation to economic actors or participants and how the economic exchanges are carried out. This includes procurement-phase costs of bidding and contract negotiation; costs related to contract observance and renegotiation in the construction and functional stages [5]. According to [5], the factors that affect construction project transaction cost, portraying their descriptions as related to construction activities are: Project Uncertainty; Relationship with other parties; Integration of design and construction; Experience in similar type of projects; Payment on time; Organizational efficiency; Change orders; Qualifications of the contractor; Material substitution; Frequency of claims; Incentive/Disincentive clauses; Leadership; Bidding behavior; Quality of decision making; Quality of communication; Technical competency; Conflict management; Project complexity; Fair risk allocation; Experience in similar type of projects; Completeness of design; Relationship with subcontractors; Competition between bidders; Relationship with previous clients; Bonding requirement; and Early contractor involvement.

2 Research Methodology

The research utilized a quantitative research method, adopting convenience sampling technique. A survey of registered construction professionals affiliated with their respective professional bodies was carried out in Lagos state, Nigeria, the study area. The professionals were namely: Architects, Quantity Surveyors, Builders and Engineers. In total, three hundred and thirty-seven (337) questionnaires were distributed while two hundred and sixty-four (264) were retrieved and suitable for analysis, while the others were either not retrieved or incorrectly filled. Quantity Surveyors, Architects, Builders and Engineers were administered with 83, 86, 78 and 90 questionnaires with respectively. The methods of data analysis deployed are Mean Item Score, ANOVA, Kruskal-Wallis and Student Newman Kaul (SNK) Post-Hoc Test.

Data Analysis

Table 1 shows the ranking of factors that affects construction projects transaction costs for the different professionals making up the respondents. Architects ranked 'Completeness of design' as the most factor that affects construction project transaction costs with a mean score of 3.82, trailed by 'Project complexity' with a score of 3.77. For Quantity Surveyors, the factors are 'Frequency of claims' and 'Relationship with subcontractors' with mean scores of 3.73 and 3.72 respectively. Builders ranked 'Quality of decision making' the highest with a mean score of 3.62 followed by 'Incentive/Disincentive clauses' with a score of 3.42. While Engineers' most ranked factors are 'Project uncertainty' and 'Relationship with previous clients' with mean scores of 3.76 and 3.58 respectively. Overall, the most ranked factor was 'Leadership' with a mean score of 3.52.

ANOVA test generated F-ratio and significance value so as to examine the degree of diverging opinions of the different construction professionals with respect to each of the identified factors. Nineteen of the twenty-five variables had p-value less than 0.05; hence, it reveals a significant divergence between the construction professionals with respect to the nineteen factors.

Test of Hypothesis

Kruskal-Wallis H-test was undertaken to ascertain if there is a differing stance among the construction professionals. The analysis indicates an H value of 6.455, Asymptotic Significance p-value of 0.038 which is less than 0.05; hence, the **Null Hypothesis** stating that there is no significant difference in professionals' perceived factors that affects construction project transaction costs is **rejected**, while the **Alternate Hypothesis** stating that there is a significant difference in professionals' perceived factors that affects construction project transaction costs **accepted**.

Table 2 outlines the post-hoc test (multiple comparisons) undertaken using the Student Newman Kauls (SNK) Post-Hoc technique which was deployed to show underlying difference in professionals' perception of the factors affecting construction transaction costs. Result indicates that there is a discovered difference in the factors that affects construction project transaction costs between three groups of Respondents. Builders having a value of 3.0130 make up the first group. Engineers with a value of 3.2446 fall under the second group while Architects Quantity Surveyors with values of 3.3983 and 3.4918 respectively makes up the third group.

Table 1. Factors that affects construction project transaction costs

Identified factors	Architect		QS		Builder		Engineer		Overall		F-ratio	Sig. (p-value)
	Mean	Rank	Mean	Rank	Mean	Rank	Mean	Rank	Mean	Rank		
Leadership	3.55	4	3.61	10	3.38	3	3.53	3	3.52	1	3.458	0.000**
Quality of decision making	3.54	5	3.66	6	3.62	1	3.24	12	3.51	2	17.355	0.003**
Competition between bidders	3.28	20	3.46	9	2.85	18	3.24	12	3.43	3	1.409	0.142**
Incentive/disincentive clauses	3.49	7	3.62	16	3.4	2	3.23	14	3.42	4	2.345	0.034**
Frequency of claims	3.52	6	3.73	1	2.89	15	3.38	7	3.38	5	2.655	0.000**
Completeness of design	3.82	1	3.21	5	3.11	6	3.11	19	3.37	6	3.902	0.833
Payment on time	3.49	7	3.72	3	2.77	20	3.4	6	3.34	7	8.125	0.000**
Relationship with previous clients	3.46	11	3.56	13	2.72	22	3.58	2	3.33	8	0.714	0.039**
Integrating design and construction	3.49	7	3.67	8	2.86	17	3.45	5	3.33	8	0.031	0.000**
Material substitution	3.36	15	3.48	17	3.1	8	3.32	8	3.32	10	0.048	0.025**
Fair risk allocation	3.42	12	3.63	24	2.65	24	3.27	10	3.31	11	1.557	0.0614
Bonding requirement	3.37	14	3.58	7	2.94	12	3.23	14	3.3	12	7.897	0.001**
Unclear method of measurement	3.3	18	3.25	23	3.11	6	3.49	4	3.29	13	11.752	0.714
Early contractor involvement	3.26	22	3.5	11	2.71	23	3.01	21	3.28	14	19.301	0.000**
Experience in similar type of projects	3.32	17	3.55	14	3.01	10	3.23	14	3.28	14	9.762	0.001**
Bidding behavior	3.27	20	3.43	20	3.21	5	3.17	17	3.27	16	15.339	0.041**
Conflict management	3.38	13	3.57	12	3.03	9	3.12	18	3.27	16	14.721	0.004**
Relationship with parties	3.29	19	3.47	18	2.91	13	3.25	11	3.23	18	3.345	0.023**
Relationship with subcontractors	3.36	15	3.72	2	2.66	23	2.59	24	3.21	19	1.119	0.974
Project uncertainty	3.61	3	3.49	19	2.82	19	3.76	1	3.21	19	12.225	0.036**
Project complexity	3.77	2	3.66	15	2.76	21	3.02	20	3.12	21	0.179	0.000**
Quality of communication	3.12	24	3.18	25	3.27	4	2.87	22	3.11	22	4.788	0.852
Technical competency	3.15	23	3.27	22	2.9	14	2.99	21	3.08	23	1.189	0.532
Qualifications of the contractor	3.01	25	3.08	27	2.99	11	2.6	23	2.92	24	23.478	0.032**
	3.4		*3.49*		*3.01*		*3.24*		*3.29*			

Table 2. SNK post-hoc test

Groups	N	Subset for alpha = 0.05		
		1	2	3
Builders	27	3.0130		
Engineers	27		3.2446	
Architects	27			3.3983
Quantity surveyors	27			3.4918
Sig.		1.000	1.000	.113

Means for groups in homogeneous subsets are displayed.
a. Uses Harmonic Mean Sample Size = 27.000.

3 Discussion of Results

The study revealed that among the identified factors that affect construction project transaction costs, the most ranked factor was Leadership. This is in support of the study [6], who opined that quality leadership entails projection of vision so as to align project participants with their goals, propel their team to take cooperative actions and to attain project set out goals, in return optimizing transaction costs. Good leadership has been proven to be indispensable quality for successful delivery in any human endeavor. With the right leadership in place for construction projects, transaction costs will be most likely optimized which turn serves as a pointer to successful project delivery. Also, 'Competition between bidders' ranked as the next highest factor affecting construction project transaction costs. [7] asserted that a reduction in transaction is most likely with the involvement of low bidders which may create a low procurement process. Next highest ranked factor was 'Incentive/Disincentive clauses'. As noted by Broome and Perry [8], Incentive clauses may inspire the contractor to minimize contract timing, thereby leading to a reduction in transaction costs.

4 Conclusion and Recommendations

Construction transaction costs can be influenced by a whole lot of varying factors. The study has established the factors that affect construction transaction costs thereafter examining these factors with statistical scrutiny. Among the factors has Leadership as the most ranked by professionals closely followed by Quality of decision taken. Inference derived from the study equally revealed that Architects and Quantity Surveyors have similar underlying notion of construction transaction costs, with both professionals agreeing to similar factors influencing construction transaction costs. This was proven quite different from Builders and Engineers. This goes to reveal that the professionals in the construction sphere have differing perception of probable influencing factors to construction transaction costs.

It is imperative that collaborators of construction projects should as a matter of utmost importance give credence to transaction costs during project execution as much emphasis has been placed on production costs. Equally, stakeholders should as place emphasis on actions that aid the optimization of transaction costs.

References

1. Hillebrandt, P.M., Hughes, W: What are the costs of procurement and who bears them? In: Proceedings of 2nd International Conference on Construction in Developing Countries. Botswana National Construction Industry Council and University of Botswana, Gaborone, Botswana (2000)
2. Winch, G.M.: The construction firm and the construction project: a transaction cost approach. Constr. Manage. Econ. **7**(4), 331–345 (1989)
3. Rahman, M.M., Kumaraswamy, M.M.: Minimising transaction costs, maximising relational benefits and optimizing risk management—through partnering in Hong Kong projects. Int. J. Constr. Mark. **3**(2), 51–72 (2002)
4. Joskow, P.L.: Vertical integration and long-term contracts: the case of coal-burning electric generating plants. J. Law Econ. Organ. **1**(1), 33–80 (1985)
5. Li, H., Arditi, D., Wang, Z.: Transaction costs incurred by construction owners. Eng. Constr. Archit. Manage. **21**(4), 444–458 (2014)
6. De Meyer, A.: Collaborative leadership: new perspectives in leadership development. Working Paper, Cambridge Judge Business School, University of Cambridge, Cambridge, UK (2010)
7. Farajian, M.: Transaction cost estimation model for us infrastructure public private partnerships. M.Sc. thesis, University of Maryland, College Park, MD (2010)
8. Broome, J., Perry, J.: How practitioners share fractions in target cost contracts. Int. J. Proj. Manage. **20**(1), 59–66 (2002)

4IR Leadership Effectiveness and Practical Implications for Construction Business Organisations

Kehinde Alade(✉) and Abimbola Windapo

Department of Construction Economics and Management, University of Cape Town, Upper Campus, Rondebosch, Cape Town 7701, South Africa
Aldkeh001@myuct.ac.za, Abimbola.Windapo@uct.ac.za

Abstract. The construction industry has generally been labelled as rigid and slow in its response to change. This is not unconnected with certain features of construction products such as their project-based nature, multiple-stakeholders and professionals, and lengthy processes involved in the construction supply chain. The fourth industrial revolution (4IR) presents another moment of change in the global business arena and is dynamically influencing different organisations and their strategic positioning. There is evidence that leadership drives the change agenda. However, scholars also suggest that construction business leaders are unhurried in their pace, strategic decisions and flexibility. This study employed a systematic literature review in identifying the leadership traits, style and intelligence required for construction leaders to lead effectively in the 4IR. The leadership framework developed gives an indication as to the practical implications of 4IR for construction business leaders and concludes that a combination of factors will lead to leadership effectiveness and high levels of 4IR implementation.

Keywords: 4IR · Change intelligence · Construction organisations · Leadership · Sustainability

1 Introduction

In a global marketplace that has presented several surprises and a high degree of uncertainty recently, disruption and change have become a common word. It has been widely acknowledged that the fourth industrial revolution (4IR), categorized by disruptive and rapid technological changes and digitization is new territory for the world. The 4IR, defined as a quick and revolutionary change to the world characterized by the fusion of technologies, blurring the lines between the physical, digital, and biological spheres, is fast changing the way organisations function and how they provide services [1–3]. The introduction of new business models and re-shaping of socio-economic systems are evidence of the profound changes taking place across all industries, thus making leadership and opportunities within 4IR more challenging [4]. As [5] and [6] claim, leadership plays a crucial role in ensuring success in almost any change initiative within an organization. It is the ultimate key to organisations successfully embracing a change initiative into their daily business operations [7]. Leadership is the process of

C. Aigbavboa and W. Thwala (Eds.): CIDB 2019, *The Construction Industry in the Fourth Industrial Revolution*, pp. 62–70, 2020.
https://doi.org/10.1007/978-3-030-26528-1_7

social influence that attributes causation to individual social actors [8–12]. A leader's job is to identify the drivers of performance and follow through to the flow of activities at the process level [6]. This leadership process drives change in the organization, and it involves various factors such as developing a vision for the organization; aligning people with the idea through communication; and motivating people to action through empowerment and basic need fulfillment [9, 11, 13–16]. Given the Fourth Industrial Revolution's rapid pace of change and broad impacts, business leaders are being challenged to take advantage of the opportunities and efficiently use digital programs to improve their efficiency and organisation performance [2]. However, a key trend also noted by scholars is the reluctance of construction business leaders to cope or learn how to make rapid changes that will positively impact their organisations [17–19]. According to Ofori and Toor [20] and Suresh *et al.* [6] there is a greater need for effective leadership in the construction industry than any other sector. The complexity of construction organisations stems from the management of project-based activities, many stakeholders involved in its process, different procurement methods, projects uncertainty, unpredictable nature of the environment and a fragmented supply chain [21].

As Oosthuizen [4] argues, there are leadership challenges and opportunities peculiar to 4IR. Likewise, the construction leader's inability to lead change, innovation culture, and cope with the challenges of technological changes and innovation is evident and a different approach to leadership is required in construction organisations, to take advantage of 4IR and the digital era [17–19]. Also, Gaultier [18] notes that most of the CEOs in construction firms have worked in an executive role for decades and have not been directly involved in the latest technological evolution, thereby leading to organisational disconnect. These challenges establish the need to understand and investigate the leadership change intelligence, traits, and styles that are required by the CEOs in construction firms for an effective 4IR in construction businesses. Therefore this study investigates the requirements for 4IR leadership effectiveness in construction organisations, towards establishing a leadership framework for driving organizational change and improving performance in construction business organisations. The paper comprises of five sections. In Sect. 2, background information on leadership change intelligence, traits, and styles are provided. Information on the research methodology is provided in Sect. 3. Section 4 presents the conceptual framework for effective 4IR leadership, while discussions, conclusions, and limitations of the study are presented in Sect. 5.

2 Theoretical Background for the Study

2.1 Leadership Change Intelligence (LCQ)

Leadership Change intelligence (LCQ) is the awareness of the leader of his or her change leadership style, and the ability to adapt the style to be optimally effective in leading change across a variety of people and situations [22]. In the light of disruption, complexity and uncertainty that the 4IR conveys, [2, 4, 23] agree that a special kind of intelligence is required for leaders to adapt, harness and maximize their potentials. Whereas [2] propagates a 4-type intelligence for leaders in 4IR (contextual, emotional, inspired and physical intelligence), [23] expanded on this to include entrepreneurial intelligence, to increase the potential of 4IR realisation for impact and value-creating solutions [4] however developed a 4IR-leadership intelligence framework, shown in Fig. 1, for leaders in general.

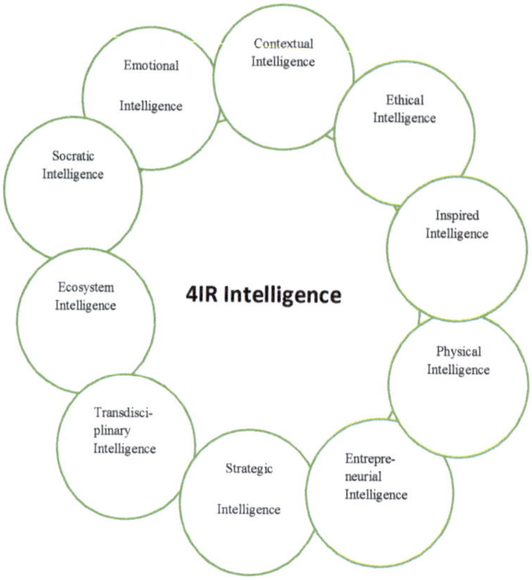

Fig. 1. 4IR leadership intelligence framework (Source: [4])

2.2 4IR Leadership Traits (4LT)

Scholars have argued whether leaders are born or made and if some characteristics or qualities distinguish leaders from followers. There are also debates as to whether leadership attributes can be developed. Current leadership trait theorists suggest a consensus is emerging around the "HEXACO" (Honesty-Humility, Emotionality, Extraversion, Agreeableness, Conscientiousness, and Openness to Experience) personality framework and domains [24] and [25]. In relation to 4IR and leadership in construction business organisations, [18] notes four key attributes of a digitization-ready leader in the construction industry which include disruption-preparedness, agility, focus on talent development, and collaborative approach.

2.3 Change Leadership Styles (CLS)

While the underlying concept of leadership is the premise on having followers, scholars submit that the relationship between a leader and follower will differ based on the style of the leader [8, 9, 26]. As a result, many leadership styles are documented in the literature. Leaders have been classified as autocratic, democratic, transactional, transformational, charismatic and strategic. Among these leadership styles, the change leadership styles developed by Trautlein and Trowbridge [22] appears to be the most appealing since it reveals the attitude of leaders to change for effectiveness in different situations. assume that leaders' function as a coach, champion, visionary, driver, executor, facilitator, and adapter when leading change in an organisation. Trautlein and Trowbridge [22] also note that each leader has a unique change leadership style shown in Fig. 2 below, depending on how strong the leader is at heart (people-oriented), head (purpose-oriented) or hands (process-oriented). It is believed that by combining these functions, leaders will effectively drive 4IR.

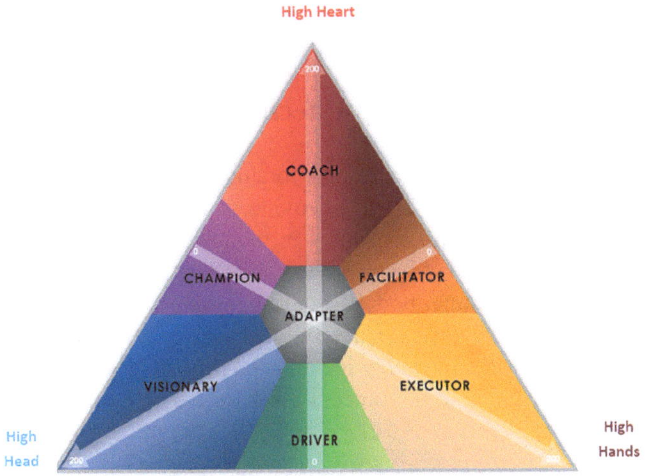

Fig. 2. Change leadership styles (Source: [22]

3 Research Method

A systematic literature review approach is used for this study. The relevant literature on 4IR, change leadership, leadership style and traits for effectiveness in construction organizations were examined based on the objective of the study employing journal articles, conference proceedings, online articles and other related texts on 4IR from the Science direct, Web of Science, Scopus and the Google scholar database. The search process covered from year 1995–2018. Whereas the research articles in the domains of 4IR and leadership is quite broad, generating an initial output of about 19,000 articles, the objective of this study- identifying the leadership traits, style and intelligence required for construction leaders to lead effectively in the Fourth Industrial Revolution, was used to narrow down the dataset to the appropriate fields of the research to an outcome of about 300. Top journals in the field of Management and leadership were considered. These include *Harvard Business Review, Academy of management review, Journal of Quality Progress, International Journal of Financial Research, International Journal of Management, Business, and Administration.* Others that relates specifically to the Engineering and Construction leadership such as *the Procedia Engineering, Journal of Construction in Developing Countries, Journal of Architectural Engineering Design Management and Journal of Architectural Engineering Technology* were also included. Top Conference papers such as those from *Conference on Responsible Leadership, Leadership and management in construction (LEAD) conference, Project Management Institute Global Congress* and *Southern African Institute of Management Scientists*. It is noteworthy that some duplicates were removed from this dataset leaving a total of 58 journal articles and 12 conference proceedings. The most relevant Journal articles and conference proceedings to the objective of the study from these were the seventeen (17) and four (4) used respectively for the study.

Using the results of the literature review, a framework is developed, and the implications are discussed the subsequent sections.

4 Developing 4IR Leadership Effectiveness Framework

The 4IR Leadership effectiveness framework for construction organisations is shown in Fig. 3. Studies of [4–7, 18, 22–24, 27] provide support for this framework. These studies concluded that leadership especially in moments of uncertainty and change is beyond a position or title; instead it is a process of influencing others to understand and agree on what needs to be done, how it should be done, whilst facilitating individual and collective efforts to accomplish shared objectives, thus signifying that a leader's job is to foresee the future and quickly identify the drivers of performance and follow through the flow of activities at the process level.

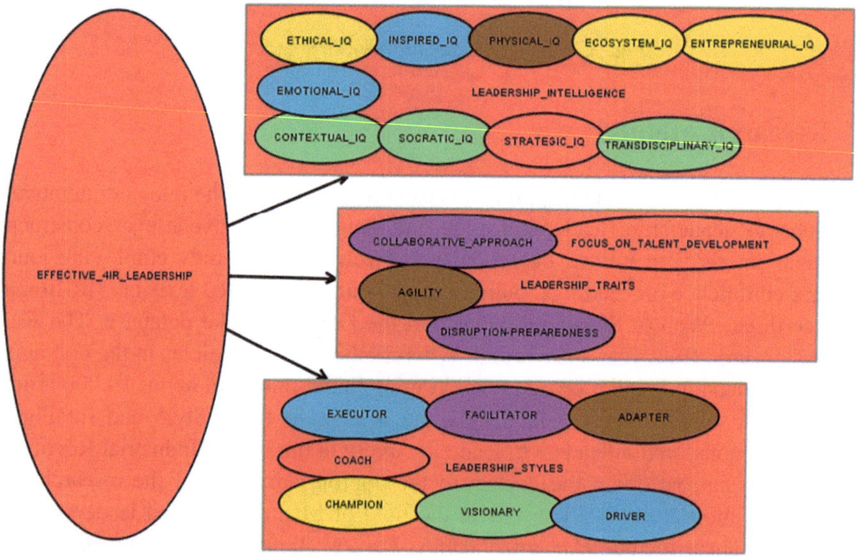

Fig. 3. 4IR leadership effectiveness in construction business organisations

The framework shown in the Fig. 3, links effective 4IR leadership to ten-type leadership intelligence, four-type leadership traits and change leadership styles. The ten leadership intelligence featured, namely, Contextual Intelligence, Emotional Intelligence, Inspired Intelligence, Physical Intelligence, Entrepreneurial Intelligence, Strategic

Intelligence, Transdisciplinary Intelligence, Ecosystem Intelligence, Socratic Intelligence, and, Ethical Intelligence require leaders to develop themselves broadly in the areas identified, to survive the 4IR. Disruption-preparedness, Agility, Focus on talent development, and Collaborative approach was identified as the leadership traits. In the framework, Disruption-prepared refers to the ability of the construction leader to embrace disruption.

Some scholars have argued that the concept of disruption is threatening. However, acknowledging its reality and putting in place a strategy for multiple futures to constant hanges with significant impact on the organization will help cushion the effect. Agility as a leadership trait refers to the ability of construction executives to respond quickly to change or risk losing forward-thinking employees to new industry players with fresh thinking. Focus on talent development is the aptitude of construction executives to look for ways to upskill their existing workforce in line with the requirements of the 4IR. It is important that construction executioves develop their workforce to take up jobs as drone operators, robot resource managers and augmented reality trainers, to name a few. The last leadership trait is a Collaborative approach, and it refers to the ability of construction executives to network not only with their internal team members but also with the, academia, industry players and other relevant bodies for mutual learning. This attribute is critical for digitization-ready leaders in construction because 4IR is a new environment that requires networking to maximize benefits.

The change leadership styles required that 4IR oblige construction CEOs to be people-oriented, purpose-oriented, and process-oriented. The change style of leadership demands that the CEOs function as a Visionary, Coach, Champion, Executor, Driver, Facilitator, and Adapter in their organizations. The visionary leader uses the head, to set a compelling vision of an exciting future. The leader here envisions the future using head focus and sees new situations and possibilities.

Operating as a coach require the leader to engage colleagues and subordinates whenever possible and provide support for people, thereby taking everyone in the organization through the change process. As a champion however, the change leader combines the strength in the heart and head to get others to change. The leader in this case not only envisions the future, but is also able to communicate effectively and encourage others warmly so that they consent to change.

The leader as an executer gets things done quickly often using execution approach backed up by comprehensive, step-by-step plans of the future. Leaders who operates as rivers are usually strong on both head and hands. Not only do they envision the future, they also use execution abilities to drive toward the vision, laying out clear strategies and tactics along the way. Leaders who operates as facilitators support people on a day to day basis using their strong heart and hands capabilities. Having identified tasks that need to be accomplished, they usually check measurable progress and succeed in motivating others to work together on those tasks. Finally, as an adapter, a change leader engages the head, heart, and hands. They tend to be generally flexible, politically savvy, and willing collaborate with others.

5 Conclusion, Recommendations and Limitations of the Study

Construction leaders driving the 4IR agenda will ideally possess a combination of *Change leadership intelligence* (LCQ), *Change leadership styles* (CLS) and the *4IR leadership traits* (4LT) to be effective. Knowledge and understanding of the 4IR landscape and strategic positioning will facilitate effectiveness. At the personal level, construction executives will ideally be change intelligent (see Fig. 3) to embrace the 4IR. Construction executives will also be conversant with their change leadership style and adopt the 4IR leadership traits for effectiveness. Besides, a firm awareness of all other change leadership styles (CLS) identified in this study will help in adapting as situation warrants to promote effectiveness. At the team/Group level: This second level is concerned with how leaders impact others and drive employees through the 4IR change. This level is also concerned with building relationships by persuading supporters, detractors, and fence-sitters to get onboard with the required changes. As such, a caring attitude and concern for developing the required 4IR skills are essential. A disposition to engage employees and be supportive will also inspire the team members to be 4IR driven. As earlier stated, the construction industry and organization are a part of the larger society and the global community and does not exist in isolation. Therefore, collaboration with other organisations such as academia, government agencies will foster effectiveness. Also, benchmarking 4IR activities of the organization with other organisations will promote efficiency. This is both an internal and external role for 4IR leadership effectiveness. In practical terms, 4IR leadership effectiveness in construction organisations will reflect all the dimensions of leadership unraveled and discussed in the previous sections. Although there are theoretical justifications for the framework advanced in this paper, this study recommends that they are empirically tested to prove its validity. Future studies will examine the effectiveness of South African construction executives' Leadership change intelligence, change leadership style and adaptability, and 4IR leadership traits employing semi-structured interviews for the testing. Data will be collected from Level 1 leaders: Chief Executive Officers (CEO), Board members senior executives and business unit managers of building and civil engineering contractors listed in Grades 7–9 on the cidb Register of Contractors in South Africa. The data obtained will be analysed using descriptive statistics and content analysis. This will form the foundation for testing the framework developed for construction leaders' effectiveness and successful deployment of 4IR initiatives in the construction business organisations.

Acknowledgements. Funding from TETFUND, the National Research Foundation (NRF) and UCT-ARUA Collaborative project towards this research is hereby acknowledged. Opinions expressed and conclusions arrived at, are those of the authors and are not necessarily to be attributed to TETFUND, NRF or UCT-ARUA.

References

1. Xu, M., Jeanne, M., Kim, S.: The fourth industrial revolution: opportunities and challenges. Int. J. Financ. Res. **9**(2), 90–95 (2018)
2. Schwab, K. The fourth industrial revolution: Currency, New York (2017)
3. Gamechangers Information. https://www.thegeniusworks.com/2017/07/disruptive-technologies-exactly-applications-mean-business
4. Oosthuizen, J.: The determinants of fourth industrial revolution leadership dexterity: a proposed framework for 4IR intelligence and subsequent 4IR leadership development. Paper presented at the 4th International Conference on Responsible Leadership, At GIBS–Johannesburg, Conference paper (2017)
5. Dinwoodie, D., Pasmore, W., Quinn, L., Rabin, R.: Navigating change: a leader's role. Greensboro (2015)
6. Suresh, S., Egbu, C., Olomolaiye, A., Renukappa, S.: Leadership roles for implementing knowledge management in construction organisations. Paper Presented at the Leadership and Management in Construction (LEAD) Conference (2008)
7. Center for Creative Leadership Information. https://www.ccl.org/wp-content/uploads/2016/09/navigating-change-a-leaders-role-center-for-creative-leadership
8. Antonakis, J., Day, D.: The Nature of Leadership. Sage publications, Los Angeles (2017)
9. Bass, B., Bass, R.: The Bass Handbook of Leadership: Theory, Research, and Managerial Applications. Free Press, New York (2009)
10. Merritt, E.A.: Leadership and Strategy: Understanding, Planning and Implementing. Day Press, USA (2017)
11. Kruse, K.: What is leadership. Forbes Mag. **3** (2013)
12. Pfeffer, J.: The ambiguity of leadership. Acad. Manag. Rev. **2**(1), 104–112 (1977)
13. Lunenburg, F.C.: Leadership versus management: a key distinction—at least in theory. Int. J. Manage. Bus. Adm. **14**(1), 1–4 (2011)
14. Goleman, D.: What makes a leader? Harv. Bus. Rev. **82**(1), 82–91 (2004)
15. Goleman, D.: Leadership that gets results. Harv. Bus. Rev. **78**(2), 78–90 (1998). HBR Boston. What makes a leader, pp. 93–102 (2000)
16. Kotter, J.P.: Leading change: why transformation efforts fail. Havard Bus. Rev. (1995)
17. Matarneh, R., Hamed, S.: Exploring the adoption of building information modeling (BIM) in the jordanian construction industry. J. Archit. Eng. Technol. **6**(1), 189 (2017)
18. Crown Publications Information. http://crown.co.za/latest-news/construction-world-latest-news/8194-a-leadership-reset-in-construction-for-the-fourth-industrial-revolution
19. Goulding, J.S., Pour Rahimian, F., Arif, M., Sharp, M.: New offsite production and business models in construction: priorities for the future research agenda. J. Archit. Eng. Design Mgt. **11**(3), 163–184 (2015)
20. Ofori, G., Toor, S.: Leadership and construction industry development in developing countries. J. Constr. Dev. Ctries. **17** (2012)
21. Liphadzi, M., Aigbavboa, C., Thwala, W.: Relationship between leadership styles and project success in the South Africa construction industry. Proc. Eng. **123**, 284–290 (2015)
22. Trautlein, B.A., Trowbridge, C.: You know your IQ, but what's your CQ? Develop your change intelligence to lead successful and sustainable transformation and results. Paper presented at PMI Global Congress North America, New Orleans, LA. Project Management Institute, Newtown Square (2013)

23. Oosthuizen, C.: Entrepreneurial intelligence: expanding schwab's four-type intelligence proposition to meaningfully address the challenges of the fourth industrial revolution. Paper presented at the Proceedings of the 28th Annual Conference of the Southern African Institute of Management Scientists (2016)
24. Waldman, D.A., Ramirez, G.G., House, R.J., Puranam, P.: Does leadership matter? CEO leadership attributes and profitability under conditions of perceived environmental uncertainty. Acad. Manag. J. **44**(1), 134–143 (2001)
25. Weller, J., Aysel, T.: Predicting domain-specific risk taking with the HEXACO personality structure. J. Behav. Decis. Mak. **24**(2), 180–201 (2011)
26. Çelik, S., Özkul, E., Tuna, P.: The relationship between strategic decision-making and leadership styles: an application in 4 and 5-star hotels in Istanbul. İşletme Araştırmaları Dergisi **8**(1), 240–264 (2016)
27. Edgeman, R., Dahlgaard, S., Dahlgaard, J., Scherer, F.: On leaders and leadership. Qual. Prog. **32**(10), 49 (1999)

Evaluation of Safety and Security Measures: Preliminary Findings of a University Student Housing Facility in South Africa

Solomon Adisa(✉), Fredrick Simpeh, and Julius Fapohunda

Department of Construction Management and Quantity Surveying,
Cape Peninsula University of Technology,
Bellville Campus, Cape Town, South Africa
solomon.akinwumi@gmail.com

Abstract. Safety and security management of tertiary institutions have become a global concern. Studies show that the safety of student housing facilities (SHFs) are not giving the necessary attention in tertiary institutions. The neglect of the safety aspects of SHF exposes students to several risk and dangers such as accidents, theft, fire outbreak, sexual harassment, assault and, worse of all, death. Consequently, the paper investigates students' perception about the level of provision of security and fire safety measures and their satisfaction level with the performance/functionality of those measures in a post-graduate on-campus SHF. The study is exploratory in nature. The methodology used for the study is mixed method. Data was collected from students who are registered member of the residence through a closed-ended questionnaire and personal observation was carried out by the researchers to validate the result. Descriptive statistics was used to analyse the data. The results obtained revealed that measures such as: water sprinkler system, and smoke extractor were not provided in the SHF, while measures including fire extinguisher, fire assembly point and fire alarm were poorly provided. Other risk factors identified include: absence of CCTV, lack of weapon detector and poor access control. It is evident that students demonstrate a high level of dissatisfaction with the provision and performance level of security and fire safety measures at their residence.

Keywords: Evaluation · Facilities · Safety · Security · Student housing

1 Introduction

The university education boosts world development, hence, enriches people intellectual competence and stimulate knowledge development among young and older generations. [12] described tertiary institutions as a social system that has both input and output cycle, the inputs of schools are comprised of human, facilities, financial and material resources. All of these inputs are transformed to produce the desired results. In consequence, tertiary education brings positive changes to the social and economic development of any nation [6].

Generally, facilities that are required in universities may differ from one institution to another depending on the nature of academic activities and population of students

C. Aigbavboa and W. Thwala (Eds.): CIDB 2019, *The Construction Industry in the Fourth Industrial Revolution*, pp. 71–81, 2020.
https://doi.org/10.1007/978-3-030-26528-1_8

[11, 13]. Some of the physical facilities required in universities include lecture theaters, administrative buildings, laboratories, library, student housing, offices, sport fields, roads, furniture and general infrastructure [16, 19]. [19] commented that a sustainable student housing can be regarded as an integral component of facilities provided at any tertiary institutions with the aim of assisting students to have a quick access to learning facilities and lecture venues. As a matter of fact, SHFs play a crucial support role towards a quality higher education and as well provides a fulfilling living experience for the students.

[3] described SHFs as a unit of an en suite single room or multiple single rooms in a shared apartment, where amenities are also shared among room and flat mates which are often located on the campus or off campus. Additionally, [11] acknowledged the important of SHFs in any tertiary institutions as such facilities are designed to meet the needs of present students without compromising the ability of future generations to meet their own needs. [1, 6, 18] elaborated on the contribution of SHFs on student academic achievement as it provides students with a sense of home away from home and influences a desirable educational outcome.

However, in recent years, safety and security have become a major challenge in SHFs globally. This has been a topic of interest among researchers and universities in SA [9]. Despite a national call for the provision of accessible, decent, safe and academically conducive student accommodation, safety and security in student housing facilities in South Africa still remains a serious challenge [4]. [7] explained that the lapses in the issue of safety and security in the management of student housing in South Africa has led to; increase number of murder cases in the student housing, fire outbreak, accident report, theft, poor lighting condition at night, sexual harassment and assaults in the university SHFs.

Although quite a number of studies on SHFs in the broader scope has been conducted [3, 6, 13, 18, 19], studies which primarily focuses on the evaluation of safety and security measures in a SHFs from the perspective of students is quite lacking. Thus, the aim of the paper is to evaluate the level of provision of security and fire safety measures as well as students' level of satisfaction with the performance/functionality of those measures in an on-campus SHFs.

2 Safety and Security Measures in Student Housing Facilities

Safety and security have been a critical challenge in the students housing facilities in both on-campus and off campus residences nationwide [6, 14]. Studies show that safety and security is not giving adequate attention in South Africa student housing. [20] argue that situation has arisen in tertiary institutions in SA, where facilities managers, university health and safety unit, hostel managers and institution management ignore the role of safety and security in the management of institution infrastructure such as student housing.

[17] stressed that one of the few studies investigating safety and security in South Africa on-campus SHF demonstrates that students feel unsafe at their respective institutions with several incidences of crime and injuries. [10] revealed another security gap/lapses in the management of SHF in South Africa tertiary institutions owing to

another incident that took place on 1st of May, 2018 at a particular university in Durban a coastal city in KwaZulu-Natal Province of South Africa where a first year Quantity Surveying student was murdered in an undergraduate residence. [8] further revealed that South Africa also has the highest level of recorded robbery with 149.4 incidents recorded per 100,000 of the population both in the campus residence and off campus accommodation.

Similarly, a study conducted by [17] identified certain security and safety risk factors in the SHFs in South Africa tertiary institutions such as lack of CCTV, poor access control, absence of burglar bars on the doors, inadequate lighting at night, absence of weapon detector, lack of security patrol around the hostel, absence of emergency help line, lack of written policy prohibiting vandalism, faulty fire alarm, absence of smoke detector, inadequate fire assembly point, inadequate fire hydrants, insufficient fire extinguisher, leaking showers, absence of signage, inadequate emergency exit and lack of water sprinkler system.

According to [6] inadequate provision of fire safety measures in the SHF is a major factor responsible for fire related incidences in the hostels and further stressed that a SHF is a typical example of buildings that accommodate a large number of occupants and such facilities are particularly vulnerable because of their high fire load which can be described as amount of fuel within the building, thus required adequate fire safety measures. [17] commented that fire and electrocution incidents are key risk factors in South Africa student housing, with 93 fires incidences recorded in 2009 at educational institutions nationwide in South Africa with 8% of these fires occurring as a result of faulty extension cord, open flame and smoking materials such as cigarettes.

The security and fire safety measures included in the questionnaire for this study are: Security measures such as: CCTV, security alarm, security check point, fencing around the hostel, smart-card access control, lighting at night, weapon detectors, security patrol around the hostel, emergency help line, written policy prohibiting vandalism and notice board. Fire safety measures include: fire alarm, smoke detectors, evacuation fire drills, water sprinklers, fire hose reels, fire assembly point, fire hydrants, fire extinguishers, signage and emergency exit.

The National Building Regulation and Building Standards Act No 103 of 1977 of South Africa highlighted safety measures that must be prioritised in buildings include: escape door, escape route, feeder route, fire extinguisher, fire hydrant, foam inlet, fire main, fire pump connection, fire stopping and smoke/heat detector [15].

3 Research Methodology

This is an exploratory study (ongoing research project) that focuses on a post graduate SHF. The strategy adopted for the study is a survey. Data was collected quantitatively with the use of questionnaire. A closed-ended questionnaire was structured and administered to students. The total number of registered students was provided by residence manager as 75. However, for the purpose of this study, questionnaires were distributed to the total number of students residing in one on-campus post graduate residence. Only 31 questionnaires were duly completed by the respondents representing a response rate of 41.3% of the total population. All questionnaires were distributed and collected in person.

During the distribution and collection period, the researchers observed and took note of security and fire safety measures provided in the SHF. A 5-point Likert scale was used to determine the level of provision of security and fire safety measures and satisfaction level of students with performance/functionality of those measures provided in the SHF. The scale reads as follows, 1 = Well provided, 2 = Provided, 3 = Somewhat provided, 4 = Poorly provided, 5 = Not provided and the Unsure option. Likewise, for the level of satisfaction, 1 = Very satisfied, 2 = Satisfied, 3 = Somewhat satisfied, 4 = Unsatisfied, 5 = Not at all satisfied and the Not applicable option. The quantitative data were analysed using Statistical Package for the Social Sciences (SPSSv25) with the use of descriptive statistic and researchers carried out observational study to validate the result.

4 Findings and Discussion

Profile of Respondent. Table 1 indicates that 54.8% (17) of the respondents are female while 45.2% (14) are male. 51.6% (16) are Masters students and 48.4% (15) are PhD students. 12.9% (4) of the respondents are Colored, 80.6% (25) are Black and 6.5% (2) represent White. 80% of the questionnaire respondents are South African and 20% are international students. 32.3% (10) of the students responded that they have been living in the post-graduate hostel for about period of 2years, 12.9% (4) for 4years, 22.6% (7) for 3years, 19.4% (6) for 1year, and 12.9% (4) for the period of 5years.

Table 1. Profile of respondents

Profile of participants Gender	N	%
Female	17	54.8
Male	14	45.2
Total	31	100
Level		
Masters	16	51.6
PhD	15	48.4
Total	31	100
Race		
Colored	4	12.9
Black	25	80.6
White	2	6.5
Total	31	100
Duration of stay		
5 yrs	4	12.9
4 yrs	4	12.9
3 yrs	7	22.6
2 yrs	10	32.3
1 yrs	6	19.4
Total	31	100

Provision of Security and Fire Safety Measures in the SHF

Table 2 below presents students' perceptions of the security measures in the SHF. It is evident from the percentages recorded that the majority of the respondents are of the opinion that; CCTV (77.4%), weapon detector (71.0%), fencing around the hostel (48.4%), emergency help lines (61.3%), security patrol around the hostel (48.4%), security alarm (45.2%) and emergency protocol poster (41.9%) are the security measures that are not provided in the selected on-campus post-graduate SHF.

However, no possible explanation can be provided for the few students who indicated that CCTV and Weapon detector was poorly provided as no CCTV and weapon detector could be found during the observation carried out in the selected post-graduate SHF. Fencing around the hostel was perceived by the majority of students as not provided, though as low as 4 respondents (12.9%) indicated that fencing was provided. However, it was observed that there is a fencing but very low and actually meant for demarcating campus boundary and not directly erected for the SHF which gives a free access to both students and complete strangers to the SHF which could be of high risk to the residents.

The findings also revealed that smart-card access control 15 respondents (48.4%), security guard on post 12 respondents (38.7%) and security check point 11 respondents (35.5%) responded that these measures were poorly provided in the SHF. Lighting at night 14 respondents (45.2%) perceived that it was provided in the SHF. It is quite surprising that as many as 7 of the respondents indicated that lighting at night was well provided and 14 students also indicated that it was provided. However, it was observed that lighting was poorly provided at night as surround and rear side of the hostel was not well illuminated mostly at night. The remaining security measures ranged from somewhat provided and poorly provided which is an indication that improvement should be made in the provision of security measures in the SHF.

Table 2 below also presents the responses from the respondents on the level of provision of fire safety measures. Smoke detector 11 respondents (35.5%), water sprinkler system 12 respondents (38.7%), fire assembly point 10 respondents (32.3%) and evacuation fire drills 10 respondents (32.3%) were ranked by the majority of the respondents as the most fire safety measures that were not provided. However, it was discovered during the observation that although there was no fire assembly point within the selected SHF, there was a nearby fire assembly point within the campus field. Furthermore, the findings show that fire alarm 10 respondents (32.3%) and fire safety signage 10 respondents (32.3%) were indicated by the students as fire safety measures that are poorly provided. While fire extinguisher 14 respondents (45.8%), fire hydrants 13 respondents (41.9%) and fire hose reels 10 respondents (32.3%) were perceived as fire safety measures that are somewhat provided in the SHF.

It can be deduced from the findings that emergency exit 11 respondents (35.5%) indicated that it was provided. Observation also revealed that some of these fire safety measures were faulty and most students do not know how to operate fire extinguisher in case of minor fire occurrence. More than 70% of the respondents responded in the range of somewhat provided and poorly provided. Thus, it is evident from the analysis that improvement is needed in the provision of fire-safety measures in the SHF. University facilities/maintenance departments should focus their attention on security and fire-safety measures in the SHFs.

The few studies on SHF safety revealed the lack of such measures in SA institutions, for example the study of [17] also identified the lack of CCTV, absence of smoke detector, poor access control, absence of burglar bars on the doors, inadequate lighting at night, absence of weapon detector, lack of security patrol around the hostel, absence of emergency help line, lack of written policy prohibiting vandalism, faulty fire alarm, faulty electrical outlets, inadequate fire assembly point, inadequate fire hydrants, insufficient fire extinguisher, absence of signage, inadequate emergency exit, and lack of water sprinkler system were key SHF safety lapses in SA universities.

Table 2. Provision of security and fire safety measures in the SHFs.

Security measures	Well provided		Provided		Somewhat provided		Poorly provided		Not provided		Unsure		Total
	N	%	N	%	N	%	N	%	N	%	N	%	
CCTV	0	0	0	0	0	0	5	16.1	24	77.4	2	6.5	31
Weapon detector	0	0	1	3.2	2	6.5	2	6.5	22	71.0	4	12.9	31
Fencing around hostel	0	0	4	12.9	3	9.7	3	9.7	15	48.4	6	19.4	31
Emergency help lines	1	3.2	0	0	5	16.1	6	19.4	19	61.3	0	0	31
Security patrol around the hostel	1	3.2	2	6.5	5	16.1	8	25.8	15	48.4	0	0	31
Security alarm	0	0	4	12.9	6	19.4	6	19.4	14	45.2	1	3.2	31
Emergency protocol poster	0	0	4	12.9	4	12.9	10	32.3	13	41.9	0	0	31
Policy against vandalism	5	16.1	3	9.7	5	16.1	8	25.8	8	25.8	2	6.5	31
Smart-card access control	0	0	6	19.4	8	25.8	15	48.4	2	6.5	0	0	31
Security on post	0	0	8	25.8	9	29.0	12	38.7	1	3.2	1	3.2	31
Security check point	2	6.5	8	25.8	5	16.1	11	35.5	5	16.1	0	0	31
Notice board	4	12.9	8	25.8	6	19.4	6	19.4	5	16.1	2	6.5	31
Safety & security sign	2	6.5	8	25.8	10	32.3	8	25.8	3	9.7	0	0	31
Lighting at night	7	22.6	14	45.2	3	9.7	4	12.9	3	9.7	0	0	31
Fire safety measures													
Smoke detector	0	0	0	0	5	16.1	12	38.7	11	35.5	3	9.7	31
Water sprinkler	0	0	7	22.6	9	29.0	1	3.2	12	38.7	1	3.2	31
Fire assembly point	0	0	6	19.4	7	22.6	5	16.1	10	32.3	3	9.7	31
Evacuation fire drills	1	3.2	4	12.9	8	25.8	5	16.1	10	32.3	3	9.7	31
Fire alarms	0	0	7	22.6	6	19.4	10	32.3	6	19.4	2	6.5	31
Signage	0	0	6	19.4	8	25.8	10	32.3	4	12.9	3	9.7	31
Electrical outlets & switches	1	3.2	7	22.6	6	19.4	7	22.6	9	29.0	1	3.2	31
Fire hydrants	0	0	6	19.4	13	41.9	3	9.7	8	25.8	1	3.2	31
Fire hose reels	1	3.2	7	22.6	10	32.3	7	22.6	5	16.1	1	3.2	31
Emergency exit	1	3.2	11	35.5	7	22.6	2	6.5	10	32.3	0	0	31
Fire extinguisher	2	6.5	8	25.8	14	45.8	3	9.7	4	12.9	0	0	31

Students' Satisfaction Level with Security and Fire Safety Measures in the SHF
This section portrays the extent at which students are satisfied with different security and fire safety measures in the SHF. Table 3 below, depicts students' satisfaction level with the security measures in the SHF. It can be deduced from the percentage obtained that the majority of the respondents expressed a feeling of dissatisfaction with the provision/performance of security measures at their residence.

The majority of respondents were either unsatisfied or not at all satisfied with measures including: weapon detector, CCTV, emergency help line and security patrol around the hostel. It can be established from the percentage obtained that none of the security measures actually met the satisfactions of students in terms of their performances except for lighting at night where 10 respondents (33.3%) students expressed a feeling of satisfaction and 7 respondents (23.3%) revealed that they are somewhat satisfied. At least more than 75% of the respondents indicated that they are unsatisfied with the security measures at their residence especially with CCTV, weapon detector, access control to the hostel, fencing around the hostel, emergency help lines and security alarm.

It could be assumed from the percentage in the table that notice board and lighting condition at night fell within the expectations of students as more than 60% of the student's responses fell within very satisfied, satisfied and somewhat satisfied which indicate a feeling of satisfaction.

Furthermore, from Table 3 below, the response obtained indicates that electrical outlets and emergency exit are the two fire safety measures that actually met the satisfaction of the students in the SHF. According to the percentage of responses for electrical outlets which range from 15 respondents (48.4%) who responded with satisfied and somewhat satisfied and emergency exit 18 respondents (58.6%) responded with satisfied and somewhat satisfied respectively which indicate a level of satisfaction.

Moreover, it is evident from the percentages recorded that the majority of the respondents are of the opinion that smoke detector 15 respondents (48.4%), fire alarm 10 respondents (33.3%) and signage 10 respondents (33.3%) were fire safety measures which fell within not at all satisfied. The majority of the percentage obtained for the other fire safety measures such as: water sprinkler system, fire assembly point, evacuation fire drill, fire hydrant and fire hose reels fell within somewhat satisfied and unsatisfied respectively. Students demonstrate a high level of dissatisfaction with fire safety measures provided in the SHF, only with exception of fire extinguisher where there is a dispersion of data around the mean. It is evident from the responses that less than 30% of the total respondents are satisfied with fire safety measures at their residence and more than 70% are unsatisfied.

Studies on safety, security and sustainable SHFs by [2, 5, 6, 17] all support that safety and security measures are crucial to occupants and often influences students' choice of tertiary institution which is an important and first factor that contributes to students' satisfaction. In addition, students generally expressed dissatisfaction with their SHFs if safety and security measures are lacking or not adequately provided as learning and peaceful living experience can only take place in a safe and conducive living environment [19].

Table 3. Students' satisfaction level with security and fire safety measures in the SHF.

Security measures	Very satisfied		Satisfied		Somewhat satisfied		Unsatisfied		Not at all satisfied		Not applicable		Total
	N	%	N	%	N	%	N	%	N	%	N	%	
Weapon detector	0	0	0	0	1	3.3	12	38.7	15	48.4	2	6.7	31
CCTV	1	3.3	0	0	2	6.7	10	33.3	16	53.3	1	3.3	31
Emergency help lines	0	0	2	6.7	2	6.7	13	43.3	13	43.3	0	0	31
Security patrol around the hostel	1	3.3	0	0	2	6.7	15	50.0	8	26.7	1	3.3	31
Fencing around hostel	1	3.3	4	13.3	4	13.3	8	26.7	11	36.7	2	6.7	31
Security check point	1	3.3	1	3.3	9	30.0	8	26.7	11	36.7	0	0	31
Security alarm	2	6.7	3	10.0	4	13.3	11	36.7	10	33.3	0	0	31
Emergency protocol poster	1	3.3	5	16.7	3	10.0	13	43.3	7	23.3	1	3.3	31
Access control	1	3.3	2	6.7	11	36.7	8	26.7	8	26.7	0	0	31
Policy against vandalism	1	3.3	5	16.7	6	20.0	12	40.0	5	16.7	1	3.3	31
Security signs	3	10.0	6	20.0	7	23.3	7	23.3	7	23.3	0	0	31
Notice board	3	10.0	5	16.7	9	30.0	9	30.0	3	10.0	1	3.3	31
Lighting at night	5	16.7	10	33.3	7	23.3	5	16.7	3	10.0	0	0	31
Fire safety measures													
Smoke detectors	1	3.3	2	6.7	4	13.3	7	23.3	15	48.4	1	3.3	31
Water sprinklers	1	3.3	7	23.3	7	23.3	8	26.7	3	10.0	4	13.3	31
Evacuation fire drill	4	13.3	2	6.7	8	26.7	8	26.7	6	20.0	2	6.7	31
Fire assembly point	1	3.3	2	6.7	9	30.0	9	30.0	6	20.0	3	10.0	31
Fire alarms	2	6.7	4	13.3	5	16.7	8	26.7	10	33.3	1	3.3	31
Signage	1	3.3	4	13.3	6	20.0	7	23.3	10	33.3	2	6.7	31
Electrical outlets & switches	1	3.3	10	33.3	5	16.7	4	13.3	8	26.7	2	6.7	31
Fire hydrants	1	3.3	5	16.7	9	30.0	4	13.3	8	26.7	3	10.0	31
Fire hose reels	1	3.3	5	16.7	9	30.0	7	23.3	6	20.0	2	6.7	31
Emergency exit	1	3.3	9	30.0	9	30.0	5	16.7	3	10.0	3	10.0	31
Fire extinguishers	4	13.3	6	20.0	11	36.7	3	10.0	5	16.7	1	3.3	31

Unfortunately, in most SHF satisfaction studies, provision of space and amenities was their main research focus but the few conducted such as student satisfaction in university residence hall by [5] and performance evaluation of sustainable student housing by [6] revealed that students feel satisfied when security and fire- safety measures are adequately provided with other residential amenities especially when they are in a good working condition. On the other hand, students get disappointed (dis-satisfied) if these measures are not well provided and do not perform to their expectation.

Therefore, it is paramount that university health and safety unit, facility managers, university maintenance department and hostel managers make every possible effort to ensure that security and safety measures are prioritised in the SHFs by making necessary and adequate provision for these safety measures so as to eradicate security and safety gap/lapses in the SHF. Furthermore, universities ought to accept that, safety and security of student, staff and visitors to the campus is their legal obligation therefore it becomes a crucial aspect of management.

5 Conclusion

Safety, security and environment are integral component of facility management e.g. SHFs. However, studies show that it has been utterly neglected and seen as a separate function by facility managers as they only focus on maintenance activities which has brought huge safety problems in many residential buildings most especially SHFs in South Africa.

SHFs are one of the most important facilities to students and is an essential facility in the university setting. Thus, it becomes important that safety and security measures are well provided, and ensured they are in good working condition in order to ensure that students are safe and secured at their various residences which would consequently increase their satisfaction level and improve their learning experience. Therefore, the study sort to evaluate safety and security measures on sustainable student housing facilities by investigate students' perceptions about the level of provision of security and fire-safety measures required in an on campus SHF.

It is evident, that students have different opinion and perception about the level of provision of those measures and their satisfaction level varies. The majority of security and fire-safety measures were ranked as not provided by the respondents and some of these measures are not functioning well as established from the observation. Very few safety measures were perceived to be well provided such as fire extinguishers.

6 Limitation

Data used for this study was obtained from one post-graduate SHF of a University in the Western Cape Province, South Africa. Thus, this may reduce the representation of other SHFs in the country.

7 Recommendation

SHFs should be well provided with security and fire safety measures such as CCTV, weapon detector, security alarm, access control with smart-card, security personnel should be on post regularly, written policy prohibiting vandalism, security patrol around the hostel, moderately high fencing, adequate lighting at night, smoke detector, water sprinkler system, fire safety alarm and emergency exit should be well provided. A regular safety inspection should be carried out in the SHFs by a competent health and safety officer/facilities manager and a written report should be prepared based on the his or findings for an immediate action as maintenance lapses may result in safety risk.

Acknowledgement. This work is based on the research supported wholly by the National Research Foundation of South Africa (Grant Numbers: 116844).

References

1. 'Ulyani Mohd Najib, N., et al.: Student residential satisfaction in research universities. J. Facil. Manag. **9**(3), 200–212 (2011)
2. Atkin, B., Brooks, A.: Total Facility Management. Wiley Blackwell, Hoboken (2015)
3. Cyprain, O., Omunagbe, B.: Drivers and consequences of residents' satisfaction with off-campus student housing in South-south, Nigeria. Nelson Mandela Metropolitan University (2015)
4. Department of Higher Education and Training: Report on the ministerial committee for the review of the provision of student housing at South African universities. Department of Higher Education and Training (2011)
5. Foubert, J.D., et al.: Predictors of student satisfaction in university residence halls. J. Coll. Univ. Stud. Hous. **27**(January), 41–46 (1998)
6. Hassanain, M.A.: On the performance evaluation of sustainable student housing facilities. J. Facil. Manag. **6**(3), 212–225 (2008). Article information
7. Hemingway, J.A.: Can we do better when students are a threat to self? A review of legal and policy implications for current practices on college campuses (2015)
8. Institute for Security Studies and Africa Check: South Africa's official crime statistics for 2013/14. https://africacheck.org/factsheets/factsheet-south-africas-official-crime-statistics-for-201314/
9. Kern, G.M., et al.: College housing fire safety directorate for engineering sciences, December 2004
10. Lungani, Z.: University student #ZolileKhumalo murdered at res, ex-student arrested | Sunday Tribune. https://www.iol.co.za/sunday-tribune/news/university-student-zolilekhumalo-murdered-at-res-ex-student-arrested-14742549
11. Mustafa, F.A.: Performance assessment of buildings via post-occupancy evaluation: a case study of the building of the architecture and software engineering departments in Salahaddin University-Erbil. Iraq. Front. Archit. Res. **6**(3), 412–429 (2017). https://doi.org/10.1016/j.foar.2017.06.004
12. Nutt, B.: Four competing futures for facility management. Facil. Emerald **18**, 124–132 (2000)
13. Olanrewaju, A.L.A., et al.: Quantitative analysis of defects in Malaysian university buildings: providers perspective. J. Retail Leis. Prop. **9**(2), 137–149 (2010). https://doi.org/10.1057/rlp.2010.2
14. Rasmussen, C., Johnson, G.: The ripple effect of Virginia Tech: Assessing the nationwide impact on campus safety and security policy and practice. Midwest. High. Educ. Compact. 1–32 (2008)
15. Regulations, N.B.: National building regulations and building standards act 103 of 1977. Building 1977, 1–20 September 1985 (1996)
16. Rinn, A., Rinn, A.N.: Academic and social effects of living in honors residence halls (2004)
17. Rodriguez, L.: Investigating risk and protective factors to mainstream safety and peace at the University of South Africa. Afr. Saf. Promot.: J. Injury Violence Prevent **11**(1), 39–60 (2013)

18. Too, L., Bajracharya, B.: Sustainable campus: engaging the community in sustainability. Int. J. Sustain. High. Educ. **16**(1), 57–71 (2015)
19. Ulyani, N., et al.: Measuring satisfaction with student housing facilities. Am. J. Eng. Appl. Sci. **4**(1), 52–60 (2011)
20. Xaba, M.: An investigation into the basic safety and security status of schools' physical environments. South Afr. J. Educ. **26**(4), 565–580 (2006)

Work-Life Balance Practices in the Construction Industry of Swaziland

Bhekinkosi Dlamini, Olalekan Shamsideen Oshodi(✉), Clinton Aigbavboa, and Wellington Didibhuku Thwala

SARChl in Sustainable Construction Management and Leadership in the Built Environment, Faculty of Engineering and the Built Environment, University of Johannesburg, Johannesburg, South Africa
os.oshodi@gmail.com

Abstract. The construction sector plays a significant role in the economy through the provision of constructed space for productive activities. The process of executing projects in the construction sector is labor intensive and often workers need to work for long hours. Evidence found in literature shows that family conflict, emotional exhaustion and job dissatisfaction are prevalent in the construction sector. These problems have negative impact on organization and project performance. Previous studies have shown that the adoption work-life balance practices lead to significant improvements in organizational performance. The purpose of this investigation is to assess the level of adoption of work-life balance practices among construction firms in Swaziland. Questionnaire survey approach was utilized in the current study. It was revealed that a large majority of the respondents are aware of work-life balance practices. Also, leave (causal, parental [paternity/maternity], and study) and wellness programs are the most commonly used work-life practices. However, work from home and part-time work are the least used. The findings of this study shed more light on the current work-balance practices in the construction sector of Swaziland. Further research is required to understand the link between work-life balance and job satisfaction.

Keywords: Systems engineering construction industry · Job performance · Interpersonal relationship · Turnover intention · Work-life balance practices

1 Introduction

The construction industry is faced with several challenges, which affects its performance. Shortage of workers, ageing construction workforce, poor project performance and low productivity are examples of the problems facing the construction sector [1–3]. Inefficient use of scarce resources play a major role in poor performance of construction projects. This outcome is due to the vast amount of financial resources used in procuring the products of the construction sector. To improve the performance of the construction sector, stakeholders (academics and practitioners) have recommended the adoption and utilization of numerous innovative processes, practices and techniques. Work-life balance is an example of such practices.

C. Aigbavboa and W. Thwala (Eds.): CIDB 2019, *The Construction Industry in the Fourth Industrial Revolution*, pp. 82–89, 2020.
https://doi.org/10.1007/978-3-030-26528-1_9

Work-life balance practices has received considerable attention from researchers in the academic field of management. A review of work-life balance literature shows that its impacts on work outcomes seem to vary across various sectors. Bambra et al. [4] found that the use of compressed working week does not always lead to improvements in the health of shift workers. In contrast, Bradley et al. [5] found that work-life balance practices has significant influence on job satisfaction among workers in the construction industry. The differences in the findings of these studies suggest that the impact of work-life balance could vary due to differences in work environment, job schedules and nature of the work. Despite these variances, the adoption of work-life balance practices by construction benefits could yield beneficial outcomes.

In the field of construction management, a number of scholars have focused on work-life balance practices. Studies have explored work-life balance practices in the Australia [7] and US [8]. However, little is known about work-life balance practices in the construction sector of countries in the global south. The aim of this study is to examine the level of implementation of work-life balance practices in the construction industry of Swaziland. To achieve this aim two objectives were investigated: (i) to assess the level of adoption of work-life balance among construction firms in Swaziland, and (ii) to examine if the implementation of work-life balance practices has an impact on job performance, quality of interpersonal relationship among employees and job turnover.

2 Literature Review

2.1 Work-Life Balance: Understanding the Terminology

Work-life balance practices refer to various interventions targeted at creating a healthy equilibrium between personal life and job tasks [8]. Greenhaus et al. [9] view work-life balance as a situation when employees can achieve the right balance between work and personal responsibilities. Research has shown that achieving this balance could lead to higher quality of life and job satisfaction and reduction in employee turnover [5, 9, 10]. Due to the long work-hours associated with jobs in the construction sector [5], the utilisation of work-life balance practices can lead to improved organisational outcomes.

2.2 Work-Life Balance Practices

A considerable amount of literature has been published on work-life balance practices. A review of previous research shows that numerous work-life practices are utilised in various companies [11, 12]. Bardoel et al. [11] identified 36 types of work-life balance practices. In contrast, [12] identified 13 types of work-life practices implemented by various organisations. Despite the existence of various approaches used to classify work-life practices, there is no generally accepted approach.

The approaches used to classify work-life balance practices are numerous. Bardoel et al. [11] classified work-balance practices into five groups, namely defendant benefits, flexible working conditions, leave options, information and personnel policies and organisational cultural issues. Cegarra-Leiva et al. [13] categorised work-life balances into four groups, namely flexible work practices, work leave, spatial flexibility and work-life balance information. Although work-balance practices were labelled using different nomenclature, research has shown that its adoption is beneficial to business organisations.

Literature shows that the use of work-life balance practices yields lot of benefits to employees, organisations and the society. The tangible benefits identified include reduction in attrition, lower stress levels, improved job satisfaction, career advancement, higher levels of productivity, better quality of life, increased commitment among employees [6, 10, 14–16]. Improvement in employees participation in family activities is among the advantages of using work-life balance practices [6]. Based on the findings emanating from literature, it is evident that the implementation of work-life balance practices in companies would improve organizational performance. Based on this evidence, the current study seeks to assess the extent of implementation of work-life balance practices among construction firms operating in Swaziland.

3 Research Method

To date various research methods have been developed and used to measure the extent of usage of work-life balance practices in different companies. Although each research method has its advantages and drawbacks, questionnaire survey remain as an appropriate method for collection of data from a large number of respondents [17]. These criteria serve a justification for using questionnaire survey in the current study. The questionnaire was used to collect data relating to the extent of usage of work-life balance practices in construction firms operating within Swaziland.

Based on the information available on the Construction Industry Council of Swaziland (http://www.cic.co.sz/), 566 contractors are registered in South Africa. Questionnaires were administered to middle management employees of construction firms operating in Swaziland between September and October 2018. The questionnaires were designed with an objective of assessing the extent of adoption of work-life practices in the construction firms. The respondents were assured of the anonymity and confidentiality of their response to the survey instrument. In addition, the aim of the study was explained to the study's respondents. Convenience sampling was utilized. Thus, the respondent firms were drawn from the industry contact of the first author.

Respondents were asked to rate the extent of adoption of work life practice on a 5-point Likert scale ranging from "1" ("not at all") to "5" ("to a large extent"). Work-life balance practices have been identified in several previous studies [11, 15]. The respondents were asked about the level of adoption of 19 work-life balance practices. Respondents were asked questions about the background information and firm characteristics. Also, responses were sought relating to the respondents perception on the impact of work-life balance practice on intention to quit their jobs. Five point Likert scale ranging from "1" ("strongly disagree") to "5" ("strongly agree") was used to rate the intention to quit.

Of the 100 questionnaires sent, the first author collected 58 filled questionnaires. From the 58 questionnaires, 54 were valid for further analysis. These four questionnaires were removed due to missing data (i.e. the respondents did not provide responses to some questions). This yields a response rate of 54%. The overall response is higher than those obtained in similar previous studies [12, 18]. The response rate in the previous studies on work-life balance practices were below 40%. Thus, it is considered that the response rate for the current study is satisfactory. In addition, the study reported in this paper forms part of a final year research project for an undergraduate degree. The first author would extend the study to a large population in the future. The data collected from the survey was analysed using Statistical Package for Social Sciences (SPSS).

4 Results

4.1 Demographic Profile of Respondents

In terms of gender, 66.7% of respondents were males and 33.3% were females. Over half (51.9%) of respondents were either married or living with a partner. Regarding the years of working experience, a large majority (63%) of respondents had less than 10 years of experience in the construction section and 9.3% had more than 20 years of experience. The average work hours per day was also measure. 31.5% of respondents indicated that they worked for more than 8 h per day.

4.2 Reliability Analysis

The reliability of the questionnaire used in the current study was assessed by estimating internal consistency using Cronbach's coefficient alpha. This coefficient ranges between zero and one, and a value close to one signifies greater reliability [18]. According to Nunnally [19], the value of Cronbach alpha coefficient should not be less than 0.7. The Cronbach alpha coefficient was found to be 0.871 and this indicates that the scales are reliable.

4.3 Work-Life Balance Practices

Table 1 shows descriptive statistics relating to the extent of utilisation of work-life balance practices among construction firms operating in Swaziland. Mean score for each practice indicates the level of adoption in the organisations that participated in the survey. Parental leave received the highest mean score of 4.24.

This indicates that parental leave is the most commonly adopted work-life balance practice among construction firms operating in Swaziland. In addition, maternity, health and wellness program, study leave and organisational understanding and support were ranked second, third, fourth and fifth with corresponding, mean scores of 4.04, 3.91, 3.85 and 3.52, respectively. In contrast, compressed week, work from home and part-time work received the lowest rank, which indicate low level of adoption among Swaziland's construction firms.

Table 1. Work-life balance practices adopted by construction firms

Work-life balance practices	Mean	Std. deviation	Rank
Parental leave or leave for other family emergencies	4.24	0.989	1
Maternity leave	4.04	1.303	2
Health and wellness program	3.91	1.336	3
Study leave	3.85	1.235	4
Organisational understanding and support	3.52	1.177	5
Job share	3.28	1.204	6
Bereavement leave	3.17	1.514	7
48/52 working year (annual hours)	3.06	1.338	8
Flexible starting and finishing times	3.04	1.453	9
Relocation assistance	2.89	1.61	10
Telecommuting	2.69	1.515	11
Shorter workdays for parents	2.65	1.556	12
Use flex days or rostered days off as half days	2.56	1.396	13
Career break	2.35	1.43	14
Childcare arrangements/benefits (crèche)	2.28	1.524	15
Non-paid extra holidays	2.24	1.516	16
Compressed week	2.2	1.25	17
Work from home on an ad hoc basis	2.19	1.402	18
Part-time work	2.17	1.397	19

4.4 Relationship Between Variables

In the current study, Spearman Rank Order correlation was used to evaluate the strength of relationship among study's variables. Table 2 shows the correlation between the top five work-life balance practices, job performance, quality of interpersonal relationship among employees and job turnover.

Table 2. Bivariate correlation between work-life balance practices and some variables

	Job performance	Interpersonal relationship	Turnover intention
Parental leave	0.113	0.194	−0.08
Maternity leave	0.223	0.143	−0.101
Health and wellness program	0.135	0.186	−0.081
Study leave	0.089	−0.008	−0.024
Organisational support	0.225	0.194	−0.03

As mentioned in Pallant [18], the size of sample affects the statistical significance of the results of a survey. It was suggested that the purpose of correlation is to evaluate the strength of a relationship and amount of shared variance. The correlation between work-balance practices and turnover intention was very small and negative. This finding indicates that job turnover decreased with the adoption of work-life balance practices. However, the value of the correlation is very small and it is difficult to determine the direction of causality between the variables. Also, a weak positive relationship exist among work-life balance practices, interpersonal relationship and job performance. Except for the relationship between study leave and interpersonal relationship among workers, which is negative. This result indicates that increase in the level of adoption of work-life balance practices is associated within an increase in job performance and quality of interpersonal relationship among employees. The strongest of these relationships is between organisational support and job performance.

Discussion of Results

Very little was found in literature on work-life balance practices in the construction industry in developing countries. The current study provide insights into the extent of usage of work-life balance practices among construction firms operating in Swaziland. The current study found that parental leave, maternity leave and wellness program are ranked as the most commonly implemented work-life balance practice in Swaziland's construction companies. Also, the implementation of work-life balance practices contributes to improvement in job performance and interpersonal relationship. However, it reduces the employees turn-over intentions. Surprisingly, it was found that an increase in the use of study leave has a negative relationship with interpersonal relationship. This finding may be due to unhealthy competition among employees.

The findings emanating from the current survey do not support those reported in previous research [12, 14]. Study leave and part-time are the most predominantly used work-life balance practices among companies operating in Australia [12]. In addition, European business organisations have preference for implementing flextime and work from home [14]. There are several possible explanations for this observed variance. First, the study area for each research was different. Second, the survey instrument in this study was administered to respondents drawn from the construction sector. In contrast, the sample in the previous studies were selected from various organisations in Australia [12] and Europe [14]. Research has shown that the usage of work-life balance practices lead to an increase in the number of females working in an industry [14]. Hence, it is possible to hypothesis that the low level of adoption of work-life practices among construction companies may be responsible for the male dominance in the sector.

The relationship among study's variables, i.e. turnover intention, job performance, interpersonal relationship and work-life balance practices, was assessed. Correlation was used to address the second objective of this study and this finding was compared with those presented in previous research. In agreement with the present study, previous studies [22, 24], have shown that reduction in employee turnover is linked to the use of work-life balance practices. Previous research have shown that less work-life conflict lead to improve job performance and the quality of interpersonal relationships [20–22]. However, it must be noted that these results differ from those described previously in literature. For example, a review of existing knowledge indicates the need for more research into the effect of work-life balance on performance [23]. On the overall, the use of work-life balance practices has been responsible for improved outcomes in terms of job satisfaction, organisational commitment and quality of life, among others [6, 24, 25]. This finding, while preliminary, suggests that construction firms need to adopt the use of work-life balance practices in their organisations.

5 Conclusion

The current study set out to evaluate the work-life balance practices among construction companies operating within Swaziland. This study has shown that parental leave and maternity leave are the prevalent work-life practices used by construction firms in Swaziland. Also, it was found that the use of work-life balance practices results in reduced job turnover, increased job performance and improvements in the interpersonal relationship among employees. Based on the results of this current study, it is suggested that construction companies need to implement work-life balance practice to improve the overall performance of their business operations.

References

1. Abdel-Wahab, M., Vogl, B.: Trends of productivity growth in the construction industry across Europe, US and Japan. Constr. Manage. Econ. **29**, 635–644 (2011)
2. Flyvbjerg, B., Skamris Holm, M.K., Buhl, S.L.: How common and how large are cost overruns in transport infrastructure projects. Transp. Rev. **23**, 71–88 (2011)
3. Thomas, H.R., Horman, M.J.: Fundamental principles of workforce management. J. Constr. Eng. Manage. **132**, 97–104 (2006)
4. Bambra, C., Whitehead, M., Sowden, A., Akers, J., Petticrew, M.: A hard day's night? The effects of compressed working week interventions on the health and work-life balance of shift workers: a systematic review. J. Epidemiol. Commun. Health. **62**, 764–777 (2008)
5. Bradley, L., Brown, K., Lingard, H., Townsend, K., Bailey, C.: Talking the talk and walking the walk: how managers can influence the quality of work-life balance in a construction project. Int. J. Manage. Proj. Bus. **3**, 589–603 (2010)
6. Lingard, H., Brown, K., Bradley, L., Bailey, C., Townsend, K.: Improving employees' work-life balance in the construction industry: project alliance case study. J. Constr. Eng. Manage. **133**, 807–815 (2007)
7. Malone, E.K., Issa, R.R.: Work-life balance and organizational commitment of women in the US construction industry. J. Prof. Issues Eng. Educ. Pract. **139**, 87–98 (2012)
8. Lingard, H.C., Yip, B., Rowlinson, S., Kvan, T.: The experience of burnout among future construction professionals: a cross-national study. Constr. Manage. Econ. **25**, 345–357 (2007)
9. Greenhaus, J.H., Collins, K.M., Shaw, J.D.: The relation between work–family balance and quality of life. J. Voc. Behav. **63**, 510–531 (2003)
10. Brough, P., Timms, C., O'Driscoll, M.P., Kalliath, T., Siu, O.-L., Sit, C., et al.: Work–life balance: a longitudinal evaluation of a new measure across Australia and New Zealand workers. Int. J. Hum. Resour. Manage. **25**, 2724–2744 (2014)
11. Bardoel, E.A., Tharenou, P., Moss, S.A.: Organizational predictors of work-family practices. Asia Pacif. J. Hum. Resour. **36**, 31–49 (1999)
12. De Cieri, H., Holmes, B., Abbott, J., Pettit, T.: Achievements and challenges for work/life balance strategies in Australian organizations. Int. J. Hum. Resour. Manage. **16**, 90–103 (2005)

13. Cegarra-Leiva, D., Sánchez-Vidal, M.E., Gabriel Cegarra-Navarro, J.: Understanding the link between work life balance practices and organisational outcomes in SMEs: the mediating effect of a supportive culture. Pers. Rev. **41**, 359–379 (2012)
14. Straub, C.: A comparative analysis of the use of work-life balance practices in Europe: do practices enhance females' career advancement. Women Manage. Rev. **22**, 289–304 (2007)
15. Yu, H.H.: Work-life balance: an exploratory analysis of family-friendly policies for reducing turnover intentions among women in US federal law enforcement. Int. J. Public Adm. **42**, 345–357 (2019)
16. Zheng, C., Molineux, J., Mirshekary, S., Scarparo, S.: Developing individual and organisational work-life balance strategies to improve employee health and wellbeing. Empl. Relat. **37**, 354–379 (2015)
17. Bryman, A., Bell, E.: Business Research Methods. Oxford University Press, Oxford (2011)
18. Pallant, J.: SPSS Survival Manual: A Step by Step Guide To Data Analysis Using SPSS. McGraw-Hill Education, Maidenhead (2010)
19. Nunnally, J.: Psychometric Theory. McGraw-Hill, New York (1978)
20. Smith, K.T.: Work-life balance perspectives of marketing professionals in generation. Serv. Mark. Q. **31**(4), 434–447 (2010)
21. Harrington, J.: Shift work and health–a critical review of the literature on working hours. Ann. Acad. Med. Singapore **23**, 699–705 (1994)
22. Van Steenbergen, E.F., Ellemers, N.: Is managing the work–family interface worthwhile? Benefits for employee health and performance. J. Org. Behav. **30**, 617–642 (2009)
23. Beauregard, T.A., Henry, L.C.: Making the link between work-life balance practices and organizational performance. Hum. Resour. Manage. Rev. **19**, 9–22 (2009). 24
24. Deery, M.: Talent management, work-life balance and retention strategies. Int. J. Contemp. Hospit. Manage. **20**, 792–806 (2008)
25. Sang, K.J.C., Ison, S.G., Dainty, A.R.J.: The job satisfaction of UK architects and relationships with work-life balance and turnover intentions. Eng. Constr. Archit. Manage. **16**, 288–300 (2009)

Apprenticeship: A Narrative Review of Factors Influencing Career Choice of Young People

Olalekan Shamsideen Oshodi[1(✉)], Clinton Aigbavboa[1], Oluwayomi Kayode Babatunde[2], and Bamidele Temitope Arijeloye[3]

[1] SARChl in Sustainable Construction Management and Leadership in the Built Environment, Faculty of Engineering and the Built Environment, University of Johannesburg, Johannesburg, South Africa
os.oshodi@gmail.com, ciagbavboa@uj.ac.za

[2] School of Construction Economics and Management, University of the Witwatersrand, Johannesburg, South Africa
oluwayomi.babatunde@wits.ac.za

[3] Department of Quantity Surveying, Federal University of Technology, Akure, Nigeria
btarijeloye@futa.edu.ng

Abstract. The ageing workforce is a significant problem affecting the construction sector in many countries. The increase in the average age of workers reduces the predictability of the outcome of construction projects. Existing literature suggests that ageing construction workforce has a significant negative impact on labor productivity and safety at construction sites and could lead to poor outcomes of construction projects. A substantial body of knowledge indicates that young people need to be attracted to the construction sector as an effective replacement of its ageing workforce. However, a comprehensive review of the scientific literature focused on this problem is lacking. A narrative literature review was conducted to examine the current state of knowledge on factors influencing career choice of young people. The findings indicate that family, salary and quality of career advice are the most common factors influencing career choice among young people. The articles selected for review were those published in academic journals. Based on the review, the directions for future research were proposed.

Keywords: Apprentice · Career choice · Construction industry · Review · Young adults

1 Introduction

The construction sector is one of the main drivers of economic growth in several countries. Over the years, research has shown that economic growth and the volume of activities in the construction sector are positively related [1, 2]. Despite the successes recorded in the construction sector, productivity is still a concern to researchers and construction professionals. Previous studies have shown that productivity is affected by weather, project characteristics, unit weight of materials, workers benefits, and age of construction workers, among others [3–5]. The decline in the productivity level has a

C. Aigbavboa and W. Thwala (Eds.): CIDB 2019, *The Construction Industry in the Fourth Industrial Revolution*, pp. 90–99, 2020.
https://doi.org/10.1007/978-3-030-26528-1_10

negative impact on the outcome of construction projects. There is an obvious need to address factors affecting productivity in the construction sector.

Ageing is one of the main factors influencing labour productivity in the construction sector. The construction workforce in Hong Kong is constituted with about 43% of workers who are above 50 years of age [6]. Similarly, McNair and Flynn [7] found that 40% of construction workers in the UK are above 45 years of age. The disadvantage of utilizing ageing workforce in construction projects include: an increase in the number of site accidents, low labour productivity and a rise in labour cost are risk associated with ageing workforce [8–10]. These risk factors lead to uncertainties in project cost and schedules. Given the concerns about ageing workforce, it has become imperative to develop strategies for recruiting young people to the sector and this might improve the outcomes of construction projects.

Several apprenticeship programs are been implemented in several countries across the globe. Enhanced Construction Manpower Training Scheme (ECMTS) of Hong Kong and WorldSkills. South Africa are typical examples of apprenticeship programs for the construction sector [11]. Despite the existence of these apprenticeship programs, a decrease in the average age of the construction workforce has not yet been achieved. In the existing literature, review studies focused on different problems within the field of construction management is acknowledged. For example, Darko et al. [3] summarized studies on green buildings. However, a review study focused on factors influencing career choices among young people is lacking. Thus, the current study addresses this gap in current knowledge. The purpose of this study is to carry-out a narrative review of studies on factors influencing career choice of young people. The study would make the following contributions to literature and practice: (i) identify the factors influencing career decisions; (ii) based on the current literature directions for future research were suggested; and (iii) a concise report on studies focused factor influencing career decisions would aid the development of strategies targeted at attracting young individuals to the construction sector.

2 Research Method

The use of a literature review is a well-established approach in the field of construction management. Reviews have been conducted to understand the drivers of green building [13], factors influencing labour productivity [5], and methods for forecasting manpower demand [14], among others. The review methods used in previous studies include systematic review, critical review and narrative review. A narrative review was used to describe the state of knowledge relating to a topic. According to Cronin et al. [15], narrative reviews are useful for providing an evidence-based solution to practice problems. In the current study, the narrative review approach was used to examine the state of knowledge relating to factors influencing career choices among young people.

The SCOPUS database was used for the search. SCOPUS database was selected due to the extensive number of construction journals indexed on the databases and its adoption by similar previous studies [16], [17]. The search was conducted using the three most appropriate keywords; (i) career choice; (ii) career selection and (iii) career decision. The search was limited to: (i) articles published in the English language;

(ii) articles published in journals; (iii) full-text availability; and (iii) focus (i.e. young people and/or construction).

The search for the study was carried out in three distinct stages namely, database search, filtering, and qualitative analysis. Upon completion of the initial search on the SCOPUS database, 101 papers were found. The search results were filtered to remove irrelevant papers (i.e. studies that did not meet the inclusion criteria). Also, articles categorised as "book reviews", "book chapter", "book", "editorial", "editor's notes", "letter to the editor" and other papers not based on empirical work were excluded. Care was also taken to ensure that duplicates were removed. At the end of the search and screening process, 16 relevant articles were subjected to further analysis.

3 Results

A narrative review was carried out to identify empirical evidence relating to the factors influencing career choices among young people. The findings of this study are based on a review of 16 empirical studies published between 1967 and 2017 (Table 1). An overview of the selected studies is presented in Table 1. The study participants are drawn from 5 countries, including 12 (75%) from the USA and UK. It is worth noting that 2 previous studies have focused on the South African context. The 16 selected studies emanated from research conducted in several countries across the globe (n = 9 USA; n = 4 Europe; 1 = Asia; 2 = Africa). The spread of the selected studies demonstrates the global relevance of the topic. The sample for most of the studies was drawn from a population of students, such as undergraduate and high school students.

Table 1. Summary of selected studies

S/N	Author(s)	Country	Study participants
1	Swinhoe (1967)	United Kingdom	Students
2	Friedman (1977)	United States of America	Students
3	Haworth et al. (1986)	United Kingdom	Students
4	Emmett and Minor (1993)	United States of America	Students
5	Federle et al. (1993)	United States of America	Construction workers
6	Bronzini et al. (1995)	United States of America	Mixed
7	Borg (1996)	Malta	Students
8	Wilkinson (1996)	United Kingdom	Students
9	Greene and Stitt-Gohdes (1997)	United States of America	Tradespeople
10	Constantine et al. (2005)	United States of America	Students
11	Singaravelu et al. (2005)	United States of America	Students
12	Bojuwoye and Mbanjwa (2006)	South Africa	Students
13	Chileshe and Haupt (2010)	South Africa	Students
14	Ling and Ho (2012)	Singapore	Students
15	Escamilla et al. (2016)	United States of America	Students
16	Bigelow et al. (2017)	United States of America	Tradespeople

Note: Mixed = students, faculty and construction professionals

3.1 Factors Influencing Career Choices of Young People

The main objective of the study is to understand the factors influencing career choice among young people. Thirty factors were identified from sixteen previous studies. The factors were identified and summarized in Table 2. It is evident that the factors influencing career choice of young people are numerous. An understanding of these factors would facilitate the development of strategies for attracting young people to apprenticeship training programs for craftspeople in the construction industry.

Table 2. Factors influencing career decisions

Label	Factor	References
1	Family influence	[3–5, 7, 10–13, 15, 16]
2	Salary upon graduation	[1, 5–8, 11, 13, 15, 16]
3	School (career advice from teachers)	[3, 5, 7, 11–13, 15, 16]
4	Personal interest and ability	[1, 2, 4, 7, 9, 11, 12, 14]
5	Job prospects (career opportunities)	[1, 2, 4, 6, 7, 13, 16]
6	Gender (male dominance)	[2, 8–10, 14, 15]
7	Industry's image (media)	[2, 5, 13, 14, 16]
8	Peers	[3, 7, 11–13]
9	Working conditions (dangerous and dirty)	[1, 5, 13–15]
10	Training opportunities	[7, 8, 13, 16]
11	Challenges associated with work	[4, 6–8]
12	Job autonomy	[1, 4, 7, 9]
13	Career advancement opportunities	[5, 8, 13]
14	Opportunity to do varied work	[1, 6, 8]
15	Role model within the profession	[2, 3, 9]
16	Family friendly nature of work	[8, 14]
17	Opportunity to help others	[1, 7]
18	Opportunity to work on new projects/network	[1, 8]
19	Perceived lack of professionalism	[14, 15]
20	Race discrimination	[10, 15]
21	Safety concerns	[1, 14]
22	Travel associated with work	[1, 8]
23	Academic performance	[12]
24	Adversarial relationship within the sector	[5]
25	Benefits (car)	[8]
26	Choice of subjects taken at school	[12]
27	Finance (Government/Sponsors)	[11]
28	No other work opportunities	[16]
29	Pension plan	[8]
30	Teachers	[11]

A total of 30 factors influencing career choice was identified. These factors were categorised into manageable grouping. Based on the factors identified in the current study, the factors were classified into five groups namely, 'family and friends', 'job and industry characteristics', 'benefits and job opportunities', 'school', and 'personal'. The findings presented in Table 2 shows that the most frequently reported factors are family influence, salary upon graduation, school, personal interest and job prospects (i.e. the availability of jobs at the completion of studies). In this study, the factors were classified into these groupings based on (1) similarities between factors and previous research. For example, Greene and Stitt-Gohdes [18] classified factors into personal and external factor grouping.

3.2 Family and Friends Factor Grouping

The main reasons influencing the career choice of young people are family and peers. The factors classified in this group include family influence (F1), peers (F8), availability of role models within the profession (F15), and family-friendly nature of work (F16). Out of the 16 selected studies, the family influence was reported in 10 studies. The influence of parent vary based on the gender and characteristics of student population. For example, Haworth et al. [19] showed that a females students' choice of an engineering profession is greatly influenced by their fathers. In addition, the family had a significant influence on the career choice among non-Asian international students [20].

A close examination of literature seems to suggest that the impact of parents on career choice tend to vary based on family income. Chileshe and Haupt [21] found that the impact of family on career choice is insignificant among South African students. In contrast, [22] reported that family members influenced career choice among South African students. The characteristics of the respondents are the main difference between these two studies carried out in South Africa. The respondents for Chileshe and Haupt [21] were drawn from High School students in Western Cape, and [22] selected students from a disadvantaged segment of the population. This suggests that the economic status of parents could be the main reason for the contrasting findings.

3.3 Job and Industry Characteristics Factor Grouping

The job and industry characteristics factor grouping encompasses factors such as gender (F6), industry's image (F7), working conditions (F9), training opportunities (F10), challenges associated with work (F11), job autonomy (F12), career advancement opportunities (F13), opportunity to do varied work (F14), opportunity to help others (F17), opportunity to work on new projects/network (F18), perceived lack of professionalism (F19), race discrimination (F20), safety concerns (F21), and travel associated with work (F22). Of all the factors reported within this group, gender discrimination is the most commonly identified factor (This factor was found in 6 of the 16 relevant studies). It is evident that male dominance in the construction sector and gender discrimination is an important factor influencing career decision among young people. In addition, Berik and Bilginsoy [23] found that gender is one of the main reasons responsible for apprentice attrition in training programs. Gender remains as an important factor influencing career decision among young people.

3.4 Relationship Between Variables

The job and industry characteristics factor grouping encompasses factors such as gender (F6), industry's image (F7), working conditions (F9), training opportunities (F10), challenges associated with work (F11), job autonomy (F12), career advancement opportunities (F13), opportunity to do varied work (F14), opportunity to help others (F17), opportunity to work on new projects/network (F18), perceived lack of professionalism (F19), race discrimination (F20), safety concerns (F21), and travel associated with work (F22). Of all the factors reported within this group, gender discrimination is the most commonly identified factor. This factor was found in six of the 16 relevant studies). It is evident that male dominance in the construction sector and gender discrimination is an important factor influencing career decision among young people. In addition, Berik and Bilginsoy [23] found that gender is one of the main reasons responsible for apprentice attrition in training programs. Gender remains as an important factor influencing career decision among young people.

3.5 Benefits and Job Opportunities Factor Grouping

Benefit grouping consists of salary upon graduation (F2), job prospects (F5), benefits associated with work (F25), no other work opportunities (F28) and pension plan (F29). This grouping refers to financial benefits, non-financial benefits and job prospects (i.e. availability of jobs upon graduation). Studies have shown that career decision among young people is heavily influenced by benefits and its impact vary across gender [24–27]. For instance, male students tend to be more influenced by salary [18]. The desire to achieve financial independence and the ease of securing jobs upon graduation significantly influence career decision among young people.

3.6 School Grouping

The school grouping consists of quality of career advising (F3), choice of subjects available to students (F26) and teachers (F30). School background has a significant influence on the career decision among young people. The quality of career advising received from teachers and guidance counselor influences the career decision among young people, however, the impact tend to vary across characteristics of respondents. For example, Chileshe and Haupt [21] found that the career decision among female students tends to be influenced by teachers and counselors. In contrast, Bojuwoye and Mbanjwa [22] reported that teachers do not influence career decision. This finding demonstrated that the quality of career advising influences the choice of career and subjects taken by students.

3.7 Personal Grouping

According to the finding of this study, personal interest (F4), academic performance (F23), adversarial relationship (F24) and finance (F26) were categorised into personal grouping. From the perspective of young people, personal interest is one of the most common factors influencing career decision among young people.

4 Data Collection Method Utilised in Selected Studies

The data collected methods used in selected studies are numerous. Five data collection methods were utilised in previous empirical studies. The data collection methods are questionnaire survey, interview, case study, focus group and mixed. Table 3 provides information on the data collection methods used in the relevant studies. It was found that the questionnaire survey is the most frequently utilized method and it accounts for 62.5% of the total number of published studies. The findings agree with those reported in previous studies, which have shown preference for the questionnaire [28].

Table 3. Data collection methods used in selected studies

Data collection methods	Number of papers	Percentage
Questionnaire survey	10	62.50%
Interview	3	18.75%
Case study	1	6.25%
Focus group	1	6.25%
Mixed	1	6.25%

5 Discussion of Results and Directions for Future Research

A narrative review was carried out to identify the factors influencing career decision among young people. An understanding of these factors would provide guidance for attracting young people to the construction sector, especially in blue-collar job positions. Shortage of construction workers has been linked to ageing construction workforce. In addition, evidence suggests that uncertain in construction cost is largely due to the competition to attract workers among employers. The findings of this study reveal that family, salary upon graduation and school (quality of career advice) are the predominant factors influencing career decision among young people. In order to increase the number of young people seeking career opportunities in the construction sector, there is a need to address these important factors (see Table 2).

Based on the findings from the review, multiple gaps within the current knowledge are identified. These shortfalls in the current knowledge serve as justification for further researcher in this area. First, a large number of studies have focused on the developed countries (i.e. USA and UK). Despite the difficulties experienced in recruiting construction workers in developing countries, there has been no study focused on factors influencing the career decision of young people. Future studies could focus on countries in Africa, South America, Asia, and Australia. The process of knowledge creation can be classified into four phases, namely description, explanation, prediction and control [29, 30]. Knowledge of factors influencing career decision among young people is at the description phase. The impact of intervention strategies, for example, parent's exposure to career opportunities in the construction, on the career choice of young people is unknown. The use of a longitudinal approach would provide an opportunity of identifying effective strategies for attracting young people to the construction sector.

Finally, there has been calls for greater use of qualitative research methods in social science research [31]. The use of qualitative methods, such as ethnography, would provide deeper insight on factors influencing career decisions, which may not be, emerge in quantitative studies.

6 Conclusion

Shortage of construction worker is a major problem affecting the construction sector in several countries across the globe. The shortage of workers has been largely due to the unattractiveness of jobs in the sector to young people. This study seeks to provide information on factors influencing career decisions among young people. The findings of this study provide relevant stakeholders with information for developing strategies to attract young people to the construction sector. Sixteen relevant studies were reviewed and 30 factors influencing career decisions were identified. Furthermore, these factors were categorised into five groups.

The study revealed that family is the most frequently reported factor influencing career decisions among young people. The identified factors were classified into five groups namely, family, industry, benefits, school and personal. The findings of the current study is a concise report on the factors influencing career decision among young people. The findings of the current study provide guidance to researchers on the current state of knowledge in this area and relevant stakeholders could also use this information to develop strategies for attracting young people to the construction sector. For example, enlightenment programs on the job opportunities and salary of construction workers could be disseminated through media outlets, such as television and social media. These programs could be televised during weekends and after office hours to gain the attention of parents and other family members. Three main gaps in the current knowledge were identified namely, state of knowledge creation, low usage of qualitative research methods and study area. Researchers should focus on addressing the gaps identified in the current knowledge on factors influencing career choice among young people.

References

1. Pietroforte, R., Bon, R., Gregori, T.: Regional development and construction in Italy: an input-output analysis, 1959–1992. Constr. Manag. Econ. **18**, 151–159 (2000)
2. Chiang, Y.H., Tao, L., Wong, F.K.: Causal relationship between construction activities, employment and GDP: the case of Hong Kong. Habitat Int. **46**, 1–12 (2015)
3. Tang, J., MacLeod, C.: Labour force ageing and productivity performance in Canada. Can. J. Econ. **39**, 582–603 (2006)
4. Goodrum, P.M., Zhai, D., Yasin, M.F.: Relationship between changes in material technology and construction productivity. J. Constr. Eng. Manag. **135**, 278–287 (2009)
5. Naoum, S.G.: Factors influencing labor productivity on construction sites: a state-of-the-art literature review and a survey. Int. J. Prod. Perform. Manag. **65**, 401–421 (2016)

6. Hong Kong Construction Association: Hong Kong's Construction Industry Vision 2020. Hong Kong Construction Association, Hong Kong (2013). http://www.hkca.com.hk/index.php. Accessed 20 Sept 2016
7. McNair, S., Flynn, M.: Managing an Ageing Workforce in Construction. Age Partnership Group, Department for Work and Pensions, London (2006)
8. Conen, W.S., van Dalen, H.P., Henkens, K.: Ageing and employers' perceptions of labour costs and productivity: a survey among European employers. Int. J. Manpower **33**, 629–647 (2012)
9. Dong, X.S., Wang, X., Largay, J.A., Platner, J.W., Stafford, E., Cain, C.T., et al.: Fatal falls in the U.S. residential construction industry. Am. J. Ind. Med. **57**, 992–1000 (2014)
10. Jebens, E., Mamen, A., Medbø, J.I., Knudsen, O., Veiersted, K.B.: Are elderly construction workers sufficiently fit for heavy manual labour? Ergonomics **58**, 450–462 (2015)
11. Ho, P.H.K.: Labour and skill shortages in Hong Kong's construction industry. Eng. Constr. Architectural Manag. **23**, 533–550 (2016)
12. Construction Industry Development Board Construction Monitor - October 2017. Construction Industry Development Board, South Africa (2017)
13. Darko, A., Zhang, C., Chan, A.P.C.: Drivers for green building: a review of empirical studies. Habitat Int. **60**, 34–49 (2017)
14. Wong, J.M.W., Chan, A.P.C., Chiang, Y.H.: A critical review of forecasting models to predict manpower demand. Constr. Econs. Build. **4**, 43–56 (2012)
15. Cronin, P., Coughlan, M., Smith, V.: Understanding Nursing and Healthcare Research. Sage, London (2015)
16. Ke, Y., Wang, S., Chan, A.P.C., Cheung, E.: Research trend of public-private partnership in construction journals. J. Constr. Eng. Manag. **135**, 1076–1086 (2009)
17. Yi, H., Yang, J.: Research trends of post disaster reconstruction: the past and the future. Habitat Int. **42**, 21–29 (2014)
18. Greene, C.K., Stitt-Gohdes, W.L.: Factors that influence women's choices to work in the trades. J. Career Dev. **23**, 265–278 (1997)
19. Haworth, G., Povey, R., Clift, S.: Girls and engineering: perceived influences on career choice amongst female craft and technician engineers and non-engineers. Br. J. Guidance Counselling **14**, 196–204 (1986)
20. Singaravelu, H.D., White, L.J., Bringaze, T.B.: Factors influencing international students' career choice: a comparative study. J. Career Dev. **32**, 46–59 (2005)
21. Chileshe, N., Haupt, T.C.: An empirical analysis of factors impacting career decisions in South African construction industry: male and female high school students' perspectives. J. Eng. Des. Technol. **8**, 221–239 (2010)
22. Bojuwoye, O., Mbanjwa, S.: Factors impacting on career choices of Technikon students from previously disadvantaged high schools. J. Psychol. Africa **16**, 3–16 (2006)
23. Berik, G., Bilginsoy, C.: Do unions help or hinder women in training? Apprenticeship programs in the United States. Ind. Relat. J. Econ. Soc. **39**, 600–624 (2000)
24. Emmett, J.D., Minor, C.W.: Career decision-making factors in gifted young adults. Career Dev. Q. **41**, 350–366 (1993)
25. Bronzini, M.S., Mason Jr., J.M., Tarris, J.P., Zaki, E.: Choosing a civil engineering career: some market research findings. J. Prof. Issues Eng. Educ. Pract. **121**, 170–176 (1995)
26. Escamilla, E., Ostadalimakhmalbaf, M., Bigelow, B.F.: Factors impacting Hispanic high school students and how to best reach them for the careers in the construction industry. Int. J. Constr. Educ. Res. **12**, 82–98 (2016)
27. Bigelow, B.F., Zarate, V., Soto, J., Arenas, J., Perrenoud, A.: Attracting and retaining tradespeople, an evaluation of influencers on construction workers in two different trades in texas. Int. J. Constr. Educ. Res., 1–16 (2017)

28. Li, Z., Shen, G.Q., Xue, X.: Critical review of the research on the management of prefabricated construction. Habitat Int. **43**, 240–249 (2014)
29. Runeson, G.: The methodology of building economics research. In: De Valence, G. (ed.) Modern Construction Economics. Spon Press, Oxon (2011)
30. Grove, S.K., Gray, J.R., Burns, N.: Understanding Nursing Research: Building an Evidence-Based Practice, 6th edn. Elsevier, Missouri (2015)
31. Fellows, R.: New research paradigms in the built environment. Constr. Innov. **10**, 5–13 (2010)

Relevance of Competencies of Construction Project Risk Managers to Quantity Surveying Practice

Edosa Mark Osazuwa[1(✉)] and Ayodeji Emmanuel Oke[2]

[1] Department of Quantity Surveying, University of Benin, Benin City, Nigeria
osazuwaedosamark@gmail.com
[2] Department of Quantity Surveying,
Federal University of Technology, Akure, Nigeria
emayok@gmail.com

Abstract. This paper aim to examine the relevance of construction project risk manager's competencies to quantity surveying practice in Nigeria and developing countries with a view to improving the overall performance and competencies of quantity surveyors in the construction industries. The study adopt a survey design and construction professionals in Edo State. Nigeria were assessed using convenience sampling with the aid of structure questionnaire. Frequency, percentage, mean score, sample *t*-test and factor analysis were used to analyse the data gathered, It was revealed that the knowledge of risk management by construction professionals and their involvement in risk management program were quite high, findings reveals that the areas of competencies of project risk managers were found to be highly relevant to quantity surveying practice. Among these areas, include the ability to draft contract agreement, effective human resource management, organization of financial operation, ability to present and write report, preparation of claims and litigation.

Keywords: Quantity surveyors · Construction industry · Competency project risk manager · Risk management

1 Introduction

The construction industries are seen as a key driver in the socio economic development of nations [1]. It contributes between 5 to 10% of the gross domestic product (GDP) in a country and employ about 10% of a country working population, which is accountable for half of the gross fixed capital formation [2]. The development of the industry is therefore the goal of many developing countries in order to stop companies from collapsing. Its development requires knowledge of risk management policies. It has been establish that at every stage of the construction process from initial investment appraisal, the built environment is subject to risk for all the parties involved. Project success requires effective risk management where it identifies risks and create plans to deal with the threats. Although, very few project risk managers have the competencies to manage them within their initiatives. Roggema-van Heusden opined that competencies is the capacity to perform task in a proficient state which involve the success of a certain duty

C. Aigbavboa and W. Thwala (Eds.): CIDB 2019, *The Construction Industry in the Fourth Industrial Revolution*, pp. 100–111, 2020.
https://doi.org/10.1007/978-3-030-26528-1_11

or mission in a manner that can be judged by others [3]. Hassal, Dunlop and Lewis asserted that competent quantity surveyors must have variety of skills and knowledge that can be used in any kind of situation and establishment [4]. Hore, O'Kelly, and Scully noted that the provision of advice on risk analysis and management are one of the functions of a quantity surveyor [5]. This collaborated Dada and Jagboro study that quantity surveyors in Nigeria construction industry and in developing countries have over the years gained significant competencies in risk management [6]. Rahman and Kumaraswamy highlighted that a competent quantity surveyor can effectively carryout the roles of project risk managers [7]. The study therefore brings to light that the competencies of project risk manager are not just peculiar to construction project risk managers but are also very relevant to other construction professionals such as Quantity surveyors.

2 Literature Review

2.1 The Concept of Competency

Entrepreneurs, administrative practitioners and researchers alike have deliberated the concept of competencies since the 1960s. However, the rational on competencies remained undeveloped during the 1970s and early 1980s, the reason being that management consultants and scholars paid more attention to other methods of strategy [8]. In the late 1990 s, the idea of competence had gained greater weight and was widely used by businesses. Rankin reported that a significant number of companies in the UK are using competencies to increase individual and commercial performance [9]. Baker, Mapes, New, and Szwejczewski observed that companies were starting to look into staffs' competencies when they recognised that providing a structure, which brings greater clarity to the idea of competence, operation and technology management in the industry in general, would lead to competitive improvement within their organisation [10]. Competency has been linked with individual performance and job performance; it is therefore the knowledge, skills, abilities and behaviours needed for a particular task [11]. It is the state of being qualified as well as having the ability to perform a specific role [12]. It can also be seen as being capable, possessing certain skills and the knowledge to do what one is supposed to do.

2.2 Competences of Project Risk Manager

Project risk managers are professionals who are responsible for identifying, analysing and responding to risk in other to achieve the objectives of the project [13]. A competent project risk manager is vital to the success of a project. To manage the project successfully, there is need for project risk manager to acquire the required competencies [14]. Successful project focusses on ensuring that project risk managers possess the important competencies needed to be successful in their project. For this reason, construction project risk manager must retain the required competency in term of skills, behaviour and knowledge [15]. They therefore must ensure that individual who works on construction site must have the necessary competencies to function effectively on their assigned roles. Some of the basic functions of project risk managers include; documenting, evaluating, managing, monitoring, reporting and scheduling. Competencies of project risk managers were grouped into primary and secondary competencies to include managerial, communication, technical and financial [16].

2.3 Quantity Surveyors Competence

The functions of quantity surveyors in the construction industry "is to ensure that the resources of the industry are utilized to the best advantage of society by providing financial management for project and cost consultancy services to the client and designers during the whole construction process [17]. This role provides the basis for proper cost management function of quantity surveyors in the construction industry within the context of forecasting, analysis, planning, controlling and accounting. With improvements in education, advancement in technology and increase in affluence of society, clients have become more discerning as well as demanding on the quality and timeliness of the service they received. There are several options in quantity surveying field for a quantity surveyor. A qualified quantity surveyor can be gainfully employed with quantity surveying firms, construction firms, or property developers. Some large public or private organisations that deal with a significant amount of building or construction procurement as part of their activities normally employ quantity surveyors among other construction professionals to become their project managers [18]. Quantity surveyors may work as a consultant in a private practice, for a developer or in the development arm of a major organisation or a multinational organisation (e.g. retailer, manufacturer, utility company or airport) and a public sector body. On the contracting side, quantity surveyors can work for a major national or international contractor, a local or regional general contractor, for a specialist contractor or sub-contractor, or for a management style contractor.

According to Royal Institution of Chartered Surveyors, some of the competencies of a quantity surveyor may include the following, Preparing feasibility studies or development appraisals, assessing capital and revenue expenditure over the whole life of a facility. Others may include advising clients on ways of procuring the project, advising on the setting of budgets, monitoring design development against planned expenditure conducting value management and engineering exercises managing and analysing risk and managing the tendering process. In addition, there are; preparing contractual documentations, controlling cost during the construction process, managing the commercial success of a project for a contractor, valuing construction work for interim payments, valuing change, assessing or compiling claims for loss of expense and agreeing final accounts, negotiating with interested parties and giving advice on the avoidance and settlement of disputes [11].

3 Research Methodology

This study set out to assess the relevance of construction project risks manager's competencies to quantity surveying practices in Nigeria. The target population comprises of registered construction professionals in Edo state, Nigeria, namely: Architects,

Quantity Surveyors, Builders and Engineers as obtained from their various professional body. The research instrument used was structured questionnaire designed based on information gathered from the review of related literatures. The questionnaires was designed in sections with the first section designed to gather information on the respondent background and level of knowledge on project risk management. Respondent were provided with the competencies of project risk managers in the second section, and were ask to rate the level of their relevance to quantity surveying practice using a 5-point Likert scale, with 5 being very high, 4 being high, 3 being average, 2 being low and 1 being very low. The questionnaire were self-administered through the help of field assistance and through the aid of electronic questionnaire. One hundred and ninety seven (197) questionnaires were distributed while one hundred and sixty (160) were returned and deemed fit for analysis, while the others were either not retrieved or incorrectly filled. Cronbach's alpha test was used to test the reliability of the research instrument. This method is used to measure the reliability of the questionnaire between each field and the mean of the whole fields of the questionnaire. The closer the reliability coefficient is to 1.0, the better the reliability of the measure. This research study therefore adopted Cronbach's alpha value of 0.7. The data gathered was analysed using frequency and percentage for analysing the background information while Mean Item score was used in ranking the relevance of competencies and one sample t-test was used to determine the importance attached to each of the identified competencies and their associated level of significance.

4 Findings and Discussion

4.1 Characteristics of Respondents

Result in Table 1 shows the information elicited with respect to the background information of respondents, which shows that 33.8% are Engineers, 29.4% are Architects, and 23.8% of the respondents are Quantity surveyors, while 13.1% are Builders. In terms of experience, 37.5% of the respondents have less than 6 years working experience, 23.8% have between 6 to 10 years' experience, 16.9% have above 20 years' experience, 14.4% have 16 to 20 years' experience while 7.5% have 11 to 15 years' experience.

Table 2 shows that the respondents are of the opinion that they possess vast knowledge in construction risk management for whatever form of project that can be embarked upon. The mean value of 4.78 indicates that out of a maximum 5.00, the majority of the professionals are well educated as to the existence of risk management in construction project; this implies that the reliability of the responses as to the validity of the research can be relied on for this study.

Table 3 reveals that 64.4% have been involved in construction project risks management program while 35.6% indicated that they have not been involved in any of such.

Table 1. Background information of respondents

Respondents characteristics	Frequency	Percentage
Designation		
Quantity surveyor	38	23.8
Architect	47	29.4
Engineer	54	33.8
Builder	21	13.1
Total	**160**	**100**
Years of experience		
0–5	60	37.5
06–10	38	23.8
11–15	12	7.5
16–20	23	14.4
Above 20	27	16.9
Total	**160**	**100**

Table 2. Respondents level of knowledge of construction project risks management

Responses	Very high (5)	High (4)	Average (3)	Low (2)	Very low (1)	Mean value
Level of knowledge	129	27	4	0	0	4.78

Table 3. Respondents involvement in construction project risks management program

Response	Frequency	Percentage
Yes	103	64.4
No	57	35.6
Total	**160**	**100.0**

4.2 Relevance of Construction Project Risk Managers Competencies to Quantity Surveying Practice

In determining the relevance of construction project risk manager's competencies to quantity surveying practice, respondents were given certain competencies of project risk managers identified from literatures and were asked to rank them based on their relevance to quantity surveying practice. One sample t-test was then used to ascertain the significance of each of the identified competence including their associated standard deviation and standard error. The mean item score for each of the items in the table is ≥ 3.00. Based on a 5 point Likert scale, a 2.50 score is usually considered as average but a score tending towards 3.00 is thus indicating that the 17 items in the Table 4 are important to project risk management.

Table 4. Relevance of project risk managers' competencies to quantity surveyors

Variables	Mean	Std. Dev	T	Mean difference (test value = 3)	Sig. p-value
Ability to draft contract agreement (CP14)	4.63	.874	23.514	1.625	.000
Effective human resources management (CP17)	4.56	.759	25.950	1.556	.000
Ability to present and write report (CP12)	4.52	.945	20.332	1.519	.000
Adequate planning and resources allocation (CP4)	4.45	.716	25.602	1.450	.000
Conversant with the use of information technology and its application (CP5)	4.44	.829	21.923	1.438	.000
Preparation of claims and litigation (CP15)	4.42	.928	19.337	1.419	.000
Good public and human relations (CP16)	4.41	.961	18.596	1.419	.000
Effective organizational financial operation (CP11)	4.39	.824	21.298	1.388	.000
Knowledge of project SWOT (strength, weakness, opportunities and threat) analysis (CP6)	4.32	.667	25.008	1.319	.000
Possess legal qualities (CP13)	4.21	1.111	13.736	1.206	.000
Protecting the employee against project hazard (CP10)	4.07	1.016	13.301	1.069	.000
Understand the impact of demand and supply on project delivery (CP2)	4.01	.735	17.417	1.013	.000
Effective management of organizational risk (CP3)	3.93	.714	16.383	.925	.000
Understanding project quality control (CP7)	3.87	.880	12.130	.884	.000
Understand data collection in project management (CP8)	3.80	.830	12.118	.800	.000
Understand the legal components of the risk management industry (CP1)	3.70	.460	19.261	.700	.000
Securing the physical assets of the enterprise (CP9)	3.50	.673	9.399	.500	.000
Average Mean	**4.19**	**0.338**			

4.3 Factor Analysis

Factor analysis is necessary for the reduction of the competencies into smaller number of logical subscales. For factor analysis to be carried out, the data must be analysed to determine its suitability for analysis. Pallant recommended the use of larger samples for factor analysis. Hence, the 17 variables used in this study coupled with 160-sample size are considered adequate for factor analysis.

Kaiser–Meyer–Olkin measure of sampling adequacy (KMO) and Bartlett's test of sphericity were also employed to test the factorability of the data. This was done using the SPSS version 21 software. KMO index ranges from 0 to 1, with 0.6 suggested as the minimum value for a good factor analysis [19]. Bartlett's test of sphericity shows whether the correlation matrix is an identity matrix. Pallant suggested that the Bartlett's test of sphericity should be significant ($p < 0.05$) for the factor analysis to be considered appropriate [20]. The KMO measure of sampling adequacy achieved a value of 0.611, exceeding the recommended minimum value of 0.6 and the Bartlett's test of sphericity was also statistically significant at 0.000 (less than 0.05), thus supporting the use of factor analysis for the data gathered which was appropriate (Table 5).

Table 5. KMO and Bartlett's test

Kaiser-Meyer-Olkin measure of sampling adequacy		.611
Bartlett's test of sphericity	Approx. Chi-Square	4262.652
	Df	136
	Sig.	.000

Factor analysis was therefore conducted for the relevance of construction project risk managers' competencies to quantity surveying practice in Nigeria using principal component analysis (PCA) with varimax rotation. Table 6 shows that 2 component with eigenvalues greater than 1.0 were extracted using the factor loading of 0.30 as the cut-off point as suggested by [20]. Hence, two competency factors with Eigen values exceeding 1 were retained, resulting in 9.195 and 2.386 selected which explains 54.087 and 14.035% of the variance respectively. This infers that the first cluster of competency factors accounted for 54.087% while the second cluster of competency factors accounted for 14.035%. These two clusters of competency factors together have a total cumulative percentage of 68.122% of the total importance, which highlights their significance from the seventeen competency factors shown.

Table 6. Total variance explained

Factor	Initial eigenvalues			Extraction sums of squared loadings			Rotation sums of squared loadings		
	Total	% of Variance	Cumulative %	Total	% of Variance	Cumulative %	Total	% of Variance	Cumulative %
1	9.195	54.087	54.087	9.195	54.087	54.087	9.065	53.321	53.321
2	2.386	14.035	68.122	2.386	14.035	68.122	2.516	14.801	68.122
3	1.506	8.861	76.984						
4	1.086	6.387	83.370						
5	.810	4.762	88.132						
6	.564	3.317	91.449						
7	.473	2.781	94.231						
8	.314	1.849	96.079						
9	.211	1.241	97.320						

(continued)

Table 6. (*continued*)

Factor	Initial eigenvalues			Extraction sums of squared loadings			Rotation sums of squared loadings		
	Total	% of Variance	Cumulative %	Total	% of Variance	Cumulative %	Total	% of Variance	Cumulative %
10	.142	.834	98.154						
11	.125	.733	98.886						
12	.065	.381	99.267						
13	.059	.347	99.615						
14	.035	.206	99.821						
15	.017	.100	99.921						
16	.010	.058	99.979						
17	.004	.021	100.000						

Extraction Method: Principal Component Analysis

The scree plot and component matrix were observed in other to determine the factors to be retained as suggested by, [20]. In analysing the scree plot, a change (or elbow) in the shape of the plot is identified and only components above this point are retained. An inspection of the scree plot on Fig. 1 reveals a break after the second factor (Understand the impact of demand and supply on project delivery (CP2)). The two large cluster factors, which are positioned on the steep slope, were retained. To aid the interpretation of these two factors, varimax rotation was performed which gave rise to the pattern matrix as shown in Table 7.

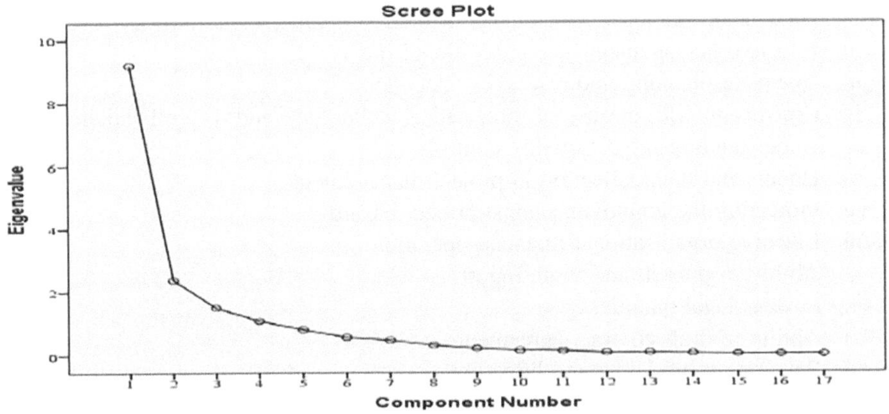

Fig. 1. Scree plot for relevance of competencies of project risk managers' to QS

Table 7 shows the factor analysis reporting the two cluster relevance factors of competence. It shows the summary of cluster relevance factor groupings for project risk managers' competence:

Table 7. Pattern matrix

	Factor	
	1	2
CP1	**0.824**	
CP2		**0.619**
CP3	**0.545**	
CP4	**0.85**	
CP5	**0.761**	
CP6		**−0.745**
CP7	**0.638**	
CP8	**0.636**	
CP9		**0.548**
CP10	**0.802**	
CP11	**0.805**	
CP12	**0.909**	
CP13	**0.927**	
CP14	**0.869**	
CP15	**0.829**	
CP16	**0.922**	
CP17	**0.869**	

Extraction Method: Principal Component Analysis

FACTOR 1 – Competences in Project Management and Its Legal Aspects

i. Understand the legal components of the risk management industry
ii. Effective management of organizational risk
iii. Adequate planning and resources allocation
iv. Conversant with the use of information technology and its application
v. Understanding project quality control
vi. Understand data collection in project management
vii. Protecting the employee against project hazard
viii. Effective organizational financial operation
ix. Ability to present and write report
x. Possess legal qualities
xi. Ability to draft contract agreement
xii. Preparation of claims and litigation
xiii. Good public and human relations
xiv. Effective human resources management

FACTOR 2 – Competences in Business Aspect of Project Risk Management

i. Understand the impact of demand and supply on project delivery
ii. Knowledge of project SWOT (Strength, Weakness, Opportunities and Threat) analysis
iii. Securing the physical assets of the enterprise

4.4 Discussion of Findings

The competencies of project risk managers were ranked to find if the competencies are relevant to quantity surveying practice and findings reveals that the identified competencies of project risk managers were highly important and significantly relevant to quantity surveying practice. This findings agrees with Dada and Jagboro study that quantity surveyor competencies include; financial management, contract administration, contract documentation, project management and communication [6]. This implies that the competencies identified through the literature, which are performed by construction project risk managers, are of great relevance to quantity surveyors. The findings from Sabana concluded that accounting principles and procedure and project evaluation are the most relevant to quantity surveyors, which is not in agreement with this study [21]. The reason for this is that the study looks at the perceived quality of procurement officers in Uganda while this study looks at the capability of quantity surveyors in Edo state Nigeria.

FACTOR 1 – Competences in Project Management and Its Legal Aspects
Factor 1 has fourteen (14) competencies of project risk managers, which were found to be highly relevant and significant to quantity surveying practice. These competencies include; *Understand the legal components of the risk management industry, Effective management of organizational risk, Adequate planning and resources allocation, Conversant with the use of information technology and its application. Others are; Understanding project quality control, Understand data collection in project management, Protecting the employee against project hazard, Effective organizational financial operation, Ability to present and write report, Possess legal qualities, Ability to draft contract agreement, Preparation of claims and litigation, Good public and human relations, Effective human resources management.* Therefore, this set of competencies was labelled "Competencies in project management and its legal aspects" and has a total variance of 54.087. This is in tandem with Dada and Jagboro study, which opined that one of the competencies of construction project risks managers, which is relevant to quantity surveying profession, is project management [6].

FACTOR 2 – Competences in Business Aspect of Project Risk Management
Factor 2 has three (3) competencies of project risk managers, which were found to be highly relevant to quantity surveying practice. These competencies include; *understand the impact of demand and supply on project delivery, Knowledge of project SWOT (Strength, Weakness, Opportunities and Threat) analysis, securing the physical assets of the enterprise.* Therefore, this set of competencies was labelled "Competencies in business aspect of project risk management" and has a total variance of 14.035. Fotwe and McCaffer collaborated this finding that one of the competencies of project risk managers, which is highly relevant in quantity surveying profession, is in technical and business or financial competencies [16]. This can be seen in the three competencies found in this category.

5 Conclusion

The study assessed the relevance of construction project risks manager's competencies to quantity surveying practice in Nigeria and developing countries in order to find out if the competencies of project risk managers are important and relevant in quantity surveying practice in construction industries using survey design. The study revealed the professionals have a high level of knowledge on risks with a mean value of 4.78 out of a maximum of 5 mean value and 64.4% of the professionals were said to have been involved in construction project risk management program. From the competencies of project risk managers, the research reveals that these competencies are highly relevant to quantity surveying practice. Among these areas, include the ability to draft contract agreement, effective human resource management, organization of financial operation, ability to present and write report, preparation of claims and litigation.

It was observed that the least competencies of project risk managers, which are relevant to quantity surveying practice, are Effective management of organizational risk, understanding project quality control, understand data collection in project management, understand the legal components of risk management industry and securing the physical assets of the enterprise. The study therefore recommend that quantity surveying practitioners should pay more attention to these areas of competencies in other to be successful in delivering the overall projects. Awareness of project risk among practitioners will help aid professionals in discharging their duties to curb risk in the industry.

References

1. Hillebrandt, P.M.: Economic Theory and the Construction Industry, 2nd edn. Macmillan, Basingstoke (2000)
2. Ofori, G.: The construction industries in developing countries: strategic review of the book'. In: Ofori, G. (ed.) New Perspectives on Construction in Developing Countries, pp. 1–15. Spon, Abingdon (2012)
3. Roggema-van Heusden, M.: The challenge of developing a competence-oriented curriculum: an integrative framework. Libr. Rev. **53**(2), 98–103 (2004)
4. Hassal, T., Dunlop, A., Lewis, S.: Internal audit education: exploring professional competence. Manag. Auditing J. **11**(5), 28–36 (1996)
5. Hore, A.V., O'Kelly, M., Scully, R.: Seeley and Winfield's Building Quantities. Explained Irish Edition. Palgrave Macmillan, Basingstoke (2009)
6. Dada, J.O., Jagboro, G.O.: Core skills requirement and competencies expected of quantity surveyors: perspectives from quantity surveyors, allied professionals and clients in Nigeria. Australas. J. Constr. Econ. Build. **12**(4), 78–90 (2012)
7. Rahman, M., Kumaraswamy, M.: Risk management trends in the construction industry: moving towards joint risk management. Eng. Constr. Architectural Manag. **9**(2), 131–151 (2002)
8. Anjana-Kak, S.: Sustainable competitive advantage with core competence: a review. Global J. Flex. Syst. Manag. (2002)
9. Rankin, N.: Annual Survey of Competency Frameworks, Competency and Emotional Intelligence, IRS (2000)

10. Baker, J.C., Mapes, J., New, C.C., Szwejczewski, M.A.: Hierarchical model of business competence. Integr. Manuf. Syst. **8**, 265–272 (1997)
11. RICS: APC and pathway guide, Quantity Surveying and Construction. Requirement and Competencies, Royal Institution of Chartered Surveyors, London (2014). http://www.rics.org/Global/APC_Requirements_and_Competencies_Guide.pdf. Accessed 27 Sept 2014
12. Chan, I.Y.S., Liu, A.M.M., Cao, S., Fellows, R.: Competency and empowerment of project managers in China. In: Smith, S.D., Ahiaga-Dagbui, D.D. (eds.) 29th Annual ARCOM Conference, Reading, UK, Association of Researchers in Construction Management, 2–4 September 2013, pp. 383–392 (2013)
13. PMI: Project Manager Development (PMCD) Framework, 2nd ed. Project Management Institute, Pennsylvania (2007). http://www.tecdigital.itcr.ac. Accessed 24 Oct 2014
14. Hwang, B., Ng, W.: Project management knowledge and skills for green construction: overcoming challenges. Int. J. Proj. Manag. **31**, 272–284 (2013)
15. Crawford, L.: Senior management perceptions of project management competence. Int. J. Proj. Manag. **23**(1), 7–16 (2005)
16. Fotwe, F.T.E., McCaffer, R.: Developing project management competency; perspectives from the construction industry. Int. J. Proj. Manag. **18**(5), 19–24 (2000)
17. Dada, O.J.: A principal component analysis of skills and competencies required of quantity surveyors: Nigerian perspective. Organ. Technol. Manag. Constr. **6**(8), 1020–1032 (2014)
18. Said, I., Shafiei, M.W.M., Omran, A.: The competency requirements for quantity surveyors: enhancing continuous professional development. ACTA Technica Corniensis Bull. Eng. **3**, 105–112 (2010)
19. Tabachnick, B.G., Fidell, L.S.: Using Multivariate Statistics, 5th edn. Pearson Education, Boston (2007)
20. Pallant, J.: SPSS Survival Manual: A Step-By-Step Guide to Data Analysis Using SPSS for Windows (Version 12), 2nd edn. Allen & Unwin, Crows Nest NSW (2005)
21. Sabana, O.S.: Competences of procurement officers, self-efficacy, Accountability and perceived service quality of Procurement and disposal entities within central Government in Uganda. An MSc dissertation, Department of Human Resource Management, Makerere University, Uganda (2010)

Killing What We Intend to Grow: The Challenges of the Local Construction Industry with the Public Procurement Systems in Ghana

Christopher Amoah[1] and Winston Shakantu[2(✉)]

[1] Department of Quantity Surveying and Construction Management, University of the Free State, Bloemfontein 9300, South Africa
amoahc@ufs.ac.za

[2] Department of Construction Management, Nelson Mandela University, Port Elizabeth, South Africa
Winston.Shakantu@mandela.ac.za

Abstract. Countries have used their public procurement systems, as a policy tool to grow their indigene construction industry by reviewing sections that are not favourable to the local construction firms. Ghanaian public procurement systems have gone through various reforms to date. This study, therefore, investigated the challenges faced by the local construction industry concerning the public procurement systems in Ghana. The qualitative research approach was used for this study. A structured survey questionnaire made up of closed and open-ended were distributed to randomly selected 65 construction firms registered and operating from Accra; of which 36 of them responded. Data received were analyzed by using content analysis method whereby similar responses were grouped together and sorted in terms of their frequencies and magnitude. The findings indicate that the local construction firms have major issues with the public procurement systems, some of which include; funding, bribery and corruption, foreign competition, lack of capacity to tender for some project, payment delays, cumbersome tender process, sole sourcing, and political influence. The implication is that the procurement systems in Ghana are still having major challenges in addressing issues faced by the local construction thereby limiting their growth. The limitation of this study is that primary data were solicited from only construction firms operating from Accra. It is therefore recommended that government of Ghana should institute measures such as local content policy, prompt payment, skill development and encourage joint venture to address these concerns in order to achieve the domestication and Ghana without aid policy in the near future.

Keywords: Construction industry · Government · Local · Public procurement

1 Introduction

The importance of public procurement for developing countries is progressively been cherished by development agencies globally, knowing that the social and economic costs [1] of the flaws in public procurement governance increases the risk for foreign

C. Aigbavboa and W. Thwala (Eds.): CIDB 2019, *The Construction Industry in the Fourth Industrial Revolution*, pp. 112–123, 2020.
https://doi.org/10.1007/978-3-030-26528-1_12

investment. Public procurement is naturally a politically delicate activity, as it involves momentous amounts of public money of a country's economy.

The OECD Development Assistance Committee puts the size of global public sector procurement at 8% (US$3.2 trillion) of the worldwide GDP of US$40 trillion [2], while others put the figure at 15% of GDP for developing countries [3]. Public procurement in Ghana is estimated at 24% of total imports and aside from personal remunerations; public procurement is valued at 50–70% of the national budget and 14% of GDP [4]. Studies done by the World Bank shows that in Ghana, the public procurement for goods, works, and consultancy services is valued at US$600 million, representing about 14% of GDP [5]. The bulk of this expenditure according to the World Bank was spent on procurement for works, goods, technical and consultancy services by Ministries, Departments, Agencies (MDAs) and District Assemblies (DAs) as well as Public Hospitals, Universities, Schools and Colleges [5]. The public procurement system has major implications on the local construction industry as a viable local construction industry would be able to execute government projects and therefore impact substantially on the economy of Ghana as government expenditure on construction projects would stay in the country, thereby creating wealth and boosting the country GDP and reduce poverty.

Ghana, since independence, has tried to address the shortcomings in its public procurement procedure through the passage of financial regulations such as the Contracts Act, 1960 (Act 25); Ghana Supply Commission Act; the Ghana National Procurement Agency Decree, 1976 (SMCD 55); and the Financial Administration Decree (SMCD 221) passed in 1976 [6]. All these regulations were instituted in the public sector to provide an extensive framework of administrative powers to standardize procurement activities. However, continuous reviews of the public procurement management in Ghana reveals considerable inadequacies, fraud and opaqueness in the procurement practices within the government agencies due to the absence of clear legal framework, coherent procedures and guidelines as well as unclear institutional and organizational provisions needed to administer the public procurement process [7]. This, therefore, prevented the attainment of value for money in procurement financed by the government and donor agencies [7].

It, therefore, became necessary to re-examine the public sector procurement systems to safeguard the effectiveness and capability of institutions in addressing the procurement related problems. The Ghana government following substantial appraisal of its public expenditure system in 1993, therefore resolved to institute an all-inclusive reform in its Public Financial Management System. This process gave birth to the Public Financial Management Reform Programme (PUFMARP), which was effective from 1995. The main goal of PUFMARP was to boost efficiency, transparency, and answerability in the management of public finances [7]. To achieve this, PUFMARP suggested the evaluation of the Public Procurement system and subsequently, instituted Public Procurement Oversight Group (PPOG) in 1999 to help formulate an all-embracing public procurement reform policy [7]. The PPOG drafted the public procurement bill in September 2002, which was passed into law as the Public Procurement Act, 2003 (Act 663) by the Ghana government. The ultimate goal of Act 663 is to address the flaws in the public procurement of goods, works, and services, to operationalize the concept of good governance, and to drive the notion of "zero tolerance" for corruption in the public procurement process.

With all these reforms to make public procurement more successful and achieve the objectives of government, the local construction industry is not benefiting from these reforms. Challenges facing the local contractors in Ghana have been identified as (1) delayed payments for works completed; (2) contractor's managerial and technical incapability to generate profit from their projects; (3) inability to secure more profitable contracts; (4) inability to get contracts; and (5) inability of obtain funds and to repay loans [8]. Therefore, more and effective funding arrangements for local contractors should be encouraged [8]. This study, therefore, investigated the challenges of the local construction industry with the public procurement systems in Ghana.

2 What Is Public Procurement?

Public procurement originated from the fiduciary responsibility of government administrations to provide services and infrastructure, such as education, health care, harbours and roads to the public or people in a particular area, town or city [9]. Public procurement can also be defined as any government action aimed at acquiring services and goods required to execute government functions [10]. Public procurement is generally described as the acquisition of goods, construction works and services by the public sector through a contract [9]. It also involves the acquisition of construction works and services if state or local authority uses its budgets, domestic loans, foreign aid and revenue for such acquisitions.

Public procurement can also be referred to as the government's activity of procuring goods and services, which are needed for the government to function effectively [10]. Public procurement is a practice whereby public sector organizations purchase goods, services, and works from third parties, including the provision of key services such as education, social care, health care, and welfare, to the citizens by private sectors. In some instances, the public sector directly provides these services directly or provides them through procurement mechanisms.

2.1 Objectives of Public Procurement Systems and Regulatory Rules

There are numerous purposes of public procurement that can be acknowledged, which are also common to public procurement systems in many countries. These objectives are implemented through various legal and regulatory rules of conducting the procurement procedures. The effective implementation of the public procurement objectives can be effected through consistent and comprehensive procedures based on the fundamental principles, and where compliance with the procedures is appraised and enforced. Different countries have different objectives for regulating their public procurement procedures. For examples, it is believed that, with the procedures prescribed in the national legislation, an enacting country will create an environment in which the public is assured that the government will expend public funds with responsibility and answerability, and thus attain value for money [10].

Different analysts also embrace different categorizations. For example, the advancement of social and political goals, economic efficiency, trade objectives as the most distinguishable strategic goals are recognized and decreasing corruption are considered as part of efficiency objective [11]. Generally eight common objectives in public procurement systems are identified, which include; value for money in acquiring services, works, and goods; truthfulness; accountability; same treatment for suppliers; efficiency implementation of industrial, social and environmental objectives; making public markets accessible to international trade; and effectiveness in the procurement practices [10].

It is vital to note that there exist different objectives within different procurement systems and the weight devoted to the numerous objectives varies quite remarkably. For example as Arrowsmith *et al.* put it, *"some systems attach much more important than others to policies of fair and actual treatment of providers to the use of procurement to promote social objectives or accountability with the result that the government may be willing to pay higher prices for goods or services or to accept greater process cost to implement these values"* [10]. The variances in the objectives among different public procurement systems, private and public organizations explain to some degree the variances in the method to procurement and the rulebooks that administer it [11]. For example, a system that dwells heavily on accountability is more likely to offer a comprehensive rule which permits for excessive public scrutiny of the procurement practices than one that does not [11].

In Ghana, however, the objectives of reforming the public procurement systems are to ensure professionalism; transparency, competitiveness, and fairness; value for money in the procurement systems; efficiency in the procurement systems; ethical approach in procurement processes, and introduction of technology in the procurement systems [6, 7].

2.2 The Public Procurement Act, 2003 (Act 663 of 2003)

The Public Procurement Law, 2003 (Act 663 of 2003) [12] is an all-inclusive regulation formulated to eradicate the inadequacies and administrative flaws which were evident in Ghanaian public procurement. The government of Ghana, having discussed with other development partners, recognized that the public procurement system needed critical consideration in view of the prevalent public perception of corrupt practices and inadequacies in the procurement system.

A study by the World Bank [13] estimated the national budget for public procurement to be about 50–70%. Therefore, an effective public procurement system could guarantee value for money in government spending, which is indispensable to a country experiencing massive challenges in its developmental agenda. To safeguard saneness and value for money in the public procurement arena, the government of Ghana through PUFMARP and Public Procurement Oversight Group in 1999, drafted public procurement bill in September 2002. In 2003, the Parliament of Ghana enacted the Public Procurement Act to help in procurement activities in Ghana. Act 663 has contributed some level of sanity into the construction sector [7].

2.3 Criticism of the Public Procurement Law (PPL) of Ghana (Act 663)

Among the findings of the research are that there are still various shortcomings that need improvement in the PPL. There are a number of criticisms enumerated against the PPL [14], which include the following:

- The PPL allows sole sourcing and restricted tendering which is not ideal in public procurement as it may promote system abuse and corrupt practices despite the laws regulating their use, including preceding approvals by the PPA.
- The PPL gives excessive prominence to events primary to contract award to the detriment of happenings after the contract is awarded.
- The PPL does not give appropriate weight on ethical practices and conflict of interest policy in public procurement and nose-dives to provide comprehensive ethical standards for defining the difference between private and public interests.
- There are contradictions in the guidelines governing the advertisement time and methods for the National Competitive Tendering solicitation documents.
- The PPL does not make adequate room for civil society and the media to participate in the public procurement process. This restricts their significance in scrutinizing public activities to ensure fairness, transparency, and non-discrimination.
- The margin of preference provided under section 60 of PPL for domestic bidders, when approved by PPA, gives advantage to domestic suppliers and contractors, which could perpetuate inefficiency and increased cost and compromise on fair competition principles and value-for-money.
- The PPL is mute on evolving public procurement models such as Public Private Partnership and Framework Agreement.

A study examining the influence of the public procurement law on the Ghanaian educational sector revealed that out of 35 respondents interviewed, 33 representing 94% agreed that they experience many challenges [15]. Some of the challenges respondents stated were; burdensome process thus delaying contract award; lack of space for negotiating contract price; and postponements in funds allocations thus leading to purchase price fluctuations. These holdups hamper the effective operation of their school's procurement systems [15]. It is further stated that the foremost causes for these holdups are unavailability of supplier's database; inadequately experienced staffs and cumbersome approval processes [15]. In addition to unavailability of suppliers database, the respondents criticized the difficulty in adhering to the procurement processes, because, in most cases, they had to provide three (3) quotations from different suppliers for textbooks and appoint the supplier that provides them with quality textbooks at a reduced cost and in some instances some of the suppliers were legally incapable to enter into a contract [15].

2.4 Problem of the Construction Industry in Ghana

Studies demonstrate that many developing countries construction industries, including that in Ghana, encounter many difficulties [16]. There are three central reasons for these challenges [16]. First, the economic weaknesses experienced by these countries ultimately affect the adequacy of resources earmarked for improving the industry thereby stifling job opportunities as well as the market forces which support innovation. Secondly, many of these countries governments do not recognise the significance of the construction industry, and therefore do not institute and implement programmes for improving the industries. Finally, as a result of these countries inherent underdeveloped construction industries, they are powerless in dealing with their flaws, make a robust argument for assistance or contribute to the efforts making by the government to improve the industries.

Due to these difficulties highlighted, the performance of developing countries, including Ghana, construction industries on projects, is perceived to be unsatisfactory in many aspects including quality, cost, and output [16]. It has been established that developing countries miss the target set by themselves for most construction projects executed in relation to schedules (time), budgets (cost) and specifications (quality) [16]. Again the construction output in these countries is believed to be substandard in terms of their sustainability and resilience. As constructed objects are major investments, they are projected to last for years and their durability and maintainability, therefore, has momentous social and economic significance. Furthermore, the construction industries performance in these developing countries looks unsatisfactory when compared to their counterparts in industrialised nations.

The features of the construction industry in Ghana resemble the features of the construction industries in developing countries stated above. Whereas governments have a propensity to control the economy through the use of the investment in the construction industry, hold-ups in the public-sector projects management often lead to abysmal performance than the targets forecasted, hence restraining the effect of the government's unrelenting determinations [16]. If a country wants to derive a supreme benefit from its construction activities, then it is important that these activities do not lead to an increment in importation or transfer of incomes by foreign firms, but must rather engender activity in other sectors of the local economy including the indigenous construction industry. Ghana must, therefore, institute policies to address the challenges facing its domestic construction industry to realise the benefits of the construction sector in its GDP growth.

Studies have revealed many challenges faced by contractors in Ghana [17]. Among these challenges are the presence of foreign contractors and the lack of contractor's capacity [17]. It has been identified that foreign contractors in Ghana are awarded most major government projects and this necessitated a suggestion from the Ghana Association of Road Contractors that 25% of the contracts awarded to foreign construction firms by the government be sublet to the local contractors with recognized capability [17]. The Association of Road Contractors argued that most of the works executed by personnel of the oversea contractors could be equally done by local contractors and professionals. However, the call has not been heeded by the government [17]. Findings from studies indicate that some contractors protested that foreigners engaged to work for foreign construction firms in some cases do not have suitable experiences in construction works [17]. Among challenges faced by local contractors in building and road industry are the presence of foreign contractors and low capacity of indigenous Ghanaian contractors and that local contractors do not oppose competition; however, the inadequate capacity of the local contractors makes it more challenging for them to compete [17].

A systematic and synchronized method in the administration improvement of the construction industry in Ghana is thus necessary for the country, and the current administrative system for this objective should be supported and the required resources be made available on a sustainable basis [16]. It has been identified that the preferential policy in the current public procurement law is inadequate in facilitating capacity growth of the local construction industry and hence the government should institute policies to address these deficiencies in the procurement law [18].

3 Research Methodology

In order to attain the purpose of this study, a combination of an in-depth literature review and semi-structured survey questionnaire were adopted. The qualitative research approach was therefore adopted for this study. The random sampling method was adopted to select respondents from the lists of contractors and consultants operating from Accra, Ghana. In this type of sampling, [19] state that every member of the target population has the same probability of being involved in the sample and can be chosen. Also, two steps that are essential for drawing a random sample are suggested: Firstly, all the units of analysis in the sampling frame should be identified and be given numbers and secondly, the method used to select the units of analysis should ensure that every number has the same chance of being selected [19]. The researcher first obtained the addresses of contractors and consultants operating from Accra, Ghana from the Ghana Telkom directory. The researcher then randomly selected the firms to be included in the study by taking into account the location of the company. The location of the companies was taken into account in order to prevent the selection of more firms operating from the same locality. A prior arrangement with the respondents (firm owners or representatives) was then done via telephone. The techniques adopted for this study was the use of a semi-structured survey questionnaire made up of both closed and open-ended questions.

Ambiguity is reduced when answer choices are provided in the instrument and respondents are more likely to answer the question that is needed by the researcher and closed-ended questions are conclusive in nature as they are intended to generate data that is easily measurable [20]. Again, it is suggested that researchers should make response choices exhaustive, offering at least one possibility with no room for vagueness and if respondents do not find a response option that matches their answer to the question, they may avoid the question entirely or select a response option that does not represent what they are thinking about [21]. Open questions are generally used when there is insufficient understanding about a particular topic, and you want to learn as much as possible without limiting the responses and also open-ended questions allow for rich qualitative data as they provide the researcher with an opportunity to gain understanding of all the opinions on a topic they are unfamiliar with [20]. The researcher personally visited each of the respondent's firm, delivered the questionnaire and waited for the respondents to fill in the questionnaire and hence the researcher was able to clarify issues with the respondents where needed. In order instances, the researcher left the questionnaire with the respondents and collected it at a later date. This technique was adopted to reduce the non-responsive rate and come-backs.

The questionnaire was distributed to 68 construction firms and consultants of which 36 were successfully received. In qualitative research, by the time you reach a sample size of 30 participants, the research would have reached the theoretical capacity point, where minimal or no fresh information can be added [22]. Considering these views [22], the 36 responses received is enough for descriptive statistical analysis. The Public Relation Officer of the Ghana Public Procurement Authority was also interviewed. Descriptive statistical analysis tools such as excel were used to analyse the data through the use of contents analysis where similar responses were grouped together and their frequencies determined.

Out of 36 respondents, 30% were managing directors of the company, 24% were project managers, 3% were Architects, 22% were Quantity surveyors, 16% projected Engineers and 5% were Operation managers for their respective organizations. The analysis of the respondent's industry profile indicates that 69% of the respondents are in the building construction industry, 19% are professional consultants in the construction industry and 12% are in the road construction sector. Forty-one percent (41%) of the respondents organizations have been in business for 1 to 5 years; 35% have been operating for 6 to 10 years; 11% for 16 to 20 years; 8% has been in business for 11 to 15 years and 5% have been in operation for 21 to 25 years. This indicates that most of the firms selected for the interview have been in business for quite a long time. Fifty-seven percent (57%) of the respondents have worked between 6 to 10 years, 22% of them have worked for 11 to 15 years, 14% have worked in the construction industry for 1 to 5 years, 5% have worked for 16 to 20 years and 3% has worked for 21 to 25 years. Thus a substantial number of the respondents have experience of working in the construction industry for quite a long time and hence are well placed to know the challenges of the procurement systems.

4 Findings and Discussions

4.1 The Challenges the Local Construction Industry Has with the Current Public Procurement System

Respondents were asked to mention the challenges they have with the current public procurement systems. The challenges reported by respondents during the interviews are as shown in Table 1. The responses were tallied to show how often that particular challenge was mentioned by the respondents and then ranked to show the seriousness of the challenge.

From Table 1, it is clear that the main challenges of the local construction industry with respect to the public procurement practices in Ghana was funding issues (ranked 1). This was followed by lack of capacity to tender (ranked 2), bribery and corruption (ranked 3), high-interest rate and foreign competition (ranked 4 respectively). The next issues are payment delays, political influence, cumbersome tender processes and lack of transparency (ranked 5 respectively). This is followed by sole sourcing, non-enforcement of local content law, and high taxes (ranked 6 respectively). The major challenges facing the players in the construction and consulting industry are funding issues, lack of capacity, bribery and corruption, high-interest rate, foreign competition,

Table 1. Respondent's challenges with the current procurement systems

Challenges	Frequency	Percentages	Rank
Funding issues	16	30.77%	1
Lack of capacity to tender for some projects	12	23.08%	2
Bribery and corruption	7	13.46%	3
Foreign competition	3	5.77%	4
High interest rates	3	5.77%	4
Payment delays	2	3.85%	5
Cumbersome tender processes	2	3.85%	5
Political influence	2	3.85%	5
Lack of transparency	2	3.85%	5
Sole sourcing	1	1.92%	6
No law to enforce local content	1	1.92%	6
High taxes	1	1.92%	6

political influence, payment delays and lack of transparency. If the government and procurement authorities address these issues, it would be of tremendous help to the local construction industry. This finding validates the findings of [8, 17, 15, 23] with respect to the difficulties confronting Ghanaian contractors as stated in Sect. 2.4. The minor concerns of the local construction industry in Ghana about the current procurement system are no law to enforce local contents, sole sourcing, cumbersome tender processes, and high taxes. This also supports the findings of Laryea [17].

4.2 Respondent's Suggestions on What Government Should Do in the Public Procurement Process to Help Grow the Local Construction Industry

Respondents were asked to suggest what the government should do about the current public procurement systems to help grow the local construction industry. The responses of the interviewees are shown in Table 2.

From Table 2, the players in the construction industry would be happy if the government would take measures to address their challenges with respect to the current public procurement processes. The following are the suggestions to the government; financial support and strict compliance with the procurement law (ranked 1 respectively), followed by local content policies, tax reduction on materials and accountability in the procurement system (ranked 2 respectively), lower interest rate (ranked 3). This is followed by an incentive package to local industry (ranked 4), Create pools of construction equipment for local contractors, prompt payment for work done and set aside policies in the procurement law (ranked 5 respectively), discourage sole sourcing, skills development from foreign to local contractors, equal treatment to all bidders and encouragement of joint ventures (ranked 6 respectively). The findings of the study, therefore, show that contractors and consultants in the construction industry mainly want the government to tackle the same problems they have with the public procurement systems.

Table 2. Respondent's suggestions to the government on the current public procurement systems

Respondent's suggestions	Frequency	Percentages	Ranking
Financial support to local contractors	7	13.73%	1
Strict compliance to the law	7	13.73%	1
Local content policies in the procurement	6	11.76%	2
Tax reduction on construction materials	6	11.76%	2
Accountability in procurement system	6	11.76%	2
Lower interest rate on loans	5	9.80%	3
Incentive package to local industry	4	7.84%	4
Prompt payment for work done	2	3.92%	5
Create a pool of construction equipment for local contractors	2	3.92%	5
Set aside policies in the procurement law	2	3.92%	5
Discourage sole sourcing	1	1.96%	6
Skills development from foreign to local contractors	1	1.96%	6
Equal treatment to all bidders	1	1.96%	6
Encourage joint ventures	1	1.96%	6

4.3 An Interview with Ghana Public Procurement Authority

An interview with the Ghana procurement authority revealed that local contractors and consultants and their foreign counterpart have the same chance of winning a government contract if all of them are tendering for the first time. This buttresses the concerns of the local construction industry regarding unrestricted foreign competition and lack of enforcement of local contents law in the procurement systems. In public procurement systems where the growth of the local construction industry is paramount, local contractors should have upper hand over the foreign contractors and consultants and the government should institute measures in the procurement process to restrict foreign competition [24, 25].

In Malawi, for example, the government instituted measures in the public procurement system to safeguard profit to the local firms, by only accepting bids that were within a certain percentage of the projected price and bonus clauses were also introduced for completing on schedule. The end result of this policy in favour of the local construction industry was the 40% cost savings realised over the use of international contractors and the improvement of the local construction industry that enabled other clients to use [25].

5 Conclusion and Recommendations

It was revealed through data analysis that, the challenges the local construction industry is facing currently are the same challenges they faced before the public procurement reforms. The major challenges are funding issues, lack of capacity, bribery and corruption, high-interest rate, foreign competition, political influence, payment delays, and lack of transparency. Other concerns identified are no law to enforce local content, sole sourcing, cumbersome tender processes, and high taxes. It can, therefore, be concluded that public procurement reforms in Ghana have failed to address the challenges of the local construction industry.

The Ghanaian government should therefore emulate the examples of countries such as Malawi, Singapore and South Africa by instituting measures such as financial support to local contractors, enforcement of local content policies in the procurement, tax reduction on imported construction materials, lower interest rate on loans, incentive package to local construction industry, creation of construction equipment pool for local contractors, set aside policies in the procurement law, prompt payment for work done, discouragement of sole sourcing, skills development from foreign to local contractors and equal treatment of all bidders in the public procurement processes to address concerns of the local construction industry.

References

1. Schapper, P., Veiga Malta, J., Gilbert, D.: An analytical framework for the management and reform of public procurement. J. Pub. Procurement **6**(1&3), 1–26 (2006)
2. OECD/DAC: Mainstreaming the Procurement Function into the Public Expenditure Policy and Effectiveness Dialogue (OECD/DAC). World Bank Roundtable, Paris, 22–23 January 2003. www.oecd.org
3. Nicol, W.: Mainstreaming the procurement function into the public expenditure policy and effectiveness dialogue. Paper presented at the World Bank Roundtable, Paris, France, 22–23 January 2003. www.oecd.org
4. Oppong, D.: The impact of the implementation of public procurement act 2003 (Act 663), on the timely delivery of goods and services: a case study of Ghana Water Company Limited (GWCL). Int. J. Appl. Innov. Eng. Manag. (IJAIEM) **2**(8) (2013)
5. World Bank: Economic Report on Ghana, Washington, DC: Ghana Country Department, The World Bank (2003)
6. Adjei, A.B.: Message from the Chief Executive. Public Procurement Board, June 2006. http://www.ppbghana.org/about/ppbboard.asp. Accessed 20 June 2015
7. Kotoka, F.A.: Assessing the Level of Compliance with the Public Procurement Act 2003, (ACT 663) in Public Entities in Ashanti Region of Ghana (2012)
8. Eyiah, A.K., Cook, P.: Financing small and medium-scale contractors in developing countries: a ghana case study. Constr. Manag. Econ. **21**, 357–367 (2003)
9. Odhiambo, W., Kamau, P.: Public Procurement: Lessons from Kenya, Tanzania and Uganda. OEDC. Development Centre, PDF file (2003)
10. Arrowsmith, S., Treumer, S., Fejo, J., Jiang, L.: Public Procurement Regulations: An Introduction (2010)
11. Trepte: Regulating Procurement: Understanding the Ends and Means of Public Procurement Regulations, pp. 18–19 (2004)
12. Public Procurement Board, Manuals - Public Procurement Act, 2003 (Act 663), Ghana
13. World Bank: Ghana Country Procurement Assessment Report Number 29055, Vol. 2. World Bank, Ghana Country Department, Africa Region (2003)
14. Ameyaw, C., Mensah, S., Osei-Tutu, E.: Public procurement in Ghana: The Implementation Challenges to the Public Procurement Law 2003 (Act 663). Int. J. Constr. Supply Chain Manag. **2**(2), 55–65 (2012)

15. Gnanih, E.L.: Investigating into the effect of the Public Procurement Law on the Ghanaian Education Sector. Thesis submitted in partial Fulfilment of Bachelor of Science Degree in Business Administration, Ashesi University College (2012)
16. Ofori, G.: The Construction Industries in Developing Countries: Strategic Review of the Book, pp. 1–15. Spon, Abingdon (2012)
17. Laryea, S.A.: Challenges and opportunities facing contractors in Ghana, West Africa Built Environment Research Conference, Accra, Ghana, 27–28 July 2010. http://centaurreading.ac.uk/1682. Accessed 18 July 2014
18. Amoah, C., Shakantu, W.: The effectiveness of the preferential policies in the Ghanaian procurement law in promoting the local construction industry. J. Constr. Proj. Manag. Innov. **7**(SI1), 1859–1876 (2017). ISSN 2223-7852
19. Welman, C., Kruger, S.J., Mitchell, B.: Research Methodology, 3rd edn. Oxford University Press, Oxford (2005)
20. Reja, U., Manfreda, K.L., Hlebec, V., Vehovar, V.: Open-ended vs. Close-ended Questions in Web Questionnaires (2003)
21. Burgess, T.F.: A general Introduction to the Design of Questionnaires for Survey Research (2001)
22. Cassim, L.: An overview of research design and methodology. Chapter 3, ERS (2011)
23. Osei-Tutu, E., Badu, E., Owusu-Manu, D.: The level of compliance with the Public Procurement Act in Ghana. Int. J. Manag. Innov. Sustain. Built Environ. (2011)
24. Thai, K.V.: Public procurement re-examined. J. Pub. Procurement **1**(1), 9–50 (2001)
25. Wells, J., Hawkins, J.: Increasing local content in infrastructure procurement in low income countries, unpublished (2008). Accessed 3 Sept 2013

Mentoring Practices in Construction Professional Firms in Nigeria

Osamudiamen Kenneth Otasowie(✉) and Ayodeji Emmanuel Oke

Department of Quantity Surveying,
Federal University of Technology, Akure, Nigeria
otasowiekenneth@gmail.com, emayok@gmail.com

Abstract. The paper assessed the exhibited mentoring practices in construction consulting firms with a view to maintaining high skill work force that can sustain the competitive advantage of these firms in the construction industry. The involvement and role of professionals in mentoring relationships, as well as the exhibited mentoring practices in professional firms within the built environment were assessed through a survey design. Structured questionnaire was used to survey quantity surveying firms from Abuja, the capital of Nigeria. Percentage, Mean Item Score and one-sample t-test were used in the analysis of the gathered data. The study reveals a good level of involvement in mentoring practices among construction professionals in the firms. However, some construction professionals are not sure of their participation in mentoring relationships. The study also revealed the mentoring practices exhibited in construction professional firms and the practices that are not exhibited. The findings show the exhibited mentoring practices in construction professional firms as well as recommendations for maintaining high skill work force and sustain the competitive advantage of the construction professional firms within the country. This paper highlights the exhibited mentoring practices in professional firms in the Nigerian construction industry and made known the practices that are not exhibited. This practices when exhibited will help in the proper psychosocial and career development of protégées in the industry.

Keywords: Mentoring relationship · Mentoring practices · Construction professional firms · Quantity surveying · Consultancy firms

1 Introduction

The building construction industry over the last few decades has witnessed both institutional and organizational shift across the world that has resulted in the continuous complex nature of professional services within the built environment. [1] opined that professional firms in the construction industry must maintain high skill work force to sustain competitive advantage in a manner that will speed up the work. This is because they depend on their experience and knowledge rather than plant and equipment to meet the ever-changing needs of clients. According to [2], professional firms can maintain high skilled work force, sustained performance by engaging some performance intervention programs, and further identified mentoring as one of such programs that can sustain both performance and competitive advantage of these professional firms.

C. Aigbavboa and W. Thwala (Eds.): CIDB 2019, *The Construction Industry in the Fourth Industrial Revolution*, pp. 124–132, 2020.
https://doi.org/10.1007/978-3-030-26528-1_13

Mentoring has become fundamental in performance improvement of employees and promotion of organization efficiency and productivity thereby placing the organization in a more competitive and advantageous position. [3] opined that mentoring has developed into a critical service for professional support and still developing into an intervention for business adversary. [4] posited that mentoring can be used as a tool of communication to those employees in terms of their performance that are not open, to break down the ice in an employee. [5] opined that mentoring practices are supported by firms for the purpose of increasing the knowledge and productivity of its employees and thereby improving the performance of the firm.

However, despite the successes of mentoring practices recorded in various firms and on individuals in other industries, such as checking youth behavior which is seen as a threat, helping graduate students develop personal and professional relationships with their academic advisors [6], the review of literature shows that there are no adequate research studies that focused on the concept in Nigerian construction industry. [2] corroborated this by revealing that insufficient attention has been given to the concept of mentoring in the nation's built environment. Another significant disclosure from such findings is that researches have basically concentrated on knowledge management and human resource management in Nigerian construction professional firms without particular attention to mentoring practices, which is core to improving competitiveness, driving performance and ensuring sustainability. It is therefore disturbing that there has not been sufficient research in this vital strategy, as it is evident that we live in a world where the construction industry plays a vital role in the development of any nation and as such, the need for mentoring in the construction industry cannot be overemphasized. This lack of study is a motivator to this study, which is to assess the mentoring practices in construction professional firms in Nigeria. The presence of several construction sites in Abuja, the nation's capital means the availability of various construction professional firms for the study to be conducted. The findings of the study can therefore be inferred for Nigeria and other Africa nations. The later parts of this paper include the review of literature, the methodology, and the findings of the study. Furthermore, conclusions drawn and necessary recommendations made from the findings.

2 Concept of Mentoring and Mentoring Practices

According to [7], central to the concept of mentoring and the word "mentor" is the Greek mythology, when Telemachus, Odysseus son, was entrusted to a close friend of his father, in other to not only oversee his professional development but also his social and personal growth.

[8] posited that the preponderance of mentoring in several settings and issues of wide range has seen scholars struggle to formulate a general definition of the term. A prove to this can be noticed in that over 50 different definitions have emanated from the social science literatures [9]; however, despite the various definitions from several studies, there exists a common feature of mentoring which includes a learning relationship between an experienced and a novice or better still a less experienced individual, a process that involves emotional and instrumental functions and that becomes more impactful over time.

The need for human capital sustainability for the purpose of economic growth and social development as noted by [10] has given importance to the concept of mentoring. Mentoring has gained some ground positively and has been linked with many successful careers particularly for the argument as opined by [11], that undergraduate programs continuously fail to produce graduates who possess the right professional and lifelong learning skills it takes to have a successful career. Learning from the experience of others is the purpose of mentoring and to this end, a way of offering training that can suit entrepreneurs' penchant for just-in-time, experiential and applied learning [12].

[13] identified role modelling, acceptance/confirmation, counselling, and friendship as mentoring practices that help in the psychosocial development of a mentee. He forwarded that the confidence, effectiveness, and competence of one involved in mentoring could be improved upon through these practices. [14] opined that role modelling help guide individuals on the required behaviours for success, while [15], posited that when the skills and knowledge of a mentee have been validated by the mentor, it shows in the support and acceptance shown to the mentee. Furthermore, counselling according to [16] avails an individual the opportunity to explore a difficult situation and dissatisfaction of another and thereafter render help that addresses the issue, while [17] opined that friendship, that is to say friends of virtue as categorised by Aristotle aims at achieving goals.

In addition to psychosocial development, mentoring can also aid in the development of the careers of individuals through practices such as sponsorship, coaching, protection, exposure/visibility, and challenging work assignments [13]. [15] opined that professionals and even the firms they are attached to witness advancement in the level of service delivery through sponsorship, while [18] sees sponsorship as a relationship that benefits parties involve and helps in professional development. Coaching which aims at improving competency in work place helps parties involve in mentoring particularly mentees according to [19, 20] attain their potentials. [15] argued that protection which is meant to shield a protégée from damaging contacts is the most overlooked and neglected practice of the available mentoring practices. This is despite its necessity in the development of young professionals. Furthermore, [21] put forward that protégées advance when they are exposed and visible to developmental relationships which in turn leads to the advancement of the firms they work with, while [22] posited that protégées are pushed outside their comfort zone to development through challenging work assignments which increases their skills and ability.

3 Research Methodology

This study assessed the exhibited mentoring practices in construction professional firms in Nigeria; hence survey research was employed and a quantitative approach in data gathering from respondents therefore used through questionnaires for the purpose of objectivity. The entire practicing Quantity Surveying consultancy firms resident in Abuja, Nigeria were targeted for the population of this research. The Information gathered from related literature review informed the design of the research instrument used which was structured questionnaire. The questionnaire was designed in sections with the preliminary section of the questionnaire focusing on the background information of the respondents such as involvement in mentoring and role in mentoring. Respondents were invited and provided with the mentoring practices obtained from literature and were asked to indicate the degree of exhibition of each of the practices based on a five-point Likert rating scale (very high = 5, high = 4, average = 3, low = 2 and little or none = 1). The questionnaires were administered through personal delivery. A total of 149 questionnaires were distributed to Quantity Surveying Consultancy firms in Abuja. 142 usable questionnaires were retrieved indicating 95.30% effective response rate.

In the analyses of the gathered data, frequency and percentage were used for the background information, Mean Item Score for ranking the practices, and one samples t-test for determining the relative level of exhibition of each of the identified practices and their significance level.

3.1 Background Information of Respondents

Table 1 presents the distribution of the ranks, working experience, professional membership, mentoring involvement and mentoring relationship role of respondents. Base on the result, Principal Partners of the surveyed firms make up 9.9% of the respondents while 18.3% are Partners in the firms surveyed. Analysis also indicated that Senior Quantity Surveyors makes up 32.4% of the respondent whereas (39.4%) were Quantity Surveyors in their respective firms. 36.60% of the respondents have 0 to 5 years of experience, while 2.10% have 6 to 10 years. 18.30% of the respondents have between 11 to 15 years, while 9.20% are within 16 to 20 years of experience and 17 out of the total number of respondents indicating (33.80%) have over 20 years of experience. On Registration with the Nigerian Institute of Quantity Surveyors of respondents, 54.9% of the respondents had Corporate membership with the Institute and (12.7%) were Fellow members. furthermore, (32.4%) were Probationer members in the Institute. Relating to the involvement in mentoring relationship by respondents, boxes were presented to the respondents to tick “Yes” if they have participated in a mentoring relationship, “No” if they have not, and “Not sure” if they are not certain they have participated. Result shows that 78.2% of these respondents have been involved in mentoring relationships, while 21.8% of the respondents are yet to participate in mentoring relationships. On role in mentoring relationships, Results shows that 19% of the respondents played the role of a mentor in the mentoring relationships they have been involved in, while 21.1% played the role of a protégée in the mentoring relationships. 38% of the respondents have been opportune to play both the role of a mentor and a protégée in the mentoring relationships involved in, while 21.8% were not sure.

3.2 Mentoring Practices in Construction Professional Firms

Pertaining to the mentoring practices exhibited in construction professional firms, respondents were presented with mentoring practices identified from literature so as to rank them based on their level of exhibition. A scale of between 0–30% was set for very low, 31–50% for low, 51–70% for average, 71–90% for high while 91–100% for very high. A one sample t-test was then performed to ascertain whether the respondents regarded a particular mentoring practice to be exhibited or otherwise. The ranked mean

Table 1. Background information

Respondents information	Frequency	Percentage
Ranks		
Principal	14	9.9
Partner	26	18.3
Senior QS	46	32.4
QS	56	39.4
Total	**142**	**100**
Years of Experience		
1–5 years	52	36.6
6–10 years	3	2.1
11–15 years	26	18.3
16–20 years	13	9.2
21 years & above	48	33.8
Total	**142**	**100**
Professional Membership		
Probationer	46	32.4
Corporate	78	54.9
Fellow	18	12.7
Total	**142**	**100**
Mentoring Involvement		
Yes	111	78.2
No	31	21.8
Not Sure	0	0
Total	**142**	**100**
Mentoring Relationship Role		
Mentor	27	19
Mentee	30	21.1
Both	54	38
Not Sure	31	21.8
Total	**142**	**100**

of each practice was tabulated to offer a clear picture of the response of the respondents. The significance level was set at 95% in compliance with conventional level of risk and based on the five-point Likert rating scale, a mentoring practice was deemed exhibited if it had a mean of 3.5 or more. In the situation where there are two or more practices with the same mean, the mean with the lowest standard deviation was assigned the highest ranking of exhibition [23]. The normality test was not carried out as according to the central limit theorem, the sampling distribution tends to be normal in large samples of hundreds irrespective of the data shape [24, 25]. Table 2 shows the mean of each of the identified practice together with their associate standard deviation and standard error.

Table 2. T-test showing one sample statistics for exhibited mentoring practices

Mentoring practices	N	Mean	Std. deviation	Std. error mean
Sponsorship	142	3.176	.708	.059
Coaching	142	3.507	.788	.066
Protection	142	3.711	.637	.053
Exposure/Visibility	142	3.683	.811	.068
Role modelling	142	4.268	.798	.067
Challenging assignments	142	3.817	.847	.071
Acceptance	142	3.824	.765	.064
Counselling	142	3.599	.817	.069
Friendship	142	3.810	.662	.056

The result in Table 2 shows that the standard error of the respective means was relatively close to zero which indicates that the population is accurately reflected in the chosen sample. This is due to the fact that a small standard error between the sampled mean and the population mean as opined by [23], represents a true reflection of the population while large standard error suggest otherwise. Also, the standard deviations are all less than 1.0 which intimates there exist a little difference in the data and respondents consistency in agreement [23].

Table 3 reveals the p-values which represent the significance of each identified mentoring practices. The p-values which show the significance for a two-tailed test was half as shown in Table 4 so as to get the significance for a one-tailed test. The null hypothesis for each practice was that the it is not exhibited (H_0:U = U_0) and the alternative hypothesis was that the practice is exhibited (Ha:U > U_0), where U_0 is the population mean which was set at 3.5.

Table 3. One sample test for exhibited mentoring practices

Mentoring practices	Test Value = 3.5					
	T	df	Sig. (2-tailed)	Mean difference	95% confidence interval of the difference	
					Lower	Upper
Sponsorship	−5.455	141	.000	−.324	−.441	−.207
Coaching	.107	141	.915	.007	−.124	.138
Protection	3.954	141	.000	.211	.106	.320
Exposure/Visibility	2.690	141	.008	.183	.049	.318
Role modelling	11.460	141	.000	.768	.635	.900
Challenging assignments	4.457	141	.000	.317	.176	.458
Acceptance	5.044	141	.000	.324	.197	.451
Counselling	1.438	141	.153	.099	−.037	.234
Friendship	5.574	141	.000	.310	.200	.420

Table 4. Mentoring practices exhibited in QS firms showing rankings

Mentoring practices	Mean	Std. deviation	Ranking	Sig. (1-tailed)
Role modelling	4.268	.798	1	0.000
Acceptance	3.824	.765	2	0.000
Challenging assignments	3.817	.847	3	0.000
Friendship	3.810	.662	4	0.000
Protection	3.711	.637	5	0.000
Exposure/Visibility	3.683	.811	6	0.004
Counselling	3.599	.817	7	0.077
Coaching	3.507	.788	8	0.458
Sponsorship	3.176	.708	9	0.000

4 Discussion of Findings

The ranks, years of experience, and professional membership of respondents shows the cogent nature of this research. It is evident from findings that a good number of the respondents have participated in mentoring relationships. This indicates that despite the lack of attention given to the concept of mentoring in construction industry both in the Nigeria and in Africa in general according to [2], respondents have been involved in mentoring relationships either as a mentor, a mentee or as both. However, there still exist those that are yet to or not sure of their participation.

Considering how the respondents rated the mentoring practices exhibited in Nigerian QS firms, the top three ranked practices by the quantity surveyors are role modelling (4.268), acceptance (3.824) and challenging assignments (3.817) while sponsorship (3.176) was ranked as the lowest exhibited mentoring practice. However, according to the hypothesis (Ha:$U > U_0$ and $U_0 = 3.5$), it can be said that only first six practices are exhibited in the quantity surveying firms. These include role modelling (p-value = 0.000), acceptance (p-value = 0.000), challenging assignments (p-value = 0.000), friendship (p-value = 0.000), protection (p-value = 0.000), and exposure/visibility (p-value = 0.004).

Role modelling is the most exhibited mentoring practice in Nigerian QS firms and followed by acceptance. According to [13] these practices will help in the psychosocial development of the protégées in these firms. Role modelling as opined by [14] will help guide these young professionals on the required behaviours for success. Furthermore, among the six identified mentoring practices exhibited in the QS firms are challenging assignments, protection, and exposure/visibility which according to [13] aid in the development of the career of protégées.

5 Conclusion

The need to assess the exhibited mentoring practices in construction professional firms has become necessary as a result of the need to maintain high skill work force that will sustain competitive advantage in these firms. The study determined the involvement of construction professionals in mentoring relationship and the role they played in the relationship in their respective firms through the survey research carried out. Furthermore, the study has been able to ascertain the exhibited mentoring practices in the respective firms that will help maintain high skill work force and sustain competitive advantage of these firms in the Nigerian construction industry. Base on finding, the study can conclude that most construction professionals have been involved in mentoring relationships either as a mentor, a mentee or even as both. Nevertheless, there are still some construction professionals who for one reason or the other have not been involved in any mentoring relationship neither were they sure of their role. Mentoring practices such as role modelling, acceptance, challenging assignments, friendship, protection, and exposure and visibility are exhibited in these construction professional firms whereas counselling, coaching, and sponsorship are not exhibited as such. Exhibiting these practices in the firms will help maintain high skill work force. The study therefore recommends that construction firms pay maximum attention to counselling, coaching, and sponsorship the same way attention is being given to role modelling, acceptance, challenging assignments, friendship, protection, and exposure and visibility. In addition, the various professional bodies of construction professionals should ensure maximum implementation of the mentoring practices not exhibited in the firms. This will guarantee that necessary skills and knowledge is sustained in the firm and the profession in general. Furthermore, despite this study was confined to quantity surveying firms, studies on the involvement and role in mentoring relationships, and the exhibited mentoring practices should be extended to other professional firms in the construction industry where such study is yet to be carried out.

References

1. Babalola, M.O., Ojo, G.K., Bello, W., Adafin, J.K., Musa, N.: Training tools and needs for quantity surveying education and training. A paper Presented at the 2011 Quantity Surveying Assembly and Colloquium held at Shehu Musa Yar'Adua Centre, Abuja, Nigeria, 28–29 September (2011)
2. Oke, A., Aigbavboa, C., Odia, O.: Human resource management practices in quantity surveying firms. In: Creative Construction Conference 2016, pp. 150–155 (2016)
3. Verster, J.J.P., Haputfleisch, A.C.: Project management maturity in the South African construction industry. In: 2nd Built Environment Conference. Association for Schools of Construction of Southern Africa (ASOCSA) and the Council for the Built Environment (CBE), Port Elizabeth, 17–19 June (2007)
4. Nkomo, M., Thwala, W.: Assessment of mentoring of construction workers in the South African construction industry: a literature review. In: International Conference on Education, Law and Humanities, 27 (2013)
5. Kahle-Piasecki, L.: Making a mentoring relationship work: what is required for organizational success. J. Appl. Bus. Econ. **12**(1), 46–56 (2011)
6. Roegman, R., Reagan, A., Goodwin, L., Yu, J.: Support and assist: approaches to mentoring in a yearlong teacher residency. Int. J. Mentoring Coaching Educ. **5**(1), 37–53 (2016)
7. Agumba, J.N., Fester, F.C.: Participation in formal mentoring programme in South Africa Construction Industry: a perspective of new knowledge workers. Afr. J. Bus. Manage. **4**(10), 954–1903 (2010)

8. Coles, A.: The role of mentoring in college access and success. Res. Pract. Brief. **1**(1), 1–11 (2011)
9. Crisp, G., Cruz, I.: Mentoring college students: a critical review of the literature between 1990 and 2007. Res. High. Educ. **50**, 525–545 (2009)
10. Slaus, I., Jacobs, G.: Human capital and sustainability. J. Sustain. **3**, 97–154 (2011)
11. Shafie, H., Mazlina, S., Khuzzan, S., Mohyin, N.: Soft skills competencies of quantity surveying graduates in Malaysia: employers' views and expectations. Int. J. Built Environ. Sustain. **1**(1), 9–17 (2014)
12. St-Jean, E., Audet, J.: Factors leading to satisfaction in a mentoring scheme for novice entrepreneurs. Int. J. Evid. Based Coaching Mentoring **7**(1), 48–161 (2009)
13. Kram, K.E.: Phases of the mentor relationship. Acad. Manag. J. **26**(4), 608–625 (1983)
14. Lockwood, P.: "Someone like me can be successful": do college students need same-gender role models? Psychol. Women Q. **30**, 36–46 (2006)
15. Murrel, A.J.: Five key steps for effective mentoring relationships. Kaitz Q. **1**(1), 1–9 (2007)
16. Aldridge, S.: A Short Introduction to Counselling. Sage Publications, London (2014)
17. Pahl, R.E.: On Friendship. Polity, Cambridge (2000)
18. Amoako, G.K., Dartey-Baah, K., Dzogbenuku, R.K., Kwesie, J.S.: The effect of sponsorship on marketing communication performance: a case study of Airtel Ghana. Afr. J. Mark. Manag. **4**(2), 65–79 (2012)
19. Antonioni, D.: Leading, Managing and Coaching. Ind. Manag. **42**(5), 27–34 (2000)
20. Vander, S.K.: Exec coaching hones competitive edge. Credit Union Executive Newsletter **26**(32), 2–4 (2000)
21. Özkalp, E., Kirel, C., Sungur, Z., Ozdemir, A.A.: Mentoring relations in the aircraft industry: a case study in Turkey. Int. J. Evid. Based Coaching Mentoring **6**(2), 67 (2008)
22. Davies, K.: Stretch assignments give employees room to grow. Gov. Financ. Rev., 25–28 (2016)
23. Field, A.: Discovering Statistics, Using SPSS for Windows. Sage Publications, London (2005)
24. Field, A.: Discovering statistics using SPSS, 3rd edn, p. 822. SAGE publications Ltd., London (2009)
25. Elliott, A.C., Woodward, W.A.: Statistical Analysis Quick Reference Guidebook with SPSS Examples, 1st edn. Sage Publications, London (2007)

A Model of Network Communication in Building Information Modelling Supply Chain

Oluseye Olugboyega(✉) and Abimbola Windapo

Department of Construction Economics and Management,
University of Cape Town, Cape Town, South Africa
OLGOLU005@myuct.ac.za, oolugboyega@yahoo.com,
abimbola.windapo@uct.ac.za

Abstract. The exact nature of communication among the BIM supply chain members and the significant influence of BIM process requirements on the nature of communication among the BIM supply chain members is yet to be understood. This study explores the influences of BIM process requirements on the nature of the communication process among the BIM supply chain members. The study employed a Meta-Analysis methodology for interpreting the concepts of network communication in BIM supply chain, and for developing the conceptual model. Results indicated that the two dimensions of network communication (mixed communication and structured communication) showed consistent relations with network communication. The findings provide empirical insights into the nature of network communication and highlight the importance of communication channels, communication protocols and platforms, and structured e-meeting in BIM supply chain within collaborative working processes among the supply chain stakeholders.

Keywords: BIM · BIM supply chain · Communication in BIM · Network Communication · Mixed Communication · Structured Communication · Communication theory

1 Introduction

Communication refers to the transmission of resources from one party to another using shared symbols and media [9]. Communication is essential to the efficient performance of the project participants because of the difference in their skills, competence, interests, and organization structure [35]. [4] build on this conclusion, stating that communication is crucial for problem-solving, decision-making, integration and collaboration, understanding, dispute-resolution, project objectives determination, and efficient coordination.

Traditionally in the construction industry, communication among the project participants has been about the transmission of project information only; while other important resources such as data and knowledge (opinions, ideas, skills, technology, and feedback) were not given adequate consideration for transmission. This was because of the organizational fragmentation and lack collaborative work processes in the construction industry [33]. Non-communication of data and knowledge alongside

C. Aigbavboa and W. Thwala (Eds.): CIDB 2019, *The Construction Industry in the Fourth Industrial Revolution*, pp. 133–143, 2020.
https://doi.org/10.1007/978-3-030-26528-1_14

information was responsible for ineffective communication practices, which has manifested in the inefficiency of project participants and poor performance of projects [3]. Additionally, the multi-disciplinary nature of project participants is also a challenge to the communication process and practice.

According to [12, 15], the solution to these challenges is to adopt Building Information Modeling (BIM) on construction projects. BIM provides structure and control for communication processes and practice; thereby ensuring communication effectiveness [12]. BIM also turns multi-disciplinary project participants into a BIM supply chain by subjecting the participants to BIM processes and principles [23, 34]. Thus, this study describes the BIM supply chain as short-term or long-term networks of clients, subcontractors, main contractors, suppliers, and consultants participating in a BIM-based project [16, 25, 31].

Pieces of evidence abound for the effectiveness of communication in BIM-based projects [8, 12, 17]. Nevertheless, the true nature of communication among the BIM supply chain and the significant influence of BIM process requirements on the nature of communication among the BIM supply chain members (BIM-SCM) is yet to be understood [15, 38]. As a result, this study aims to understand the nature of communication among the BIM-SCM and establish whether the requirements of the BIM process influence this form of communication. This paper is presented in four sections. Section 2: theoretical background on the nature of communication among the BIM-SCM. Section 3: research methodology for the study. Section 4: findings and discussion of the findings and Sect. 5: conclusions and limitations of the study.

2 Nature of Communication Among the BIM-SCM and the Influence of BIM Process Requirements

2.1 Network Communication

Communication theory is required to provide insight into the nature of the communication process. Until now, communication theory has provided explanations for the nature and pattern of communications that emerged in the construction project supply chain [6, 13, 22, 32, 36, 37]. However, recent studies on communication process among the BIM supply chain have shown that the existing communication theories provide little or no insight into the nature of communication among the BIM supply chain [10, 15, 26–28, 30]. For example, [15] assessed the relevance of communication theories and media in the construction industry following the introduction of BIM in the sector. The study concluded that BIM has a significant impact on communication plans. [26] theorized that BIM improves the communication process among the BIM supply chain by enhancing the interoperability of information between the various software applications deployed by the supply chain members.

BIM process requires the communication of resources such as data, information, and knowledge [35]. This requirement complicates the communication process and alters the nature of communication among the BIM-SCM [12]. Thus, ensuring effective communication of resources among the BIM-SCM means that the members must be interconnected in all directions of the chain (network) [14]. This interconnection is achieved in the BIM process with an electronic collaborative network (e-mail, teleconferencing, mobile phones, internet, multimedia, virtual reality, and intranet) and network structures (contractual relationships, coordination, integration and collaboration requirements) [8, 9]. The above descriptions indicate a unique form of communication (that is, an organized and controlled communication among the BIM-SCM that takes place via an electronic collaborative network and network structures). This study describes this form of communication as *Network Communication*. The following subsections identify the features and dimensions of *Network Communication*.

2.1.1 Mixed Communication

For communication to be effective among the BIM-SCM, it must take place across the project lifecycle, across different communication facets, and must compose of different methods of transmitting resources among members of the BIM supply chain [5]. [2] inferred that communication across facets such as interpersonal, inter-group, intra-group, inter-professional, intra-organization, inter-organizational, and among team members is essential to integration. Communication across different facets becomes essential among the BIM-SCM owing to the need for integration and collaboration in BIM processes and because groups and teams are generally formed from several specialist organizations in BIM supply chain to undertake BIM-related tasks. [9] found that communication across different facets among the supply chain members is essential to carrying out of the normal practices of individual organizations such as information generation and management, planning and control, and decision-making. According to [11], cross-functional and cross-organizational communication are some of the ways by which the BIM process alters the nature of communication among the BIM-SCM. Also, [21] established that the BIM process requires the BIM-SCM to keep open communication and described open communication as a form of communication that permits a multi-directional flow of resources. [29] observed that in BIM, the multi-directional flow of resources among the BIM-SCM means that each member grants the other members a non-exclusive license to reproduce, distribute, display, or otherwise use that member's contributions for the needs of the project.

Based on the above discussion, *Mixed Communication* emerged as a term to describe a dimension of the nature of communication among the BIM-SCM. *Mixed Communication* implies a pattern of communication among the BIM-SCM that consists of different communication facets, takes place across the project life-cycle, contains a mixed methods approach to transmitting resources, occurs at different places at the same time or same place at different times, as well as demonstrates the collaboration and integration requirements of BIM processes [5, 8].

2.1.2 Structured Communication

BIM process requires the planning for communication channels, communication platforms, communication protocols, structured e-meeting, real-time communication, and joint problem-solving techniques among the BIM-SCM [5, 7, 9, 19]. This serves as a manifestation of a clearly defined communication process. According to [9], the use of communication channels such as meetings, discussion, and communication technologies enhance communication flow and promote better understanding among the BIM-SCM. Communication process becomes effective when the flow and understanding of resources are enhanced among the members [2]. In the BIM process, the effectiveness of the communication process is made possible by the BIM process requirements such as usage of communication protocols, which gives structure and consistency to the communication process [8]. A case study of communication among the BIM-SCM by [1] confirmed that BIM process requirements ensure the occurrence of overlapping interaction, flexible formalities, and transparency in the communication process. In the traditional work process, conflicting interests and lack of cooperation occur because of the lack of requirements for the establishment of a secured line of communication and the use of a formal communication channel [33]. However, in BIM, the requirements for the communication process ensures cooperation among the BIM-SCM. Studies by [19, 20] confirm this claim. [20] demonstrated how the use of wireless network and collaborative platforms for real-time communication support cooperation among the BIM-SCM. [19] confirmed the flexibility of formalities among the BIM-SCM through real-time communication and structured e-meetings as enabled by the use of Wireless Fidelity, cameras, laptops, mobile phones, and web-based communication tools.

The above explanation induced a pattern of the communication process and this communication pattern is termed *Structured Communication*. The term *Structured Communication* can, therefore, be described as a form of communication that takes place across clearly defined channels, platforms, protocols, time-space, and objectives.

3 Method

3.1 Concept Formulation

This study sought to understand the nature of communication among the BIM-SCM and establish whether the requirements of the BIM process precipitate the nature of communication. The study described the nature of communication among the BIM-SCM as *Network Communication* based on the influence that the BIM process has on communication among the BIM-SCM. *Network Communication* was developed into a conceptual model comprising two dimensions and twenty-one components. These dimensions were termed *Mixed Communication* and *Structured Communication*. The dimensions in the model and the related components are shown in Fig. 1. The model integrated BIM process requirements with communication processes among the BIM-SCM. The model also gives an adequate representation of the nature of communication among the BIM-SCM as influenced by the BIM process requirements. This gives an appropriate communication process in BIM-based projects. The model explains the various mechanisms that differentiate communication among the BIM-SCM from other forms of communication among the construction supply chain. These differences are captured as *Network Communication* (*Mixed Communication* and *Structured Communication*). Section 2 has elaborated on *Mixed Communication* and *Structured Communication*. A meta-analysis of the database developed was also conducted to provide further support for these dimensions and their components.

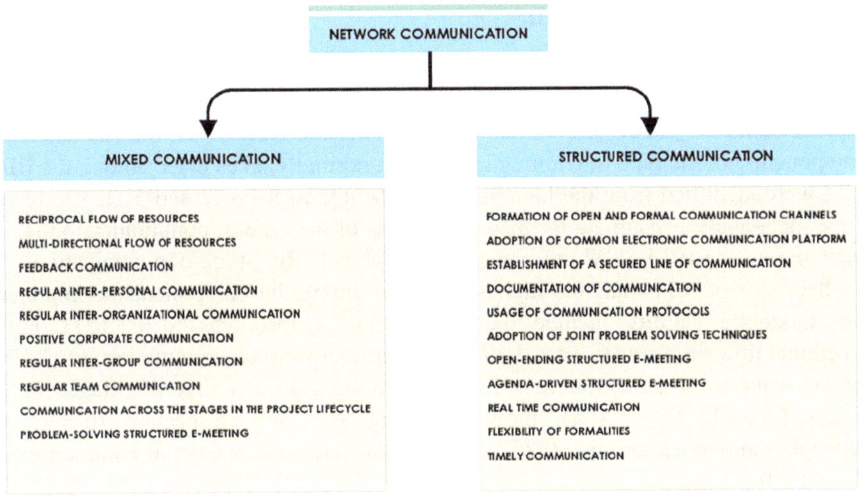

Fig. 1. A model of Network Communication in BIM supply chain.

3.2 Database

To understand the nature of communication among the BIM-SCM; a literature database was compiled from Scopus, Engineering Village, Ebsco, Google Scholar, and Web of Science. The search was limited to peer-reviewed articles that were published from 1992–2019. The search was done using keywords such as *communication among BIM project team, communication among BIM stakeholders, communication among BIM supply chain, communication in BIM process, communication tools and methods in BIM, effective communication and BIM*, and *effective communication among the project team*. These efforts resulted in 617 articles after the elimination of articles that were not published in the English Language and not construction-related. The database for the meta-analysis review was created using these articles.

3.3 Meta-analytic Procedures

The study used the meta-analytic procedures of [18]. Meta-analytic estimates such as correlation coefficients, relative weight, effect size, variance, credibility interval, and confidence interval were carried out in this study. The study reported both the confidence interval and credibility interval around the estimated population correlation in order to get detailed information about the nature of the correlation coefficients. While credibility intervals provide an estimate of the variability of individual correlations across the meta-analytic database; confidence intervals provide an estimate of the variability around the estimated population mean and correlation coefficient that is due to sampling error. Besides, the standard error of the mean was estimated. This was done in order to measure the accuracy with which a sample represents a population.

4 Results and Discussion

4.1 Nature of Communication Among the BIM-SCM

To understand the nature of communication among the BIM-SCM, the characteristics (components) of the type of communication that normally takes place among the BIM-SCM were identified from the literature (as presented in Sects. 2 and 3.1). The results of the meta-analysis estimate for the components of the type of communication taking place among the BIM-SCM are presented in Table 1. Real-time communication with an effect size of 3.114 has the highest validity among the components. Components with moderate validity include reciprocal flow of resources (d = 1.151), multi-directional flow of resources (d = 1.177), regular interpersonal communication (1.126), regular team communication (d = 1.087), communication across the stages in the project lifecycle (d = 1.164), problem-solving structured e-meeting (d = 1.190), feedback communication (d = 0.802), and regular inter-organizational communication (d = 0.880).

Table 1. Meta-analysis estimate of the validities of the components of network communication.

Components of network communication	Frequency (f)	Relative weight (w)	Effect size (d)
Mixed communication			
Problem-solving structured e-meeting	16	0.014	1.190
A multi-directional flow of resources information	17	0.015	1.177
Communication across the stages in the project life-cycle	18	0.016	1.164
Reciprocal flow of resources information	19	0.017	1.151
Regular inter-personal communication	195	0.180	1.126
Regular team communication	192	0.177	1.087
Regular inter-organizational communication	176	0.162	0.880
Feedback communication	170	0.157	0.802
Regular inter-group communication	170	0.157	0.802
Positive corporate communication	109	0.100	0.012
Structured communication			
Real-time communication	210	0.507	3.114
Establishment of a secured line of communication	6	0.014	0.579
Documentation of communication	15	0.036	0.416
Formation of open and formal communication channels	16	0.038	0.398
Usage of communication protocols	17	0.041	0.380
Adoption of common electronic communication platforms	18	0.043	0.362

(continued)

Table 1. *(continued)*

Components of network communication	Frequency (f)	Relative weight (w)	Effect size (d)
Adoption of joint problem-solving techniques	18	0.043	0.362
Open-ending structured e-meeting	23	0.055	0.271
Flexibility of formalities	23	0.055	0.271
Agenda-driven structured e-meeting	24	0.057	0.253
Timely communication	44	0.106	0.108

In general, the relative weight and effect size values show a piece of overwhelming evidence in support of the components of *Network Communication*. These findings suggest the nature of communication among the BIM-SCM through the dominating components. The nature of communication among the BIM-SCM relates to *Network Communication*, and it is characterized by regular communication among the team and groups within the supply chain organizations; as well as across the stages in the project lifecycle. This regular communication must be reciprocated, allow feedback, be multi-dimensional, and happen in real time.

4.2 Influence of BIM Process Requirements on the Nature of Communication

The analysis of the theoretical background in Sect. 2 yields a conceptual model, but the model was further synthesized using meta-analysis. The results of the meta-analytic estimates were used to establish the dimensions and components of the model. The use of meta-analytic estimates to validate the model was necessitated by the need to evaluate how well the model represents communication among the BIM-SCM in the real world. As presented in Table 2, the two dimensions of the model of *Network Communication* in BIM supply chain [*Mixed Communication* ($r = 0.489$) and *Structured Communication* ($r = 0.105$)] were significant predictions of the nature of communication among the BIM-SCM as influenced by BIM process requirements. The findings of this study show that the form of communication among the BIM-SCM has collaborative attributes and that these attributes can be explained or summarized as *Mixed communication* and *Structured communication.*

Table 2. Meta-analysis estimate of the validities of the dimensions of network communication.

Meta-analytic estimates	Dimensions of network communication	
	Mixed communication	Structured communication
Correlation coefficient (r)	0.489	0.105
Population mean (M)	108.2	37.6
Sample standard deviation (s)	81.44	57.91

(continued)

Table 2. (*continued*)

Meta-analytic estimates	Dimensions of network communication	
	Mixed communication	Structured communication
Variance (sample standard deviation) (s^2)	6633.73	3354.25
Population standard deviation (μ)	77.26	55.22
Variance (population standard deviation) (μ^2)	5990.36	3049.32
Total population (N)	10	11
The standard error of the mean (SEx)	25.76	17.46
95% Confidence Interval • Lower • Upper	0.011 0.785	−0.354 0.524
Fisher's Zr	0.535	0.105
95% Credibility Interval • Lower • Upper	0.011 1.060	−0.370 0.582

The findings of this study corroborate the conclusions of [14, 38] who indicated that the influence of BIM process requirements on the nature of communication among the BIM-SCM is reflected in need to communicate data and knowledge apart from information. The communication of information characterizes the traditional form of communication among the CPSC. The communication of data, knowledge, together with information among the BIM-SCM is due to the BIM process requirements. This becomes necessary because of the need to overcome the problems of disorganized and uncontrolled communication as found in the traditional form of communication. For example, [17, 21] concluded that the traditional form of communication among the CPSC is not usually open and lack protocols.

5 Conclusions, Practical Implications, and Limitations

The purpose of this study was to understand the nature of communication among the BIM-SCM towards establishing the influence of BIM process requirements on this form of communication. The collaborative attributes of BIM are responsible for *network communication* among the BIM-SCM. This was presented as a two-dimensional model. These two dimensions were termed as *mixed communication* and *structured communication*. *Mixed communication* consisted of nine components, while *structured communication* consisted of ten components. The study leveraged theoretical grounding to conceptualize the model; while meta-analysis was used to validate the dimensions and components of the model. The findings of this study provide information that can guide the understanding of how BIM process requirements have changed the nature of communication among the BIM-SCM.

Additionally, the information provided in this study is useful for the effective coordination of the communication process in BIM-based projects. An empirical case study of the nature of communication among the BIM-SCM will benefit significantly from the insights provided in this study. Understanding the nature of communication among the BIM-SCM will enable the understanding of the communication process in BIM and enable BIM managers to establish expectations in the communication process to set the foundation for a truly collaborative BIM project. The insight will also provide a sufficient framework and depth to support the development of a generally applicable model.

The study was unable to produce vigorous quantitative findings because of a lack of robust analysis and primary data. The studies used as the database for this study did not have the same research questions. These incompatibilities serve as limitations for the findings of this study. Nevertheless, this study has presented a conceptual model that can be validated in an in-depth and rigorous future study. Also, for future research, it is essential to validate the components of network communication among the BIM-SCM.

Acknowledgment. The financial assistance of the University of Cape Town (UCT), TETFUND and the National Research Foundation (NRF) towards this research is hereby acknowledged. The opinions expressed, and conclusions arrived at, are those of the authors and are not necessarily to be attributed to UCT, TETFUND or NRF.

References

1. Al Hattab, M., Hamzeh, F.: Information flow comparison between traditional and BIM-based projects in the design phase. In: Proceedings for the 21st Annual Conference of the International Group for Lean Construction, pp. 761–770 (2013)
2. Anumba, C.J., Evbuomwan, N.F.O.: A taxonomy for communication facets in concurrent life-cycle design and construction. Comput. Aided Civil Infrastruct. Eng. **14**(1), 37–44 (1999)
3. Baiden, B.K., Price, A.D.F.: The effect of integration on project delivery team effectiveness. Int. J. Proj. Manage. **29**(2), 129–136 (2011)
4. Barlish, K., Sullivan, K.: How to measure the benefits of BIM—a case study approach. Autom. Construct. **24**, 149–159 (2012)
5. Bassanino, M., Fernando, T., Wu, K.-C.: Can virtual workspaces enhance team communication and collaboration in design review meetings? Architectural Eng. Des. Manage. **10**(3–4), 200–217 (2014)
6. Bechtel, C., Jayaram, J.: Supply chain management: a strategic perspective. Int. J. Logistics Manage. **8**(1), 15–34 (1997)
7. Buvik, M.P., Rolfsen, M.: Prior ties and trust development in project teams–a case study from the construction industry. Int. J. Project Manage. **33**(7), 1484–1494 (2015)
8. Charalambous, G., Thorpe, T., Demian, P., Yeomans, S.G., Doughty, N., Peters, C.: Collaborative BIM in the Cloud and the Communication tools to support it (2013)
9. Cheng, E.W.L., Li, H., Love, P.E.D., Irani, Z.: Network communication in the construction industry. Corp. Commun. Int. J. **6**(2), 61–70 (2001)
10. de Blois, M., Herazo-Cueto, B., Latunova, I., Lizarralde, G.: Relationships between construction clients and participants of the building industry: structures and mechanisms of coordination and communication. Archit. Eng. Des. Manage. **7**(1), 3–22 (2011)

11. Dike, I.U., Kapogiannis, G.: A conceptual model for improving construction supply chain performance. In: Proceedings of the 30th Annual Association of Researchers in Construction Management Conference (ARCOM 2014), Portsmouth, UK (2014)
12. Dubas, S., Pasławski, J.: The concept of improving communication in BIM during transfer to operation phase on the Polish market. Procedia Eng. **208**, 14–19 (2017)
13. Erik Eriksson, P.: Improving construction supply chain collaboration and performance: a lean construction pilot project. Supply Chain Manage. Int. J. **15**(5), 394–403 (2010)
14. Fiala, P.: Information sharing in supply chains. Omega **33**(5), 419–423 (2005)
15. Fox, S., Leicht, R.M., Messner, J.I.: Assessing the relevance of media synchronicity theory to the use of communication media in the AECO industry. J. Archit. Eng. **16**(2), 54–62 (2009)
16. Getuli, V., Ventura, S.M., Capone, P., Ciribini, A.L.C.: A BIM-based construction supply chain framework for monitoring progress and coordination of site activities. Procedia Eng. **164**, 542–549 (2016)
17. Goh, K., Goh, H., Toh, S., Peniel Ang, S.: Enhancing communication in the construction industry through BIM. In: 11th International Conference on Innovation and Management, pp. 313–324 (2014)
18. Hunter, J.E., Schmidt, F.L.: Dichotomization of continuous variables: the implications for meta-analysis. J. Appl. Psychol. **75**(3), 334 (1990)
19. Lee, J., Bernold, L.E.: Ubiquitous agent-based communication in construction. J. Comput. Civil Eng. **22**(1), 31–39 (2008)
20. Leung, S., Mak, S., Lee, B.L.P.: Using a real-time integrated communication system to monitor the progress and quality of construction works. Automation Constr. **17**(6), 749–757 (2008)
21. Meng, X.: The effect of relationship management on project performance in construction. Int. J. Project Manage. **30**(2), 188–198 (2012)
22. Min, J.U., Bjornsson, H.C.: Agent-based construction supply chain simulator (CS 2) for measuring the value of real-time information sharing in construction. J. Manage. Eng. **24**(4), 245–254 (2008)
23. Nawi, M.N.M., Haron, A.T., Hamid, Z.A., Kamar, K.A.M., Baharuddin, Y.: Improving integrated practice through Building Information Modeling-Integrated Project Delivery (BIM-IPD) for Malaysian Industrialised Building System (IBS) construction projects. Malaysia Constr. Res. J. (MCRJ) **15**(2), 1–15 (2014)
24. Papadonikolaki, E., Verbraeck, A., Wamelink, H.: Formal and informal relations within BIM-enabled supply chain partnerships. Const. Manage. Econ. **35**(8–9), 531–552 (2017)
25. Papadonikolaki, E., Vrijhoef, R., Wamelink, H.: The interdependences of BIM and supply chain partnering: empirical explorations. Architectural Eng. Des. Manage. **12**(6), 476–494 (2016)
26. Pauwels, P., Van Deursen, D., De Roo, J., Van Ackere, T., De Meyer, R., Van de Walle, R., Van Campenhout, J.: Three-dimensional information exchange over the semantic web for the domain of architecture, engineering, and construction. Ai Edam **25**(4), 317–332 (2011)
27. Peansupap, V., Walker, D.H.T.: Information Communication Technology (ICT) implementation constraints: a construction industry perspective. Eng. Constr. Architectural Manage. **13**(4), 364–379 (2006)
28. Perumal, V.R., Bakar, A.H.A.: The needs for standardization of document towards an efficient communication in the construction industry. Acta Technica Corviniensis-Bull. Eng. **4**(1), 23 (2011)
29. Porwal, A., Hewage, K.N.: Building Information Modeling (BIM) partnering framework for public construction projects. Autom. Constr. **31**, 204–214 (2013)

30. Rankin, J.H., Luther, R.: The innovation process: adoption of information and communication technology for the construction industry. Can. J. Civ. Eng. **33**(12), 1538–1546 (2006)
31. Robson, A., Boyd, D., Thurairajah, N.: UK construction supply chain attitudes to BIM. In: 50th ASC Annual International Conference (2014)
32. Vrijhoef, R., Koskela, L.J., Howell, G.: Understanding construction supply chains: an alternative interpretation. In: Proceedings of 9th International Group for Lean Construction Conference, pp. 185–199 (2001)
33. Wikforss, Ö., Löfgren, A.: Rethinking communication in construction. J. Inform. Technol. Constr. (ITcon) **12**(23), 337–346 (2007)
34. Wu, C., Xu, B., Mao, C., Li, X.: Overview of BIM maturity measurement tools. J. Inform. Technol. Constr. (ITcon) **22**(3), 34–62 (2017)
35. Wu, G., Liu, C., Zhao, X., Zuo, J.: Investigating the relationship between communication-conflict interaction and project success among construction project teams. Int. J. Project Manage. **35**(8), 1466–1482 (2017)
36. Xue, X., Shen, Q., Ren, Z.: Critical review of collaborative working in construction projects: business environment and human behaviors. J. Manage. Eng. **26**(4), 196–208 (2010)
37. Xue, X., Li, X., Shen, Q., Wang, Y.: An agent-based framework for supply chain coordination in construction. Automation Constr. **14**(3), 413–430 (2005)
38. Zhang, S., Teizer, J., Lee, J.-K., Eastman, C.M., Venugopal, M.: Building information modeling (BIM) and safety: automatic safety checking of construction models and schedules. Automation Constr. **29**, 183–195 (2013)

A Building Information Modelling-Integrated Model of Construction Project Performance Indicators

Oluseye Olugboyega(✉) and Abimbola Windapo

Department of Construction Economics and Management, University of Cape Town, Cape Town, South Africa
OLGOLU005@myuct.ac.za, oolugboyega@yahoo.com

Abstract. The traditional construction project performance indicators are no longer the sole determinant of the performance of construction projects, firms, and actors because other construction performance indicators and factors such as Building Information Modelling are increasingly becoming important and relevant. This paper identifies a comprehensive set of construction project performance indicators, and it establishes a new, integrated, model for evaluating construction performance. The model uses ten types of performance indicators and four types of performance factors for Building Information Modelling-based construction projects. The result of the study is a set of comprehensive construction project performance indicators that are useful in enhancing the performance of construction firms, projects, and actors. The paper concludes with recommendations for the development of relative weightings and measurement models for the indicators and the development of measurement models for the various components of construction project performance indicators and Building Information Modelling performance indicators in future studies.

Keywords: Construction industry indicators · Construction project performance indicators · Project success criteria · Key performance index · Building Information Modelling · Construction performance · BIM performance · Construction performance evaluation

1 Introduction

Construction project performance indicators (CPPIs) are measures of construction project success and systems of measuring construction performance [4]. Different terminologies such as construction industry indicators (CIIs), key performance indicators (KPIs), and project success criteria (PSC) have been used to describe CPPIs. For this study and clarity of discussion, the term CPPIs will be adopted in this paper. According to [11], CPPIs are useful for the decision-making process and promoting changes in organizations. Besides, [2] reported that CPPIs are useful for promoting best practices and establishing business objectives in the construction industry. Another notable importance of CPPIs is their usefulness in improving the application of construction management techniques and identifying the strengths and weaknesses in the

C. Aigbavboa and W. Thwala (Eds.): CIDB 2019, *The Construction Industry in the Fourth Industrial Revolution*, pp. 144–157, 2020.
https://doi.org/10.1007/978-3-030-26528-1_15

performance of the various sub-sections of the construction industry [26]. There are emerging pieces of evidence to suggest that the existing models of CPPIs are inadequate [15, 19, 33]. [15] concluded that the existing models of CPPIs do not represent the various aspects of construction project performance. [33] maintained that the dynamic nature of the construction industry requires regular updating of the CPPIs. In the opinion of [19], construction performance evaluation systems do not reflect the effect of increased globalization, improvement in competition and efficiency of construction firms, technological evolution, and increased regulation. These evidences prove that the existing CPPIs models are not comprehensive and do not incorporate the impact of performance factors (project success factors) such as innovative construction project management systems (sustainable construction system, lean construction system, concurrent construction system), Building Information Modelling (BIM), and other Fourth Industrial Revolution technologies [31].

Construction project performance is a multifaceted concept, and a comprehensive model of its indicators must address a unique dimension of project performance and reflect the impact of performance factors on the indicators [3, 13]. This is important because a small change in any of the performance factors will result in a significant effect on the indicators and the estimated construction project performance. This affirms the conclusion of [34] that the level of performance of construction projects depends mainly on the obtainable value of performance factors as indicated by the quality of the construction project management system. BIM is a construction project management system with ample performance factors that are capable of improving the processes of project design and construction as well as increasing the level of competence of construction firms, workers and professionals [14]. Hence, there is a need to extend the dimensions of the existing models of CPPIs in order to safeguard their adequacy for ensuring superior performance, priorities of each of the indicators, and integration of the indicators of common performance factors.

There are justifications for a BIM-integrated model of CPPIs. First, a BIM-integrated model of CPPIs will enhance construction project management and enable a more comprehensive explanation of construction performance. Second, a BIM-integrated model of CPPIs will provide a clear and unambiguous evaluation system for construction performance. The purpose of this study is to identify a comprehensive set of CPPIs in order to establish a BIM-integrated evaluation system for construction project performance.

To achieve the aim of this study, the study developed a theoretical background in Sect. 2 to identify a comprehensive set of CPPIs and performance indicators for BIM. In Sect. 3, the study categorized the CPPIs into dimensions and components, integrated the CPPIs with BIM's performance indicators, and then used meta-analytic techniques to test the validities of the dimensions and components empirically. In this way, the study strived to synthesize the literature on CPPIs and BIM's performance indicators. The findings from the meta-analytic estimate are presented in Sect. 4; while Sect. 5 was used to present conclusions, limitations of the study, and directions for future research.

2 Theoretical Background

2.1 Existing Models of CPPIs

Table 1 shows the various models of CPPIs that are available globally. As distinct as these models appear to be, they also share some similarities such as productivity (time and cost performance), quality, health and safety, functionality, profitability, and client satisfaction. Also, there are emerging CPPIs that are targeted towards technological advancement and the continuous improvement of the construction industry. Examples of the emerging CPPIs are employees' satisfaction, innovation, the rate of employees trained, sustainability, growth, technological capability, impact on environment and biodiversity, energy use, water use, and waste. Collectively, these CPPIs address industry satisfaction [11, 23, 25]. The emerging CPPIs are derivatives of lessons learned from the performance of past projects and the industry. The best practice is to update the model of CPPIs with the emerging CPPIs owing to the dynamics of construction projects and industry. However, only a few models are being updated regularly. The United Kingdom model of CPPIs represents the best form of CPPIs as it is updated annually [29], while the South African model of CPPIs developed in 2003 is among the redundant models. Its development has enabled the South African government to evaluate the impact of contemporary interventions for timely and pro-active execution of revised legislation, strategies and development agendas to act as an updated roadmap for the future wellbeing and growth of the South African construction industry every year [22].

Table 1. Models of CPPIs.

CPPIs model by country	Dimensions	References
United Kingdom	Productivity (time and cost performance), profits, quality, safety, client satisfaction with the product, client satisfaction with service, cost and time predictability, health and safety, business performance	[8]
United State of America	Cost, time, value, effectiveness, safety, change, rework	[1]
	Project cost growth, project budget factor, project schedule growth, project schedule factor, total project duration, change cost factor, recordable incident rate, lost workday case incident rate, total field rework factor phase cost, factor phase cost growth, phase duration factor, construction phase duration	[9]
Croatia	Quality, cost, number of investor interferences, changes in project support, time increases, client satisfaction, employees' satisfaction, innovation and learning, time and identification of client's interest	[26]
Malaysia	Revenue and profits, the benefit to stakeholder, cost, time, quality, safety, achieving scope, customer satisfaction, technical specifications, functional requirements, market share, competitive advantage, reputation	[16]
Brazil	Cost deviation, time deviation, degree of client satisfaction (user and owner), average time for selling unit, contracting index, ratio between the number of accidents and total man-hour input, non-conformity index in the unit delivery, percentage of plan completed, construction site best practice, supplier performance, number of non-conformity in audits, degree of employee satisfaction, rate of training courses, rate of employees trained	[11]

(*continued*)

Table 1. (*continued*)

CPPIs model by country	Dimensions	References
Scotland	Product, service, quality, time, cost, safety, environment, people, business	[32]
Chile	Safety, productivity, quality, the efficiency of labor, rework, training, planning effectiveness, cost variation, schedule variation	[27]
	Productivity, planning, effectiveness, urgent orders, risk rate, accident rate, the efficiency of direct labor, subcontracting rate, cost client, change in amount contracted, deviation of construction due date, cost deviation by project	[10]
South Africa	Client satisfaction, contractor satisfaction, procurement indicators, health and safety	[22]
Saudi Arabia	Client satisfaction, communication, safety, closeness to budget, profitability, payment, claims, staff experience, the planning period	[19]
India	The benefit, risk, project status, decision effectiveness, production, cost-effectiveness, customer commitment, stakeholders, project management	[24]
China	People, cost, time, quality, safety, client satisfaction, communication, environment	[7]
Canada	Cost, time, quality, safety, scope, innovation, sustainability, client satisfaction	[6, 28]
Vietnam	Construction cost and time, client satisfaction, quality management, team performance, change management, material management, safety	[21]
Thailand	On time, under budget, specifications, efficiency, effectiveness, safety, defects, stakeholders, disputes	[35]
Korea	Profitability, growth, stability, customer satisfaction, market share, development, technological capability, business efficiency, information, organizational competency	[38]
Portugal	Productivity, profitability, growth, safety, customer satisfaction, predictability (cost and time), rework index, direct labor efficiency, employee satisfaction, defects, impact on environment and biodiversity, energy use, water use, waste	[25]
Ghana	Client satisfaction, cost, time, quality, health and safety, business performance, productivity, people, environment	[23]
Denmark	Time, defects remediation, number of defects during handing-over, accident frequency, contribution ratio, work intensity, labor productivity, change in project price during construction, square meter price, customer satisfaction	[5]

Even though the model of CPPIs is expected to be unique in terms of the national context and vision for the construction industry; the model must also reflect the international best practices and advances in the construction industry [31]. Many models of CPPIs do not measure up in this regard [7, 9, 16, 21, 23–25, 27, 38]. For instance, most of the existing models employed a single indicator to measure the satisfaction of all agents and did not consider the functionality of construction projects. Additionally, the time and cost performance of the project was not considered in some of the existing models of CPPIs. Other drawbacks of the existing models of CPPIs are the use of percentage satisfaction level for evaluation, the lack of components to capture the exact reasons for all the problems experienced by project stakeholders, the lack of indicator for monitoring and evaluating project impact on the industry, non-separation of project functionality from project quality, and the use of a single indicator to measure the satisfaction of project stakeholders. A comprehensive model of CPPIs is therefore required to address these drawbacks.

2.2 BIM and Construction Project Performance

Construction project performance is determined by the performance indicators of the construction project management system employed for the project and the effectiveness of the project stakeholders (denoted by the construction project performance indicators) [36]. Building Information Modelling (BIM) is a significant construction project management system that provides efficient integration and collaboration tools and platforms for construction management [14]. The performance indicators of BIM on construction projects are signified by the level of network communication, knowledge sharing and transfer, information sharing and exchange, and trust-based relationships among the project stakeholders [17]. According to [30], BIM has a multiplier effect on CPPIs through its performance indicators. Thus, there is a need for a new model of CPPIs that will incorporate BIM performance indicators (BIM-PIs) [20, 37].

3 Method

3.1 Concept Formulation

A conceptual model of integrated construction project performance indicators (CPPIs) and BIM performance indicators (BIM-PIs) was developed in this study based on the theoretical background presented in Sect. 2. The model integrated two sets of performance indicators (a comprehensive set of CPPIs and BIM-PIs) to conceptualize an effective, systematic, and comprehensive method of evaluating BIM-based construction projects. The model uses sixty-three components across ten dimensions of CPPIs and four dimensions of BIM-PIs. The components and dimensions for the model were identified by a critical examination of global best practices and by addressing the contextual limitations of the industry-specific performance measurement practices.

As shown in Fig. 1, the dimensions of CPPIs include contractor satisfaction, client satisfaction, consultants' satisfaction, subcontractors' satisfaction, suppliers' satisfaction, health and safety performance, time performance, cost performance, industry satisfaction, and quality performance. The dimensions for BIM-PIs are network communication, knowledge sharing, and transfer, information sharing and exchange, and trust-based relationships among the project stakeholders. The level of application of the model corresponds to whether a project to be evaluated was delivered using BIM or conventional project delivery systems (CPDS). As presented in Fig. 1, the performance of BIM-based construction projects must be evaluated using the ten dimensions of CPPIs and the four dimensions of BIM-PIs. This is necessitated by the need to evaluate the performance of the construction projects and the performance of BIM on the construction projects. it is Expected that the BIM-PIs will indirectly boost the performance of the construction projects through their direct effects on the CPPIs.

[30] describes this phenomenon as the multiplier effect of BIM on construction projects. Also, the model applies to projects delivered using CPDS. For such projects, the performance evaluation must be done using the ten dimensions of CPPIs.

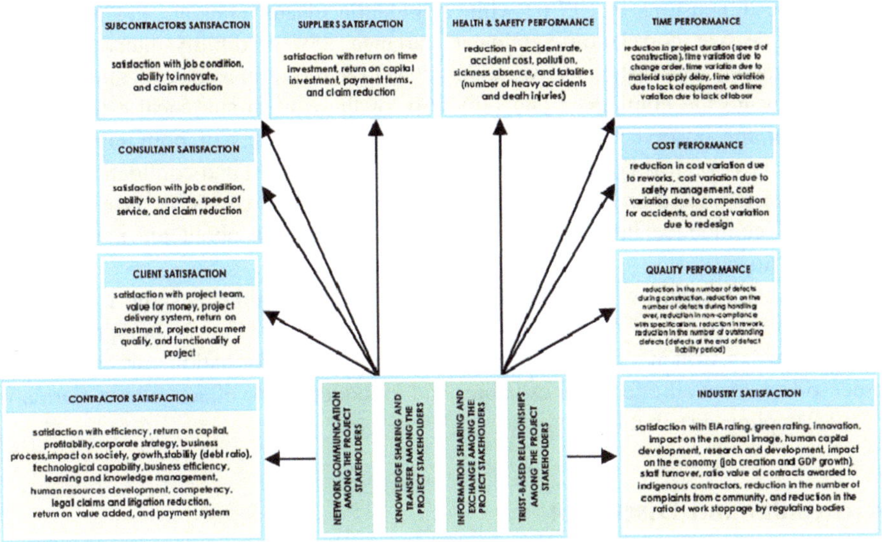

Fig. 1. A BIM-integrated model of CPPIs.

Although, the selection of appropriate performance indicators depends on the objectives and expectations of project stakeholders or industry regulators. However, [12] acknowledges that it is usually best to evaluate projects using a wide, project delivery system-based and globally relevant performance indicators.

3.2 Database

Databases such as Scopus, Engineering Village, Ebsco, Google Scholar, and Web of Science were searched for peer-reviewed articles from 1962 to 2019. The terms that were used to identify the articles include *project success, project performance, construction industry indicators, construction key performance indicators, and BIM performance indicators*. The search results were restricted by language (English Language), document type (abstract of published items), timespan, and exclusion of patents and citations. The search efforts resulted in the identification of 1,331 abstracts. Employing electronic search engines, articles that are not construction project specific and that are not useful for meta-analysis (that is, articles containing information needed to calculate correlation coefficient) were eliminated. Articles that mentioned the dimensions of CPPIs and BIM-PIs, as well as their components, were selected for inclusion in the database for the meta-analysis. Finally, a total of thirty-one (31) articles were identified for the database.

3.3 Meta-analytic Procedure

Meta-analytic procedures defined by [18] was used in this study. Meta-analysis was employed to estimate the validities of the dimensions and components of CPPIs. The sample means, weight and effect size was computed for the components identified for each of the dimensions of CPPIs; while population means, standard deviation, combined sample size, correlation coefficient, variance, 95% credibility interval and confidence interval, and Fisher's Z were computed for the dimensions of the CPPIs. In order to ensure the confidence and reliability of these meta-analytical estimates, the variance of the observed individual estimates for the effects of both sampling and measurement error was corrected. Credibility intervals (95%) and confidence intervals (95%) were estimated and reported for the correlation coefficients because both credibility and confidence intervals convey unique information about the nature of the precise estimates.

The sample weight was estimated to ensure that correlations ensuing from studies with larger sample sizes have greater weight. The effect size for the components was estimated using Cohen's d effect size. The Fisher's Z transformation was estimated in order to transform all effect size estimates and to moderate the effects of non-normality of the sampling distribution.

4 Results and Discussion

4.1 Comprehensive Sets of Construction Project Performance Indicators

To achieve the objective of this study, a conceptual model of BIM-integrated CPPIs was developed from a meta-analysis study of literature. The validities of the CPPIs were determined using meta-analytical estimates as discussed in Sect. 3. The meta-analysis results presented in Table 2 show that all the dimensions are valid indicators of construction project performance. Health and safety performance shows the highest overall validity ($r = 0.594$) but is directly followed by suppliers' satisfaction ($r = 0.375$). Time performance ($r = 0.335$), client satisfaction ($r = 0.302$), contractor satisfaction ($r = 0.261$), and industry satisfaction ($r = 0.261$) show moderately strong validities. The dimensions with averagely strong validities are quality performance ($r = 0.152$), consultants' satisfaction ($r = 0.135$), cost performance ($r = 0.073$), and subcontractors' satisfaction ($r = 0.043$). None of the dimensions has a 95% credibility and confidence interval of 0. This shows that more than 95% of the individual correlation coefficient for CPPIs dimensions are greater than zero. The findings of this study imply that all ten dimensions of CPPI are relevant in evaluating the performance of construction projects.

Table 2. Meta-analysis estimate of the validities of the dimensions of CPPIs.

Dimension	M	SD	N	r	95% C.I	Fisher's Zr	95% C.I	v
Health and safety performance	379.80	370.62	5	0.594	0.22 0.818	0.683	0.215 1.152	0.057
Suppliers' satisfaction	105.75	13.62	4	0.375	−0.049 0.685	0.394	−0.049 0.838	0.051
Time performance	168.60	218.93	5	0.335	−0.082 0.652	0.348	−0.082 0.779	0.048
Industry's satisfaction	113.45	82.73	11	0.261	−0.109 0.568	0.267	−0.109 0.644	0.037
Quality performance	9.14	22.00	5	0.152	−0.261 0.518	0.153	−0.267 0.573	0.046
Consultants' satisfaction	90.20	17.24	5	0.135	−0.276 0.505	0.136	−0.284 0.556	0.046
Cost performance	87.00	10.23	4	0.073	−0.341 0.463	0.073	−0.355 0.502	0.048
Subcontractors' satisfaction	85.67	16.44	3	0.043	−0.376 0.448	0.040	−0.395 0.482	0.050
Contractor's satisfaction	82.91	23.97	16	−0.261	−0.568 0.109	−0.267	−0.644 0.109	0.037
Client's satisfaction	103.00	38.17	6	−0.302	−0.106 0.623	0.312	−0.106 0.730	0.045

Note: *M = population mean, SD = Standard deviation, n = combined sample size, r = correlation coefficient, v = variance, 95% credibility and confidence interval.*

4.2 BIM-Integrated Model for Construction Project Performance

The results of the meta-analysis for the components of the ten CPPIs dimensions are presented in Table 3. The meta-analysis was done in order to validate further the conceptual model of BIM-integrated CPPIs that was developed in Sect. 3. As the results in Table 2 indicate, Cohen's effect size values suggest a small (d = 0.2) to large (d = 0.8 and above) practical significance for most of the components. Only eight components (satisfaction with return on capital (d = 0.170), satisfaction with profitability (d = 0.121), satisfaction with project delivery system (d = 0.000), satisfaction with return on investment (d = 0.130), satisfaction with claim reduction (d = 0.091), reduction in accident cost (d = 0.078), satisfaction with staff turnover (d = 0.102), and reduction in the number of complaints from community (d = 0.079) have low practical significance. However, these components were not eliminated from their respective dimensions in the model owing to the high relative weights they exhibited (Table 3). The established BIM-integrated model of CPPIs was formulated mathematically in Eqs. 1–4 to show the multiplier effect of BIM-PIs on the CPPIs.

Table 3. Meta-analysis estimate of the validities of the components of CPPIs.

Components of CPPIs	Frequency (n)	Relative weight (w)	Cohen's effect size (d)
Contractor satisfaction			
Satisfaction with growth	5	0.003	3.438
Satisfaction with stability (debt ratio)	68	0.051	0.622
Satisfaction with impact on society	97	0.072	0.587
Satisfaction with corporate strategy	96	0.072	0.546
Satisfaction with business process	94	0.070	0.462
Satisfaction with technological capability	94	0.070	0.462
Satisfaction with efficiency	93	0.069	0.420
Satisfaction with business efficiency	92	0.069	0.379
Satisfaction with return on capital	87	0.065	0.170
Satisfaction with profitability	80	0.060	0.121
Satisfaction with human resources development	98	0.073	0.629
Satisfaction with the payment system	95	0.071	0.504
Satisfaction with competency	72	0.054	0.455
Satisfaction with learning and knowledge management	93	0.069	0.420
Satisfaction with return on value added	93	0.069	0.420
Satisfaction with legal claims and litigation reduction	74	0.055	0.371
Client satisfaction			
Satisfaction with the project team	178	0.288	1.964
Satisfaction with value for money	79	0.127	0.628
Satisfaction with the functionality of the project	79	0.127	0.628
Satisfaction with project document quality	81	0.131	0.576
Satisfaction with return on investment	98	0.158	0.130
Satisfaction with project delivery system	103	0.166	0.000
Consultants' satisfaction			
Satisfaction with the reliability of service	64	0.141	1.519
Satisfaction with claim reduction	108	0.239	1.032
Satisfaction with job condition	102	0.226	0.684
Satisfaction with the ability to innovate	84	0.186	0.359
Satisfaction with the speed of service	93	0.206	0.162
Subcontractors' satisfaction			
Satisfaction with job condition	67	0.260	1.134
Satisfaction with claim reduction	98	0.381	0.749
Satisfaction with the ability to innovate	92	0.357	0.384
Suppliers' satisfaction			
Satisfaction with return on capital investment	124	0.293	1.338
Satisfaction with return on time investment	92	0.217	1.008
Satisfaction with payment terms	100	0.236	0.421
Satisfaction with claim reduction	107	0.252	0.091

(continued)

Table 3. (*continued*)

Components of CPPIs	Frequency (n)	Relative weight (w)	Cohen's effect size (d)
Health and safety performance			
Reduction in fatalities (number of heavy accidents and death injuries)	1000	0.526	1.673
Reduction in the accident rate	82	0.043	0.803
Reduction in sickness absence	122	0.064	0.695
Reduction in pollution	286	0.150	0.253
Reduction in accident cost	409	0.215	0.078
Time performance			
Reduction in project duration (Speed of construction)	560	0.664	1.787
Reduction in time variation due to change order	63	0.074	0.482
Reduction in time variation due to lack of labor	63	0.074	0.482
Reduction in time variation due to material supply delay	78	0.092	0.413
Reduction in time variation due to lack of equipment	79	0.093	0.409
Cost performance			
Reduction in cost variation due to redesign	102	0.293	1.466
Reduction in cost variation due to safety management	79	0.227	0.782
Reduction in cost variation due to reworks	83	0.238	0.391
Reduction in cost variation due to compensation for accidents	84	0.241	0.293
Industry satisfaction			
Satisfaction with the green rating	351	0.281	2.871
Satisfaction with the impact on the economy (i.e. job creation, GDP growth)	64	0.051	0.597
Satisfaction with a ratio value of contracts awarded to indigenous contractors	65	0.052	0.585
Reduction in the ratio of the work stoppage by regulating bodies	71	0.056	0.513
Satisfaction with an impact on the national image	78	0.062	0.428
Satisfaction with research and development	79	0.063	0.416
Satisfaction with human capital development	147	0.0117	0.405
Satisfaction with EIA rating	83	0.066	0.368
Satisfaction with innovation	85	0.068	0.344
Satisfaction with staff turnover	105	0.084	0.102
Reduction in the number of complaints from the community	120	0.096	0.079
Quality performance			
Reduction in rework (Rework index)	119	0.260	1.254
Reduction in the number of defects during construction (defects remediation)	70	0.153	0.972

(*continued*)

Table 3. (*continued*)

Components of CPPIs	Frequency (n)	Relative weight (w)	Cohen's effect size (d)
Reduction in the number of outstanding defects (i.e., defects at the end of the defect liability period)	111	0.242	0.890
Reduction in the number of defects during handing-over	78	0.170	0.609
Reduction in non-compliance with specifications	79	0.172	0.563

$$Z = [\varphi] * [\gamma] \tag{1}$$

$$\varphi = \beta_1 + \beta_2 + \beta_3 + \beta_4 = \sum_{i=1}^{n=4} \beta \tag{2}$$

$$\begin{aligned}\gamma = & \sum_{i=1}^{n=16} \omega_1 + \sum_{i=1}^{n=6} \omega_2 + \sum_{i=1}^{n=5} \omega_3 + \sum_{i=1}^{n=3} \omega_4 + \sum_{i=1}^{n=4} \omega_5 \\ & + \sum_{i=1}^{n=5} \omega_6 + \sum_{i=1}^{n=5} \omega_7 + \sum_{i=1}^{n=4} \omega_8 + \sum_{i=1}^{n=11} \omega_9 + \sum_{i=1}^{n=5} \omega_{10}\end{aligned} \tag{3}$$

$$\begin{aligned}Z = & \left[\sum_{i=1}^{n=4} \beta\right] * \left[\sum_{i=1}^{n=16} \omega_1 + \sum_{i=1}^{n=6} \omega_2 + \sum_{i=1}^{n=5} \omega_3 \right. \\ & + \sum_{i=1}^{n=3} \omega_4 + \sum_{i=1}^{n=4} \omega_5 + \sum_{i=1}^{n=5} \omega_6 + \sum_{i=1}^{n=5} \omega_7 \\ & \left. + \sum_{i=1}^{n=4} \omega_8 + \sum_{i=1}^{n=11} \omega_9 + \sum_{i=1}^{n=5} \omega_{10}\right]\end{aligned} \tag{4}$$

where $\mathbf{Z}$ = *Performance Score for a BIM-based Construction Project;* $\boldsymbol{\varphi}$ = *Total Performance Score for BIM on the Construction Project;* $\boldsymbol{\gamma}$ = *Total Performance Score for CPPIs;* $\boldsymbol{\beta_1}$ = *Performance Score for Network Communication;* $\boldsymbol{\beta_2}$ = *Performance Score for knowledge sharing and transfer;* $\boldsymbol{\beta_3}$ = *Performance Score for Information Sharing and Exchange;* $\boldsymbol{\beta_4}$ = *Performance Score for Trust-based Relationships;* $\boldsymbol{\omega_1}$ = *Performance Score for Contractor Satisfaction;* $\boldsymbol{\omega_2}$ = *Performance Score for Client Satisfaction;* $\boldsymbol{\omega_3}$ = *Performance Score for Consultants' Satisfaction;* $\boldsymbol{\omega_4}$ = *Performance Score for Subcontractors' Satisfaction;* $\boldsymbol{\omega_5}$ = *Performance Score for Suppliers' Satisfaction;* $\boldsymbol{\omega_6}$ = *Performance Score for Health and Safety Performance;* $\boldsymbol{\omega_7}$ = *Performance Score for Time Performance;* $\boldsymbol{\omega_8}$ = *Performance Score for Cost Performance;* $\boldsymbol{\omega_9}$ = *Performance Score for Industry Satisfaction;* $\boldsymbol{\omega_{10}}$ = *Performance Score for Quality Performance.*

5 Conclusion

This study identified a comprehensive set of CPPIs and established a BIM-integrated evaluation system for construction project performance. In particular, the study contributes to the literature on construction project performance evaluation in at least four ways. First, it shows that there are ten distinct dimensions of CPPIs for projects that are based on CPDS. These ten dimensions meaningfully and comprehensively explain the concept of construction project performance. This finding has weighty practical implications in that the model is useful in ensuring organizational productivity and projects' physical output. Second, the findings provide further support for the extension of CPPIs dimensions. Third, the findings show the relative importance and components of industry satisfaction such as satisfaction with a green rating, satisfaction with innovation, satisfaction with impact on the national image, and satisfaction with the impact on the economy. This contribution is of great consequence in that the literature has not previously identified the full components of industry satisfaction with construction project performance. Fourth, the findings (as depicted in Eq. 4) speak to the relative contribution of BIM in evaluating the performance of construction projects.

In sum, this meta-analytic review has helped to identify and synthesize the dimensions and components of CPPIs across studies and countries. Given the evidence put forward in this study, future research should concentrate on the following gaps: empirical investigation of the significance and relationships of the dimensions and components of the BIM-integrated model of CPPIs, development of software technology to simplify the application of the model, and development of standardized formulas and measurement models for the components of the model.

6 Limitations of the Study

Considering that the study is meta-analytic research, it contains the limitations of the studies on which the database for the study was built. The study could not test the correlation coefficients of the BIM-PIs dimensions and components identified for CPPIs dimensions, owing to a lack of research on them. A final limitation of this study is that the study was not able to investigate boundary conditions that might apply to the BIM-integrated model of CPPIs. The study recognizes that components are moderating the dimensions of BIM-PI, but these were not covered in this study.

Acknowledgments. The financial assistance of the University of Cape Town (UCT), TETFUND and the National Research Foundation (NRF) towards this research is hereby acknowledged. The opinions expressed, and conclusions arrived at, are those of the authors and are not necessarily to be attributed to UCT, TETFUND or NRF.

References

1. Alarcon, L.F., Ashley, D.B.: Modeling project performance for decision making. J. Constr. Eng. Manag. **122**(3), 265–273 (1996)
2. Ali, H.A.E.M., Al-Sulaihi, I.A., Al-Gahtani, K.S.: Indicators for measuring the performance of building construction companies in the Kingdom of Saudi Arabia. J. King Saud University Eng. Sci. **25**(2), 125–134 (2013)
3. Bassioni, H.A., Price, A.D.F., Hassan, T.M.: Performance measurement in construction. J. Manag. Eng. **20**(2), 42–50 (2004)
4. Beatham, S., Anumba, C., Thorpe, T., Hedges, I.: KPIs: a critical appraisal of their use in construction. Benchmarking Int. J. **11**(1), 93–117 (2004)

5. BEC: Benchmarking Danish Construction. The Benchmark Centre for the Danish Construction Sector Copenhagen (2014)
6. CCIC: Measuring the performance of the Canadian construction industry: pilot project final report. Canadian Construction Innovation Council, Toronto, Ontario (2007). www.ccic-ccic.ca/reports.html. Accessed Feb 2018
7. Cheung, S.O., Suen, H.C.H., Cheung, K.K.W.: PPMS: a web-based construction project performance monitoring system. Autom. Constr. **13**(3), 361–376 (2004)
8. Constructing Excellence: UK Construction Industry: Key Performance Indicators. Constructing Excellence, UK (2004)
9. Construction Industry Institute: Benchmarking and Metrics (2001). www.constructioninstitute.org/. Accessed 12 July 2018
10. Corporación de Desarrollo Tecnológico (CDT): National benchmarking system for the construction industry (2002)
11. Costa, D.B., Formoso, C.T., Kagioglou, M., Alarcón, L.F., Caldas, C.H.: Benchmarking initiatives in the construction industry: lessons learned and improvement opportunities. J. Manag. Eng. **22**(4), 158–167 (2006)
12. Costa, D.B., Lima, H., Formoso, C.T.: Performance measurement systems for benchmarking in the Brazilian construction industry. In: Performance Measuring System for Benchmarking Conference, Rotterdam, Netherlands, pp. 1029–1040 (2004)
13. Dawood, N., Sikka, S.: Development of 4D based performance indicators in the construction industry. Eng. Constr. Architectural Manag. **16**(5), 438–458 (2009)
14. Fazli, A., Fathi, S., Enferadi, M.H., Fazli, M., Fathi, B.: Appraising effectiveness of Building Information Management (BIM) in project management. Procedia Technol. **16**, 1116–1125 (2014)
15. Furneaux, C.W., Hampson, K.D., Scuderi, P., Kajewski, S.L.: Australian construction industry KPIs. In: CIB World Congress Proceedings–Building a Better World, pp. 1–12. CIB (2010)
16. Garbharran, H., Govender, J., Msani, T.: Critical success factors influencing project success in the construction industry. Acta Strutilia **19**(2), 90–108 (2012)
17. Hartmann, T., Van Meerveld, H., Vossebeld, N., Adriaanse, A.: Aligning building information model tools and construction management methods. Autom. Constr. **22**, 605–613 (2012)
18. Hunter, J.E., Schmidt, F.L.: Dichotomization of continuous variables: the implications for meta-analysis. J. Appl. Psychol. **75**(3), 334 (1990)
19. Jastaniah, Y.R.: Performance evaluation and benchmarking of construction industry projects using data envelopment analysis. Ph.D. dissertation, Southern Methodist University (1997)
20. Love, P.E.D., Matthews, J., Simpson, I., Hill, A., Olatunji, O.A.: A benefits realization management building information modeling framework for asset owners. Autom. Constr. **37**, 1–10 (2014)
21. Luu, T.-V., Kim, S.-Y., Cao, H.-L., Park, Y.-M.: Performance measurement of construction firms in developing countries. Constr. Manag. Econ. **26**(4), 373–386 (2008)
22. Marx, H.: Construction industry indicators affecting contractors. J. Adv. Perform. Inf. Value **4**(1) (2012)
23. Ofori-Kuragu, J.K., Baiden, B.K., Badu, E.: Key performance indicators for project success in Ghanaian contractors. Int. J. Constr. Eng. Manag. **5**(1), 1–10 (2016)
24. Pillai, A.S., Joshi, A., Srinivasa Rao, K.: Performance measurement of R&D projects in a multi-project, concurrent engineering environment. Int. J. Proj. Manag. **20**(2), 165–177 (2002)
25. Pinheiro, J.P.C.: Key performance indicators applied to construction: sector performance and benchmarking. Extended Abstract: Instituto superior tecnico (2011)

26. Radujković, M., Vukomanović, M., Dunović, I.B.: Application of key performance indicators in South-Eastern European construction. J. Civil Eng. Manag. **16**(4), 521–530 (2010)
27. Ramırez, R.R., Alarcón, L.F.C., Knights, P.: Benchmarking system for evaluating management practices in the construction industry. J. Manag. Eng. **20**(3), 110–117 (2004)
28. Rankin, J., Fayek, A.R., Meade, G., Haas, C., Manseau, A.: Initial metrics and pilot program results for measuring the performance of the Canadian construction industry. Can. J. Civ. Eng. **35**(9), 894–907 (2008)
29. Roberts, M., Latorre, V.: KPIs in the UK's construction industry: using system dynamics to understand underachievement. Revista de la Construcción **8**(1), 69–82 (2009)
30. Rokooei, S.: Building information modeling in project management: necessities, challenges, and outcomes. Procedia Soc. Behav. Sci. **210**, 87–95 (2015)
31. Sarhan, S., Fox, A.: Performance measurement in the UK construction industry and its role in supporting the application of lean construction concepts. Constr. Econ. Build. **13**(1), 23–35 (2013)
32. Scottish Construction Centre: Scottish KPIs (2009), From the Scottish Construction Centre website www.scocon.org. Accessed 17 April 2008
33. Sibiya, M., Aigbavboa, C., Thwala, W.: Construction projects' key performance indicators: a case of the South African construction industry. ICCREM **2015**, 954–960 (2015)
34. Takim, R., Akintoye, A.: Performance indicators for successful construction project performance. In: 18th Annual ARCOM Conference, vol. 2, pp. 545–555 (2002)
35. Toor, S.-ur-R., Ogunlana, S.O.: Beyond the «iron triangle»: stakeholder perception of key performance indicators (KPIs) for large-scale public sector development projects. Int. J. Proj. Manag. **28**(3), 228–236 (2010)
36. Tsiga, Z.D., Emes, M., Smith, A.: Critical success factors for the construction industry. PM World J. **5**(8), 1–12 (2016)
37. Won, J., Lee, C.: How to tell if a BIM project is successful: a goal-driven approach. Autom. Constr. **69**, 34–43 (2016)
38. Yu, I., Kim, K., Jung, Y., Chin, S.: Comparable performance measurement system for construction companies. J. Manag. Eng. **23**(3), 131–139 (2007)

Graduating Female Students' Long-Term Career Decisions and Underrepresentation of Women in South Africa's Construction Industry

Yolanda Moraba(✉) and Oluwayomi Babatunde

School of Construction, Economics, and Management,
Wits University, Johannesburg, South Africa
mmanoko.moraba@students.wits.ac.za

Abstract. The underrepresentation of women in South Africa's construction industry has not improved over the past decade despite the increase of the graduation of females in the university's faculty of the built environment. This study investigates the career decisions made by graduating female students based on their perceptions of the construction industry, and how their long-term career decision affects the underrepresentation of women therein. This pilot study used an online questionnaire distributed to female students registered for BSc Honours in Quantity Surveying at a university in South Africa. The pilot study was extended with interviews conducted with recent QS graduates. The findings from the study report on the factors that motivated female students and recent graduates to study a QS degree, the challenges experienced during their studies, and how the experienced challenges influence their long-term career decisions. To sustain the industry's business development and growth, graduating female students need to be attracted and retained by the construction industry.

Keywords: Barriers · Career decisions · Construction education · Factors · Female postgraduates

1 Introduction

For years, the persistent underrepresentation of women in South Africa's construction industry has been a critical area of concern [21]. The general characteristic of the construction industry is that it is male favoured with the lack of presence of females, especially in top influential positions [15]. The traditions, organizational cultures and the sexist attitudes that exist within the industry play a vital role when employing females in the industry [14, 21]. As a result, females currently employed in the construction industry are found more in administrative roles than in technical roles [2, 21].

In universities, the faculty of the built environment has experienced a steady increase in the number of female students registering and graduating from construction-related programmes [8, 9]. This increase, however, is not reflected in the industry's workplace and women remain underrepresented [8]. The long-term career decisions of graduating female students within construction-related degrees programmes are

C. Aigbavboa and W. Thwala (Eds.): CIDB 2019, *The Construction Industry in the Fourth Industrial Revolution*, pp. 158–167, 2020.
https://doi.org/10.1007/978-3-030-26528-1_16

reported to be influenced by the barriers found in the construction industry [13]. These barriers include the perceived unsafe nature of construction, challenging labour requirements and the dangerous work environment [13]. This study explores factors that motivated graduating female student to join the construction industry and how their long-term career decision within the industry influences the under-representation of females in South Africa's construction industry.

2 Literature Review

2.1 Factors Motivating Construction Education

Career paths are often decided on the final years of a person's high school career [14]. However, Bigelow et al. [15] report that the career-decision making process is often complicated for high school students and more than 80% of students in university believe that they made their career decision too early, with some reporting that they would change careers given a chance. In high school, females report that they face gender-based career stereotyping, which makes it harder for them to choose a career in the construction industry. There is also a sense of isolation that exists amongst young girls in high school and universities because of the limited networking opportunity afforded to them to meet other female professionals working in the construction industry [14, 15].

After high school, female students have a choice to either study a construction-related degree or a non construction-related degree. Of the students that chose to study a construction-related degree, the factors that have been identified as influential to their career decision making process decision include the poor image of the construction industry [15], the impacts of family members, friends, teachers, and counselors [18]. More factors that enhance the retention of female students include, mentoring, fellowships, scholarships, internships, career opportunities available to them after graduation and industry experience [5].

2.2 Internal Barriers

While the number of female students registering and graduating from construction-related degree programmes has increased over the years [1, 9, 16], underrepresentation of women within the construction industry still perpetuates. During the construction education process, female students face several barriers that have been identified as internal barriers in this study. These barriers have been found to be the lack of networking opportunities for graduating female students to meet and interact with other female professionals in the industry [6, 22], the male-dominated learning environments and dominated male academic staff found in universities [6, 22]. The male-dominated environment is argued to breed gender-based discrimination and harassment against graduating female students by their male colleagues [10, 14]. These barriers result in female students considering careers in other sectors such as health, commerce and IT over the construction industry because of the barriers they have faced during construction education [14]. These barriers further create perceptions of how the construction industry's work environment is like, and the contributions to the long-term career decision of graduating female students.

2.3 External Barriers

External barriers are the perceived challenges for graduating female students face when they enter the construction industry [13]. External barriers are the factors that have been in the industry for many generations which graduating female students have no control over. The nature of the construction industry, working conditions, and sexist attitudes have been identified by [13] as external barriers found in the construction industry. These barriers may be the primary source of the low participation rates we see of women in South Africa's construction industry [8].

Nature of the Construction Industry
The image that the construction industry has built over decades is that of male dominance [11, 14, 21], with difficult working conditions [8]. This plays a role in graduating female students not wanting to join the industry after graduation [8, 13]. However, the industry is becoming more high-tech because of the unfolding trends of the 4th industrial revolution, which requires more mental strength than brute physical strength [11]. The macho image of the industry is evolving with the 4th industrial revolution; therefore, graduating female students ought not to make their long-term career decisions based on a macho image.

Working Conditions
The construction industry is a place which expects employees to work long hours, neglecting their family responsibilities in the process [1]. The consequence of this model is that women who chose to work and have a family are prejudiced against [1]. The "all boys club" created by organisations requires long working hours, competition and isolation [8]. Females are therefore expected to adopt the male culture that exists in organisations or leave the industry. If they do not do either, they remain in unimportant positions [8]. This culture is a barrier to graduating female students [13] because they will eventually be regarded as unreliable for taking time off to fulfill their family responsibilities [21].

Sexist Attitudes
Gender stereotypes, glass ceiling, organisational cultures made from the male-dominance of the industry and bias recruitment practices are the leading sexist barriers found in the construction industry. The recruitment of graduates is conducted informally [1] using unstructured interviews [8, 9]. The successful female candidates enter the industry and start on equal footing with their male counterparts. Their career advancement is however different for both because of the sexist attitudes that exist within organisations [12]. The female graduate is often faced with a glass ceiling, preventing her from advancing in the organisation. Females are further faced with isolation and limited access to mentorship, which may help move up the organisation's ladder [2].

3 Conceptual Framework

Figure 1 presents the relationship between the early days when female students start deciding on the career path they want to pursue. After high school, female students either register for a construction-related degree or a non construction-related degree. For those that register for a construction-related degree, their decision is influenced by the factors motivating construction education under the literature review Sect. 2. The circle represents female students' duration as undergraduate students at a university of choice in South Africa. During their undergraduate studies, female students have reported being faced with challenges that were identified broadly by Ling and Poh [13] as internal and external barriers. Despite having faced barriers during their undergraduate years, some female students decide to continue with their studies by enrolling for a postgraduate degree. This set of students could have decided at this point to pursue another career path or find a job because they have a bachelor's degree. Instead, they decided to improve their academic knowledge of the construction industry, showing interest in the industry. A previous study by Ling and Poh [13] report that graduating students are aware that females in the construction industry are faced with external barriers relating to the nature of the construction industry, working conditions found in the construction industry and sexist attitudes but they still choose to pursue a postgraduate degree. This group's long-term career decision is therefore of importance which this study will explore.

4 Research Methods

The fieldwork investigated factors that motivated graduating female students to study a construction-related degree, their experience of construction education and the construction industry. Using convenience sampling, an online questionnaire was distributed to female students registered for BSc Honours (QS) at a university in South Africa. Graduates who have been working in the construction industry for two years were also asked about what motivated them to study a construction related degree and their experiences of the construction industry for the two years they have been working. The questions were adopted from a previous study by Ling and Poh [13] and the Career Construction Theory was adopted for questions relating to career decisions and factors influencing their decisions. A summary of the results is presented below and will is discussed in the subsequent sections.

4.1 Results and Discussion

Factors Motivating Construction Education
The results in Table 1 show that career opportunities available (F1) to graduating students after graduation and the prospective financial stability (F2) that comes with it were the leading factors that motivated them to study a QS degree. People often choose to follow in the footstep of people they look up to when deciding on a career path to pursue [6]. As a result, the perceived image of the construction industry (F3) which was

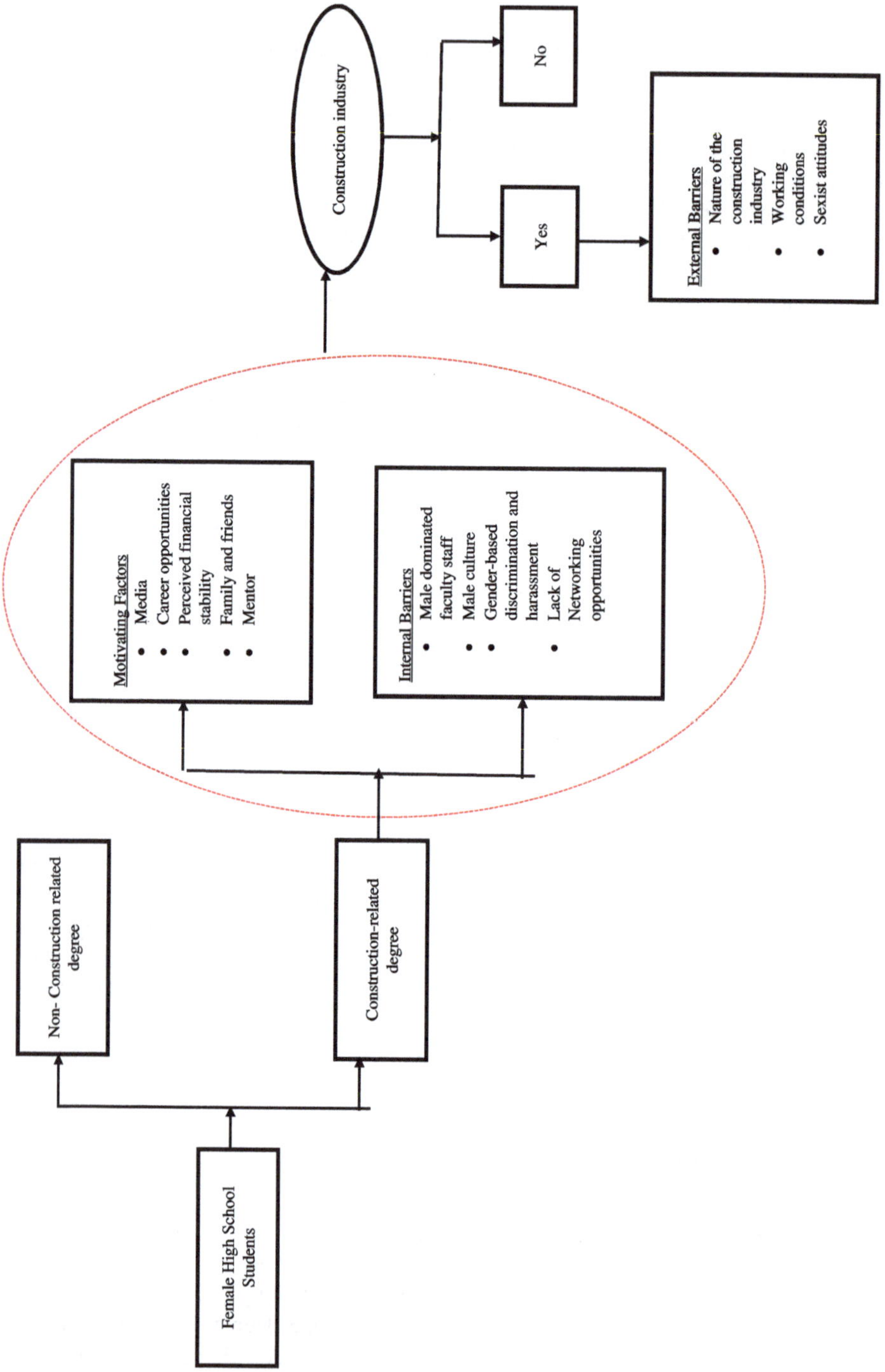

Fig. 1. Conceptual Framework for construction education motivation, barriers, and retention, [17]

Table 1. Summary of results

Barriers/Factors	Rankings	Mean	Median	SD
Factors motivating construction education				
F1 Career opportunities available	1.00	4.00	3.60	3.32
F2 Prospective financial stability	2.00	4.00	3.60	3.82
F3 Image of the construction industry	3.00	3.67	1.50	3.66
F4 Family, friends and teachers	4.00	3.67	1.50	3.82
F5 Mentor and high school career counselors	5.00	3.11	4.00	3.60
F6 Media	6.00	2.56	4.60	3.31
F7 Having a father in the industry	7.00	2.00	1.50	2.23
Internal barriers				
I1 Male-Dominated faculty staff and lecturers	1.00	4.44	2.00	3.19
I2 Male culture exists in construction education	2.00	4.00	3.60	3.20
I3 Discrimination against female students	3.00	3.67	3.00	2.93
I4 Lack of Networking Opportunities	4.00	3.33	3.00	3.49
I5 Female students face gender-based harassment in construction education	5.00	3.33	3.00	3.31
I6 Female students are not readily accepted by their male counterparts	6.00	3.11	4.00	3.31
External barriers				
E1 Construction jobs are masculine	1.00	4.33	1.50	3.82
E2 Construction jobs occur at the expense of family responsibilities	2.00	4.22	1.00	3.60
E3 Gender stereotyping occurs in the construction industry	3.00	4.11	0.50	3.88
E4 Construction jobs have long working hours	4.00	3.89	1.00	3.31
E5 Male cultures exist in construction	5.00	3.89	2.00	2.23
E6 The construction industry has a poor image of	6.00	3.89	0.50	3.66
E7 Female graduating students face glass ceiling	7.00	3.56	0.50	3.38

drawn by family members, friends, high school teachers (F4), mentors, high school career counselors (F5) and the media (F6) played a role in the decision-making process. No graduating female student reported to have a father in the construction industry for this study; therefore (F7) was not a motivating factor for construction education.

In the interview female graduates were asked what the key motivating factors were in studying a construction-related degree.

> ***Respondent 1**: "My uncle was an executive at a construction company and advised me to study Quantity Surveying."*

> ***Respondent 2**: "When I was in grade 10, we went to a Construction and Engineering Expo. I learned a lot about the construction industry, and I became interested in pursuing a career in the industry. At the time, it made sense to me because I have always had an interest in houses and architecture. I knew that I wanted to be a Quantity Surveyor after my first year because that was my favourite module in the first year."*

This shows that the active involvement by industry professionals, directly or indirectly, play a significant role in enhancing the construction industry as a career choice for female students. This suggests that professionals who speak to high school girls about the construction industry have a positive influence on the career choice of females in high school.

Internal Barriers

The results in Table 1 suggest that the number of male faculty staff members and lecturers (I1) is the leading barrier found in construction education. It was reported that male culture exists in lecture halls (I2) where female students are not readily accepted by their male counterparts (I6). This often results in discrimination against female students (I3) and gender-based harassment (I5). This is no different from the construction workplace where women are found mainly in administrative roles.

Graduating female students report that they are not given networking opportunities (I4) with other females working in the industry for mentoring and ways to deal with the barriers they are facing as female students. Active involvement from female industry professionals for mentorship and an increase in the presence of female lecturers could help boost graduating female students' confidence in their decision to continue with a career in the construction industry. Fielden et al. [9] suggest that it is essential for universities to recognise women as a source of new workers, and to provide them with appropriate training to fill the gap.

The interviews with the recent female graduates outlined barriers female graduates have faced during construction education.

> **Respondent 1**: "*One of the internal barriers I have experienced is the lack of interaction between female students and female professionals. The industry is dominated by males who dictate and limits interactions for females within the profession. I have not been exposed to females who have high positions, which is unfortunate.*"

> ***Respondent 2***: *"During my undergraduate and postgraduate as a Quantity Surveying student, my class was female dominant. I, therefore, did not experience any gender-based challenges. The only challenges I faced were the regular challenges every student face, regardless of the degree they are enrolled in".*

The interviews suggest that female graduates do not face challenges in university that prevents them from entering the construction industry. As noted, the number of female students registering in and graduating from university have increased over the past years. These results show that the problem does not lie in university.

External Barriers

Graduating female students reported that the construction industry has a poor image (E6) which was built on the traditional model of being masculine (E1) and requires employees to work long hours (E4), which occurs at the expense of family responsibilities (E2). The industry traditionally expects employees to work overtime. The more willing an employee is to working overtime, the more committed the employee seems to be to their work. For a person who bears the primary role of family responsibility, working overtime is not suitable for their family needs. However, the 4th Industrial Revolution will enable the construction industry to be high-tech [12].

The male culture found in the construction industry (E5) is said to give rise to gender stereotyping (E3). The culture of the industry is that of a male-dominance, requiring brute strength and good tolerance for outdoor conditions, severe weather and bad language. Fielden et al. [9] reported that the construction industry is characterised as an "all-boys club" represented by "bad" language and behaviour. This macho culture is regarded as hostile and discriminatory to women. As a result, women who seek entry and career advancement into the male-dominated industry must adapt to the male culture if they want to have a successful career, leave the industry, or remain in unimportant positions within the organization. Females are also not given equal opportunities with their male counterparts and are faced with a glass ceiling quicker than their male counterparts (E7). Kolade and Kehinde [12] found that the above reasons are the major contributors to women's barriers to career advancement and the glass ceiling. These conditions are barriers that further deter graduating female students from entering the construction industry after graduation and deciding on having a long-term career within the industry [13].

In the interviews with the recent female graduates, they were asked about the barriers they faced while working in the construction industry.

> ***Respondent 1**: "I never felt safe on site. It was male dominated, and an environment that created a persistent sense of discomfort. There was an incident where a subcontractor's employee invaded my personal space and touched me inappropriately. I immediately reported him to his supervisor; I would not allow anyone to touch or be too close to me physically".*

> ***Respondent 2**: "The main challenge I face in the industry is the chauvinistic behavior from my male colleagues. My male colleagues also always wanted to have the last say on any matter under discussion, assuming that they know better, regardless of their level of experience. I would sometimes follow what they say to avoid conflict, and keep projects going".*

5 Conclusion

Factors that motivated graduating female students to study a construction-related degree include career opportunities available, perceived financial stability, the influence of mentors, the media and family members. The lack of female lecturers was found to be the significant barrier experienced during construction education. Graduating female students are faced with gender-based discrimination during construction education. The nature of the industry is seen as masculine and requiring long working hours. The male culture and gender stereotypes that exist within the industry often subject women to a slow career progression than their male counterparts. Despite the barriers, graduating female students have shown interest in pursuing long-term careers within the construction industry. This is a positive contribution towards the current state of underrepresentation of women in South Africa's construction industry because universities are continuously injecting female who wants to have long-term career decisions in the construction industry. However, graduating female students remain a wasted and untapped resource within South Africa's construction industry. If the industry continuously fails to attract and retain female graduates, the industry will face a labour shortage which will result in a slow economic and infrastructural development of the country. South Africa's construction industry and universities, therefore, need to transform and diversify so that it can sustain its business development and growth in the 4th Industrial Revolution.

References

1. Adogbo, K., Ibrahim, A., Ibrahim, Y.: Development of a framework for attracting and retaining women in construction practice. J. Constr. Dev. Countries **20**(1), 99–115 (2015)
2. Agherdien, S., Smallwood, J.: Women in South African construction. In: Proceedings of the 19th CIB World Building Congress, pp. 1–11. Construction and Society, Brisbane (2013)
3. Ayalp, G., Öcal, M.: Determining construction management education qualifications and the effects of construction management education deficiencies on turkish construction. Creative Educ. **7**(1), 254–268 (2016)
4. Aulin, R., Jingmond, M.: Issues confronting women participation in the construction industry. In: Advances in Engineering and Technology- Contribution of Scientific Research in Development, Entebbe, pp. 312–318 (2010)
5. Bigelow, B.F., Bilbo, D., Ritter, L., Mathew, M., Elliott, J.W.: An evaluation of factors for retaining female students in construction management programs. Int. J. Constr. Educ. Res. **12**(1), 18–36 (2016)
6. Bigelow, B.F., Saseendran, A., Elliot, J.W.: Attracting students to construction education programs: an exploration of perceptions by gender. Int. J. Constr. Educ. Res. **14**(3), 179–197 (2017)
7. Del Puerto, C., Guggemos, A.C., Shane, J.S.: Exploration of strategies for attracting and retaining female construction management students. In: 47th ASC Annual International Conference Proceedings, Nebraska, pp. 6–9 (2011)
8. Fernando, N.G., Amaratunga, D., Haigh, R.: The career advancement of professional women in the UK construction industry: the career success factors. J. Eng. Des. Technol. **12**(1), 53–70 (2014)
9. Fielden, S.L., Davidson, M.J., Gale, A., Davey, C.L.: Women, equality, and construction. J. Manag. Dev. **20**(4), 293–305 (2001)
10. Francis, V., Prosser, A.: Exploring vocational guidance and gender in construction. Int. J. Constr. Educ. Res. **10**(1), 39–57 (2014)
11. Ginige, K., Amaratunga, R.D., Haigh, R.P.: Effect of attitudes towards creating women managers in the construction industry. Research Institute for the Built Environment, University of Salford, pp. 169–179 (2007)
12. Kolade, O., Kehinde, O.: Glass ceiling and Women career advancement: evidence from the Nigerian construction industry. Iran. J. Manag. Stud. **6**(1), 79–99 (2013)
13. Ling Yean Yng, F., Pei Poh, Y.: Encouraging more female quantity surveying graduates to enter the construction industry in Singapore. Women Manag. Rev. **19**(8), 431–436 (2004)
14. Madikizela, K., Haupt, T.: Influences on women's choices of careers in construction: a South African study. Constr. Econ. Build. **10**(1), 1–15 (2010)
15. Martin, P., Barnard, A.: The experience of women in male-dominated occupations: a constructivist grounded theory inquiry. SA J. Ind. Psychol. **39**(2), 1–12 (2013)
16. Morganson, V.J., Jones, M.P., Major, D.A.: Understanding women's underrepresentation in science, technology, engineering, and mathematics: the role of social coping. Career Dev. Q. **59**(1), 169–179 (2010)

17. Moraba, M.Y., Babatunde, O.: Graduating female students' long-term career decisions and underrepresentation of women in South Africa's construction industry. In: SACQSP 10th International Research Conference, Johannesburg, South Africa, pp. 67–76 (2018)
18. OstadaliMakhmalbaf, M.: Identifying and prioritizing the fundamental factors that lead to inaccurate perceptions of the construction industry. Master's thesis, Texas A & M University (2014)
19. Savickas, M.L., Porfeli, E.J.: Career adapt-abilities scale: construction, reliability, and measurement equivalence across 13 countries. J. Vocat. Behav. **80**(3), 661–673 (2012)
20. Savickas, M.L., Porfeli, E.J., Hilton, T.L., Savickas, S.: The student career construction inventory. J. Vocat. Behav. **106**(1), 138–152 (2018)
21. Sewalk, S., Nietfeld, K.: Barriers preventing women from enrolling in construction management programs. Int. J. Constr. Educ. Res. **9**(4), 239–255 (2013)
22. Shane, J., del Puerto, C.L., Strong, K., Mauro, K., Wiley-Jones, R.: Retaining women students in a construction engineering undergraduate program by balancing integration and identity in student communities. Int. J. Constr. Educ. Res. **8**(3), 171–185 (2012)

Applied Systems Analysis for Analysing Challenges in Construction Projects: A Methodological Approach

Dillip Kumar Das(✉)

Central University of Technology, Bloemfontein, Free State, South Africa
ddas@cut.ac.za

Abstract. Construction projects constitute a complex set of activities and engagement of stakeholders. The activities and stakeholders are interlinked often in a non-linear way. Therefore, it becomes difficult to comprehend and analyse the challenges of construction projects adequately and evolve apposite policy or strategic interventions to meet these challenges. Applied Systems Analysis (ASA) is argued to be one of the methods by which the construction related challenges can be comprehended and analysed. Therefore, the objective of the paper is to provide a methodological approach of how ASA and System Dynamics modelling approach can be employed to examine the challenges in construction projects, and understand the mechanisms that cause construction challenges based on the non-linear causal feedback relationships, and to evolve policy interventions to meet the challenges in construction projects. The study was relied on literature review and case studies on how ASA has been applied on various aspects of construction projects. The paper is expected to contribute to the methodological approaches to analyse the challenges in construction projects and also assist the project and construction mangers in practice to comprehend the challenges and take remedial actions.

Keywords: Applied Systems Analysis · Causal feedback · Construction · Methodological approach · Project management

1 Introduction

Construction projects are complex challenges involving a multitude of activities and stakeholders (Abd El-Razek et al. 2008; Bon Gang and Lay Peng 2013; Doloi et al. 2012). The activities and the stakeholders in construction projects are observed to be interlinked and influence each other (Aiyetan and Das 2016; Das 2015; Das and Emuze 2017; Han et al. 2013; Love et al. 2008; Lyneis and Ford 2007). Consequently, it is found that the non- performance of one aspect has a cascading effect on the other aspects resulting into challenges of overrun of schedules, increase in cost, poor quality of work, rework, etc. Although, methods such as Construction Management, Project Management, Lean Construction, Last Planner System, Building Information Management Systems, etc., are being used largely, sometimes, the non-linearity and dynamicity associated with the various aspects in the construction projects make it difficult to comprehend and analyse the challenges of construction projects adequately

C. Aigbavboa and W. Thwala (Eds.): CIDB 2019, *The Construction Industry in the Fourth Industrial Revolution*, pp. 168–178, 2020.
https://doi.org/10.1007/978-3-030-26528-1_17

and evolve apposite policy or strategic interventions to prevent or meet the challenges. Applied Systems Analysis (ASA) because of its ability- to comprehend the dynamicity and nonlinearity, to elicit the causal feedback mechanisms among various relevant variables, for flow of information and materials and to consider delay (Beck et al. 2018) is argued to be an appropriate method to analyse and understand the construction challenges and the mechanisms that cause such challenges in construction projects (Han et al. 2013; and Lyneis and Ford 2007). The paper therefore offers a methodological approach of how ASA can be employed to examine the challenges in construction projects, and understand the mechanisms that cause the construction challenges based on the non-linear causal feedback relationships and also to emanate policy measures to resolve the issues in construction projects. The study relied on a number of case studies that have been carried out within the ASA framework on various aspects of construction projects such as delay, skilled labour challenges and health and safety in developing countries. The case studies were augmented by the findings that are accrued from the established main stream literature. Based on the case studies and findings from literature it is found that ASA could be a very relevant and useful method to understand and examine the various challenges in construction projects and enable evolving of policy interventions to meet the challenges. The paper is envisaged to contribute in two forms such as (1) it offers a methodological approach to analyse the challenges in construction projects and also assists the project and construction mangers in practice to comprehend the challenges and take remedial actions, and (2) it provides a modelling approach for the post graduate students to deal with research challenges in construction.

2 Approach of Investigation

This investigation is based on literature review, and case study analyses of empirical researches. ASA and System Dynamics (SD) modelling were considered as precursors to the case study analysis. However, prior to that the framework of ASA and SD modelling approach was reviewed and discussed in the general context and specifically in the context of construction. Construction challenges that includes rework, delay and health and safety from developing countries were used for the case study analysis. The case studies were used from the published main stream literature and unpublished works of the author relating to construction.

3 Understanding of ASA and SD Modelling Approach

Use of model in analysis and forecasting in the applied science field is pervasive. The reason being that the environment, ecological, economic, social and more so infrastructural systems are too complex to reason through the needs of policy or strategy without models (Beck 2009). They are also likely to remain an integral part of the process of deriving of policy interventions or strategies to shape the decision making. According to Batty (2009), models represent the realities in a simplified form, which represent systems in such a way that crucial features of theory and its application are identified and emphasized. In this context, models enable experimentation and testing with theory in a prognostic sense. It also enhances the understanding of the situations that have not been realized, for example, what is expected to happen in future (Das 2016). However, the challenge remains with the appropriate choice of models and their veracity to replicate the real world to avoid any ills that may come because of the failure of the models to predict rightly and place the trust of stakeholders in the models (Beck 2009; Das 2016).

Applied System Analysis (ASA) is argued to be one of the such methods- modelling principles that can be appropriately used to understand the behaviour of the complex problems, that include challenges relating to construction. According to many scholars and modellers ASA is more of quantitative in nature, that is relied on computational or mathematical models. However, on the other hand, it can also be used qualitatively to comprehend the challenge conceptually (Aiyetan and Das 2018; Beck et al. 2018). Essentially, ASA is a modelling principle that is multidisciplinary and cross disciplinary and has the ability to deal with the complex problems having (or without having) causal (feedback or no feedback) relationships and provides a holistic approach to find solutions to complex problems (Laszlo and Krippner 1998). However, while dealing with the complex problems, the environment is considered as the system, which might be either closed or open.

System Dynamics (SD) modelling approach is a modelling techniques that is essentially rests on ASA. Ideas from Systems theories form the basis for the SD modelling. Essentially, it is an outcome of the interaction of ideas that are emanated from traditional management, cybernetics, and computer simulation. According to Forrester (1968), it represents a theory of structure and behaviour of system, which can be easily used. Yet, it uses sophisticated mathematical algorithms while undertaking practical systems investigation. It has been adapted and used in understanding the behaviour of complex systems that include but not limited to urban development, industrial systems, construction, tourism, land use, and so on (Bon Gang and Lay Peng 2013; Coyle 1977; Forrester 1969, 1971; Shen et al. 2009). According to Sterman (1982), it is a problem evaluation approach in which essential elements of a system are linked in the form of a structure. It generates the behaviour of the system, and suitable to analyse the problems that is premised upon the feedback relationships. Moreover, the model offers the advantage of simulating the impacts of the policy interventions or measures on both the problem under inquiry as well as the whole system.

SD has been used to understand the behaviour of different aspects of construction such as rework, project delay, improving the effectiveness of the decision-making process, etc. (Aiyetan and Das 2016; Das 2015; Das and Emuze 2017; Han et al. 2013; Love et al. 2008; Lyneis and Ford 2007). Also, according to scholars such as Han et al. (2013) and Lyneis and Ford (2007), application of SD has not been confined to analysis of challenges of construction projects, rather moved beyond the projects to examine the behaviours of different phenomena in construction and project management.

SD model has several elements such as stock, rate, flow (physical and materials) and delay. In a system, there is always a stock (known as level variable). Stock is the accumulation of the difference between input and output (Sterman 2000). The stock is governed by the rate variables, in other words the decision rules or policies. They are the criteria that are used to regulate flows, which try to drive the system to a preferred state. Flows represent the movement of materials or information in the system. The chain of causality is denoted by the feedback loops, which influence the current and future scenarios. However, delays cause the dynamics that create disturbances, instability or oscillations in the system. The basic structure of a dynamic system is made by the stock, rate variables and the flows (Coyle 1996; Guimarães et al. 2009).

4 Overall Procedure of ASA for Construction Challenges

Based on the premise of ASA paradigm the overall process of ASA to analyse and comprehend construction challenges is presented below. While discussing the process, examples from the case studies of construction challenges dealt by use of ASA linked SD modelling principles both qualitative and quantitative have been used.

4.1 Initial Crafting of the "Problem"

The important thing is to first understand the problem and then initially craft the problem to substantially represent the actual or real challenge. The problem belongs to a system or environment in which the behaviours of the important variables or the system as a whole are to be comprehended or analysed. In other words, something is to be resolved or limited or enhanced and the behaviour of that variable is simulated or predicted under different options of interactions (non-interactions) of related variables. According to Beck et al. (2018), the important aspect is to craft the problem appropriately that should be based on the critical review of the literature combined with probing and understandings that engender out of the experiences of others. Further, the problem should be transformed into the specific challenges to be resolved, which may not be unique. However, first challenge is to understand whether ASA is relevant or not and if it is relevant then develop a coherent storyline based on the causal logic.

Case Study: Delay is pervasive in construction industry and it can happen because of any stakeholders and any components of construction projects. For example, clients, contractors, and consultants as stakeholders contribute to cause delay. The case study of "A Dynamic Model of Contractor Induced Delays in India", offers an insight of how delay occurs because of contractor related challenges and how the project can behave based on different policy or strategic interventions (Das and Emuze 2017). In this respect the initial problem is that contractors essentially can influence delay because of various challenges such as lack of funding, commitment, lack of planning, etc., which can contribute to the overall cost and time overrun of the project. These factors might influence each other or related factors to reinforce the causation of delay in a construction project. Although contractor related factors also influence other stakeholders and other activities in project, the system boundary is limited to contractor factors only.

4.2 Identifying the System, Its Environment, and the Sub-systems of the System

Usually the problem that is sought to be resolved or solved drive the choice and definition of the system. In a system there are sub systems and each sub-system might have its sub-systems. Also, the sub-systems could be systems in themselves. Relatively more detailed features are implicitly essential to define or identify a sub-system; however, some features that are not so important may be excluded. In other words, some features or parameters of the sub-system are considered important and some are not essential and thus can be ignored. Further, a system may be a part of a larger system called a supra-system, and a supra-system may be contained in a supra-supra-system (Beck et al. 2018). However, often the subjective judgement and choice of the Systems Analyst or modeller play a vital role in delineating the various systems and sub-systems. This choice is influenced by a mixture of the initial crafting of the problem to be addressed (as mentioned in Step 1) and some expectation about the nature and type of the computational analysis to follow (refer Step 4 in Sect. 4.4), in particular. However, the outcomes of the model will be based on the right or wrong or good or bad choices (Beck et al. 2018).

Case Study: Occurrence of work place accidents in construction is ubiquitous. In the case study of "Developing Safety Archetypes of Construction Industry at Project Level Using System Dynamics" (Mohammadi et al. 2018), although accidents occur in construction industry, a project is considered as the system (the environment). Aspects related to design, cost, contractors and supervisors are taken as the various sub-systems that influence the safety in a construction project (Mohammadi et al. 2018). These sub-systems apparently influence safety independently as well as also influence other factors such as delay, re-work, budget, safety control, etc., which accordingly either contribute to the occurrence or reduction of accidents in construction projects. The challenge is however the delineation of the 'system - the project' and the sub-systems.

4.3 Preliminary Analysis of Logic, and Conceptualization - Causal Loop Diagrams (CLD)

Development of causal loop diagrams (CLDs) for the system generally preludes computational SD modelling. Essentially CLDs are largely the visual and qualitative representation of the behaviour of the system, in respect of the interactions among the variables and entities, which are essential to resolve the problem (that is crafted in Step 1) (Beck et al. 2018). The CLDs generally offer dynamic hypothesis (theory) on which the systems are believed to work or behave and identify the leverage points on which decisions can be made or policy interventions can be evolved. The development of CLDs forms one of the most crucial elements for building models as the behaviour of the system would largely depend on the way the various parameters interact and play in mechanism. The important thing to consider is the identification of feedback loops. When the CLDs are developed the cause effect relations should properly have comprehended and appropriate polarities (negative or positive) are assigned. The loops can be reinforcing or balancing. When an increase (decrease) in a variable causes a tendency for the variable to be increased (decreased) due to the feedback through other variables in the loop, a reinforcing loop is created. However, when an increase (decrease), in a variable lead to a tendency for the same variable to decrease (increase) on account of the feedback of other variables in the loop, a balancing loop is formed (Mohammadi et al. 2018; Sterman 2000). However, before developing the CLDs, a mental map on the cause and effects can be developed by the modeller and the one way causalities mapped. Once one way causalities are established, the feedbacks relations are checked and established. Also, while developing the loops only 'nouns' should be used and polarities are decided based on the effect of a variable on the succeeding variable. The preceding or the succeeding variables do not affect the polarities between the two consecutive concerned variables. The analysis of the CLDs can enable the identification of the feedback loops, which may govern the behaviour of the system as a whole. These feedback loops are considered as the dynamic hypotheses, which consequently might outline the policy measures to be tested through quantitative modelling and analysis (Step 4). The CLDs form the base of the structure of the computational model (Step 4). Besides, it can also identify leverage points. Moreover, while developing the loops operational thinking – the way the variables function such as – information - decision - action – environment may be also used (Olaya 2012).

Case Study: In the case study of 'System Dynamics Approach to Mitigating Skilled Labour Shortages in the Construction Industry: A South African Context' (Olatunji and Das 2018), as shown in Fig. 1, the CLDs relate to investment, productivity and skilled labourer availability in the construction industry. It is hypothesized that if the investment is not adequate, it leads to poor labour wages, which influence the availability of skilled labourers in the construction industry negatively due to either few people opt for the profession or higher attrition rate as people leave the profession in search of better paying jobs. This is represented by a balancing loop B1. This situation gets worse or exacerbated as the lack of skilled labourers results to creation of challenges in the project delivery such as delay, re-work or cost overrun leading to negative or lesser contribution to the investment situation in the industry or projects because of poor productivity- represented by B1A. Thus, B1A complements B1 and aggravates the skilled labour shortage situation. There will therefore disturbances in the construction project- 'the system' However, productivity remains the vital factor that is dependent on the availability of skilled labourers. So, if the investment is increased in the labour wage sector, it will reinforce the availability of skilled labourers, leading to higher productivity and investment through a feedback relationship CLD R1. Higher productivity also positively influences the labour wages leading to higher availability of skilled labourers because of either reduction in attrition rate or more people opting for this profession (R1A), which reinforces R1 (Olatunji and Das 2018). Thus, the CLDs in this case created two dynamic hypotheses - one causes skilled labour shortage and the other provides ways to alleviate the challenge.

4.4 Model Formulations for Computational Analysis

ASA linked computational models can be built based on the CLDs developed in the previous step (Step 3). A number of mathematical, statistical or econometric models may be built to analyse the behaviour of the system. However, care should be taken to

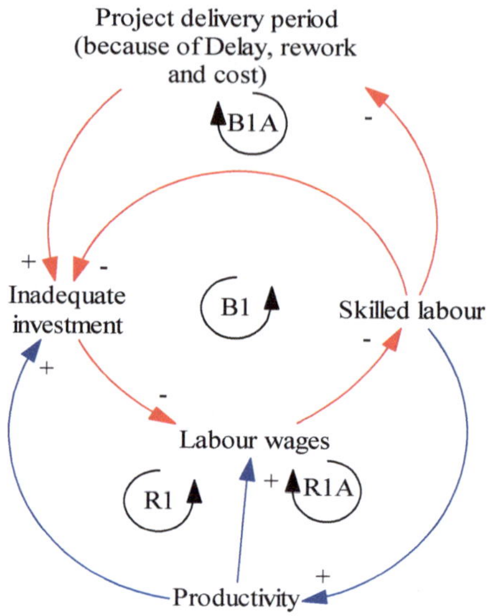

Fig. 1. Example of CLD

observe if the model realistically represents the causal relations and specifically the feedback relationships. In this particular case, a SD model was found relevant, which can be built by using any appropriate software platform such as 'Powersim', 'Stella', 'Vensim', or 'Any logic' and so on and writing appropriate algorithms to fit to the causal relations. However, while developing the model, the variables such as stock, rate, flow and auxiliary variables should be appropriately identified and used, and a stock flow (structure of the model) diagram should be built and the variables are initialized. The stock flow diagram or the structure of the models should be built based on the CLDs identified. The model then is initially simulated and then validated by using relevant validation principles. There are several ways to validate the model which include structure verification test, algorithm check, behaviour analysis, comparing with real life scenarios, etc. The validated model is then simulated to examine the behaviour of the system under varied simulated scenarios which can be used to develop policy interventions.

Case Study: In the case study of 'A Dynamic Model of Contractor Induced Delays in India' (Das and Emuze 2017), project period was considered as the stock variable, which is observed to be increasing or decreasing at a variable rate under different scenarios of finance delay, ineffective planning, and combined scenarios of finance delay and ineffective planning in a project. The stock variable - project period is observed to be showing an increasing trend compared to business as usual scenario (normal scenario - when everything goes as planned) at varying degrees (Das and Emuze 2017) (Fig. 2).

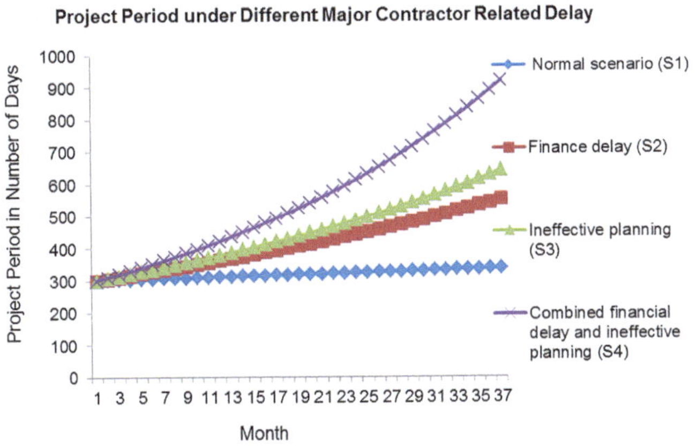

Fig. 2. Simulated scenarios

4.5 Screening and Analysis of Computational Model Simulated Results

The objective of the computational analysis of the model - in this case SD model is, to examine or evaluate what could occur in the future, and policy interventions can be evolved and implemented in the system to get the desired behaviour. The model can generate innumerable simulated scenarios with unique or not so unique behaviour under each scenario of questions such as "What if?". However, it is essential to limit the scenarios for policy analysis. Largely, it depends on the analysts' experience and understanding of the real system and phenomena that are occurring or might occur. However, having generated all the responses to all the "What If?" questions related to the identified problem in the system, the policy interventions and consequent future outcomes should be examined or evaluated based on a set of decision criteria, in order to arrive at appropriate and explicit interventions or how best the situation can be resolved and desired effects are obtained (Beck et al. 2018).

Case Study: In the case study of 'A Dynamic Model of Contractor Induced Delays in India' (Das and Emuze 2017), from a number of scenario analysis, it is found that a scenario that is based on the combination of effective communication, the availability of fund, provision for rework and appropriate use of technology and construction methods, the project period is expected to reduce significantly. For instance, the project period will be marginally higher (9.2%) than the project period under business as usual scenario.

4.6 Implementing the Decision and Handling of Uncertainty

There are decisions to make and also to meet the challenges of uncertainties, which are generally integral parts of any construction. So while simulating the model the worst case scenarios might be evaluated and see what if such worse cases arise. In this context, an understanding of the uncertain occurrences and ideas and experiences might help. While making decisions, the appropriate behaviour under different scenarios might be considered and discussed. In this respect as a rule of thumb, the scenarios such as the business as usual scenario, worst case scenario, optimistic scenarios and a set of likely or perceived scenarios should be considered. Besides, the decisions might be considered based on the leverage points that are identified in the step 3.

5 Conclusion

Construction is a complex challenge. Construction projects have a number of interrelated components and stakeholders which work through causal relations including causal feedback relations. So, resolving the challenges of construction necessitate appropriate mathematical/computational models to comprehend the behaviour of the project under different complex and complicated scenarios of effect of individual or combination of multiple variables. ASA is argued to be one of the such modelling principle which can meet the challenge. Therefore, in this paper the process of developing ASA linked SD model to understand the construction related challenges, evolve policy interventions and take appropriate decisions is explained with case studies. The process was explained in a linear fashion with examples drawn from the literature and works of the author. However, while developing the model(s), it could be very messy and it may not represent the real life scenario. So, two important things need to be carefully considered, that are: first operational thinking as how in reality the system works and what precisely is the problem/challenge to be solved or resolved and second, perhaps a number of repetitions (iterations) among the different stages that are actually required along with continual adaptation of the real life scenario. Besides, a lot depends on the modellers and analysts, their experience and ingenuity. Besides, validation remains a crucial challenge to be resolved by the modellers before using the model for simulation and developing policy interventions.

The paper has some limitations such as it is based on the literature review and case studies depicted from published and unpublished works. Besides, due to paucity of space, the detailed explanations and sub steps were not provided. However, yet it is envisaged that ASA linked models (particularly SD model) can be a viable option to comprehend and analyse the construction challenges. The modellers can find the modelling process useful and can adapt to their modelling needs to analyse and understand the challenges of construction.

References

Abd El-Razek, M.E., Bassioni, H.A., Mobarak, A.M.: Causes of delay in building construction projects in Egypt. J. Constr. Eng. Manage. **134**, 831–841 (2008)

Aiyetan, O.A., Das, D.: Resolving contractor commitment challenges in project delivery by using conceptual system dynamics models. J. Constr. Proj. Manage. Innov. **6**(1), 1563–1582 (2016)

Aiyetan, O.A.: Das, D: System dynamics approach to mitigating skilled labour shortages in the construction industry: a South African context. Constr. Econ. Build. **18**(4), 45–63 (2018). https://doi.org/10.5130/AJCEB.v18i4.6041

Batty, M.: Urban Modelling. University College London, London, UK (2009)

Beck, M.B., Das, D., et al.: Cities as forces for good in the environment: a systems approach. In: Mensah, P. (ed.) Applying Systems Analysis to Complex Global Problems, pp. 9–40. Springer, Heidelberg (2018)

Beck, M.B.: Grand challenges of the future for environmental modelling (in the setting of NSF's environmental observatories initiatives. The University of Georgia, USA (2009). ISBN 978-1-61584-248-3

Bon Gang, H., Lay Peng, L.: Comparison of schedule delay and causal factors b tween traditional and green construction projects. Technol. Econ. Dev. Econ. **19**(2), 310–330 (2013). https://doi.org/10.3846/20294913.2013.798596

Coyle, R.G.: System Dynamics Modelling: A Practical Approach. Chapman and Hall, London (1996)

Coyle, R.G.: Management System Dynamics. Weiley, London (1977)

Das, D.: Development of mechanisms by using conceptual system dynamics models to resolve delay in construction projects. In: Proceedings of the 5th International Construction Specialty Conference of the Canadian Society for Civil Engineering (ICSC). Vancouver, Canada. pp. 279-1–279-10, 8–10 June 2015

Das, D.: System analysis and system dynamics modelling for social and ecological sustainability challenges. In: Pan, S., Mukherjee, S., Chakrobortty, D., Ganguly, S. (ed) Environment, People and Management, a Developmental Perspective, pp. 1–20. Renu Publishers, New Delhi (2016). ISBN: 978-93-85502-15-6

Das, D., Emuze, F.: A dynamic model of contractor induced delays in India. J. Constr. Developing Countries **22**(1), 21–29 (2017). http://web.usm.my/jcdc/vol22_1_2017/JCDC%2022(1)%20Art%202_early%20view.pdf

Doloi, H., Sawhney, A., Iyer, K.C., Rentala, S.: Analysing factors affecting delays in Indian construction projects. Int. J. Proj. Manage. **30**(4), 479–489 (2012)

Forrester, J.W.: Principles of Systems. Productivity Press, Cambridge (1968)

Forrester, J.W.: World Dynamics. Wright-Allen Press, Cambridge (1971)

Forrester, J.W.: Urban Dynamics. MIT Press, Cambridge (1969)

Guimarães, V.P., Tedeschi, L.O., Rodrigues, M.T.: Development of a mathematical model to study the impacts of production and management policies on the herd dynamics and profitability of dairy goats. Agric. Syst. **101**(3), 186–196 (2009)

Han, S., Love, P., Peña-Mora, F.: A system dynamics model for assessing the impacts of design errors in construction projects. Math. Comput. Modell. **57**, 2044–2053 (2013)

Laszlo, A., Krippner, S.: Systems theories: their origins, foundations, and development. In: Jordan, J.S. (Ed.) Systems Theories and A Priori Aspects of Perception. Chap. 3, pp. 47–74. Elsevier Science, Amsterdam (1998)

Love, P.E.D., Edwards, D.J., Irani, Z.: Forensic project management: an exploratory examination of the causal behaviour of design-induced error. IEEE Trans. Eng. Manage. **55**(2), 234–248 (2008)

Lyneis, J.M., Ford, D.N.: System dynamics applied to project management. Syst. Dyn. Rev. **23**, 157–189 (2007)

Mohammadi, A., Mehdi, T., Khosravi, Y.: Developing safety archetypes of construction industry at project level using system dynamics. J. Saf. Res. **67**, 17–26 (2018)

Olatunji, A.A., Das, D.: System dynamics approach to mitigating skilled labour shortages in the construction industry: a South African context. Constr. Econ. Build. **18**(4), 45–63 (2018). https://doi.org/10.5130/AJCEB.v18i4.6041

Olaya, C.: Models that include cows: the significance of operational thinking. In: Proceedings of the 30th International Conference of the System Dynamics Society. System Dynamics Group, St Gallen, Switzerland (2012)

Shen, Q., Qing, C., Tang, B.S., Yeung, S., Yucun, H., Gordon, C.: A system dynamics model for the sustainable land use planning and development. Habitat Int. **33**, 15–25 (2009)

Sterman, J.D.: Business dynamics: systems thinking and modelling for a complex world (No. HD30. 2 S7835 2000) (2000)

Sterman, J.: Business dynamics: systems thinking and modelling for a complex world. Boston: McGraw Hill (1982)

Effects of Land Administration on RDP Land and Buildings-A Case of Alexandra Township

Gladness Makgobi[1,2], Mashwama Nokulunga[1,2(✉)], Thwala Didi[1,2], and Aigbavboa Clinton[1,2]

[1] Department of Construction Management and Quantity Surveying, University of Johannesburg, Johannesburg, South Africa
kelebogilemakgobi@gmail.com, {nokulungam, didibhukut, Aigbavboa}@uj.ac.za

[2] SARChI in Sustainable Construction Management and Leadership in the Built Environment, Faculty of Engineering and the Built Environment, University of Johannesburg, Johannesburg, South Africa

Abstract. Land administration system plays a major role in legalising improvements outside the formal system through legislation and other measures. Hence, the study investigates the effects of land administration on existing RDP houses in South Africa. A quantitative approach was adopted. Structured questionnaires were designed and distributed to department of housing and to RDP home owners. 60 Questionnaires were distributed and 50 came back complete and eligible to used, this represented 83.3% of response rate. The findings on the effects of land administration were conflict among the owners and government, poor government planning, some of the properties are not evaluated, poor government record keeping on change of ownership, were some of the effects obtained from the study. Land management is a principal component of any country's administrative portfolios. Hence, land administration is a crucial procedure of governing land tenure, valuation of land, rights on land and development of land in order to formulate appropriate regulatory policies. The study will motivate that RDP houses must be valuated and registered so that beneficiaries could have title deeds. Also the beneficiaries would benefit from the study by knowing the land rights so they can value and appreciate it.

Keywords: Housing · Land administration · Land reform · Policies · RDP

1 Introduction

Land management is a principal component of any country's administrative portfolios, which includes the administration of land tenure, land valuation, land use, and development of land [1]. Furthermore, without a proper national land administration foundation a country will battle to be administered systematically. [2], stated that, land administration foundation is the strategic measure, a lawful system, an institutional plan, and specialised method. According to Department of housing [3], RDP beneficiaries are not aware of land information such as cadastral lines, rules of encroachment, the ownership rights, invasion of public land, land registration, approval of land for

C. Aigbavboa and W. Thwala (Eds.): CIDB 2019, *The Construction Industry in the Fourth Industrial Revolution*, pp. 179–187, 2020.
https://doi.org/10.1007/978-3-030-26528-1_18

consolidation and subdivisions, land and property value and the consequences of erecting illegal improvements and other property bylaws.

High rate of informal land attacks restrict attempts to timeously discharge appropriate land for human settlement in a planned way [4]. In South Africa most residents are obliged to secure shelter outside this formal framework [5]. Department of Land Reform and Rural development plays an important role to manage land tenure programmes by means of legislation [6]. Lack of administrative support on land right is seen as the main problem to management of rural land in South Africa. New systems and approaches to land use planning and cadastral techniques require monitoring on regular basis [7].

2 Land Reform Policies

The Department of Land Affairs [8] has developed three land reform policies namely: Land Tenure Reform, Restitution and Land Redistribution. Land tenure reform is a policy that acknowledges the individual's right to own land. The Restitution process deals with compensation of individuals who have been forcefully removed from their land.

Redistribution is the most important part of land reform in South Africa. Land is purchased from registered owners (willing buyer) by the government (willing purchaser) and redistributed, keeping in mind the end goal to keep up open trust in the land market. In spite of the fact that this framework has worked in different countries, in South Africa it has ended up being exceptionally hard to execute. This is on the basis that numerous owners don't really observe the land they obtained because they are not included in the essential choices made toward the purchasing and building.

3 Concept of Land Administration

The development and application of good land governance principles is necessary to ensure transparency, accountability, consistency, participation, access to clear and accurate information [9]. However, in South Africa there is a lack of solid national approach to land administration. Procedures and land information are often separated across municipalities, countries, provinces and states. The escalating theory of land administration requires the need for national systems and the verification for this involves more explanation [2]. Furthermore, land administration comprises a range of framework and procedures to administer, such as land tenure; Land value; land use and land development [10].

4 Reconstruction and Development Housing Program (RDP) Houses

The National Housing policy has an appropriate framework which provides free houses to poor household members who earns between R0 to R3500, 00 monthly [11]. The housing subsidy assistance is a once off capital sponsorship, which is registered in the name of the recipient household member, yet paid out to the developers responsible for developing (constructing), the subsidised houses (RDP). The Housing White Paper [12] additionally expresses that housing policy must be monetarily and financially economical in a long period.

According to the Department of Housing [3], the reconstruction and Development Programme (RDP) conceptualizes housing delivery is based on the following programmes:

i. Meeting basic needs;
ii. Developing human resources;
iii. Building the economy; and
iv. Democratizing the state and society.

5 Effects of Land Administration on RDP Land and Buildings

5.1 The Requirement for Improvement in Land Rights

Housing officials, policy makers are challenged with the breaking points of land and housing alternatives in South Africa [13]. In the past two decade the backlog in housing delivery has reached it height, furthermore, the quantities of those people who don't have housing just grew over night then legislation becomes unfit to take care of the demand [5]. There is no better time to display reasonable systems to empower advanced improvement in land rights and tenure frameworks [14]. Land tenure is understood to incorporate the delicate idea of view of security in land (social authenticity and importance) notwithstanding the hard idea of rights set up and covered in law (lawfulness) [15].

5.2 Enforcement of Policies and Support Services

[16] stated that the circumstance of the poor who live in the extra-legal or informal township, granting fundamentally appealing and well meaning, are possible just when certain exact conditions are available. Such as: (1) that the State is capable and willing to make the new land administration organizations broadly open to all nationals, including to the poorest and less literate; (2) that the dominant part of 'poor people' (not only a little extent of them) are capable and willing to enter effectively into the universe of formal rights and lawful establishments; and (3) that the State is enough staffed to go up against new frameworks of land administration, and adequately steady, solid and devoted to the general enthusiasm to have the capacity to change the

principles of access. [6] reported that in South Africa changes in ownership are just perceived in law in the event that they are registered. Off-register exchanges poses serious problems for the buyer and seller, for the honesty of registration system, and for government housing programmes.

5.3 Involvement of the Community in Decision Making Process

Residents drive and direct the establishment communal land boundary as indicated by their own particular ability and needs [17]. The community land boundaries is best when it is controlled by the vitality, inspiration, and endeavours of group individuals themselves who are completely mindful of the purposes for the mapping exercise. Full community involvement is fundamental. All residents must take part in community land limits procedures for it to be effective. Members of the community, including women, men, youth, older folks, customary pioneers, regular clients and individuals from minority groups ought to be invited to all meetings and urged to raise their concerns and or give opinion. Solid administrative program must be combined with heart-focused initiative [6].

5.4 A Reasonable Model of Provincial Land Tenure Security

Successful land administration is the way to secure land tenure [13]. The model was developed to embody, in a comprehensive way, the land tenure security in the rural communities of sub-Saharan Africa. Furthermore, land tenure security is a development of a land tenure framework, which is an element of the association of the framework's components in general [16]. A productive and powerful cadastral frameworks should be minimal effort so that expenses can be recouped decently and without unduly loading clients [7]. [18] highlighted that ICT can contribute in minimising the expenses related to land administration in spite of the fact that this may diminish openness for poor people. Available cadastral frameworks should be free of complex structures, methods and controls that would back off the framework and mitigate use thereof.

5.5 Insecure Land Ownership

This segment inspects the restrictions of spatial equity and land use rights in situations where title deeds have not been registered and additionally where responsibility for property is insecure [18]. In the two circumstances the property holder is not allowed to apply for building plan or land use endorsement since they don't hold the lawful status of ownership. In informal settlements, none of the plots are overviewed and registered in the deeds office (aside from on account of site and administration improvements). This instantly makes it incomprehensible for the landholder to apply for land use building plan authorisation [4].

Owners who have procured property informally can't lodge land use applications without authorisation of the deed holder. As indicated by a report by [18], 1.1 to 1.4 million housing endowment recipients don't have the title deeds to their properties. According to [19] 44% of South African family units (likely) have a title deed, and land use in South Africa has lost control. [9] stressed that blacks have land outside the formal structure, yet don't have title deeds.

5.6 Problem of Bureaucracy

Sub-Saharan Africa have a tedious and difficult management processes, which delays the house exchanges and stimulate deception [4]. In Ghana registering land that has been sold, includes many visits to separate government offices to receive the relevant document, to assure that an examination has been carried out and the assessment of the property has been undertaken. For various individuals, the procedure is tedious and it's very expensive, furthermore, procedure takes over a year and half even for individuals with connection in government and legal report [14].

6 Methodology

6.1 Research Area, Approach and Design

Quantitative approach method was adopted to investigate a stakeholder's perspective on the impact of land administration on RDP land and building in the township of Alexandra. The study was carried out in Gauteng Province of the Republic of South Africa. 60 Questionnaires were distributed and 50 were brought back which were all valid and usable and this represented 83.3% which indicates a credible study. A well-structured questionnaire was distributed to department of housing in Gauteng and occupants of RDP buildings in Alexandra township. The study was conducted from reliable scholarly sources such as articles, journals, books, publications, websites and site experience on the field.

Statistical Package for the Social Science (SPSS)
The quantitative data collected was analysed with Statistical Package for the Social Science (SPSS) a computer programme which is used for analysing data concerned with social phenomena. The software was used to generate various statistical, including descriptive statistic, which provides a basic summary of all variables in the data [20]. The benefits of using SPSS is that it allows for scoring and analysing quantitative data at speed and it can also be used to perform multivariate analysis. SPSS also helps to present the data in a logical format [20] thereby reducing time spent on calculating scores. However, accuracy in results is highly dependent on inputs, hence the need to accurately capture data from the questionnaire.

5 Point Linkert Scale
5- point linkert scale was adopted for the study which gave a wider range of possible scores and increase statistical analyses that are available to the researcher. The first linkert scale read is on agreement form as follows:

1 - Strongly Disagree (SD)
2 - Disagree (D)
3 - Neutral (N)
4 - Agree (A)
5 - Strongly Agree (SA)

The second linkert scale read is on likelihood as follows:

1 - Extremely Unlikely (EU)
2 - Unlikely (U)
3 - Neutral (N)
4 - Likely (L)
5 - Extremely Likely (EL)

The 5 point scales were transformed to mean item score abbreviated as (MIS).

Computation of the Mean Item Score (MIS)
The computation of the mean item score (MIS) was calculated from the total of all weighted responses and then relating it to the total responses on a particular aspect. The mean item score was adopted to rank the factors from highest to lowest. The Mean Item Score (MIS) is expressed and calculated for each item as follows:

$$\mathbf{MIS} = \frac{1n1 + 2n2 + 3n3 + 4n4 + 5n5}{\sum N} \quad (1)$$

Where;

n1 = number of respondents for strongly disagree
n2 = number of respondents for disagree
n3 = number of respondents for neutral
n4 = number of respondents for agree
n5 = number of respondents for strongly agree
N = Total number of respondents

Descriptive analyses was used to determine the impact/effect of land administration to RDP land and building in Gauteng; mean item scores and standard deviations were then tabulated in a descending manner to highlight the highest and the least impact.

7 Findings

7.1 Effects of Land Administration to RDP Land and Buildings

The respondents were asked based on their experience as to which factor has an impact on land administration regarding land on existing RDP. Conflicts among owners and government was ranked first with a (MIS = 4.12 & STD = 1.12); Poor government planning was ranked second with (MIS = 4,06 & STD = 0,85); Some of the properties are not evaluated and No government record being kept in relation to change of ownership was ranked third with (MIS = 4,03 & STD = 0,94; 0.8 respectively); No data in regards to government standards, approaches and land forms was ranked fourth with (MIS = 3,97 & STD = 0,72); Little or no understanding about land rights and confinements; Poor information about data of the land in the RDP and Most of the plots are not registered in the deeds office (MIS = 3,94 & STD = 0,92; 0.80; 0.81 respectively); Insecure land ownership was ranked sixth with (MIS = 3,88 & STD = 0,81); No reasonable model of provincial land tenure security and Property holders are not allowed to apply for building plan or land use endorsement was ranked seventh with (MIS = 3,85 & STD = 0.86;1,02 respectively); Poor tenure structure and land right; Most of the property owners don't have title deed and Lack of land policies and support services was ranked eighth (MIS = 3,74 & STD = 0.83; 0.93; 1.00); Natural disaster events and Too much land disputes among owners was ranked ninth with (MIS = 3,18 & STD = 1.03;1,17 respectively); No productive and powerful cadastral framework about the land and Lack of community involvement was ranked tenth (MIS = 2,93 & STD = 1.04 & 0.95); Negative strategic manoeuvre and The problem of bureaucracy was ranked eleventh with (MIS = 1,21 & STD = 0,69; 0.66 respectively) (Table 1).

Table 1. Effects of land administration to RDP land and buildings

Effects	MIS	S.D	Rank
Conflict among the owners and government	4,12	1,12	1
Poor government planning	4,06	0,85	2
Some of the properties are not evaluated	4,03	0,94	3
No government record being kept in relation to change in ownership	4,03	0,80	3
No data in regards to government standards, approaches and land forms	3,97	0,72	4
Little or no understanding about land rights and confinements	3,94	0,92	5
Poor information about data of the land in the RDP	3,94	0,81	5
Most of the plots are not registered in the deeds office	3,94	0.8	5
Insecure land ownership	3,88	0,81	6
No reasonable model of provincial land tenure security	3,85	0,86	7
Property holders are not allowed to apply for building plan or land use endorsement	3,85	1,02	7
Poor tenure structure and land rights	3,74	0,83	8
Most of the property owners don't have title deed	3,74	0,93	8
Lack of land policies and support services	3,74	1,00	8
Natural disaster events	3,18	1,03	9
Too much land disputes among owners	3,18	1,17	9
No productive and powerful cadastral framework about the land	2,93	1,04	10
Lack of community involvement	2,93	0,95	10
Negative strategic manoeuvre	1,21	0,69	11
The problem of bureaucracy	1,21	0,66	11

8 Conclusion

The findings demonstrate that the effects of land administration on RDP land and buildings were: conflict among the building owners and government, poor government planning; properties not properly evaluated; poor government record on change of ownership and poor data records to government standards, approach and land forms were the highest ranked impacts of land administration to RDP housing and land. In addition the best land administration requires that government standards, approaches and forms must be made openly available, accessible to the public in general.

The researcher recommends that models and standards approaches must be developed to control and monitor approval of construction and to manage conflicts between government and recipients of RDP houses. Land ownership must be formalised to black townships and information regarding land information must be clearly understood by recipients of RDP houses. The state must take reasonable legislative approaches and other measures to stimulate conditions which enable beneficiaries of RDP housing to gain access to land on just and equitable manner (application of sec 25 (5) of the Constitution). There is a need for proper establishment of new comprehensive land use and management regulations. Land policies and legislation must be monitored, amended and implemented. The available and newly implemented spatial management must benefits the poor.

References

1. Enemark, S.: Understanding the land management paradigm. In: FIG Commission 7 Symposium on Innovative Technologies for Land Administration. Madison, Wisconsin: International Federation of Surveyors (2005)
2. Williamson, I., Enemark, S., Wallace, J., Rajabifard, A.: Land Administration for Sustainable Development. ESRI Press, New York (2010)
3. Department of Housing. Housing Act, Act No. 107 of 1997. Department of Housing, Pretoria (1995–1997)
4. Simbizi, M.C.D., Bennett, R.M., Zevenbergen, J.: Land tenure security: revisiting and refining the concept for sub-saharan Africa's rural poor. Land Use Policy **36**, 231–238 (2014)
5. Cousins, B.: A Role for Common Property Institutions in Land Redistribution Programmes in South Africa (2003)
6. Roux, L., Barry, M.: Land Transactions in Social Housing in the Western Cape of South Africa: A Narrative of Kransdorp (2012)
7. FIG 1995: FIG Statement on the Cadastre. FIG Commission (1995)
8. Department of Rural and Land Reform. Communal Land Tenure Policy. Land Reform Policy workshop, 23–24 August 2013
9. Pienaar, G: Land information as a tool for effective land administration and development (2011)
10. Enemark, S.: Sustainable land governance: spatially enabled, fit for purpose, and supporting the global agenda. In: Annual World Bank Conference on Land and Poverty. The World Bank, Washington, D.C (2012)
11. Mashwama, N.X., Aigbavboa, O.C., Thwala, D.W.: Challenges of reconstruction and development (RDP) houses in South Africa. In: Proceedings of the International Conference on Conference on Industrial Engineering and Operations Management, Washington DC, USA, pp. 1695–1702, 27–29 September 2018
12. White Paper on Housing. https://www.housing.gov.za/Content/planned/Docs/Housing%20White%20Paper.pdf

13. Ziqubu, N., Cousins, T., Hornby, D.: Using local practices and records to secure individual tenure rights in common property situations – Lessons from the case studies on what might work on the ground. In: Department of Land Affairs' National Land Tenure Conference (2001)
14. Rubin, M.: Land management and democratic governance in the city of Johannesburg: Synthesis report. Centre for Urban and Environmental Studies (2008)
15. FAO 2007: Good governance in land tenure and administration. In: FAO Land Tenure Series no. 9. Rome (2007)
16. Hull, S., Sehume, T., Sibiya, S., Sothafile, L., Whittal, J.: Land allocation, boundary demarcation and tenure security in tribal areas of South Africa. S. Afr. J. Geomatics **5**(1), 68–81 (2016)
17. Mbarak: Towards Securing Community Land Tenure in Kenya A holistic approach to community-based natural resource governance: The case of Wayu Community in Tana River County and Lokapel Community of Turkana County in Kenya (2017)
18. Gordon, R., Nell, M., Di Lollo, A.: Investigation into the delays in issuing Title Deeds to Beneficiaries of Housing Projects funded by the capital subsidy. Urban Landmark (2011)
19. Gilbert, A., Mabin, A., Mc Carthy, M., Watson, V.: Low-income rental housing: are South African cities different? Environ. Urbanization **9**(1), 133–148 (1997)
20. SPSS. A guide to the processing, analyzing and reporting of (research) data. Wolters-Noordhoff BV: Netherlands (2004)
21. Constitution of the Republic of South Africa Act 108 of 1996 (Constitution). www.info.gov.za/documents/constitution/index.htm

Awareness of Green Building Prerequisites for Skill Development Among Built-Industry Professionals in Nigeria

Comfort Olubunmi Ade-Ojo(✉) and O. A. Awodele

Department of Quantity Surveying, Federal University of Technology, Akure, Ondo-State, Nigeria
ade-ojoco@futa.edu.ng, deledayo4christ@yahoo.com

Abstract. This paper assessed the level of awareness of the prerequisites for green building development among built industry professionals in South-West Nigeria. Data was collected using the Leadership in Energy and Environmental Design (LEED) v4. The LEED v4 Project checklist for New Construction and Major Renovation was adapted on a 5 point Likert's scale. The methods for the data analysis included the Mean Item Score (MIS) and the Kruskal Wallis *H* Test. The Kruskal Wallis *H* Test showed no statistically significant difference in the level of awareness of the prerequisites. The paper concluded that there is no significant difference in the level of awareness of the prerequisites for green building development among the built industry professionals in the study area. The study recommended the need for professional bodies to propagate the principles of green building for adequate skill development among their members.

Keywords: Awareness · Green building · LEED · Prerequisites · Professionals

1 Introduction

Green buildings are developed to leave lighter footprint on the environment. Developing a green building requires compliance with some prerequisites. Prerequisites are the basic criteria to be met before a building can be considered for green building certification. Green building certification refers to the extent to which a building complies or performs with specific environmental goals and requirements. A certified project is designed with the intention of reducing the impact of the built environment on both human health and the natural environment.

The Paris agreement was enacted to combat the devastating effect of human development on the physical environment with the attendant challenges to the socio economic life [1]. The Paris agreement requires every nation to set its own goal hence making it easy for everyone to effect change, to reduce greenhouse gas emission. Hence, different countries have green building assessment or rating tools meeting the peculiar need of the immediate community. The tools are used as a means of certifying the greenness (level of green) of a building [2–6]. However, studies have been carried out to assess the prospects of green building development in Nigeria by Dahiru [7].

C. Aigbavboa and W. Thwala (Eds.): CIDB 2019, *The Construction Industry in the Fourth Industrial Revolution*, pp. 188–196, 2020.
https://doi.org/10.1007/978-3-030-26528-1_19

Others include Onuoha's comparing efforts towards green building development in Nigeria and Malaysia [8]. Waniko [9] in agreement with Dahiru and Onuoaha noted that efforts towards green building in Nigeria is still at infancy. Therefore, the construction industry in Nigeria lacks the required skill for green building development.

The construction industry is complex due to the fragmented nature of its activities and requiring inputs from different professionals in the execution of its activities. The fragmented nature has made it imperative for every profession to redefine sustainable development along its professional obligation. This is to ensure that the client gets value for his investment while the users derive maximum comfort for using the property [10]. Therefore, the aim of the study is to evaluate the level of awareness of construction industry professionals of the prerequisites for green building development in Nigeria with a view to enhancing skill development and transfer. The Research Hypotheses states that there is no significant difference in the level of awareness of construction industry professionals of the prerequisites for green building development in Nigeria.

2 Literature

A building designed, built, operated and disposed of in a resource-efficient manner is referred to as green building. Green buildings are designed to minimize the overall (negative) impact of the building on the built environment, human health and the natural environment [11–13]. Green building practices include implementing construction techniques which reduces waste and promotes efficient utilization of construction resources. This also minimizes the ecological footprint of the built environment. Developing a green building requires that all of the design professionals work cooperatively towards a common goal from the inception of the project [14, 15].

2.1 Green Building Standards and Certification

Implementing green building development goals requires that some form of assessment tools be established. Assessment tools are developed by considering what is to be measured and how it is to be measured. Thus the application of rating systems to allocate sustainable points to different area of green building development [16]. In evaluating different tools for green roof system noted that since 1990 when the first green building assessment tools was developed, there has been tremendous increase in the number of sustainability assessment tools [4]. To develop these tools a weighting system is applied to the green building parameters culminating to 100 points altogether. Although these tools differ in concepts and principles depending on the nature and scope of its intended implementation, the basic sustainable parameters for green building development are the same [6].

2.2 Green Building Development in Nigeria

Construction projects are segregated along activities and professional lines. Hence, implementing green building practices are developed along same line of professional segregation resulting into lack of coordination of building development activities to enhance sustainable performance. Reviewing the challenges of sustainable building development, Salama and Alshuwaikhat [17] posits that there is need for a trans-disciplinary approach to green building development. In line with this, Shen et al. [10] proposed a matrix of framework of project sustainability performance factors for assessing sustainable projects from inception to completion. The framework as proposed is to provide a uniform basis for all professionals involved in the construction project to assess the sustainability parameters across the social, economic and environmental sustainability.

Furthermore, Dahiru, Dania, and Adejoh [7] assessed the prospects of green building practices in Nigeria. The study identified lack of awareness, lack of enabling policies and legislation to encourage prospective clients towards green building development and unfavorable economic situation of the country as challenges to green building development in Nigeria. In another vein, Onuoha et al. [8], comparing efforts towards green building development in Nigeria and Malaysia noted that Malaysia started towards the development of green buildings since the establishment of the 3rd Malaysian plan of 1976–1980. The study concluded that Malaysia has a long history of green building development policy. This has matured over the years while effort towards green building policy and program is still at infancy in Nigeria. Another point of note is that there has been more commitment to green building education in Malaysia than Nigeria. The lack of sustainable construction skills was also identified as a hindrance to green building development in Nigeria. Having considered the differences and similarities for green building development in Nigeria and Malaysia, the conclusion is that Nigeria should develop her own rating tools. This is premised on the basis that the use of South Africa Green Star may not have significant impact on green building development in Nigeria.

Amasuomo, Atanda, and Baird [18], comparatively analyzed the LEED and BREAM rating systems. The study noted that green building rating tools are developed for specific local jurisdictions. The study believes that concerns of the local architecture, socio-cultural and economic dynamics of Nigeria are important consideration in developing a building performance assessment tool for the country. Therefore, the LEED checklist will be adopted for the purpose of this study.

2.3 LEED Certification

The LEED is the most widely used green building rating tools due to its easy applicability (Mapp, Nobe and Dunbar) [19]. Promoting a whole building approach to sustainability, the LEED identifies five key performance areas: sustainable site development, water savings, energy efficiency, material selection and indoor environmental quality. The LEED scoring system comprises of three basic components namely; the requirements, the credits and the scores [20]. The LEED like other certification systems consists of three elements; the prerequisites, the credits and the points. LEED Prerequisites represent the minimum requirement for a building to be qualified for LEED certification. A building does not score any point for complying with prerequisites but once there is compliance, the building can then be assessed to earn credits which builds up towards earning higher levels of LEED certification. However, pre-requisites do not earn points like credits.

3 Methodology

The study is a mixed design with cross-sectional survey of built industry professionals in South-West Nigeria. the mixed design allows the use of qualitative research and generalizing findings to a large population using quantitative research [21, 22]. Cross-sectional surveys are carried out at one point in time. This is useful in getting definite (snapshot) information about a group at a particular time [23]. Hence, the research design consists of an exploratory survey on the built industry professionals in TEIs in South-West Nigeria. These are the Architects, Quantity surveyors, Builders, Town planners, Estate surveyors and Valuers, Engineers and Land Surveyors. A total number of 69 built industry professionals employed in the Physical Panning Units of the institutions were covered under the study.

The data collection instrument was a well-structured questionnaire adapted from the LEED scoring system. The questionnaire was used to assess background information of the respondents and to assess the level of awareness of the pre-requisites for green building development among the professionals. Using the LEED v4 project checklist for New Construction and Major Renovation developed by US Green Building Council [24]. Respondents were required to rank their level of awareness of green building requirements on a Likert's scale of "5" for very much aware to "1" not aware. The data collection instrument was administered on participants on one-on-one basis. To ensure the quality and accuracy of the responses, preliminary discussions were conducted with the respondents before going ahead to fill the questionnaire. This method allows the respondents to clear any doubts they have while filling the questionnaire.

Quantitative and qualitative analytical tools were employed for the data analysis. The Mean Item Score (MIS) was calculated to extract information on the level of awareness of the professionals of the prerequisites for green building development in the study area [25]. There are five prerequisites for the LEED certification. The mean score formula is given by:

$$\frac{5n_1 + 4n_2 + 3n_3 + 2n_4 + n_5}{5} \tag{1}$$

Using 5 (five) point Likert's scale a mean score below 3.00 is considered as not significant, while any mean score from 3.00 and above is considered as statistically significant [26].

4 Data Presentation and Analysis

The background information of the research population have the building industry professionals in the physical planning units of the selected TEIBs from Engineers (26.1%) to Quantity Surveyors (24.6%), Architects (17.4%), Town Planners (14.5%), Builders (8.8), Estate Surveyors & Valuers (4.3) and Land Surveyors (4.3) respectively (Table 1).

Table 1. Background characteristics of the respondents

S/N	Characteristics	Frequency	%
A.	Professional background		
	Architects	12	17.4
	Builders	6	8.7
	Estate Surveyor & Valuers	3	4.3
	Engineer	18	26.1
	Land Surveyors	3	4.3
	Quantity Surveyors	17	24.6
	Town Planners	10	14.5
	Total	69	100.0

In Table 2, Materials and Resources with a mean score of 3.47 ranked first on the level of awareness of the prerequisites for green building development. Sustainable Sites was ranked second with a MIS of 3.42 followed by Indoor Environmental Quality (3.15), while Water Efficiency and energy atmosphere had MIS of 3.11 and 2.80 respectively. Implicitly the rate of awareness of the prerequisite for Green building by the respondents was high with an overall mean value of 3.19. The table also reveals that all the items on each of the prerequisite had mean values above 2.5 on a 5 point scale of measurements. Table 2 also indicates that the Architects have the highest level of awareness with mean value of 3.52 followed by Land Surveyors with a mean score of 3.51 and Town Planners with 3.22. The Quantity Surveyors, Estate surveyor & Valuers and the Builders followed respectively.

Hypothesis Testing

The Kruskal-Wallis *H* test was applied to determine the significant difference in the level of awareness of the prerequisites for green buildings among professionals within the study area. The Kruskal-Wallis *H* test is a non-parametric statistical test of variance between different means. It is non-parametric because there is no need to assume a normal distribution unlike the One-Way Analysis of Variance (ANOVA). The test allows the comparison of more than two independent variables under a free distribution [27, 28].

The Kruskal-Wallis test statistic for k samples, each of size n_i is:

$$H = \frac{12}{n(n+1)} \sum \frac{T_i^2}{n_i} - 3(n+1) \quad (2)$$

Where n is the total number (all n_i) and R_i is the sum of the ranks (from all the samples) for the ith sample and the null hypothesis of the test is that all k distribution functions are equal.

Table 2. Level of awareness of prerequisite for GBD among construction industry professionals

S/N	Green building prerequisite	Mean within group	Overall mean	Rank	Professional background (Mean)						
					ARC	BLD	ES&V	ENG	L/SV	QS	TPL
1	**Materials and resources**		***3.47***	1	3.58	3.42	2.83	3.5	4.12	3.35	3.5
	Storage and collection of recyclables	3.48									
	Construction and demolition waste management plan	3.46									
2	**Sustainable sites**		***3.42***	2	3.83	3.5	3	3.28	3.33	3.24	3.6
	Construction activity pollution prevention	3.42									
3	**Indoor environmental quality**		***3.15***	3	3.63	2.83	3.33	2.94	3.83	2.97	3.2
	Minimum indoor air quality performance	3.19									
	Environmental tobacco smoke control	3.12									
4	**Water efficiency**		***3.11***	4	3.64	2.61	3	3.2	3.44	2.8	3.04
	Outdoor water use reduction	3.19									
	Indoor water use reduction	3.22									
	Building-level water metering	2.91									
5	**Energy and atmosphere**		***2.8***	5	2.91	2.26	2.46	2.92	2.75	2.51	2.76
	Fundamental commissioning and verification	2.67									
	Minimum energy performance	2.94									
	Building-level energy metering	2.68									
	Fundamental refrigerant management	2.92									
					3.52	**2.92**	**2.93**	**3.17**	**3.51**	**2.97**	**3.22**

Key: QS- Quantity Surveyors, ARC- Architects, BLD-Builders, TPL-Town Planners, ES&V-Estate surveyor & Valuers, L/SV- Land Surveyors, Eng-Engineers.

The Kruskal-Wallis *H* test showed that there was no statistically significant difference in the level of awareness of the prerequisites for green building development among construction industry professionals in South-West Nigeria, $x^2(6) = 9.888$, $p = 0.129$, with a mean rank of awareness score of 27.1 for Architect, 24.1 (Land Surveyor), 19.2 (Town Planner), 17.6 (Engineer), 13.0 (Quantity Surveyor), 12.9 (Builder) and 12.1 for Estate Surveyor and Valuer. Given the confidence interval at 0.05 alpha levels, the null hypothesis was accepted. It was therefore concluded that there is no significant difference in the level of awareness of the prerequisites for green building development among Construction Industry Professionals in South-West Nigeria (Table 3).

Table 3. Kruskal-Wallis *H* test on the level of awareness of prerequisite for green building development among construction industry professionals

Professional (Grouping variable)	Mean rank	Ranking	Chi-square	df	Asymp. Sig.
Architect	27.1	1	9.888	6	0.129
Land Surveyor	24.1	2			
Town Planner	19.2	3			
Engineer	17.6	4			
Quantity Surveyor	13.0	5			
Builder	12.9	6			
Estate Surveyor	12.1	7			

Discussion of Findings

The survey result shows that the professionals are statistically aware of the prerequisites for green building development. This is ranked in order from Materials and Resources (3.47), sustainable sites (3.42), Indoor Environmental Quality (3.15), Water Efficiency (3.11) and Energy and Atmosphere (2.80) respectively. The levels of awareness of the prerequisites were generally above average of 2.5 mean item score. However the level of awareness of the prerequisites for Energy and Atmosphere (2.80) was not significant falling below 3.00 decision rule. The awareness of Storage and collection of recyclables (3.48) followed by construction and demolition waste management plan (3.46) significantly increased the awareness for Materials and Resources.

Assessing the level of awareness of the prerequisites along professional lines showed that the professionals were generally above average with land Surveyors ranking first, followed by Architects and Engineers and having mean item score (MIS) of 3.45, 3.32 and 3.08 respectively. Three of the professionals namely, the Builders, Estate surveyor & Valuers and the Quantity Surveyors ranked from the rear with 2.92, 2.93, 2.97 respectively. The import is that the level of awareness among this group of professionals is not statistically significant. However, the result indicates that the professionals are not totally ignorant of the prerequisites for the green building development as previously observed [9, 29].

Furthermore, the Kruskal –Wallis *H* test on the awareness of the prerequisites showed no significant variation among the professionals. The result however corroborates the study by [30] which reported an improvement on the level of awareness of green building concepts as a greater percentage of the professionals surveyed were familiar with the concept.

Conclusion and Recommendation

Emanating from the data analysis, the study concludes that the level of awareness for the prerequisites is generally above average. And that some of the built industry professionals are more aware of the prerequisites (such as Architect, Land Surveyor, Town Planner and Engineers) than others like the Quantity Surveyor, Estate Surveyors & valuers and Builders respectively. This is disapproves the claim that building industry professionals are ignorant of the principles of green building. it however shows distinctively the level of awareness of each profession. The study therefore recommends that there is need to propagate the principles of green building development among the Built-industry industry professionals across the country and for each profession to develop requisite professional skills for green building development.

References

1. United Nations: "Framework Convention on Climate Change". In: Conference of the Parties on its twenty-first session, vol. 01194, pp. 1–36, January 2015
2. U.S. Green Building Council, LEED ® for New Construction & Major Renovations, October 2005
3. Vierra, S.: Green Building Standards and Certification Systems, National Institute of Building Sciences, Washington, DC, September 2016
4. Fauzi, M.A., Malek, N.A.: Green building assessment tools: evaluating different tools for green roof system. Int. J. Educ. Res. **1**(11), 1–14 (2013)
5. CIDB: Best Practice: Project Assessment Scheme (2011)
6. Pandey, S.: MIT-UTM, "Impact of Green Building Rating Systems on the Sustainability and Efficacy of Green Buildings: Case analysis of Green Building Index," Malaysia (2016)
7. Dahiru, D., Dania, A.A., Adejoh, A.: An investigation into the prospects of green building practice. J. Sustain. Dev. **7**(6), pp. 158–167 (2014)
8. Onuoha, I.J., Kamarudin, N., Aliagha, G.U., Okeahialam, S.A., Atilola, M.I., Atamamen, F.O.: Developing policies and programmes for green buildings: what can Nigeria learn from Malaysia's experience? Int. J. Real Estate Stud. **11**(2), 50–58 (2017)
9. Waniko, D.P.: Green building in Nigeria emerging opportunities for the quantity surveying profession. Academia, 1–7 (2012)
10. Shen, L., Hao, J.L., Tam, V.W., Yao, H.: A checklist for assessing sustainability performance of construction projects. J. Civ. Eng. Manage. **13**(4), 273–281 (2007)
11. Kats, G., Capital, E., Alevantis, L., Berman, A., Mills, E., Perlman, J.: The costs and financial benefits of green buildings (2003)
12. Langdon, D.: Examining the Cost of Green, Santa Monica, California (2004)
13. Langdon, D.: The cost & benefit of achieving Green buildings, Australia (2007)
14. US EPA: Green Building (2011)
15. World Green Building Council: the Business Case for Green Building (2013)

16. Poveda, C.A., Young, R.: Potential benefits of developing and implementing environmental and sustainability rating systems: making the case for the need of diversification. Int. J. Sustain. Built Environ. **4**(1), 1–11 (2015)
17. Salama, A.M., Alshuwaikhat, H.: A trans-disciplinary approach for a comprehensive understanding of sustainable affordable housing. Glob. Built Environ. Rev. **5**(3), 35–50 (2006)
18. Amasuomo, T.T., Atanda, J., Baird, G.: Development of a building performance assessment and design tool for residential buildings in Nigeria. Procedia Eng. **180**, 221–230 (2017)
19. Mapp, C., Nobe, M.C., Dunbar, B.: The cost of LEED—an analysis of the construction costs of LEED and NON-LEED banks. J. Sustain. Real Estate **3**(1), 254–273 (2011)
20. U.S. Green Building Council: LEED v4 User Guide (2013)
21. Creswell, J.W.: Research Design: Qualitative, Quantitative, and Mixed Methods Approaches, 4th edn. SAGE Publications Inc, Los Angeles (2013)
22. Mawlood, B.H.: Pragmatic approach to research. Education Slideshare, p. 13 (2017)
23. Mathers, N., Fox, N., Hunn, A.: Surveys and Questionnaires, Sheffield (2009)
24. U.S. Green Building Council, LEED v4 for Building Design and Construction (2017)
25. Sambasivan, M., Soon, Y.W.: Causes and effects of delays in Malaysian construction industry. Int. J. Proj. Manage. **25**(5), 517–526 (2007)
26. Kothari, C.R.: Quantitative Techniques, 3rd ed. Vikas Publishing House Pvt Limited, New Delhi (2009)
27. Hole, G.: The Kruskal-Wallis test, vol. 10, pp. 1–10 (2011)
28. Cheng, G., et al.: Nonparametric Statistical Methods, p. 110 (2014)
29. Abolore, A.A.: Comparative study of environmental sustainability in building construction in Nigeria and Malaysia. J. Emerg. Trends Econ. Manag. Sci. **3**(6), 951–961 (2012)
30. Nduka, D.O., Ogunsanmi, O.E.: Construction professionals' perception on green building awareness and accruable benefits in construction projects in Nigeria. Covenant J. Res. Built Environ. **3**(2), 30–52 (2015)

A Review of the Advantages and Disadvantages of the Use of Automation and Robotics in the Construction Industry

Josephine Llale(✉), Matleko Setati, Success Mavunda, Thando Ndlovu, David Root, and Paulin Wembe

University of the Witwatersrand, 1 Jan Smuts Avenue, Braamfontein, Johannesburg 2000, South Africa
{9205371p, 707344, 832137, 1062488, david.root}@students.wits.ac.za, david.root@wits.ac.za, pwembe@gmail.com

Abstract. The construction industry is hazardous and dirty. Automation and robotics in construction (ARC) were introduced to execute tasks that are difficult for humans and to reduce the number of incidents on construction sites. This study discusses the advantages and the disadvantages of the use of ARC in the South African construction industry. A narrative literature review was conducted. Given that the construction industry is a wide concept, that includes amongst others; materials, processes and site activities, the study only considers the use of ARC as it relates to site activities in South Africa. In spite of the many advantages that ARC has to offer the South African construction industry, ARC is yet to be widely adopted in South Africa. ARC can potentially reduce the number of fatalities and improve efficiency and productivity on site. However, high costs of adoption and the loss of jobs are some of the factors that are hindering adoption. The construction industry in South Africa needs structural changes in order to benefit from the use of ARC. The results of the study will benefit stakeholders in the construction industry in South Africa, particularly construction companies and the government.

Keywords: Automation · Construction industry · Robotics · Site activities

1 Introduction

The construction industry is known to be hazardous and dirty [1–3]. It is estimated that "*one in every six work-related fatal accidents occurs on a construction site*" [4: 2]. 258 accidents and 56 fatalities in the construction industry were reported in 2012 in South Africa [5]. ARC was introduced to reduce the number of incidents on site and to execute tasks that are difficult for humans [6]. In comparison to human labour, using ARC on construction sites, has other advantages such as increased productivity and efficiency, better quality, improved working environment, speed, reduced construction cost and time, speed and construction processes that are environmentally friendly or sustainable [7–16].

In spite of ARC's advantages, adoption in South Africa, is slow. High cost of ARC [11, 14] and job losses [17–20] are some of the barriers to wider adoption. [21]

C. Aigbavboa and W. Thwala (Eds.): CIDB 2019, *The Construction Industry in the Fourth Industrial Revolution*, pp. 197–204, 2020.
https://doi.org/10.1007/978-3-030-26528-1_20

estimates the unemployment rate in South Africa at around 27,5%. The South African government aims to reduce the unemployment rate to 6% by 2030 [18]. Further job losses are therefore unwanted. Wider adoption of ARC and the South African government's objective of reducing unemployment presents South African construction industry with the challenge of; how can it enjoy the benefits of wider adoption of ARC while creating jobs?

This paper discusses the advantages and disadvantages of the use of automation and robotics (ARC) in the South African construction industry. Since, the construction industry is a wide concept that includes amongst others materials, processes and site activities, the study is only concerned with the use of automation and robotics as it relates to site activities in South Africa.

A narrative literature methodology was followed to review literature on the advantages and disadvantages of the use of automation and robotics in the South African construction industry. To search for the following main terms; robotics, automation, construction industry, occupational health and safety, Google Scholar and the Wits Library were used as the main database for the search. Articles that were considered for review for this paper were those whose full text was available. The selection of relevant articles to include in this review paper was guided by the aim of the study.

The advantages and disadvantages of the use of ARC are discussed under the following sections; (Sect. 2), what is automation and robotics in construction (ARC)? (Section 3) the advantages of the use of ARC, (Sect. 4) disadvantages of the use of ARC, and (Sect. 5) Conclusion.

The findings of this review paper are that in spite of the advantages that the use of automation and robotics bring on site such as, safety and increased productivity, wider adoption in South Africa is still lagging. Wider adoption of ARC should take into consideration the South African's government's objective of creation of jobs. Labour intensive projects should incorporate ARC that work with human labour such as teleoperated machines and programmable construction machines to minimize potential job losses and to align with South African's government's objective.

In spite of its benefits, ARC is expensive and can potentially lead to job losses. For wider adoption, structural changes in the South African construction industry might be necessary. Stakeholders in the construction industry should find ways to remove structural barriers that may be preventing wider adoption of ARC such as how the construction industry is arranged, composed, and organized. This will also include (1) evaluating materials used and (2) off and on-site processes and activities. Further research in the wider adoption of robotics and automation in developing countries is recommended.

2 What is Automation and Robotics in Construction (ARC)?

There is no universal definition of ARC. [22] describe automation as "*the use of technology to facilitate the work of human or extend their physical and mental capacity.*" [23] in [11: 23] define ARC as "*a machine-centered construction factory technology for applying robotic systems on the construction field.*" [20: 169] define automation as, "*the use of control systems and information technologies to reduce the need for human work in the production of goods and services*". [24] describe ARC as the application of computer-controlled processes, mechanization concepts and technologies in construction. In this paper, ARC, is described as [24].

Different types of ARC are classified according to their interaction with humans [24]. The first type, teleoperated systems, is a type of ARC where humans use a remote control to operate the machines [24]. The second type are programmable construction machines, this type of ARC enables humans to perform certain tasks by choosing pre-programmed menu or function for them [24]. The third type are called intelligent systems, these include semi-autonomous or full autonomous machines that can operate without humans or with limited human interaction [24]. In this paper, unless otherwise stated, no distinction is made between the different types of ARC.

3 The Advantages of Using ARC

Literature provides many advantages of the use of ARC [7–16] This paper only discusses three advantages of ARC. Advantages of ARC in, (1) improving safety, (2) increasing productivity and (3) construction processes that are environmentally friendly (sustainable) and their potential implication on the South African construction industry are discussed.

3.1 Safety

The most common forms of incidence on construction site include, (1) slips and falling, (2) falls from extreme heights, (3) electrocutions, (4) structural collapses, (5) machinery failure, (6) welding emissions, (7) lead, (8) unguarded machinery, (9) being struck by heavy construction equipment and moving vehicles or falling objects, (10) asphyxiation and similar [25, 26]. Proponents of ARC claim that the use of ARC on construction site has the potential to reduce such incidents by carrying work in environments that are unsafe for humans to do so [7, 10, 14, 24].

In South Africa, being struck by (44%), falling on to different levels (14%) and striking against (10%) are the three major causes of injuries on construction sites [27]. On the other hand, motor-vehicle accidents (MVAs) (47%), struck by (17%) and falls on to different levels (17%) accounted for the most fatalities [27]. If claims from literature about ARC's ability to improve safety on construction sites are reliable, then stakeholders in the South African construction industry need to adopt ARC technologies and methods that can reduce incidents identified by [27] on construction sites. Semi-autonomous and autonomous robotics should be considered for unsafe work environments. This might potentially reduce the number of incidences on construction sites as machines instead of human labour will be exposed to dangerous situations.

3.2 Increased Productivity

There are bricklaying robotics that are able to build a complete house in two days as compared to weeks and months by human labour [12, 16]. Human labour productivity can be negatively affected by tiredness, fatigue and absenteeism [22, 28, 29]. In comparison ARC seems to be more productive than human labour. For instance, a bricklaying robot called Hadrian 105 can work 24 h a week and lay 1000 brick per hour in comparison to human labour at 100–200 bricks per hour [16].

[9] claims that by using contour crafting an entire house can be completed in a few hours. Contour crafting is a 3D printing technology that is capable of printing a house on site [16]. By using a gantry system carrying the nozzle that moves on two parallel lanes, a 200 m^2 double storey can be built in 2 days [9]. Figure 1 shows a construction of a building using contour crafting.

Using machines such as Hadrian 105 and contour crafting, the urban infrastructure backlog can potentially be reduced. To what extend can the South African housing backlog, estimated at 2,1 million [30], be reduced by using bricklaying machines such as Hadrian 105 or contour crafting which can build an entire house within 2 days?

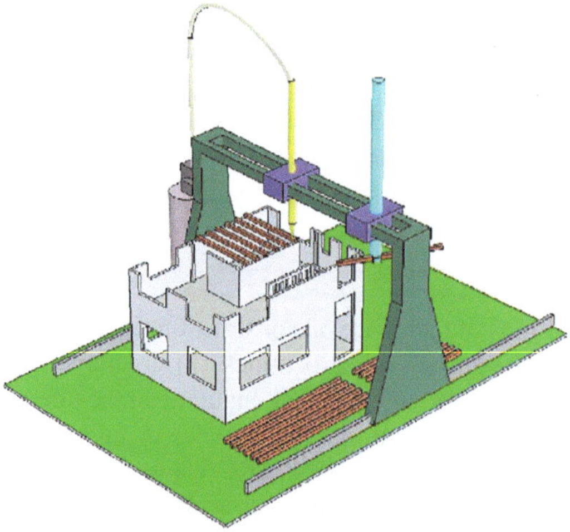

Fig. 1. Construction of a building using contour crafting [9]

3.3 Sustainability

The construction industry is the leading consumer of raw material in the world [31]. 25–40% of the world's total carbon emissions are attributed to constructed objects [31]. [32] claim buildings in South Africa account for over a quarter of all national greenhouse emissions. A potential solution to carbon and greenhouse emissions is ARC. Contour crafting (see Fig. 2a and b) is regarded as an environmentally friendly building alternative to traditional building [7, 9].

Contour crafting creates an environmentally friendly environment by reducing emissions and material waste [9]. An emission free environment is created by the use of electric contour crafting machines that are emission free [9]. Material waste is reduced by the use of computer to determine, for example, the exact amount of concrete to be deposited into the contour crafting machine for specific locations in the building [9].

(a) (b)

Fig. 2. (a) The first structure to be printed, (b) 3D printing in progress [33] in [7]

Unless, South Africa embraces environmentally friendly construction materials and processes, near zero energy buildings [32], that is buildings that contribute less greenhouse emissions, will not be achieved.

4 The Disadvantage of Using ARC

As there are advantages of ARC, there are also disadvantages to the use of ARC on construction site. Disadvantages include potential job losses, high capital and maintenance cost, unavailability of technology, constant change in technology and inflexibility [7, 11–14, 24, 34]. This paper discusses the disadvantages of ARC in relation to potential job losses and high capital cost.

4.1 Job Losses

The implementation of ARC on construction site has the potential to result in job losses. In the manufacturing industry, jobs were lost because of automation and robotics [17, 18]. The manufacturing and automotive industries are often cited as examples of job losses resulting from the use of automation and technology. For example "*some 22 million manufacturing jobs were lost globally between 1995 and 2002 as industrial output soared 30%*" [19]. Low skilled and low wage occupations jobs are usually adversely affected by automation [20].

Through the Expanded Public Works Programme (EPWP), the South African government aims to reduce unemployment by creating temporary employment for the unemployed [35, 36]. EPWP creates low-skilled job opportunities through labour-intensive projects [37]. If literature findings on the use of ARC are to be relied upon, the use of ARC on construction sites might be at odds with the job creation aims of the South African government. The possibility of job losses might also be one of the reasons for the slow adoption of ARC in the South African construction industry.

4.2 High Capital Cost

High capital cost of ARC is a major barrier to the wider adoption of ARC [14]. Small and medium size companies dominate the construction industry [11]. Many of these companies do not have the financial means to invest in ARC [11].

The Construction Industry Development Board (CIDB) grades contractors in South Africa according to the value of construction work a contractor is deemed capable of performing within a certain class of works [38]. In South Africa, it is compulsory for contractors who want to participate in public construction projects to apply for a CIDB grading [38]. Grading ranges from level 1 to level 9. A level 1 contractor, for instance, can only undertake works whose value does not exceed R200 000 [38]. There is no limit for a grade 9 contractor in terms of the value of work they can undertake [38].

Of the 94 792 contractors registered with the CIDB, 84 770 had a CIDB grading of Level 1 [39]. This means that the majority of contractors in South Africa are small contractors and may, therefore, be unable to afford to invest in ARC. How will the South African construction industry enjoy the benefits of the wider adoption of ARC while creating jobs? How can barriers to wider adoption such as high cost of ARC be resolved? These are some of the questions that need to be answered if the South African construction industry is to promote safety, increased productivity and sustainability on construction sites through the wider adoption of ARC.

5 Conclusion

In this paper the advantages and disadvantages of the use of robotics and automation in the South African construction industry were discussed. Some of the advantages include safety, increased productivity and sustainability. The disadvantages include possible job losses and high capital cost. Possible solutions to benefit from the advantages of the use of robotics and automation while creating new jobs should include structural changes in the construction industry. Wider adoption will possibly include how the South African construction industry is structured and organized. It might also include an evaluation of materials used and processes involved on and off construction sites.

Even though job creation is an important imperative for the South African government, safety of construction workers and providing adequate infrastructure for its citizens are also important imperatives that need consideration. Labour intensive projects can possibly incorporate the use of tele-operated or programmable construction machines that allow for the interaction of human and machines on construction sites. Semi- and fully autonomous machines could possibly be used in work environments that are unsafe for human labour. The use of bricklaying machines and contour crafting could possibly be used to reduce the urban infrastructure backlog. Unless, however, stakeholders in the construction find a solution for reducing high cost of ARC, wider adoption of ARC will remain a challenge.

This paper recommends further research on the wider adoption of construction robotics and automation in developing countries such as South Africa in order to draw comparisons.

References

1. Wang, J., Zou, P.X., Li, P.P.: Critical factors and paths influencing construction workers' safety risk tolerances. Accid. Anal. Prev. **93**, 267–279 (2016)
2. Li, H., Lu, M., Hsu, S.-C., Gray, M., Huang, T.: Proactive behavior-based safety management for construction safety improvement. Saf. Sci. **75**, 107–117 (2015)
3. Fang, D., Wu, H.: Development of a safety culture interaction (SCI) model for construction projects. Saf. Sci. **57**, 138–149 (2013). https://doi.org/10.1016/j.ssci.2013.02.003
4. International Labour Organization (2005). https://www.ilo.org/wcmsp5/groups/public/—dgreports/—dcomm/documents/publication/wcms_067574.pdf
5. Death and injury stalk construction sites | Safe Build Forum. http://forum.safebuild.co.za/index.php?threads/death-and-injury-stalk-construction-sites.758/
6. Cottle_Edward_M.pdf. http://repositorio.unicamp.br/bitstream/REPOSIP/286408/1/Cottle_Edward_M.pdf
7. Hager, I., Golonka, A., Putanowicz, R.: 3D printing of buildings and building components as the future of sustainable construction? Procedia Eng. **151**, 292–299 (2016)
8. Pan, M., Linner, T., Pan, W., Cheng, H., Bock, T.: A framework of indicators for assessing construction automation and robotics in the sustainability context. J. Cleaner Prod. **182**, 82–95 (2018)
9. Khoshnevis, B.: Automated construction by contour crafting—related robotics and information technologies. Autom. Constr. **13**, 5–19 (2004). https://doi.org/10.1016/j.autcon.2003.08.012
10. Elattar, S.M.S.: Automation and robotics in construction: opportunities and challenges. Emirates J. Eng. Res. **13**, 21–26 (2008)
11. Chen, Q., de Soto, B.G., Adey, B.T.: Construction automation: research areas, industry concerns and suggestions for advancement. Autom. Constr. **94**, 22–38 (2018)
12. Oesterreich, T.D., Teuteberg, F.: Understanding the implications of digitisation and automation in the context of industry 4.0: a triangulation approach and elements of a research agenda for the construction industry. Comput. Ind. **83**, 121–139 (2016)
13. Kamaruddin, S.S., Mohammad, M.F., Mahbub, R.: Barriers and impact of mechanisation and automation in construction to achieve better quality products. Procedia – Soc. Behav. Sci. **222**, 111–120 (2016). https://doi.org/10.1016/j.sbspro.2016.05.197
14. Mistri, P.S., Rathod, H.A.: Remedies over barriers of automation and robotics for construction industry. Int. J. Adv. Res. Eng. Sci. Manage. 1–4 (2015)
15. Vähä, P., Heikkilä, T., Kilpeläinen, P., Järviluoma, M., Gambao, E.: Extending automation of building construction—survey on potential sensor technologies and robotic applications. Autom. Constr. **36**, 168–178 (2013)
16. Shinde, V., Sarode, G.: Robotics application in construction industry. Int. J. Eng. Manage. Res. (IJEMR) **8**, 16–18 (2018)
17. How Technology Is Destroying Jobs. https://www.researchgate.net/publication/295614289_How_Technology_Is_Destroying_Jobs
18. Cerasis_IT: Industrial Automation: Brief History and Use in Manufacturing (2014). https://cerasis.com/industrial-automation/
19. McAfee, A.: Manufacturing Jobs and the Rise of the Machines (2013). https://hbr.org/2013/01/manufacturing-jobs-and-the-ris
20. Frey, C.B., Osborne, M.A.: The future of employment: how susceptible are jobs to computerisation? Technol. Forecast. Soc. Change **114**, 254–280 (2017). https://doi.org/10.1016/j.techfore.2016.08.019

21. P02113rdQuarter2018.pdf. http://www.statssa.gov.za/publications/P0211/P02113rdQuarter2018.pdf
22. Figueiredo, A.E., Carvalhal, R. de D., Hoeflich, S., Figueiredo, L., Pereira, S.L., Dias, E.M.: Port operation–increase of automated systems, decline of workforce jobs? In: Proceedings of 19th International Conference on Circuits, Systems, Communications and Computers (CSCC 2015), Recent Advances in Computer Science, pp. 259–266 (2015)
23. Jung, K., Chu, B., Hong, D.: Robot-based construction automation: an application to steel beam assembly (Part II). Autom. Constr. **32**, 62–79 (2013). https://doi.org/10.1016/j.autcon.2012.12.011
24. Ardiny, H., Witwicki, S., Mondada, F.: Construction automation with autonomous mobile robots: a review. In: 2015 3rd RSI International Conference on Robotics and Mechatronics (ICROM), pp. 418–424 (2015). https://doi.org/10.1109/ICRoM.2015.7367821
25. International Labour Organization. https://www.ilo.org/wcmsp5/groups/public/—ed_dialogue/—sector/documents/instructionalmaterial/wcms_161770.pdf
26. ElSafty, A., ElSafty, A., Malek, M.: Construction safety and occupational health education in Egypt, the EU, and US firms. Open J. Civil Eng. **2**, 174 (2012)
27. Construction Health and Safety in South Africa.pdf. http://www.cidb.org.za/publications/Documents/Construction%20Health%20and%20Safety%20in%20South%20Africa.pdf
28. Koehn, E.E., Kothari, R.K., Pan, C.S.: Safety in developing countries: professional and bureaucratic problems. J. Constr. Eng. Manage. **121**, 261–265 (1995). https://doi.org/10.1061/(ASCE)0733-9364
29. Hsu, P.-Y., Angeloudis, P., Aurisicchio, M.: Optimal logistics planning for modular construction using two-stage stochastic programming. Autom. Constr. **94**, 47–61 (2018). https://doi.org/10.1016/j.autcon.2018.05.029
30. African News Agency: Housing backlog at 2.1 m, says Minister Sisulu. https://www.engineeringnews.co.za/article/housing-backlog-at-21m-says-minister-sisulu-2016-04-22/rep_id:4136
31. WEF_Shaping_the_Future_of_Construction_full_report__.pdf. http://www3.weforum.org/docs/WEF_Shaping_the_Future_of_Construction_full_report__.pdf
32. Ampofo-Anti et al.: Potential for reducing greenhouse gas emissions in the South African Construction Sector. https://www.sagreenfund.org.za/wordpress/wp-content/uploads/2016/05/CSIR-Potential-for-reducing-GHG-emissions-in-the-South-African-construction-sector.pdf
33. Rudenko 3D Printer. http://www.totalkustom.com/gallery.html
34. Effects of Automation on Construction Industry Performance. https://www.researchgate.net/publication/319199428_Effects_of_Automation_on_Construction_Industry_Performance
35. EPWP_as_a_Catalyst_for_Development-Mr_Dladla.pdf. https://www.parliament.gov.za/storage/app/media/ProjectsAndEvents/2017_ncop_provincial_week_october/docs/EPWP_as_a_Catalyst_for_Development-Mr_Dladla.pdf
36. EPWP Policy Brief.pdf. https://evaluations.dpme.gov.za/images/gallery/EPWP%20Policy%20Brief.pdf
37. EPWP_Infrastructure_Guidelines_3rd_Edition_June2015.pdf. http://www.epwp.gov.za/documents/Infrastructure/Infrastructure%20incentive%20manual/2015/EPWP_Infrastructure_Guidelines_3rd_Edition_June2015.pdf
38. About the RoC. http://www.cidb.org.za/contractors/Pages/Register-of-Contractors.aspx
39. Greve, N.: CIDB Grade 1 offers 'no barrier' to industry entry, needs rehaul. http://www.engineeringnews.co.za/article/cidb-grade-1-offers-no-barrier-to-industry-entry-needs-rehaul-2014-04-08

Corporate Governance and REITs Performance

Moloi Nosipho and Akinsomi Omokolade(✉)

Faculty of Engineering and Built Environment,
School of Construction Economics and Management,
University of the Witwatersrand, Johannesburg, South Africa
0412472x@students.wits.ac.za,
Kola.akinsomi@wits.ac.za

Abstract. South African Real Estate Investment Trusts (SA REIT) is prominent and favorable to other emerging markets, since inception the returns on investments have increased with a market capitalization of R400 billion in 2018 as compared to other markets globally. South Africa (SA) has the most advanced Corporate Governance (CG) mechanism, thus contributions to literature on CG has become vital; any collapse of South Africa's CG mechanism has far dire impact beyond the African continent to the rest of the world. For analysis, the Generalised Method of Moments is in use. The data is source from McGregor. The study period is from 2008 to 2017. Results proved that CGI has a positive but statistically insignificant increase on SA-REIT performance.

Keywords: SA-REITs performance · Corporate Governance

1 Introduction

Numerous studies have compared the performance of Real Estate Investment Trust (REITs) with those of other financial assets and markets; these have reached various conclusions globally, by [1–8]. The literature concentrated on the commercial REITs and the impact of Corporate Governance (CG) on its performance, with similar outcome that better governed corporations award investors with greater performance also proven within the South African context by [9, 10]; on CG studies that compliance affords promising returns to the investor.

The Problem Statement

Overall literature has proven a positive statistically noteworthy connection between performance and decent CG practice, Rahman et al. enlightened that depending on the context and performance measures adopted; there might be variations as this was a case in China [11]. In contrast to other emerging economies Ntim attested that South African corporations with good CG practice in place tend to perform better than their counterparts, a positive link of TQ with CG [9] and Pamburai et al. found an optimistic relationship of TQ and the portion of lower level directors [10]. In addition, Manskemp et al. employed models that indicated a positive alpha for portfolios of companies containing a high CG practice [12]. In addition, Hormark, reported a positive alpha for

C. Aigbavboa and W. Thwala (Eds.): CIDB 2019, *The Construction Industry in the Fourth Industrial Revolution*, pp. 205–214, 2020.
https://doi.org/10.1007/978-3-030-26528-1_21

the JSE, but the results were different for Brazil, Russia, India and China when the CAPM was employed [13].

A comprehensive CG mechanism is mandatory is when analyzing the link between performance and CG to capture the multifaceted concept of CG as reported by [14], employing a comprehensive CG mechanism makes way for any interdependent multi-facet of CG [15]. Similar to other studies the present study also adopts the CGI covers normal CGI attributes [16]. Other non-SA studies have adopted academic indices by [17] and or market indices by [18]. The academic indices are difficult to repeat [19], with corporate governance mechanism for each organization, and the other indices are westernized. The present study is some-what dissimilar to previous studies [20], as it adopts SA setting attributes, these include non-discriminative and provisions from stakeholders, in addition to these HIV/AIDS and Employment equity are included.

Research Questions

1 How Corporate Governance policies associate with the performance of SA-REITs?

Sub-question: How does Corporate Governance's comprehensive index provisions affect the performance of both an internally and externally managed SA-REITs?

Objectives of this Study

1. To assess SA REITs variation in performance and if it is related Corporate Governance Policies.
2. To evaluate the performance of SA REITs and its relation to Corporate Governance's 50 provisions contained in the index taken from King II report for SA listed corporations.

The Hypothesis Formulation

Alreemy et al. utilized certain provisions contained in the CG index [21], with a variation in results to [24] as they adopted a comprehensive CG index which proved that corporation with a good CG in place are likely to perform better for S-REITs.

In addition, Larcker et al. contend that CG mechanism is complex, and to capture any provisions that are interdependent may be a challenge [14], thus a comprehensive CG index is mandatory [8], context specific attributes should be included in the proxy [20]. H1: The performance of the internally and externally managed SA-REITs is influence by the comprehensive CGI indices contained in the King II report.

Aim

To examine the CG policies adopted by SA-REITs with more emphases on the provisions contained therein. Thus, develop a model to examine the reasons behind the significant correlation of the performance and CG provisions.

Limitations

The sample size might require improvement, and data may not be available for all corporations before the King III and King IV reports.

Corporate Governance
The South Africa's model is unique; it is in a form of King Reports that from the Anglo-American style [9]. Similar to the UK and the US, the first CG mechanism is from the UK framework, later improved to King II, III and IV to incorporate SA specific context.

REITs
Much of the South African based CG studies challenges are embedded on the availability of information, some corporations types are excluded in various studies of [9–13]. However, on REITs, not much work in the industry. The SA REIT portfolio is smaller in proportion to other JSE portfolios, ever since inception REITs portfolio has outperformed stocks, T-Bills and so forth [23]. The SA-REIT market capitalization in 2018 was R400 billion and are committed to good governance, reported by [24]; furthermore SA-REIT is favorable compared to other emerging markets [25].

REIT is a listed property venture that consists of different corporations that obtain their earnings through administration, operation and proprietorship of the property investment; its inception was in the 1960s. In South Africa REIT was implemented in 2013 Prior to the introduction of REITs the Property unit trust (PUTs) and property loan stock (PLS), these were unevenly regulated, hence the introduction of REITs that was more flexible, transparent, simple and with tax certainty [25]. As of January 2018, there were thirty-one (31) listed REITs corporation on JSE as reported by the [24, 25].

2 Literature Review

The literature concentrated on the performance of commercial REITs and CG, better governed corporations award investors with greater performance as proven by [9, 10] and also be affording promising returns to the investor. Lacomte & Ooi and Chong et al. elaborate that CG index with all country specific provisions is mandatory to better govern organizations and thus improve performance [22, 26].

Performance of REITs and Corporate Governance
CG and REITs performance studies in developed nations by [18, 27, 28] focused on individual CG provisions or aggregate CG measures. Previously, normally board and audit committee were in previous studies as variables, which gave limited analyses on the effectiveness on CG and REITs. Emerging economies have limited information or specific measure for CG, Asian Pacific Real Estate Association (APREA) introduced this measure in 2010 that was originally introduced by the European Public Real Estate Association (EPRA), hence the framework has been applied ever since. Lecomte and Ooi adopted the APREA framework in this respect, suggested that this framework is mandatory for other markets; they found an optimistic association between CG and stock performance [22].

An Indian study was done to evaluate the effectiveness of the APREA framework, Das and Thomas mentioned that even though these studies were conducted in Asian markets this will not determine the effectiveness of the framework to measure CG and REITs in India [29]. A study in Singapore by [26]. Conflicted the results in [22] even though in a similar region as India, this study proved otherwise that CG has no noteworthy effect on the performance of REITs, value of REITs, and return on asset.

Lecomte and Ooi suggested that current organization and ownership of REIT could not bring in enough investor interest in the investment market; these devalue the reduction of the worth of REITs this implied that REITs should reconsider redefining their management arrangement [22]. Yung et al. focused on CEO over confidence lead to faults and advantage and decisions made to buy back shares [40].

Chong et al. cited that REITs executives face considerable charges when they wish to reach a balanced level to agency charges, CEO studies are important to determine buyback decision while other focused more on shareholding and managers [26]. As mentioned above managing a portfolio can be expensive, and achieving equilibrium issues is still daunting with finance models.

It has been proven that better governed South African corporations is linked with better financial performance as reported by Pamburai et al. and Ntim found a positive link between TQ an accounting measure with CG [29, 37].

The African continent has different challenges to developed nations such as the unstable political climate, poverty, HIV/AIDs, fluctuating economic climate and many more, each country in Africa is experiencing the some of these or more [30–32].

3 Research Methodology

CG financial data is from the *Perfect Information Database* (PID) to assess the connection between CG and SA-REITs performance. The price rise rate, its inverse function of exponent change of the CPI, the Treasury Bill (RTS) as its present ratio less its annual moving average, and the interest rate factor is one to 9 years. The real estate returns volatility is 12 months moving standard deviation, all from McGregor. The study period spans from 2008 to 2017.

Effectiveness of Corporate Governance on REITs

Generalized method of moments was adopted by [26] similar to the present study, as it is more robust than ordinary least squares (OLS) and panel data. Many researchers choose board and audit committee as CG measures for performance like [33, 34]. A comprehensive CG index is prepared to provide a full picture of CG incorporating SA context attribute [14–16].

The independent variable is the CGI to help understand the connection between CG and SA-REIT performance. 50 provisions are taken from the King II report which make the CGI, the 50 provisions are made from five wide-ranging sections enclosed in the King II report; thus all corporations must comply or state reasons for non-compliance. This study is unique but in line with previous studies, the CG mechanism contains board of directors and ownership [16], as it adopts variables like EE, HIV/AIDS and Affirmative Action. If a corporation has five provisions, then a value of one is given or zero otherwise. It is mandatory to note that the scoring of a corporation can vary in a year from zero to containing all 50 provisions.

To test effectively the impact of CG mechanism on SA-REIT, panel econometric analysis is applied. The key ingredient in the regression models is CG provisions of the individual SA-REITs, the below equation is from [22]:

$$R_{i,t} = \alpha + \beta_1 CGI_{i,t-1} + \beta_2 in(Size_{1,t}) + \beta_3 MTB_{i,t} + \beta_4 D/E_{i,t} + \sigma X_t + e_{i,t} \quad (1)$$

CGI index in year t-1 for SA-REIT, this was adopted by [35] furthermore Ooi and Liow suggested that also several variables should be incorporated that could influence returns [36]. For the present study, the variable revenue (*REV*) also shown in above equation as Rit represents the returns on investment.

The size of the corporation and growth prospects are controlled by the normal log of SA-REIT I's commercial capitalization in year t and the ratio of SA-REIT i's market value of assets in year t. The D/E $Ratio_{t,i}$ represents the firms use leverage, the debt asset ratio for $SA\text{-}REIT_i$ $year_t$ as a control variable.

To capture the period fluctuating effects, for instance credit availability and costs X_t shows the series.

There are two return test for performance. $R_{i,t}$, is the constantly compounded for an extra holding time on yield for SA-REIT i for time t. The initial calculation of the yearly variations in the specific SA-REIT price thereafter accustomed for the conforming market return. The equation is as follows:

$$R_{it+1} = (P_{it+1}/\ R_{it-1}) - (M_{t+1}/M_{t-1}) \quad (2)$$

Where Pi, t is the price of stock of *i*th REIT at time t, and the price of stock index at time t is represented by Mt. As a robustness test, the influence of CG on the financial advances of SA-REITs using Jensen alpha (αi), resultant from the below market model:

$$R_{it+1} - Rf_i t = \alpha_i + \beta(R_{mt} - R_{ft}) + \varepsilon_t \quad (3)$$

$R_{i,t}$ is the annual return on SA-REIT stock. Annual returns are represented by $R_{i,t,}$ $R_{f,t}$ and $R_{m,t.}$ Where $R_{i,t}$ is on the SA-REIT stock, and $R_{f,t}$ is on non-risk Treasury bill and $R_{m,t}$ is the market portfolio. If alpha is not expressively diverse from nil, then that particular portfolio is earning standard returns [36].

4 Results and Analysis

The Economic theory suggests that the relationships we are interested in are dynamic, especially in panel data [37].

Descriptive Statistics

The variable *REV* or Rit ranges from 1.000 to 4640000. With an average of 250543.5 this indicates some form of variation of our samples data. The independent variable *REV* and independent variables *SIZE, MT, DE* and *CGI* show wide variation in data, similar to [22, 26, 29]. Basic relationship between variables is evident; and to control highly skewed data by taking logarithms making them easier to interpret intuitively as in this case (Table 1).

Table 1. Descriptive statistics

	REV	SIZE	MT	DE	CGI
Mean	509355.4	5639184.	1271.792	1039.574	0.741667
Median	250543.5	4149080.	973.5000	1.109524	1.000000
Maximum	4640000.	45702000	9485.000	122372.6	1.000000
Minimum	1.000000	870.0000	1.000000	0.009014	0.000000
Std. dev.	787163.2	7144736.	1429.476	11170.04	0.439554
Skewness	3.098933	3.111877	3.189634	10.81464	−1.104211
Kurtosis	14.04764	14.44630	15.22695	117.9737	2.219282
Jarque-Bera	802.3191	848.7640	950.9666	68433.94	27.43325
Probability	0.000000	0.000000	0.000000	0.000000	0.000001
Sum	61122650	6.77E+08	152615.0	124748.9	89.00000
Sum sq. dev.	7.37E+13	6.07E+15	2.43E+08	1.48E+10	22.99167
Observations	120	120	120	120	120

Normality Test

The variables *REV, SIZE, MT, DE* are not normally distributed; however, *CGI* is close to 3.7, though still not normally distributed. The skewness test should be near zero (0) if the value is positive it means that the data is tilt to the right, as shown below. The independent variable *REV* and the dependent variables *SIZE, MT* and *DE* are tilt to the right and the independent variable *CGI* is tilt to the left (Fig. 1).

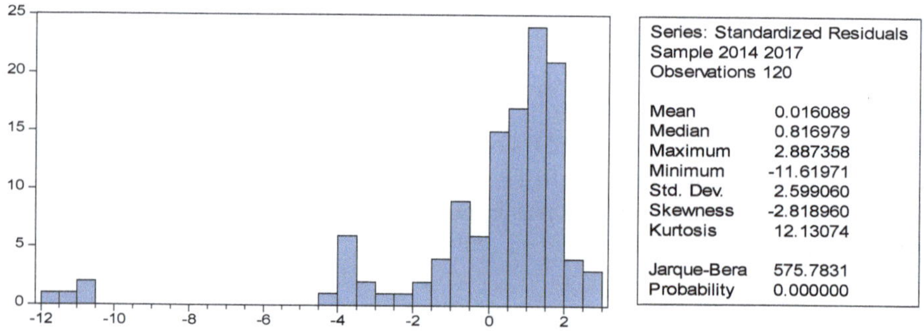

Fig. 1. Normality test

Unit Root Test

Non-stationarity in panel time-series is a natural extension and the concepts of integration and co-integration, thus variables that are non-stationary can lead to spurious regression. The Levin et al. tests are employed [38]; the p-value for *REV* is 0,000, which implies stationarity. When p- value is less than 10%, it means that variable is stationary [38]. The results at first order for variables *SIZE, MT* and *DE* are 0.000, 0.0942 and 0.0684, respectively.

To test for robustness of the above results [34], the first order results show that *REV, SIZE, MT and DE* are stationary, at 0.0001, 0.00038, and 0.3413 and 0.8458, respectively. At intercept for *MT* the result is 0.0000 and *DE* 0.0525.

Correlation Matrix

The correlation matrix shows the relationship amongst variables as stipulated below. The variable *SIZE* and *CGI* positively correlate to *REV*, respectively at 93% and 19%. However, *MT* and *DE* both negatively correlate to REV at −0.02% and −0.03%. In addition, Lecomte & Ooi proved a relationship among similar variables, with *SIZE* and *CG* provisions.

Panel Generalize Method of Moments

Nickell attests to GMM-type estimators, which make use of the optimal number of instruments, as variables tend to break down when they are believed to be I(1) [39]. With Instrumental Variable Techniques (IVS) to prevent serial correlation amongst the variables, as other techniques will lead to spurious results. If the probability coefficient value should be less than 10%, it means the variable is statistically insignificant. The variable *SIZE* has optimistic impact on *REV*. Therefore, if *SIZE* were to increase by 1% the *REV* will increase by 80%. However, *MT* has a negative impact that is statistically insignificant; a 1% increase will lead to 11% decrease in REV. On the other hand, *CGI* has an optimistic but statistically irrelevant effect on *REV*, which means a 1% increase in *CGI* would constitute a 37% increase in *REV*. However, *DE* positively influences *REV* but insignificant statistically, a 1% increase will lead to 28% increase in *REV*. R^2 should be above 50%, however this instance it is 27%; however this is not a problem, also note that Ooi and Liow has reported somewhat similar findings [36] even though Ntim had adverse results [9].

Panel generalized method of moments				
Variable	Coefficient	Std. error	t-statistic	Prob.
LOG (SIZE)	0.804353	0.077477	10.38187	0.0000
LOG (MT)	−0.111208	0.148187	−0.750459	0.4545
CGI	0.374461	0.633442	0.591152	0.5556
LOG (DE)	0.287302	0.119026	2.413774	0.0174
R-squared	0.274436	Mean dependent var		11.43033
Adjusted R-squared	0.255671	S.D. dependent var		3.051313
S.E. of regression	2.632504	Sum squared resid		803.8892
Durbin-Watson stat	0.336493	J-statistic		0.295361
Instrument rank	5	Prob (J-statistic)		0.586806

5 Conclusions

This paper considers CG and SA-REITs performance. The literature discovered that changes in economics affect the performance of REITs. Much literature had similar findings with minor variations, there is substantial work done on CG and various corporations on JSE. Literature has proved that there is statistical relationship between CG and REITs, this proved to be the similar with empirical results, even though CGI has a positive influence on SA-REITs, performance it does it is not significant. The literature has proven an optimistic association with CG and performance [22], though depending on the framework adopted and context they may be variations in results [31]. This study employed a similar framework to [9]; however, the present results proved otherwise.

Proven models are employed to test as to why there is a statically association with performance and CG practice. The SA context is different, as corporation have to comply with the SA CG mechanism that introduced in a King Reports formats. Even though the SA REIT has proven to be of interest to foreign investors, the contextual obligations like BEE, CSR, Affirmative Action and so forth. SA corporations are required to comply with the CG mechanism and JSE regulations.

References

1. Hoesli, M., Oikarinen, E.: Are REITs real estate? Evidence from international sector level data. J. Int. Money Financ. **31**(7), 1823–1850 (2012)
2. Chiang, M., Tsai, I., Sing, T.F.: Are REITs a good shelter from financial crises? Evidence from the Asian markets. J. Prop. Invest. Financ. **31**(3), 237–253 (2013)
3. Antonakakis, N., Floros, C.: Dynamic interdependencies among the housing market, stock market, policy uncertainty and the macro economy in the United Kingdom. Int. Rev. Financ. Anal. **44**, 111–122 (2016)
4. Betrand, P.: Residential real estate investment. Aix-Marseile Graduate School of Management, AIX-Marseile University (2016)
5. Ma, Y., Xu, B., Xu, X.: Real estate confidence index based on real estate news. Emerg. Mark. Financ. Trade **1**, 1 (2016)
6. Delfim, J.C., Hoesli, M.: Risk factors of European non-listed real estate fund returns. J. Prop. Res. **33**(3), 190–213 (2016)
7. Kawaguchi, Y., Sa-Adu, J., Shilling, J.D.: REIT stock price volatility and the effects of leverage. Real Estate Econ. **45**(2), 452–477 (2017)
8. Ciesielska, D., Koltuniak, M.: Outward foreign direct investments and home country's economic growth. Phys. A Stat. Mech. Appl. **482**(1), 127–146 (2017)
9. Ntim, C.G.: An integrated corporate governance framework and financial performance in South African-listed corporations. South Afr. J. Econ. **81**(3), 373–392 (2013)
10. Pamburai, H.H., Chamisa, E., Abdulla, C., Smith, C.: An analysis of corporate governance and company performance: a South African perspective. South Afr. J. Acc. Res. **29**(2), 115–131 (2015)
11. Rahman Belal, A., Rezaul Kabir, M., Cooper, S., Dey, P., Ahmed Khan, N., Rahman, T., Ali, M.: Corporate environmental and climate change disclosures: empirical evidence from Bangladesh. In: Research in Accounting in Emerging Economies, pp. 145–167. Emerald Group Publishing Limited (2010)

12. Mans-Kemp, N., Erasmus, P.D., Viviers, S.: Does investing in sound corporate governance pay? A South African study. South Afr. J. Bus. Manag. **48**(2), 33–43 (2017)
13. Hörnmark, P.: Responsible investment: should investors incorporate ESG principles when investing in emerging markets. Master's thesis. Jönköping International Business School, Jönköping, Sweden (2015)
14. Larcker, D.F., Tayan, B.: CEO attributes and firm performance. Corporate Governance Research Initiative (2017)
15. Beiner, S., Drobetz, W., Markus, M., Zimmermann, H.: An integrated framework of corporate governance and firm valuation. Eur. Financ. Manag. **12**(2), 249–283 (2006)
16. Black, B., Love, I., Rachinsky, A.: Corporate governance indices and firm's market values: time series evidence from Russia. Emerg. Mark. Rev. **7**(4), 361–379 (2006)
17. Bianco, C., Ghosh, C., Sirmans, C.F.: The impact of corporate governance on the performance of REITs. J. Portfolio Manag. 175 (2007)
18. Bauer, R., Eichholtz, P., Kok, N.: Corporate governance and performance: the REIT effect. Real Estate Econ. **38**(1), 1–29 (2010)
19. Daines, R.M., Gow, I.D., Larcker, D.F.: Rating the ratings: how good are commercial governance ratings? J. Financ. Econ. **98**(3), 439–461 (2010)
20. Klapper, L.F., Love, I.: Corporate governance, investor protection, and performance in emerging markets. J. Corp. Financ. **10**, 703–728 (2004)
21. Alreemy, Z., Chang, V., Walters, R., Wills, G.: Critical success factors (CSFs) for information technology governance (ITG). Int. J. Inf. Manage. **36**(6), 907–916 (2016)
22. Lecomte, P., Ooi, J.T.L.: Corporate governance and performance of externally managed Singapore REITs. Real Estate Financ. Eco. **46**, 664–684 (2013)
23. Ntuli, M., Akinsomi, O.: An overview of the initial performance of the South African REIT market. J. Real Estate Lit. **25**(2), 365–388 (2017)
24. Real Estate Investment Trust Association (2018). www.sareit.com. Accessed 22 Aug 2018
25. Acheampong, P., Du Plessis, P., Newell, G.: The real estate markets in South Africa. J. Real Estate Lit. **1**(2), 279–294 (2002)
26. Chong, W.L., Ting, K.H., Cheng, F.F.: The impacts of corporate governance on the performance of REITs in Singapore. J. Real Estate Lit. **24**(2), 317–344 (2016)
27. Sirmans, C.F.: Research on REIT corporate governance. Real Estate Times 75–79 (1997)
28. Ooi, J.T.: Managerial opportunism and the capital structure decisions of property companies. J. Prop. Invest. Financ. **18**(3), 316–331 (2000)
29. Das, P., Thomas Jr., C.R.: Strategic development of REITs in India. J. Real Estate Lit. **24**(1), 103–131 (2016)
30. Rossouw, G.J., Watt, A.V., Malan, D.P.: Corporate governance in South Africa. J. Bus. Ethics **37**, 289–302 (2002)
31. Markus, S.: Corporate governance as political insurance: firm-level institutional creation in emerging markets and beyond. Soc.-Econ. Rev. **6**(1), 69–98 (2007)
32. Esser, I.M., Dekker, A.: Dynamics of corporate governance in South Africa: broad based black economic empowerment and the enhancement of good corporate governance principles. J. Int. Com. L. Tech. **3**, 157 (2008)
33. Yermack, D.: Higher market valuation of companies with a small board of directors. J. Financ. Econ. **40**, 185–211 (1996)
34. Guest, P.M.: The impact of board size on firm performance: evidence from the UK. Eur. J. Financ. **15**(4), 385–404 (2009)
35. Ambrose, B.W., Pennington-Cross, A.: Economies of scale on multi-product firms: the case of REITs. Published Working Paper, Real Estate Research Institute, Bloomington, IN (2000)
36. Ooi, J., Liow, K.H.: Risk-adjusted performance of real estate stocks: evidence from developing markets. J. Real Estate Res. **26**(4), 371–396 (2004)

37. Pesaran, M., Shin, Y.: Generalise impulse responses: analysis in multivariate models. Economics **58**, 17–29 (1998)
38. Levin, A., Lin, C., Chu, C.: Unit root test in panel data: asymptotic and finite-sample. J. Econ. **8**(1), 1–24 (2002)
39. Nickell, S.: Biases in dynamic models with fixed effects. Econometrica **49**(6), 1417–1426 (1981)
40. Yung, K., Li, D.D., Sun, Q.S.: CEO overconfidence and financial policies of real estate investment trusts (REITs). J. Prop. Res. **32**(4), 384–406 (2015)

Repositioning Waste Management Architecture for Sustainable Upstream Performance in Lagos, Nigeria

Akintade Samuel Afolayan(✉)

Department of Estate Management,
University of Lagos, Akoka Yaba, Lagos 101017, Nigeria
akinafolayan@gmail.com, aafolayan@unilag.edu.ng

Abstract. The Lagos Mega City witnessed a breakdown of refuse upstream evacuation mechanism more than ever before. New and sustainable approach needs to be found for improvement, in the face of looming environmental health hazards. This study examines public policy on solid waste management in Lagos Waste Management Authority (LAWMA) enabling law, 2007 and explores human and technological resources available for a sustainable operation. It also utilises interviews with refuse evacuation stake holders; the waste-cart pushers and survey on households to appraise the current state of refuse management. The study discovers a massive displacement of grass-root mechanisms which could have been harnessed for effective waste evacuation. It also finds that scale of plant was underutilised and wrongly applied, resulting in inefficiency of Private Sector Participation (PSP) refuse collectors. The study recommends a restructuring of the evacuation mechanisms through means of the 'refuse sale' approach on three major levels.

Keywords: Refuse Cart Pushers · Refuse sale · Waste collection · Waste generation · Waste management

1 Introduction

The built environment is a product of the construction industry in its old, refined and conceptual models. The aim and a major indicator of the success of the construction industry is the wellbeing and satisfaction of man in his basic, social, recreational and economic uses of built environment. This is where design, construction and management of the created structures and infrastructure become very critical. While the technicalities of design and the construction stages determine the prospects of actualisation of the physical structures, the management and administration stage determine the level at which the aims of all inputs into the creation of built environment would be achieved. Management or/and administration was acknowledged as the 'software' aspect of built environment and thus styled it "*humanware*", as distinct from other physical components of the built environment which was referred to as 'hardware' elements [1]. The source agreed that these 'soft issues' constituted a key factor based on which users make their decisions about the overall quality of structures.

C. Aigbavboa and W. Thwala (Eds.): CIDB 2019, *The Construction Industry in the Fourth Industrial Revolution*, pp. 215–224, 2020.
https://doi.org/10.1007/978-3-030-26528-1_22

Management and administration occur at the post-occupancy stage of the value chain. Most crucial at this stage is environmental maintenance for livability which, is anchored much on performance of the structures created, as well as health and safety of users towards achieving satisfaction therein. Environmental user satisfaction and well-being is a composite of ratings from several domains. These include those of housing, health and environment [2]. [3] had also earlier established positive correlations between the well-being of man and his physical environment which, today is the focus of upgrade in Construction 4.0 framework. Environmental factors and conditions are very crucial in the determination of satisfaction and subjective well-being of man in built environment [4]. One of the major determinants of user satisfaction and well-being as a focus of construction industry is cleanliness and sustainable solid waste management (SWM). However, urbanisation and its accompanying complexities tend to weigh down our cities' environment in Africa. A major sector usually affected in this process is in maintenance of cleanliness. Lagos mega-city, Nigeria exhibits a typical failure of SWM among African cities. Unfortunately, Nigeria as a nation, was also attested to as witnessing a regime of refuse generation which outweighs the evacuation rate [5]. Even the whole of the global south was acclaimed to be bedeviled with inequitable and inconsistent solid waste collection services resulting from rapid urbanization [6].

Lagos state had landfills and dumpsites including those of Olusosun, Solous, Ikorodu, and Igando as well as new constructions in Epe to handle SWM. The Lagos State Waste Management Authority (LAWMA) in conjunction with Private Sector Participation (PSP) operators appeared working hard to rid the State of filth. Notwithstanding these, the streets are filled with refuse, creating midstream complications. The problem, however, is not a midstream issue per se. The preponderance of refuse on the streets exhibits the failure of the upstream (source-centre) evacuation mechanism. The focus of this paper is thus, on the enhancement and sustainability of upstream evacuation processes. Furthermore, only household wastes, light commercial and institutional wastes which usually find their way, albeit illegally, onto the streets are considered. Industrial and medical wastes which have specialised channels of disposal are left out. Also, the efficiency or otherwise of the downstream sector are not covered here.

2 Statement of Research Problem

The current view of refuse on the streets of Lagos mega-city is unprecedented. The efforts of LAWMA and PSP operators among others have not yielded positive results for residents' satisfaction and well-being enough to justify the aim of immense activities of the construction industry. A common feature at such refuse heaps is the batching of components in one form of container or the other ranging from cellophane wraps, baskets, sacks to other forms of characteristically poor bagging meant to be disposed of, alongside the contained refuse. The implication of this is that such refuse had been carried over a distance to be dumped there. If there had been effective evacuation machinery at the various source-centres, the illegal spots would not have received waste products.

The situation had degenerated so much that it gave birth to the whims of various schools of thought. Three of such schools could be identified here. The Public Policy School took advocacy approach through erection of signposts canvassing against dumping of refuse in some locations while, in the Political School, evolving effective SWM approach had become a campaign slogan for political aspirants. In the School of Traditions, bizarre slogans came up threatening wrath of the gods on those illegally dumping refuse in certain locations. If all efforts of the evacuation and street cleaning agents had failed to guarantee clean and healthy environment, big epidemics loom. The effects of this would be felt not only in Lagos but all over Nigeria and across West African subregion from where traders move in and out of Lagos.

3 Methodology

The research is exploratory and adopted a survey approach encompassing service of questionnaire, interview as well as direct and indirect observations. Towards achieving its objectives, the paper firstly perused the LAWMA enabling Law of 2007. The provisions therein for public policy execution and management of SWM as well as enablement under the law were identified. The in-situ operation and strategies of the SWM operatives were explored to see both conceptual and operational lacuna which led to the Cleaner Lagos Initiative (CLI) of 2017. The process further evaluated the current practices to identify shortfalls in the achievement of operational visions towards an emergence of remedial actions as well as new innovations to arrive at a truly cleaner Lagos.

Questionnaires were served on systematically sampled residents in the Lagos Mainland local government area on the observations about SWM operatives, the household attitudes and response as well as operational machinery of refuse evacuation. Interviews were also held with other stakeholders along the value chain including private sector participation (PSP) group and refuse cart pushers (RCPs). A total of 136 copies of the retrieved questionnaire from households were coded and analysed for the study.

4 Excerpts from Lagos Waste Management Authority Law 2007

The Lagos Waste Management Authority Law 2007 is a 30-section promulgation by the State House of Assembly [7]. The S.1 (1) of the law created Lagos Waste Management Authority (LAWMA) as a body Corporate with perpetual succession and common seal. It was empowered in S.1 (2) c as capable of holding, purchasing, acquiring and disposing of movable and immovable property for the purpose of its duties. S.2 (3) a-i specified the composition of the Board of Directors of the Authority as encompassing a broad spectrum of officials including the Commissioner for the Environment and even a representative of the State University's Department of Civil Engineering. The S.4 (1) a-h earmarked duties for the Authority. The Body was also empowered to prepare and update from time to time the master plans for waste collection and disposal in the State. It was further to draw agreement with, license and regulate acts of private-sector waste collectors.

The law gave the Board a somewhat blank cheque to 'do all such acts as are necessary or incidental to the proper discharge of its duties under this Law.' Section 5 spelt out powers of the Board of LAWMA. Section 5(1) (j) granted the Board the power to acquire any land for its purpose, in accordance with existing Laws of the State. It also had powers to provide other related SWM services not mentioned in the law. The S.9 to S.11 dealt with operation license for waste collectors. The procedures for licensing and prescription of fees payable for same were also contained herein. The mode of transporting and dumping of wastes were covered in S.14 and S.15 of the law. Monitoring, enforcement and stipulation of offences as well as penalties and modes of seeking redress under the law were spelt out in Sections 17–19 of the Law.

5 Findings and Discussions

The following were findings from analysed data and the discussions on them.

5.1 Demographic Details of Respondents

The demographic details of respondents in the study area are as summarised in Table 1 below. From the Table, 58.8% of the respondents were female as the male took up

Table 1. Demographic details of respondents

Characteristics		Frequency	%
A: Gender	Male	56	41.2
	Female	80	58.8
	Total	136	100.0
B: Education status	O',level	54	39.7
	NCE/OND	31	22.8
	HND/B.sc	9	6.6
	Others	42	30.9
	Total	136	100.0
C: Employment status	Employed	42	30.9
	Self employed	54	39.7
	Artisan	31	22.8
	Others	9	6.6
	Total	136	100.0
D: Length of residency in the area	Below 5 years	25	18.4
	5–10 years	58	46.6
	11–15 years	36	26.5
	21–25 yrs	17	12.5
	Total	136	100.0
E: Average number of residents in building	1–10	9	6.6
	11–20	31	22.8
	21–30	54	39.7
	Above 30	42	30.9
	Total	136	100.0

41.2%. This is particularly significant in that female members of the households were more directly involved and thus knowledgeable in refuse disposal as an appendage of home management.

The spread of educational attainment by the respondents endowed them to understand the importance of environmental cleanliness. Furthermore, 87.5% of the respondents were old residents, quite familiar with developments in refuse collection situation therein over time. The table also shows that the building (residences) were well populated with attendant implications for refuse generation. In line with [5], factors including those in the table above and related household features are germane in influencing the per capita waste generation rate in urban environment.

5.2 The Enabling Law

By provisions of S.1(2) c, the Board could acquire land within each district for primary pooling of wastes from which PSP operators could load for onward conveyance to landfills. The specification of membership of the Board was broad-based as given in S.2(3) a-i. While the inclusion of the representative from Civil Engineering Department of the State University was in order, the non-inclusion of representative from the Department of Environmental Management appeared to be an oversight. So also, was the silence about representation from Lagos State Environmental Protection Agency (LASEPA). The provisions of S.4 (1) d allowing for updates on SWM masterplan have implications for strategic management. The provisions for licensing of evacuation operators criminalised refuse cart pusher (RCP) operators. It also provided the pedestal for festering refuse-related corruption within the local government area. That the criminalisation did not deter the RCPs' actions or/and patronage by households suggests an overlooked importance of their roles in upstream evacuation and general contribution to urban economy.

5.3 Major Sources of Refuse Generation

The respondents' perception of the major sectors contributing to refuse generation in the neighbourhood was summarised in Table 2 below. Here, the general notion of population size in the city ranked first with a mean value of 3.812. The population of Lagos which was put at 22million about 2011 constitutes a formidable base for generation of wastes. [8] attested to the importance of overall population in generation of refuse in Lagos Metropolis. Households' lifestyle was identified as another contributor to refuse generation here. It ranked second with a mean figure of 3.620. Household food items were identified to include direct farm produce the consumption of which, brings into town both edible food and its waste-coverings.

Economic and trade activities ranked next with a mean value of 3.482. By the nature of land use planning in the study area, most neighbourhoods have mixed residential cum light commercial zonation. Small and medium Enterprises (SMEs) generate immense waste. Furthermore, roughages from outside the State come into town in form of packages for food and commercial items. Social gatherings were highlighted

Table 2. Identified major sources of refuse generation

Source	Opinion	Frequency	%	Cum. %	Mean	Ranking
Overall population activities	Strongly agree	40	29.4	29.4	3.812	**1st**
	Agree	52	38.2	67.6		
	Undecided	28	20.4	88.0		
	Disagree	11	8.4	96.3		
	Strongly disagree	5	3.7	100.0		
	Total	**136**	**100.0**			
Life style of households	Strongly agree	46	33.8	33.8	3.620	**2nd**
	Agree	44	32.4	66.2		
	Undecided	24	17.6	83.8		
	Disagree	15	11.0	94.8		
	Strongly disagree	7	5.2	100.0		
	Total	**136**	**100.0**			
Economic/trade activities	Strongly agree	70	51.5	51.5	3.482	**3rd**
	Agree	53	39.0	90.5		
	Undecided	13	9.5	100.0		
	disagree	00	0.0	100.0		
	Strongly disagree	00	0.0	100.0		
	Total	**136**	**100.0**			
Social gatherings/parties	Strongly agree	48	35.3	35.3	3.412	**4th**
	Agree	49	36.0	71.3		
	Undecided	21	15.4	86.7		
	Disagree	14	10.3	97.0		
	Strongly disagree	4	3.0	100.0		
	Total	**136**	**100.0**			
Institutions	Strongly agree	40	29.4	29.4	3.402	**5th**
	Agree	52	38.2	67.6		
	Undecided	20	14.7	82.3		
	Disagree	11	8.1	90.4		
	Strongly disagree	5	3.7	94.1		
	Missing item	8	5.9	100.0		
	Total	136	100.0			

with mean figure of 3.412 at the fourth position. Activities which involve bringing people together in celebrations, protests, crusades and several outdoor engagements generate considerable refuse. Also identified were wastes from institutions such as schools and colleges, business centres, offices and markets.

6 Assessment of Solid Waste Evacuation Practices in Lagos

The mechanisms for evacuation of wastes in the homes and streets included the direct official pick-ups along the streets by LAWMA, the quasi-formal arrangement of PSP operators and the informal RCPs harvesting wastes from the homes. Another approach was in the form self-help efforts by households throwing wastes into gutters and on the roads.

6.1 The PSP Operators, in Visiting the Homes Hardly Kept to Schedule

The system of poor pre-evacuation storage of refuse in the homes, and the unstandardised nature of refuse containers and vans which, necessitated manual loading of refuse, were further inhibitions to successful evacuation. The payment system was through regular monthly billings on households. However, public policy came to a head in 2017 when the status of PSP operators was changed to Waste Collection Operators (WCO) to deal with only commercial wastes. A new WCO (Visionscape Sanitation Company) took their place under a new Cleaner Lagos Initiative (CLI). LAWMA was only left to regulate the operations of the WCOs. Confusion was created in the process. Litigations came up and waste collection efficiency came to a halt.

6.2 The Unofficial Refuse Cart Pushers Were Another Set of Refuse Evacuation Operators

These were the group referred to by [6] as informal waste pickers, the local realities which conventional urban waste management largely ignored. They were persecuted by public policy for their penchant for tipping collected refuse on the streets and blocking up most drainage channels. Local governments saw them as evasive and uncontrollable. However, the interview with a group of the RCPs showed that they were plagued by problems, wriggling free from which would determine their economic survival. In the interview, they admitted it was a very viable business notwithstanding the persecution from public policy and exploitation by local government officials. However, because of its criminalisation, RCP operations could not be captured under government's local tax net. They were a critical group variously perceived either as solution providers or culprits in SWM situations in Lagos State. In fact, notwithstanding their prominence and inputs into refuse evacuation in the urban area, the RCP elsewhere, as in Lagos, were relegated in favour of foreign and imported technology [9]. The payment system was direct cash from the households though as extra burden on the households in addition to regular LAWMA bills. Based on some selected broad criteria, the summary of ratings and perception of respondents about RCP operators are as given in Table 3 below.

From the Table, it was agreed by 93.38% of respondents that RCP operators were very close to the homes. They were, in most cases, more regular in patronage than PSP operators. They could also access channels and alleys more easily than the big PSP trucks. However, they were susceptible to exploitation by local government officials. In order to make as many trips as possible to cover both payment to local government touts and also earn a living, they were always illegally tipping refuse anywhere.

Table 3. Respondents' perception of Refuse Cart Pushers (RCPs)

Characteristics		n	%
A: The closeness to households	Agree	127	93.38
	Neutral	2	1.47
	Disagree	7	5.15
	Total	136	100.0
B: Regularity of operations	Agree	129	94.85
	Neutral	7	5.15
	Disagree	0	0.0
	Total	136	100.0
C: Ease of access to households	Agree	117	86.03
	Neutral	10	7.35
	Disagree	9	6.62
	Total	136	100.0
D: Susceptibility to exploitation by local government officials	Agree	82	60.30
	Neutral	30	22.06
	Disagree	24	17.64
	Total	136	100
E: Contribution to illegal dumping	Agree	105	77.21
	Neutral	16	11.76
	Disagree	15	11.03
	Total	136	100.0
F: Overall prospects for improved refuse evacuation operation	Agree	76	55.88
	Neutral	39	15.44
	Disagree	39	28.68
	Total	136	100.0

77.21% agreed that the RCP operators contributed much to blocked gutters and drains in the Metropolis. On the whole, only 28.68% of the respondents disagreed that RCPs could, in the end, be the solution to refuse problems in the state, given strategic handling.

7 Conclusion

The rate of refuse generation was high in Lagos because of the volume of population as well as interpersonal actions in social, economic and institutional settings. The rate of household refuse generation was higher than its evacuation. The performance of PSP operators was poor. RCP's activities were also criminalised. Refuse evacuation mechanism ran into hitches due to policy somersault, misapplication of plant scale and shortfalls in the application of the enabling laws. There were also inadequate zonal refuse transfer centres. Available ones were far from what small-scale refuse collectors could access without motorised appliances. Corruption was noticed at grassroots as perpetrated by rent-seeking local government officials. This aided cart pushers' resolve to dump refuse just anywhere, in order to make marginal adjustments in earnings. The study also noted the strong survival instinct of cart pushers as an attribute which, could be harnessed for public good as entrepreneurial base for sale of refuse.

8 Recommendations

The LAWMA enabling Law contained several empowerment clauses which, the Board needed to tap into. With the finding that RCPs were not deterred by public policy persecution, and with the hindsight of trust the households had in them, they should have been considered a ready manpower block to tackle refuse evacuation. They could be organised, trained, registered and equipped for refuse evacuation from the homes into neighbourhood loading centres from where PSP trucks would load. This would curtail the waste of PSP plant time in the epileptic journey around homes.

The rewarding system for reformed RCP operations should be such that the operators would not be paid any money by the households. They should, instead, be reimbursed at the loading centres to the tune of weighted and recorded amount of sourced refuse brought in. The payment period could be weekly or any stretch of time considered equitable enough to create an atmosphere of dignified, salaried workmanship. The PSP operators and other WCOs too should not enjoy any blanket contract sum but be paid based on the weighted and recorded amount of refuse they deposited at mega landfills from the transfer zonal centres. These add up to 'refuse for sale' approach to mitigate upstream evacuation problems. The overall funding would come from the normal monthly billings on the households by LAWMA.

In this approach, there would be three levels in solid waste transfer; from homes (upstream), transfer center (Mid-stream) and unto landfills (downstream). Specific measures could then be integrated for efficacy of each level. The oragnisation, training and engagement of the RCPs would provide employment for teaming number of small-scale private sector willing operators who could also form another local income tax base for the government. The regularisation of the RCPs would eliminate the extortion by local government officials and accord RCPs human dignity in employment. Again, corruption in the form refuse round-tripping would have to be guarded against in the operations.

References

1. Auchterlounie, T., Hinks, J.: The measurement of customer satisfaction in the private house building sector. In: COBRA 2001 Conference Papers (2001). www.rics-foundation.org. Accessed 25 Sept 2016
2. van Praag, B.M.S, Frijters, P.: The anatomy of subjective well-being. Discussion Paper 265, Deutsches Institut für Wirtschaftsforschung, Berlin (2001
3. Ryff, C.D., Keyes, C.L.: The structure of psychological well-being revisited. J. Pers. Soc. Psychol. **69**, 719–727 (1995)

4. van-Hoorn, A.: A short introduction to subjective well-being: its measurement, correlates and policy uses. Paper Prepared for the International conference: Is happiness Measurable and What Do Those Measures Mean for Policy? 2–3 April. University of Rome 'Tor Vergata' (2007)
5. Abel, O.A.: An analysis of solid waste generation in a traditional African city: the example of Ogbomoso, Nigeria. Environ. Urb. SAGE J. **19**(2), 527–537 (2009)
6. Dias, S.M.: Waste pickers and cities. Environ. Urb. **28**(2), 375–390 (2016)
7. Lagos State Government: Lagos Waste Management Authority (LAWMA) Law. Government Press, Ikeja (2007)
8. Idowu, O.B.A., Omirin, M.M., Osagie, J.U.: Outsourcing for sustainable waste disposal in Lagos metropolis: case study of agege local government, Lagos. J. Sustain. Dev. **4**(6), 116–131 (2011)
9. Medina, M.: The World's Scavengers: Salvaging for Sustainable Consumption and Production. Altamira Press, Plymouth (2007)

Regulatory Distress: Architects' Perspective on Enforcement of Building Regulations in Ghana

Justice Kwame Offei-Nketiah(✉), Titus Ebenezer Kwofie, and Daniel Yaw Addai Duah

Department of Architecture, Kwame Nkrumah University of Science and Technology, Ghana Private Mail Bag, Kumasi, Ghana
jkoffeinketiah@gmail.com

Abstract. Building regulation is considered as a remarkable means for ensuring orderly and sustainable development. Unfortunately, it has become a fault line and an elusive tool for developing countries. In Ghana, the chronic collapse of buildings, haphazard developments and increased number of defective buildings are signs of weaknesses in the enforcement of building controls. Based on the perspective of architects, the study seeks to confirm the key factors that accounts for the laxity in enforcement process and the potential use of smart building regulations to improve the system. Architects present during the Annual General Meeting of architects in Ghana were purposively sampled to answer survey questionnaires. Based on response from 86 architects, institutional factors; Inefficient and Insufficient Staff, Corruption and Absence of Public-Private Partnership were established as key factors that impacts negatively on enforcement of building regulations. The study recommends the use of smart approaches for effective building controls in Ghana.

Keywords: Building regulations · Smart regulations · Sustainable development · Building control · Building collapse

1 Introduction

The building industry plays a vital role in the socio-economic development of every society. The construction industry and its related sectors accounts for about 15% of the gross domestic product of countries throughout the world [1]. About 90% of human activities occur in and around buildings, or are either affected by buildings or connected to buildings [2, 3]. Despite these significant gains, the industry is a formidable polluter of the environment, source of chronic risk and disasters such as haphazard developments, rampant building collapse and building fires that destroys properties and threatens the continuance of human life [4, 5]. Consequently, various governments worldwide have taken up the challenge to ensure orderly and safe developments in the built environment through strategies including the use of building regulations [6, 7]. Generally, building regulations (BR) play an important role in ensuring the safety, public health and welfare of people in the built environment.

C. Aigbavboa and W. Thwala (Eds.): CIDB 2019, *The Construction Industry in the Fourth Industrial Revolution*, pp. 225–240, 2020.
https://doi.org/10.1007/978-3-030-26528-1_23

The use of BR in ensuring orderly developments is particularly important to urban areas of Sub-Sahara African (SSA) countries where infrastructural development is not complementary to the rapid pace of urbanization. The population of SSA countries is increasing at a galloping rate of 2% considered as the fastest in the world [8]. This translates into a situation where the population of individual countries is expected to double by the year 2050. The major challenge is that 75% of the populace will live in urban areas where currently the scale of urban growth has outpaced the capacity of cities to provide the required services [9, 10]. As a result, there is the prevalence of unregulated and informally constructed settlements which are prone to hazards and vulnerable to climate and disaster risks such as flooding, building collapse, fires, landslide etc. Despite the use of BR as powerful tools for ensuring safety, resilience and reduction in risk by developed countries, it remains elusive to developing countries due to the absence of matured regulatory regimes [5, 11, 12]. The incessant collapse and increasing number of defective buildings in developing countries is evidence of gaps between building regulations and enforcement [13]. Ghana like most countries in SSA is not absolved from these challenges. Development of urban areas in Ghana is fraught with challenges such as inadequate infrastructure, overcrowding, poor housing, squalid environmental conditions, increase in slum conditions and crimes [14].

Although the enactment of building controls is very important to developments in the building industry, it is one thing to legislate and another to ensure compliance to the legislations [4, 15]. In Ghana, the National Building Regulation 1996 (L.I 1630) which derives its mandate from Local Government Act 1993 (Act 462), now amended as Local Governance Act 2016 (Act 936), has been in use for over 22 years without any revision contrary to international best practices. Although a National Building Code was launched in November 2018, the regulation (LI 1630) is yet to be amended to bring it into harmony with the code. Preliminary research [16–20] shows the absence of a robust building control regime that can guarantee the delivery of safe buildings to users due to non-enforcement and non-compliance to building controls.

Although [16] reported on some of the major catastrophes from building collapse in Ghana before the year 2012, the situation has not changed much from then. In 2017 alone, five years after the "Achimota disaster" in 2012 where 22 lives were lost and 78 people injured in the capital city of Accra, a number of building collapse cases were recorded and have been indicated below;

- On 24th January 2017, a market square under construction in Haatso in Accra collapsed. Although no fatalities were recorded, construction workers were severely injured [21].
- Six [6] pupils aged between 4 and 5 years died and several were injured on 31st January 2017 when a school building collapsed on over 70 pupils at Breman Jamera Methodist School in the Odoben Brakwa District of the Central region as they prepared to settle for studies [22].
- On 17th February 2017 a five storey building, owned by Fruit of Christ International Church and was constructed within six months mostly at night without building permit collapsed in Awoshie, a suburb of Accra. There were no casualties [23].

- On 8th May 2017, twenty church members of The Christlike Ministries Church at Gbawe in Accra sustained injuries when their church building collapsed on them [24].
- On 5th June 2017, two workers died when a two-storey building under construction collapsed on them at Okpoigonno in Accra [25].
- On 7th June 2017, a two-storey market store building collapsed at the Techiman Market where several people were trapped. One person died and the rest were rescued and taken to the hospital for treatment [26].

The objective of the study was in twofold: first, the study seeks to confirm the significant factors that have resulted in the laxity in the enforcement of BR in Ghana. Although this compares to studies by [16], who identified factors that affects the implementation of Ghana's BR by eliciting views from building practitioners, owners and staff of local authorities, this study is unique as it focusses on expert views of only architects. This is a prelude to a comparative study between various built environment professionals and the level of compliance in the northern, middle and southern sectors of Ghana. Second, the views of the architects about the potential use of smart regulations in enhancing the building regulatory enforcement in Ghana were sought due to the important role they play in ensuring that buildings designed and built meet the requirement of building regulations.

2 Literature Review

2.1 Overview of Building Regulatory Regime

Building regulations sets minimum acceptable quality requirement that ensures that buildings are safe, orderly and healthy for users and their immediate environment [27]. These standards contain provisions that apply to the design, construction, alteration, maintenance and demolition of buildings or structures as well as the appurtenance attached to it. It is established by [4, 28] that effective and efficient implementation of building controls is paramount to the orderly, safe and sustainable development of the built environment. That notwithstanding, there is a dearth of literature on building regulations in urban planning, urban design and general regulatory studies as compared to other regulated sectors [29].

Modern day building regulations can be traced to the 19th Century when poor environmental, housing and sanitary conditions prompted governments especially in developed countries to intervene in the building industry [29]. In contemporary times, BRs have been adapted by many countries to achieve a broad range of regulatory goals such as safety, public health, infrastructural development, sustainability and resilience. Research works such as [5, 30] have already established a direct relationship between an effective and efficient regulatory regime and urban sustainability. Regulations are as effective as their enforcement [4]. Thus, regulations work effectively when they are enforced. A robust regulatory regime is therefore a complete framework of regulations and enforcements aimed at achieving regulatory goals [29, 31]. Although regulation by itself may be comprehensive and fit for purpose, weak institutional framework for implementation may render it ineffective for achieving regulatory goals.

Ghana's building regulatory regime is made up of the LI 1630 of Act 1996 which was mandated by the Local Government Act 462 of 1993. The regulation has its genesis from the Town and Country Planning Ordinance (Cap 84) of 1945. The absence of periodic reviews has resulted in a situation where certain outdated and obsolete elements from pre-independence era are found in the legislation. Although the regulation seeks to promote the health, safety and convenience of people in the built environment, there is no emphasis on contemporary issues such as sustainability and resilience [18]. The 19 parts of the regulation consist of 187 regulations and 9 schedules that focusses on planning, design and construction of buildings. The administration of the regulation is the responsibility of Planning Authorities in the Metropolitan, Municipal and District Assemblies (MMDAs) of the Ministry of Local Government. In summary the building regulatory regime of Ghana comprises the LI 1630 together with other related regulations and their provisions for enforcement.

2.2 The State of Enforcement of Building Regulations in Ghana

There is a consensus that the traditional form of enforcement of building regulation was a bipartite process where government through state agencies acts as the regulator and the building developer (business) as a regulated entity [11, 29, 31, 32]. Despite the availability and plurality of regulatory forms coupled with the fact that compliance to regulations can be achieved through myriad strategies, enforcement of building regulations in Ghana rests in the ambit of state agencies. The Ghana National Building regulation through the decentralization policy of the Local Governance Act 462 mandates Planning Authorities of MMDAs to play an oversight role and ensure strict compliance to building regulations in Ghana. Accordingly, this task traditionally revolves around the assessment of building plans prior to permit acquisition, inspection of building under construction and buildings in use [33].

After going through the necessary planning procedures of zoning and approval of Sub-division plan as outlined by [33], assessment of building plans are carried out by Planning Authorities. When all requirements have been scrutinized, a development permit is issued for construction to commence. The BR stipulates ten work stages which must be inspected by Planning Authorities. According to the LI 1630, no construction work (work stages) shall be covered until it has been inspected and approved by the Planning Authorities at the MMDAs. Adequate checks are to be made during the inspections to ensure that buildings conform to the planning standards and construction standards before certificate of habitation is granted. Beyond this, the BR makes provision for periodic maintenance, procedures for alteration and demolition.

Preliminary studies on the BR has shown that much emphasis is placed on the permit acquisition process than the routine inspections. It must be noted that the granting of these permit alone is not a guarantee for safe, sustainable and orderly developments [18]. The enforcement of building regulations in Ghana and planning standards in general is weak (19, 20) and this has consequently resulted in increased number of slums, poor housing conditions, congestion and poor environmental conditions especially in urban areas of the country (16, 17, 34). Inadequate monitoring and enforcement of developments beyond acquisition of permit is a threat to sustainable and orderly development.

Based on studies by (16, 17, 20, 35), ten (10) factors were identified to be responsible for the weakness in enforcement of Ghana's BR. The factors shown in Table 1 were grouped based on the classification by [35].

Table 1. Factors affecting enforcement of building regulations in Ghana

Classification	Factors
Institutional	Obsolete Regulations
	Bureaucracy
	Insufficient Staff (Planning Authority)
	Inefficient Staff (Planning Authority)
	Lack of Innovation (Absence of ICT)
	Absence of Public-Private Partnership
Political	Political Interference
	Corruption
	Lack of commitment by Central Government
Socio-Economic	Lack of Awareness

Field survey (2018)

2.2.1 Institutional Factors

The low level of expertise of Planning Authorities and inefficiency due to the absence of the needed resource and logistics accounts for the ineffectiveness of enforcement mechanisms for Ghana's building regulatory regime [35]. As indicated earlier, the BR has never been reviewed since its enactment resulting in difficulties in responding to contemporary demands such as sustainability and resilience [18]. The over dependence on state agencies, bureaucratic processes and absence of innovative means for enforcement of building regulations remains a setback for achieving regulatory goals [36].

2.2.2 Political Factors

Lack of political will to enforce regulations as well as interference from government and traditional leaders poses a threat to the effective implementation of regulations in the built environment. The simmering tension between Planning Authorities and customary land owners (Chiefs) militates against orderly development in the built environment [34]. Again, the loose enforcement process has created avenues for extortion and corruption by building inspectors. Without he political will and commitment of government, it is very difficult to ensure compliance and enforcement of building regulations especially in a traditional command and control regime [6].

2.2.3 Socio-Economic Factors

The lack of education and enlightenment of the public on building regulations is considered a major factor that contravenes the effectiveness of building regulatory regimes (35). People are likely to comply with regulation when they are well informed about the regulations and the consequences of non-compliance.

2.3 Overview of Smart Building Regulatory Regime

As indicated earlier, smart regulation was presented to architects as a potential boost for building regulatory enforcement in Ghana. Gunningham et al. (1998) made the first advocate for smart regulations. The concept refers to a more flexible and innovative forms of regulatory governance that harnesses both governments and third party resources for effective and efficient implementation of regulations [2, 38]. Smart regulation is not about more or less regulations but rather, the delivery of regulatory goals in a less burdensome way through attributes that includes a mix of actors (private sector participation], enforcement styles, use of ICT and contemporary issues such as sustainability [38].

Unlike the building regulatory regime in most SSA countries where the business of regulatory governance is a bipartite process between the government agencies and regulated entities, the concept of smart regulations asserts that government agencies alone cannot shoulder the responsibilities of governance. The enforcement of regulations must go beyond the state to include the participation of third parties in building inspection, reviews, monitoring etc. The concept also embraces the use of technology as an enabler and integrator in the enforcement of regulations.

3 Methodology

Planning Authorities in developing countries like Ghana struggle to keep up with the demand for their services due to their limited resources and tight budgets [39]. These together with other reasons have accounted for the ineffectiveness of building regulatory enforcement in developing countries. This study uniquely focusses on the perspective of the architect in identifying the factors that have accounted for the laxity in the enforcement of building regulations in Ghana. Architects have been identified in many building regulatory literature as one of the key professionals who produces the most economically safe and functional buildings without any compromise to aesthetics [40, 41]. As the leader of the design-construction team, the architect gets involved with the Client first, produces designs that meets the needs and aspirations of the client and is responsible for quality control of all aspects of the building including services rendered by colleague professionals in the building industry [40].

The work of the architect at both the pre-contract (inception, sketch design, production drawings, tendering) and post-contract phases (construction and commissioning) has traditionally been conditioned by rules, regulations, standards, and governance practices [36, 37, 42]. Architects are prescribed or confronted with one regulation or the other the moment they set out to design or build. Although architects are sometimes inclined to perceive regulations as a burden on creativity and a delimitation that undermines the capacity to design, extensive work on architects and regulations suggest that regulations rather open up possibilities in the creative practice and processes in architecture. In some developed economies, architects partner with government and municipal authorities to develop and implement policies for the built environment and are directly involved in the enforcement of regulations. There is an advocate for the use of architects as intermediaries in building regulatory governance in order to meet contemporary demands of sustainable and low-carbon built environments [41]. Considering the critical stakeholders in the building regulatory regime, the expert views of architects in confirming the t important factors that have resulted in weak enforcement of building regulations in Ghana is very relevant. Hence, the focus on architects.

Data for the research was obtained at the 2018 Annual General Meeting (AGM) of the Ghana Institute of Architects (GIA) in Accra.

3.1 Population Sampling and Technique

Architects present at the 2018 GIA-AGM in Ghana formed the sample selected to answer structured questionnaires as part of the survey. Although 59 respondents were to be selected for the study based on a sample size calculation with 90% confidence level, 10% margin of error and 433 architects representing the total population of architects in Ghana as at 2018 [44], all architects present at the AGM were purposively sampled for the study. The 90% confidence level was chosen prior to the administering of the questionnaires because of the small number of the study population who were in good standing, active and available to answer the questionnaire during the AGM. However, out of the 120 questionnaires that were returned, 86 respondents which is in excess of the anticipated number of respondents (59) were found to be responsive after careful scrutiny for qualification, omissions, errors and completeness representing 71.6% response rate. This number is exclusive of all students and probationers present at the meeting and represents 30.3% of architects in good standing as at the time of the study.

3.2 Method of Data Analysis

Respondents were tasked to confirm the factors accounting for the laxity in enforcement of building regulations. The Likert scale (1–5) used for the study was labelled as 1 = Not Important, 2 = Less Important, 3 = Moderate, 4 = Important, 5 = Very Important. Respondents gave scores in accordance with their level of importance of the factors identified in literature for non-enforcement of building regulations in Ghana. Descriptive statistics were used to analyze nominal and interval data obtained from the study whilst mean item score was used to rank and confirm the importance of the factors accounting for laxity in building regulatory enforcement in Ghana.

As part of the study, Fleiss Kappa (κ) was used to test for consistency and agreement among the various respondents. Fleiss Kappa (κ) is defined as follows;

$$\text{Kappa } (\kappa) = \frac{\text{Pa} - \text{P}\varepsilon}{1 - \text{P}\varepsilon}$$

Where the factor $1 - \text{P}\varepsilon$ is the degree of agreement that is attainable above chance and $\text{P}\alpha - \text{P}\varepsilon$ gives the degree of agreement actually achieved above chance [43].

4 Data Presentation, Analysis and Discussions

4.1 Characteristics of Respondents

The data on the characteristics of respondents was limited to the experience level of the respondents and their membership status within the Ghana Institute of Architects (GIA). The purpose was to establish the fact that the more experienced respondents are, the more they are able to provide more reliable data that can be used for making policy decisions. From Table 2, architects with 5–10 years post qualification experience formed 58.1% of the respondents. A total of 13 respondents had over 15 years of experience in the profession constituting a total of 15.1% of the total respondents. Out of this number, 6 of them were Fellows of the Institute whilst the remaining together with those with less than 20 years' experience formed the Chartered Membership of the Institute as shown in Table 2.

Table 2. Post-qualification experience and membership status of respondents

Age	Frequency	Percentage
5–10 years	50	58.1
11–15 years	23	26.8
16–20 years	5	5.8
>20 years	8	9.3
Total	***86***	100%
Membership status of respondents		
Chartered members	80	93
Fellows	6	7
Total	**86**	**100**

Field survey (2018)

4.2 Presentation of Data and Analysis

Respondents were also asked to indicate how often they make reference to the Ghana's BR in their daily practice. In all, 30 respondents admitted that they do make reference to the regulations occasionally whilst the remaining 56 representing 65% of the

Table 3. Fitness of purpose of Ghana's building regulation

SN	Fitness level	Number of respondents	Percentage
1	Fit for purpose	19	22
2	Not entirely fit for purpose	59	69
3	Not fit for purpose	8	9
	Total	86	100.0

Field survey (2018)

respondents do so frequently. Table 4 shows the stage of works at which architects in Ghana makes reference to the BR. From the study, majority of architects makes reference to the BR at pre-contract stage of their works (mostly inception, sketch design and production drawings). At the tendering stage, the respondents indicated that the BR has little provision for tendering as such reference documents such as the Public Procurement Act and manuals are used for such purposes.

Table 4. Reference to building regulations at pre and post-contract phase of work

SN	Stage of work	Number of respondents (Expected score = 86)	Percentage score (Expected % = 100%)
Pre-contract stage			
1	Inception	5	6
2	Sketch design	19	22
3	Production drawings	7	8
4	Tendering	29	34
Post contract			
5	Construction	44	51
6	Commissioning	52	61

Field survey (2018)

Table 5. Notification for inspection of works

SN	Notification stage	Number of respondents (Expected score = 86)	Percentage of response (Expected score = 100%)
1	Demarcation & siting of building	31	36
2	Setting out of building foundation	28	33
3	Excavated foundation and leveling for concrete	13	15
4	Foundation concreted	12	14
5	Trenches for drainage work excavated	9	10
6	Drains laid and joined ready for testing	7	8
7	Steel rebar fixed in position before concreting	12	14
8	Concrete shuttering ready for striking	7	8
9	Walls completed to wall plate level	7	8
10	Roof framework completed before covering	10	12

Field survey (2018)

Out of the 86 respondents, only 34 representing less than 40% of the total number of respondents gave notices to Planning Authorities for inspection of the various stages of their works. Table 5 provides details on the frequency of notification to Authorities at the various stages of construction as identified in the BR. "Demarcation & Siting of Building" recorded the highest frequency of 31 respondents representing 36% followed by "Setting out of Building Foundation" which recorded 32.56% representing 28 out of 86 respondents. Testing of drains and completion of walls recorded the lowest frequency of 7 out of 86 respondents which corresponds to 8.14% of the expected number of notifications.

Table 6 also shows the mean score and ranking for the various factors that have been attributed to the weakness in enforcement of BR in Ghana. "Inefficient Staff for Planning Authorities" was ranked first with a mean score of 4.16 (approximately 4) followed closely by Corruption which was rather ranked 1st by the study conducted by [16]. The staffing situation of the Authorities is worrying as their numerical strength was ranked 4th amongst the factors by respondents. The "Absence of Private Participation" which ranked 3rd gives indication that respondents would appreciate the involvement of third parties in the enforcement of the BR. Although "lack of commitment by Government" and "Political Interference" were the least ranked factor (10th). Approximating the mean scores to a whole number, the study concludes that all the factors are important to the laxity of enforcement of the BR except "Lack of Commitment by Central Government" which received a moderate score.

As part of the study, respondents were asked to indicate if strict compliance to the existing BR will result in sustainable and resilient urban development. Out of the 86 respondents, 37 representing 43% were of the view that strict compliance and enforcement of the current regulations will automatically result in sustainable and

Table 6. Factors contribution to the laxity in enforcement of building regulations in Ghana

	Ordinal scale					∑ FX	Mean	Rank
Factors	**1**	**2**	**3**	**4**	**5**			
Obsolete Regulations	6	8	19	24	29	320	3.72	8th
Lack of Innovation (Absence of ICT)	2	4	26	25	29	333	3.87	6th
Lack of Awareness	5	5	20	22	34	333	3.87	6th
Lack of Commitment by Central Government	8	15	20	26	17	287	3.34	10th
Bureaucracy and Inordinate Delays	2	8	18	23	35	339	3.94	5th
Absence of Public-Private Partnership	2	5	20	22	37	345	4.01	3rd
Inefficient Staff (Planning Authority)	2	3	22	22	37	358	4.16	1st
Corruption	3	3	15	25	40	352	4.09	2nd
Insufficient Staff	4	2	23	20	37	342	3.98	4th
Political Interference	6	10	18	28	24	312	3.63	9th
Total	40	63	201	237	319			

1 = Not Important, **2** = Less Important, **3** = Moderate, **4** = Important, **5** = Very Important.
Field Survey (2018)

resilient urban development. Seven (7) respondents declined whilst the remaining 42 representing 48.8% of the respondents expressed the view that strict compliance to the regulation will not necessarily achieve sustainable and resilient development citing obsolete regulations and weak enforcement strategy as the reasons for stance.

Again, Respondents expressed their views on their knowledge about smart regulations. From the study, only 22 respondents admitted awareness of the concept of smart. The remaining 64 representing 74.4% declined knowledge about smart regulations. That notwithstanding, 72 respondents representing 83.7% indicated that the use of third parties (private sector and civil society group) will enhance the enforcement of regulations in Ghana. Again, 96.5% of respondents agreed that the introduction of innovative ways of governance through ICT will benefit the implementation of building regulations in Ghana.

5 Discussion of Results

The survey on the perception of architects on enforcement of building regulations revealed the following;

5.1 Perceived Weaknesses in Building Regulatory Regime

Generally, the study confirmed the weaknesses in Ghana's building regulatory regime. Although the need for inclusion of contemporary provision for sustainability, resilience and climate adaptation became apparent, the perceived weakness was mainly attributed to weak enforcement of regulations. The outcome of respondents' judgement on the fitness of purpose of the regulations was in consonance with the ranking of the factors responsible for the laxity in enforcement of regulations. Over 70% of respondents declined to accept that BR and its enforcement provisions were fit to achieve regulatory goals. This is in agreement with a study by (44) where though regulations were at least 70% fit for purpose, actual score for implementation of these regulations were below 40%. The lack of institutional capacity and expertise in the implementation of the BR as illustrated by the ranking of the factors for non-enforcement in Table 6 emphasizes the enforcement challenges of the BR.

5.2 Overreliance on Government Agencies for Enforcement

The BR regime of Ghana have not benefited much from new approaches in regulatory governances that addresses contemporary issues like sustainability and resilience. Although the institutions mandated to carry out enforcement are weak (poor ranking of institutional factors in Table 6), the framework in the LI 1630 makes no provision for other enforcement actors although private sector participation in regulatory processes has shown positive results in achieving regulatory goals in literature [2, 36, 45]. The gap in enforcement of regulations has been attributed to the tight budget and resource constraints of Municipal Authorities and the dearth of private sector participation in regulatory regimes of developing countries like Ghana.

5.3 Absence of Proper Inspection Beyond Permit Acquisition

Building permits play an important role in ensuring that developments are in conformity with planning and building regulations. In the absence of regimes to promote voluntary compliance, the ineffective inspection of work stages beyond permit acquisition by Planning Authorities undermines compliance to regulations. The study explicitly revealed that inspections were not adequately enforced as architects who act on behalf of their clients as consultants for projects fail to notify the Planning Authorities to conduct inspection at the work stages as shown in Table 5. According to the National Building Regulation 1996 (LI 1630), it is mandatory for developers to give forty-eight hours' notice in writing to Planning Authorities prior to commencement of works and provide dates for which stages of works identified in the regulations would be inspected. The study revealed serious violation of this provision by architects. It was established that Planning Authorities in most instances fail to inspect building construction unless they have been invited to do so. This undermines the crucial role played by inspection in ensuring compliance with the BR.

5.4 Level of Awareness of Smart Regulations Amongst Professionals

Although the level of awareness of smart regulations amongst architects was low, respondents agree that the introduction of third parties and other innovative forms of regulatory governance into the BR regime in addition to the efforts of State Institutions will boost enforcement of BR in Ghana.

5.5 Factors Accounting for Laxity in Enforcement of Building Regulations in Ghana

The ranking of the factors by architects reinforces the fact that the enforcement regime of the BR is weak. In addition to the fact that 9 out of 10 of the factors were deemed to be important, the institutional factors (Insufficient staff, inefficient staff, obsolete regulations, bureaucracy, lack of innovation and absence of public private partnership) were the most significant. The political and socio-economic factors cannot be overlooked as they recorded mean score closer to that of the Institutional factors. The most important factors agreed by architects for the poor enforcement of BR in Ghana in order of importance are; Inefficient Staff, Insufficient Staff, Corruption, Absence of Public-Private Partnership, Inordinate Delays and Bureaucracy, Lack of Awareness and Lack of Innovation, Obsolete Regulations, Political Interference and Lack of Commitment from Central Government

6 Inter-rate Reliability Analysis

To determine the inter rater agreement between the 86 respondents on the factors accounting for the non-enforcement of BR in Ghana, the Fleiss Kappa (**K**) inter rater reliability test was conducted. The results for the test indicated a relatively low Kappa (**K**) = 0.005 (scale of –1 to 1). This is a relatively low. However, a kappa value of

0.005 according to (44) implies that there is a slight agreement between the raters and that the rating could not be out of chance as indicated in Table 1. According to [46], any kappa above 0 is statistically significant especially for studies with large sample size such as this study. Hence, there is some consistency in the rank by the individual raters.

7 Conclusion and Recommendation

It is evident from the study that the granting of permit alone will not guarantee compliance to building regulation. Inspections are necessary to maintain standards during construction. Weaknesses in inspection of the work stages and failure to notify Planning Authorities for inspection by architects undermines compliance and enforcement. It was also established that inspections were limited to the early stages of construction (Demarcation, Setting out and Excavation of foundation).

The study also revealed that the current building regulatory regime (the regulations and enforcement provisions) cannot ensure the achievement of regulatory goals mainly due to lack of enforcement. The obsolete regulations (due to the absence of provisions for sustainability and resilience and other contemporary issue) and the weakness in the enforcement of regulations poses a threat to sustainable urban development in Ghana.

Five important factors responsible for the ineffectiveness of enforcement of building regulations according to Architects in Ghana in order of importance are (i) Inefficient Staff, (ii) Insufficient Staff, (iii) Corruption, (iv) Absence of Public-Private Partnership, (v) Inordinate Delays and Bureaucracy and Lack of Awareness. It is interesting to note that all these factors are part of the mandate of the Planning Authorities vested with the power to enforce the BR. The Planning Authorities are increasingly finding it difficult to ensure compliance to the BR due to their resource constraints (funding, logistics and expertise). Even though majority of the architects demonstrated low awareness of smart regulations, they were supportive of the introduction of third parties into BE enforcement.

In summary, the study revealed weaknesses in the enforcement of the BR and identified the potential use of smart approaches to enhance the enforcement of regulations for sustainability and resilience. The following recommendations are made to improve the enforcement gaps in the BR regime;

- There must be regular inspection of works stages by Planning Authorities
- Permit acquisition should be complemented with regular and effective monitoring of works stages
- The expertise of the private sector (especially professionals) should be leveraged for sustainable and resilient urban development
- There must be recruitment of qualified staff and in-service training of staff of Planning Authorities
- The BR should be reviewed to include provisions for contemporary issues and harmonized with the National Building Code. and
- There must be a strategic redefinition of the roles of all stakeholders in the building regulatory regime.

The study has established that the challenges of building regulatory enforcement can be attributed to deficiencies on the part of Planning Authorities. The absence of a framework that will allow for third party support and involvement in building control must be addressed. Future research must focus on the development of a robust framework for the use of smart regulations to improve compliance to building regulations in Ghana.

Acknowledgment. This paper is part of an ongoing PhD study on building regulatory regimes for sustainable urban development in Ghana.

References

1. Babatunde, O.K, Low, S.P.: Chinese construction firms in the Nigerian construction industry. Habitat Int. **40**, 18–24 (2013)
2. Van Der Heijden, J.: Smart privatization: lessons from private sector involvement in Australian and Canadian building regulatory enforcement regimes. J. Comput. Policy Anal. Res. Pract. **12**(5), 509–525 (2010)
3. Klepeis, N.E., Nelson, W.C., Ott, W.R., Robinson, J.P., Tsang, A.M., Switzer, P.: The National Human Activity Pattern Survey. Lawrence Berkeley Natl Lab [Internet] **11**(3), 231–252 (2001). http://exposurescience.org/the-national-human-activity-pattern-survey-nhaps-a-resource-for-assessing-exposure-to-environmental-pollutants
4. Owusu-Ansah, J.K., Atta-Boateng, F.: The spatial expression of physical development controls in a fast growing Ghanaian city. Land Use Policy **54**, 147–157 (2016)
5. World Bank. Building Regulation for Resilience: Managing Risks for Safer Cities. Global Facility Disaster Reduction and Recovery (2016)
6. Van der Heijden, J.: Governance for Urban Sustainability and Resilience (2014)
7. Akadiri, P.O., Chinyio, E.A., Olomolaiye, P.O.: Design of a sustainable building: a conceptual framework for implementing sustainability in the building sector. Buildings **2**(2), 126–152 (2012)
8. UN-Habitat: Urban Planning and Design at UN-Habitat, pp. 1–35
9. Mutunga, C., Zulu, E., De Souza, R.-M.: Population Dynamics, Climate Change, and Sustainable Development in Africa (2012)
10. UN-Habitat, Africa Planning Association. The State of Planning in Africa: An Overview (2013)
11. McAllister, L.K.: Dimensions of enforcement style: factoring in regulatory autonomy and capacity. Law Policy **32**(1), 61–78 (2010)
12. Goodfellow, T.: Planning and development regulation amid rapid urban growth: explaining divergent trajectories in Africa. Geoforum **48**, 83–93 (2013)
13. Ojelabi, R.A., Oyeyipo, O.O., Afolabi, A.: Built environment professionals' perceptions of the effectiveness of building control measures in Lagos state. J. Constr. Dev. Countries **22**(1), 41–54 (2017)
14. Ghana Statistical Service. 2010 Population and Housing Census Report: Urbanization (2014)
15. Dahiru, D., Abdulazeez, D., Abubakar, M.: An evaluation of the edequacy of the National Building Code for achieving a sustainable built environment in Nigeria. Res. J. Environ. Earth Sci. **4**, 857–865 (2012)
16. Ametepey, S.O., Ansah, S.K., Edu-Buandoh, K.B.M.: Assessing factors affecting implementation of the national building regulations (L. I. 1630) in Ghana. Publ. Policy Admission Res. **5**(2), 66–74 (2015)

17. Botchway, E.A., Afram, O.S., Ankrah, J.: Building permit acquisition in Ghana: the situation in Kumasi. Developing Country Stud. **4**(20), 11–22 (2014). www.iiste.org
18. Casa Associati. Africa Adaptation Programme on Climate Change Risk Reduction in Physical Development (2012)
19. Duah, D.: Building Regulation Enforcement in Ghana. eyeonafrica.matrix.msu.edu/2013/09/building-regulations-enforcement-in-ghana/
20. Agyeman, S., Assiamah, S., Abeka, H.: Re-engineering the building permits acquisition processing to control the development of unauthorized structures in ghanaian communities. China-USA Bus. Rev. **15**(4) (2016). http://www.davidpublisher.com/index.php/Home/Article/index?id=27131.html
21. Ghanaweb.com (2017). https://www.ghanaweb.com/GhanaHomePage/NewsArchive/Haatso-Market-building-collapses-construction-workers-flee-503807
22. Myjoyonline.com (2017). https://www.myjoyonline.com/news/2017/January-31st/methodist-school-building-collapses-kills-4-pupils.php
23. Ghana Broadcasting Corporation (2017). https://www.gbcghana.com/1.8554738
24. Ghanaweb.com (2017). https://www.ghanaweb.com/GhanaHomePage/NewsArchive/20-seriously-injured-in-church-building-collapse-535956
25. Ghanaweb.com (2017). https://www.ghanaweb.com/GhanaHomePage/NewsArchive/Collapsed-building-kills-two-in-Accra-544711
26. Adu-Gyamerah, E.: Techiman market building under construction collapses, killing one. Graphic Online (2017). https://www.graphic.com.gh/news/general-news/techiman-market-building-under-construction-collapses-killing-one.html
27. Pedro, J.B., dos Santos, C.P.: Building regulations and control in the face of climate change: Overview of the Portuguese situation (2011)
28. Meacham, B.J.: Toward next generation performance-based building regulatory systems. In: SFPE 11th Conference on Performance-Based Codes and Fire Safety Design Methods (2016)
29. Van der Heijden, J., De Jong, J.: A better understanding of building regulation. Environ. Plann. B **36**, 1038–1052 (2013)
30. Gunningham, N.: Compliance, Enforcement and Regulatory Excellence. https://www.law.upenn.edu/live/files/4717-gunningham-ppr-bicregulatordiscussionpaper-06
31. May, P.J.: Regulatory Regimes and Accountability. Regulatory Governance (2007). http://doi.wiley.com/10.1111/j.1748-5991.2007.00002
32. Gunningham, N., Sinclair, D.: Smart regulations. In: Drahos, P. (ed). Regulatory Theory: Foundations and Applications, pp. 133–148. Anu Press (2017)
33. Baffour Awuah, K.G., Hammond, F.N., Lamond, J.E., Booth, C.: Benefits of urban land use planning in Ghana. Geoforum (2013). http://dx.doi.org/10.1016/j.geoforum
34. Owusu-Ansah, J.K., Atta-Boateng, F.: The spatial expression of physical development controls in a fast growing Ghanaian city. Land Use Policy (2016). http://dx.doi.org/10.1016/j.landusepol
35. Somiah, M.K., Osei-Poku, G., Aidoo, I.: Relative importance analysis of factors influencing unauthorized siting of residential buildings in the metropolis of Ghana. J. Build Constr. Plan Res. **3**, 117–126 (2015)
36. Imrie, R.: The role of the building regulations in achieving housing quality. Environ. Plan B Plan Des. **31**(3), 419–437 (2004)
37. Gunningham, N., Sinclair, D.: Smart regulation. Regul. Theory **1998**, 133–148 (2017)
38. Eurofound: Smart regulation. Regulation (2004). https://www.eurofound.europa.eu/observatories/eurwork/industrial-relationsdictionary/

39. Daye, B.N., Marie, L.D., Imane, F.T.M., Keiko, S.J.S., Yanovich, Y.: Dealing with Construction Permits: Private Sector Participation in Construction Regulation. Doing Business. World Bank Group, Washington, D.C (2018). http://documents.worldbank.org/curated/en/303541513853647106/Dealing-with-construction-permits-private-sector-participation-in-construction-regulation
40. Gunningham, N., Sinclair, D.: Smart regulation. In: Drahos, P. (ed.). Regulatory Theory: Foundations and Applications, pp. 133–148. Acton ACT, Australia: ANU Press (2017). http://www.jstor.org/stable/j.ctt1q1crtm.16
41. Hamma-Adama, M., Kouider, T.: Causes of building failure and collapse in Nigeria: professionals' view. Am. J. Eng. Res. (AJER) **6**(12), 289–300 (2013)
42. Imrie, R., Street, E.: Architectural Design and Regulation (2011)
43. Landis, J.R., Koch, G.G.: The measurement of observer agreement for categorical data. Society **33**(1), 159–174 (2013)
44. Commonwealth Association of Architects. Planning for Rapid Urbanisation Survey of the Architectural Profession (2018). https://issuu.com/comarchitect.org/docs/caa_survey_of_the_architectural_pro/6
45. Osei-Kyei, R., Chan, A.P.C.: Review of studies on the critical success factors for public-private partnership (PPP) projects from 1990 to 2013. Int. J. Proj. Manage. **33**(6), 1335–1346 (2015)
46. Viera, A.J., Garrett, J.M.: Interrater agreement: Kappa statistic. Fam. Med. **37**(5), 360–363 (2005)

Making Rental Housing in the Gap-Market More Affordable Through Green Building Technology

Prisca Simbanegavi, Ziev Shani(✉), Justin Watkins, and Kimera Ramruthan

Construction and Economics Management, University of the Witwatersrand, Johannesburg, South Africa
zievshani@gmail.com

Abstract. There exists a phenomenon whereby "affordable housing" developed by the private sector has become unaffordable. The advancements of the 4th Industrial Revolution, namely Green Building Technology (GBT), provides a possible solution to making rental housing in the gap market more affordable by reducing the tenant's utility bills. The paper uses a Systematic Literature Review (SLR) to determine: Various GBT's that can be used; if the Green Building (GB) tenant benefits financially; benefits and barriers for the developer. The findings showed that the incorporation of GBTs can indeed reduce the tenants housing expenditure. The paper makes a strong case for the private sector to build green affordable housing and has found many benefits that enhance the feasibility of such development. These include: increased property value; lower tenant turnover; tax benefits; reduced operating costs among others. The development of Green Buildings should become standard practice as we enter the 4th Industrial Revolution.

Keywords: Green · Building · Gap-Market · Rental · Poverty

1 Introduction

Despite numerous Governmental innovations, South Africa is still trying to solve its decades old housing crisis [1–4]. One of the reasons for this is that most programs are home-ownership driven, which has proven to be ineffective in closing the housing gap [4]. The home-ownership model is particularly ineffective when it comes to the gap housing market: Earners who bring home between R3500 to R15000 per month and are therefore too poor to qualify for home loans, but too 'rich' to qualify for government subsidy housing [2].

Dugard et al. [5] and Tipple [6], suggest that a rental market may offer a more sustainable solution to improve the gap housing issue as owner-occupied freehold stands are not built fast enough to meet increasing demand. However, research has shown that Government may not be the best provider for this solution, and has concluded that the key to servicing the gap-housing market lies in the private sector [7, 8].

C. Aigbavboa and W. Thwala (Eds.): CIDB 2019, *The Construction Industry in the Fourth Industrial Revolution*, pp. 241–251, 2020.
https://doi.org/10.1007/978-3-030-26528-1_24

Private development companies have been taking the initiative and have developed accommodation for this low-income gap market bracket. However, this accommodation that is supplied by private companies is typically accompanied by a higher rental rate, making this 'affordable housing' less affordable [9].

One way to look at housing expenditure is as the sum of 'rent' and 'utility' bills, and so, this study explores the possibility of balancing out these high rentals by lowering the tenant's utility cost, resulting in a more affordable total housing expenditure. As Green Building Technology (GBT) is known to reduce utility bills, the creation of green affordable housing is the ultimate solution. However, it is important to note that the success of this solution highly depends on voluntary participation of the private sector developers. As the private sector is motivated by profit, the only way this solution would be viable is if the same GBTs that reduce the tenant's utility bills offer significant benefits that accrue to the building owner. Therefore, this study provides value by assessing, from the perspective of the private sector, if indeed there is a convincing case to be made in favour of building affordable rental GBs in South Africa.

This problem of unaffordable housing is a serious one. While Tissington [9] found that much of private sector supplied housing does not meet the criteria for affordability, the problem is made even worse through the tenant's energy expenditure. Energy utility costs in affordable housing add an additional 15% on rental costs, which may prevent poor households from being able to spend money on other necessities [7].

The Objectives of this study are to: 1. Give a brief overview of the types of GBTs that can be implemented 2. To determine if the tenant benefits financially by occupying a green affordable unit 3. To see if there is a case for private developers to build Green Affordable housing by assessing the benefits and barriers of doing so.

The study is *limited* because while the Systematic Literature Review (SLR) provides a qualitative analysis of the benefits of GBTs in affordable housing developments, it does not consider the quantitative cost savings or benefits. The focus of this study is only on developers/owners who build affordable housing to derive rental income and not develop-to-sell models.

2 Methodology

The methodology used by this study is that of a Systematic Literature Review (SLR). A SLR is defined as "a review of a clearly formulated question that uses systematic and explicit methods to identify, select, and critically appraise relevant research, and to collect and analyse data from the studies that are included in the review" [10]. This methodology creates a "structured protocol within which to approach a comprehensive literature review on a given topic" however it can be prone to human error [11]. To minimize this issue and ensure replicability the researchers conducted virtually the entire SLR process in the same room. Thus, any inconsistencies were clarified instantly and all participants were searching, eliminating and coding the information in the exact same manner. The process of conducting a systematic literature review can be summarized in 5 main steps (Fig. 1).

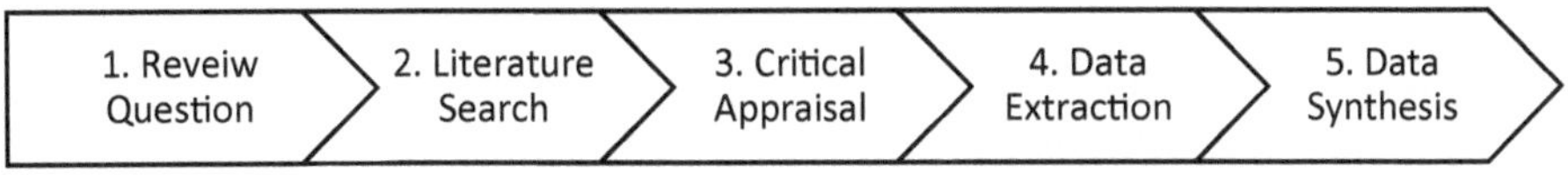

Fig. 1. Systematic literature review process [10]

Step 1 – *Develop Review question*: Based on the initial literature search, as can be seen in the introduction, the following question was formulated. Can GBTs solve the problem of unaffordable "affordable housing" developed by the private sector?

Step 2 – *Searching the literature*: A search strategy was formulated as follows:

The exclusion criteria removed articles that spoke on: Retail, office or industrial properties; Inclusionary housing; Rural housing initiatives (RDP); Non-academic sources; Articles that were not in English; Articles which focused on the health benefits of GBs; Articles that focused on home ownership.

The inclusionary criteria were considered if they explicitly discussed on the following: Green affordable/low-income housing; GBTs (residential); Relevant grey literature as excluding some of these sources may lead to publication bias [12]; Articles focusing on the financial/economic benefits of GBs; Articles focusing on rental tenure; Indirect incentives for the developer to build green.

The following keywords were derived following the guidelines set out in Xiao and Watson [12]: "green low-income housing; green affordable housing; sustainable low-income housing; sustainable affordable housing; residential green buildings and green buildings incentives; the split-incentive problem; green affordable housing financing; green buildings cost premium; affordable housing and sustainability; cost savings in green buildings; cost-benefit analysis of green buildings."

These keywords were placed into nine multidisciplinary databases, as no one database contains all published works [12]: "Taylor and Francis; Web of Science; Scopus; Science Direct; ProQuest; JSTOR; EBSCOhost; Google Scholar; Wiley online library."

Zotero software was used to collect articles hits based on the keywords. Only articles that were written in English and that fell in between the range of 2002–2018 were selected for download [11]. When entering search terms into the databases, non-academic sources were excluded in the filter settings. The search obtained a total of 2749 articles.

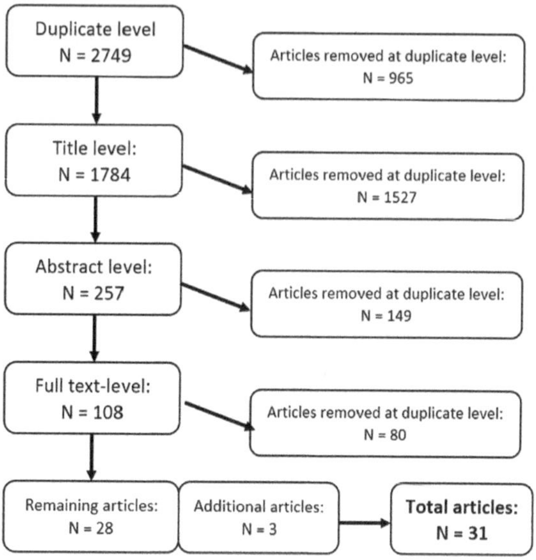

Fig. 2. Elimination process

Step 3 – *Critical appraisal*: The articles captured in the initial search then underwent a process of elimination. This process first removed 965 duplicates. The next stage, eliminated 1527 articles by assessing their titles for applicability based on our inclusion and exclusion criteria. The process was repeated looking at the abstracts and then full-texts eliminating 149 and 80 articles at each stage respectively. This cut down the number to 28. Following the process outlined in McCabe et al. [11] 3 additional articles, deemed necessary, were added to ensure the completeness of the dataset. Thus, final number of articles considered for the study was 31 (Fig. 2).

Step 4 – *Data extraction*: The remaining articles were analysed and coded using NVivo software in order to extract the necessary information. Based on the themes emerging from the literature, seven nodes were chosen, namely: "Utility costs and rental costs in housing; GBT; Benefits and incentives for the GB developer or landlord; Benefits to the GB tenant; Rent premiums and the split incentive problem; Reasons why landlords don't build green; Cost premium of GBs". The information gathered in the nodes was then used to create a narrative systematic literature review.

Step 5 – *Data Synthesis*: The information that was extracted was then summarized and synthesized into findings that are presented in the following section. The headings for each of the findings are based off the initial research aims as well as the themes that emerged from the literature.

3 Findings

3.1 Green Building Technology for Affordable Housing Developments

There are various opportunities, throughout a building's life cycle, to make a building green. From siting, planning and designing to selecting materials and construction, to the operational phase, maintenance and removal phases [13–16].

Types of GBTs that can be implemented in developments, at a reduced cost or no additional cost, are: Orienting the building to maximise natural daylight, minimize glare and heat radiation; natural ventilation; using locally available or recycled material; minimizing heat radiation by landscaping; reducing water usage through low-flow fixtures; optimizing the building shape through thermal modelling; using energy efficient appliances and light fittings; optimising, upgrading or removing HVAC systems; Lighting controls; harvesting rainwater; using recycled Greywater for flushing and having efficient drainage systems. [7, 14, 17–23].

3.2 The Benefits of GBT for the "Affordable Housing" Tenant

The benefits of building green for the tenant are material and substantial in many cases, especially in terms of energy and water utilities cost savings [13, 14, 19, 24, 25]. In EPA [26], it was found that improving energy efficiency in rented affordable housing units can contribute to making developments more affordable for residents. These benefits will become more pronounced over time with electricity prices increasing in excess of rental cost increases [7, 25].

Utility costs and rental payments are important indicators in determining affordability especially for low-income households, and therefore the reduction of utility costs will allow for affordable housing to remain affordable [26]. Hands [21], highlighted that these cost savings allow tenants to meet their rental payments more consistently, along with having additional income to use on other basic needs. This is concurred by Geng [20] who found that a decreased housing expenditure will greatly improve people's living standards and have a huge effect on the economy. In addition, GBs "positively impacts productivity, health, and welfare of human beings throughout its entire life cycle" indicating that the improved physical wellbeing of the tenant can bolster their ability to make a living [27]. Therefore, it seems as though the financial benefits of GBs can improve the tenant's economic well-being.

While most of the literature showed that the tenant will receive a positive financial benefit when living in a GB, one study cautions against a possible 'rebound effect'. As discussed, green technologies can create more disposable income for the tenant. However, a unique phenomenon may emerge that: As tenants have an increased purchasing power, they may purchase new electrical appliances that will increase their overall electricity consumption and as a result "part or all of energy saving effects vanish" in a 'rebound effect' [25].

3.3 The Case for Private Developers to Build Green Affordable Housing

3.3.1 The Split Incentive Problem

The 'split incentive' problem describes a situation in which the landlord is responsible for the energy efficiency expenditure while the tenant enjoys the benefit of a reduced utility bill, and thus the landlord is not incentivised to incur such an expenditure [13, 25, 28–31]. This problem inhibits the uptake of GB development, as the stakeholders in the position to build 'green' lack incentive to do so [24, 32].

One of the primary methods to alleviate this problem is for landlords to increase the rent that they charge through a green rent premium [24, 25, 33, 34]. However, the charging of a green rent premium should be avoided as in an environment where tenants already struggle to meet their rental obligations, as an increase in this burden is likely to lead to defaulting and tenant arrears [4]. This negatively affects the landlord as it effectively cancels out the benefit of increased rental collection and lower tenant turnover/evictions which will be discussed next.

Yet, this opportunity cost may only exist in the short-term, as the rate of utility cost escalations exceeds the rate of rental escalations [7]. The significance of this is that while a rent premium may keep the status quo in the short term, as time goes by and energy costs become an increasingly significant proportion of housing expenditure, the GB tenant will be better off than their conventional counterpart.

With regards to the purchase of assets that generate electricity or hot water through renewable means GPF [7] describes a solution. The tenants will make use of hot water and electricity that is generated onsite, remunerating the owner instead of the state utility provider [7]. Therefore, the cost of the asset will be recuperated through consumption charges to the building user. Once the initial investment price has been paid back, this then provides the owner with an additional income stream.

Ultimately, any incentive or benefit that accrues to the landlord as a result of the building being green, either financial or structural [15] serves to help solve the split incentive issue. Therefore, all of the owner's benefits and incentives identified in these findings can be thought of as ways to solve this particular problem.

3.3.2 Decreased Life-Cycle Costs, Operating Expenses and Maintenance Cost

The cost-saving ability of GBs can only be seen when it is considered from a life cycle perspective. Life-cycle costing considers costs that are incurred during the operational stage thereby not focusing exclusively on initial design and construction costs [13]. The cost savings are significant, in fact, that "an incremental green investment of about two per cent of construction costs typically [yield] life cycle savings of over ten times the initial investment" [25].

The operating costs in GBs are lower than their conventional counterparts, besides for rare cases where the building design has been compromised [7, 13, 14, 16, 20, 25, 35]. The decrease in operating expenditures, in many cases, is significant enough that it covers the additional cost in the green investment [7, 13, 14, 16, 20, 21, 36]. Zhang et al. [25] found that "on average, the aggregate operating cost (including water utilities, energy utilities, general maintenance, grounds maintenance, waste and recycling, and janitorial costs) of GBs was 19% lower than the industry average."

GBs have lower maintenance costs over their lifetime when compared to similar conventional buildings [13, 19, 21, 25, 35, 36]. This is due to the fact that "green design" includes more durable materials which can last for prolonged periods of time without significant degradation [13, 19, 21]. As maintenance costs are typically the responsibility of the building owners this benefit accrues solely to them. Geng [20], concludes that "These advantages are particularly valuable for affordable housing projects, where low operating costs are critical to affordability."

3.3.3 Increased Property Value

There is enough evidence in literature which shows that in the vast majority of cases, the 'green features' of a building increase the value of the property. Indeed, "sustainable design, which seeks to balance environmental, social, and economic needs, adds long-term value to development projects" [37]. Copiello [24] found that there was an average price premium of 3.7% across A, B and C grade GBs. Ambrose [29], found that investments in green technologies for a building increased the value of the property by 15 000 to 42 000 lb (≈R285 000 to R798 000) depending on location. Geng [20], showed that the value of a home would increase by $20 (≈R280) for every $1 (≈R14) spent on energy efficient upgrades. The reason for this is primarily that all of the benefits of GBs are "capitalized into the market value" [20].

The literature has shown that as 'green features' are integrated, property values increase which means that energy efficient homes sell for higher than average prices [20, 33]. One explanation of this phenomenon is that Net Operating Income (NOI), a key determinant in property valuation, increases as the building becomes more energy and water efficient [14, 20]. An alternative hypothesis says that there is a greater willingness to pay for GB's, and due to this increased demand, the value is driven up [32].

Either way, it is conclusive that, "the increased asset value is an incremental benefit for developers" [25]. These findings show that even though it may be difficult for affordable housing projects to directly recover the invested amounts via increased rentals or lower utility bills, they can profit if ever selling the property [34].

3.3.4 Improved Tenant Performance: Consistent Rental Payments and Lower Tenant Turnover

In Ambrose [29], landlords agreed that high utility costs threaten a tenant's ability to make the rental obligations. Therefore, as GBT reduces a tenant's utility cost, tenants are then able to pay rent more consistently [26]. This is an incentive for landlords to help reduce tenants' energy costs [29]. GPF [7] concurs, saying that GBs enhance the tenants' resilience to pay rent and avoid defaults. Nedbank [38] shows that utility cost is the second highest expense for the South African tenant and particularly with "above-inflation tariff hikes," the ability to afford rental is eroded.

Oblunmi et al. [32] found that in the commercial space, GBs see lower vacancies and tenant turnover, which creates an uninterrupted flow of returns on the investment. This correlates to the point made previously with regard to tenant benefits, as if tenants have more resilience in their ability to make rental payment then they are less likely to default [7, 29, 38]. If tenants are not defaulting, lower rate of evictions and lower tenant turnover will result.

3.3.5 Tax Benefits

Although there was an abundance of literature on various tax benefits, the paper restricts itself to tax benefits that affect the South African housing developer. Literature shows how tax policy has adjusted to incentivize developers in GBT.

In South Africa, Section 12L of the Income Tax Act, (as amended in 2015), allows for "deductions in respect of energy efficiency saving." Every kilowatt-hour saved as a result of energy efficient upgrades allows the owner to claim a 95-cent reduction in their taxes [39, 40]. This provision is only for savings due to energy efficiency upgrades i.e. insulation and shading but not for renewable energy generation [7, 40].

Section 12B of the Act provides tax incentives for assets that produce renewable energy such as biomass, wind, solar and generation [39]. It allows the owner of the asset to write it off at an accelerated period of 3 years, or 1 year if it's a solar asset generating 1 MW or more [40]. There are considerable tax benefits to the developer of affordable housing as an incentive for green building.

3.3.6 Green Building Cost Premium

Many studies agree with the common perception that GBs cost more to produce [14, 16, 19, 33, 41]. As a result of this "green and affordability are often not perceived to be achievable at the same time [and] green affordable housing is often not perceived as economically feasible" [20]. However, these studies focus heavily on the initial capital cost and fail to take into account life-cycle costing [13, 18, 37].

Sustainable construction by its very nature must incorporates life cycle costing analysis [35, 37]. In fact, by considering life cycle periods, more sustainable and affordable housing can be produced at the same, or sometimes even lower cost [20, 25, 35]. While initial costs of green projects may be slightly higher than for conventional ones, savings in operating cost over the life of the building is generally offset and may far exceed incremental initial capital costs [13, 20, 21]. Many green technologies can pay for themselves over a short period of time through lower maintenance, operation and utility costs [13, 16, 20, 21].

The perception of Green costing more came about when it was still new and so: Limited technology, lack of relevant skills, knowledge and expertise leading to a premium charged on both technology and services, meant that there was a real green price premium [18, 33]. However, that is not always the case today: the cost of green materials and technologies have decreased and more accessible, especially supply chains for green materials and equipment are maturing with time; professionals such as architects and contractors are being trained and accredited through GB forums; as the industry improves the delivery of green design and technology [18, 20, 25].

Adopting a life-cycle approach, green affordable housing has shown to be less costly, in terms of the net present value (NPV) of the development, in comparison to a conventional equivalent [20]. Green affordable housing does not automatically include a premium during construction and can, in fact, often be constructed for less than conventional affordable housing [13]. This point is crucial as the "perceived cost premium" is the main reason why developers don't build green [13, 18, 20, 25, 27, 33, 37, 42].

4 Conclusion

The paper aimed to build a case on GBT and affordable housing to incentivize private developers into seeing the benefits of building green.

In fulfilling the first objective, the paper found that GBTs do offer numerous ways of GBT that can be used by affordable housing developers.

Addressing the second objective, it was found that tenants are very likely to benefit financially from occupying a green affordable unit; for as long as tenants work against the danger of the 'rebound effect'.

In terms of the third objective, clear case was made for the private developer of affordable housing to build green: GBs do not necessarily cost more than conventional buildings and there are many measures that can be used to 'green' a building at little or no additional cost; For GBTs that do cost more, the difference can be offset by the reduced operating expenses and maintenance of the property or through consumption charges to the tenant; GBs receive favourable tax treatments; GBs enjoy increased property value; GBs tend to have better life cycle costing. They are cheaper to run and have lower maintenance costs; GBs improve tenant resiliency and improve their ability to make rental payments. Therefore, the landlord may enjoy more consistent rental payments and lower tenant turnover.

The study found that building green can improve profitability and return on investment. By highlighting financial and investment benefits, it may stimulate an increased interest by private developers to build green in the gap market.

5 Recommendations

There are five recommendations from this study. 1. Green building should be encouraged and incentivised as they can help make rental housing more affordable for low income tenants; 2. Information regarding the benefits of green buildings should be advertised in order to raise awareness of the benefits they provide to the landlord and tenant; 3. Tenants should be taught how to use their Green Buildings in order to avoid a potential 'rebound effect'; 4. Property owners should elect to build green because, if done correctly, the benefits of green buildings over the longer term will always outweigh the initial costs; 5. The landlord should not elect to charge a large green rent premium on their building, and rather charge a small premium or none at all as the increase in tenant affordability leads to huge direct and indirect benefits.

References

1. Newton, C., Schuermans, N.: More than twenty years after the repeal of the group areas act: housing, spatial planning and urban development in post-apartheid South Africa. J. Hous. Built Environ. **28**(4), 579–587 (2013)
2. CAHF.: Understanding the challenges in South Africa's Gap Housing Market and opportunities for the RDP resale market. Centre of Affordable Housing Finance in Africa, Johannesburg (2015)
3. Andersen, N.: Wait for homes: Gauteng housing unit backlog now at over a million (2018). https://www.thesouthafrican.com/wait-for-homes-gauteng-housing-unit-backlog-now-at-over-a-million/
4. Muhoro, G.N.: Towards innovative approaches for affordable housing in the gap market: a case study of Khayelitsha Township in Cape Town, South Africa. Ph. D. University of Cape Town (2015)
5. Dugard, J., Clark, M., Tissington, K., Wilson, S.: The Right to housing in South Africa. Foundation for Human Rights, Johannesburg (2016)
6. Tipple, G.: Housing policy-making in Africa: ten common assumptions. Habitat Int. **49**, 413–418 (2015)
7. Gauteng Partnership Fund.: Green Building Guideline Medium Density Affordable Housing. Johannesburg: WSP Contacts, pp. 1–30 (2014)

8. Newton, D.: Public-Private Cooperation crucial to solving SA housing crisis. Old Mutual, Johannesburg (2017)
9. Tissington, K.: Minding the Gap: An Analysis of the Supply of and Demand for Low-Income Rental Accommodation in Inner-City Johannesburg. Socio-Economic Rights Institute, Johannesburg (2013)
10. Ham-Baloyi, W.T., Jordan, P.: Systematic review as a research method in postgraduate nursing education. ScienceDirect Health SA Gesondheid **21**, 120–128 (2016)
11. McCabe, A., Pojani, D., van Groenou, A.B.: The application of renewable energy to social housing: a systematic review. Energy policy **114**, 549–557 (2018)
12. Xiao, Y., Watson, M.: Guidance on conducting a systematic literature review. J. Plann. Educ. Res. **39**, 1–20 (2017)
13. Bradshaw, W., Connelly, E.F., Cook, M.F., Goldstein, J., Pauly, J.: The costs and benefits of green affordable housing. New Ecology and The Green CDCs Initiative, Cambridge (2005)
14. Mutwiri, G.E.: An Analysis of Benefits of Green Buildings. Undergraduate. University of Nairobi (2013)
15. Shazmin, S., Sipan, I., Sapri, M.: Property tax assessment incentives for green building: a review. Renew. Sustain. Energy Rev. **60**, 536–548 (2016)
16. Wallbaum, H., Ostermeyer, Y., Salzer, C., Zea Escamilla, E.: Indicator based sustainability assessment tool for affordable housing construction technologies. Ecol. Ind. **18**, 353–364 (2012)
17. Ahmad, S.F.: Techniques for improving lives by green low-income housing. In: International Conference on Sustainable Construction Materials and Technologies (2016)
18. Al-Ali, S.Z.: Analysis of Project Costs for Green Buildings in the UAE–A Case Study. The British University in Dubai (2014)
19. Chan, C., Ma, T.: Energy efficiency housing in South Australia – a gap analysis between the expected and actual benefits. Procedia Eng. **164**, 446–452 (2016)
20. Geng, L.: The application of information and incentives as tools to promote green affordable housing development. Ph.D. Massachusetts Institute of Technology (2004)
21. Hands, E.V.: Housing can be both green and affordable|Innovative projects demonstrate how. Master's. The University of Montana (2003)
22. Manoj, P.K.: Research paper economics prospects and challenges of green buildings and green affordable homes-concept: a study with reference to Ernakulam, Kerala. Economics **2** (12), 45–49 (2013)
23. Ross, N., Bowen, P., Lincoln, D.: Sustainable housing for low-income communities: lessons for South Africa in local and other developing world cases. Constr. Manage. Econ. **28**(5), 433–449 (2010)
24. Copiello, S.: Achieving affordable housing through energy efficiency strategy. Energy Policy **85**, 288–298 (2015)
25. Zhang, L., Wu, J., Liu, H.: Turning green into gold: a review on the economics of green buildings. J. Cleaner Prod. **172**, 2234–2245 (2017)
26. Environmental Protection Agency: Energy Efficiency in Affordable Housing. Local government climate and energy strategy series (2011)
27. Dwaikat, L.N., Ali, K.N.: Green buildings cost premium: a review of empirical evidence. Energy Build. **110**, 396–403 (2016)
28. Williams, B.E.: Overcoming barriers to energy efficiency for rental housing. Master's. Massachusetts Institute of Technology (2008)
29. Ambrose, A.R.: Improving energy efficiency in private rented housing: why don't landlords act? Indoor Built Environ. **24**(7), 913–924 (2015)
30. Bird, S., Hernández, D.: Policy options for the split incentive: increasing energy efficiency for low-income renters. Energy Policy **48**, 506–514 (2012)

31. Charlier, D.: Energy efficiency investments in the context of split incentives among French households. Energy Policy **87**, 465–479 (2015)
32. Olubunmi, O., Xia, P., Skitmore, M.: Green building incentives: a review. Renew. Sustain. Energy Rev. **59**, 1611–1621 (2016)
33. Hwang, B.-G., Zhu, L., Wang, Y., Cheong, X.: Green building construction projects in singapore: cost premiums and cost performance. Proj. Manage. J. **48**(4), 67–79 (2017)
34. Chegut, A., Eichholtz, P., Holtermans, R.: Energy efficiency and economic value in affordable housing. Energy Policy **97**, 39–49 (2016)
35. Emekci, Ş., Tanyer, A.M.: New Perspective on More Sustainable and Affordable Housing for Lower Income Group in Turkey-Assessing Life Cycle Cost. Green Lines Institute for Sustainable Development, Turkey, pp. 555–564 (2016)
36. Zuo, J., Zhao, Z.-Y.: Green building research-current status and future agenda: a review. Renew. Sustain. Energy Rev. **30**, 271–281 (2014)
37. Dolata, R.D.: Addressing the barriers to sustainable design in the process of developing multifamily affordable housing in Minnesota. Master's. University of Minnesota (2011)
38. Nedbank: Nedbank Funding makes green affordable housing a reality. Nedbank (2017)
39. Income Tax Act No. 25 of 2015, pg. 50 and 54. https://www.gov.za/sites/default/files/gcis_document/201601/395888-1act25of2015taxationlawsamendacta.pdf
40. Tucker, C.: Energy efficiency initiatives in South Africa, Bowmans (2017). https://www.bowmanslaw.com/insights/environmental-law/energy-efficiently-initiatives-south-africa/
41. Chen, M., Chen, J.G., Cheng, X.X.: Life cycle incremental cost-benefit analysis of green building. Appl. Mech. Mater. **71–78**, 4645–4651 (2011)
42. Sentman, S., Del Percio, S., Koerner, P.: A climate for change: green building policies, programs, and incentives. J. Green Build. **3**(2), 46–63 (2008)

Identification of Planning and Design Factors Influencing Stakeholders' Acceptance of Public Transport Facility

Innocent Shima Azege[1(✉)], Dilip Kumar Das[1], and Bankole Awuzie[2]

[1] Department of Civil Engineering, Central University of Technology, Bloemfontein 9300, Free State, South Africa
azegeinnoc2012@gmail.com, ddas@cut.ac.za
[2] Department of Built Environment, Central University of Technology, Bloemfontein 9300, Free State, South Africa
bawuzie@cut.ac.za

Abstract. Adequate functionality and utilization of transportation infrastructure is critical to national development of a country. In view of this, South Africa's National Development Plan is set out to ensure that all transport facilities are functional and utilized by 2030. However, stakeholders reject some public transportation infrastructure projects or facilities across the nation thereby causing traffic and economic growth related problems. This paper has examined the geometric parameters of Mangaung Intermodal Transport Facility and the influence of planning and design factors on stakeholders' acceptance of public transport infrastructure. Physical observations and measurements of geometric parameters of case study facility as well as stakeholders' interview are used to obtain geometric parameters and stakeholders opinions. The geometric parameters are analysed using relevant empirical models while content analysis is used to understand the various interviews conducted. The results show that minimum turning radii, driveway width, parking bays size, proximity to trip ends, community need for a transport facility and parking facility size are function of stakeholders' acceptance.

Keywords: Design · Infrastructure · Planning · Public transportation · Stakeholders

1 Introduction

Resilient infrastructure is critical to achievement of Sustainable Development Goals (SDGs) by motivating industrialization; induce economic growth and poverty reduction of any country. Buhr [1] defined infrastructure as a platform that enables an agent to carry out economic activities. Such platforms are transportation, energy, telecommunication, water and sanitation, buildings etc. This plays important role in influencing both economic and social development of a nation. Among these infrastructure categories, it is worthy to note that transportation infrastructure is a catalyst to the development of nation's economic growth [2]. The impact of quality and adequate transportation infrastructure assets to a city or country offers it competitive advantages over others.

C. Aigbavboa and W. Thwala (Eds.): CIDB 2019, *The Construction Industry in the Fourth Industrial Revolution*, pp. 252–264, 2020.
https://doi.org/10.1007/978-3-030-26528-1_25

However, stakeholders in many parts of the world have reportedly rejected some of these public transportation infrastructure projects/assets [3, 4]. This action has resulted to increased cost of transportation, traffic congestion, costs associated with design change and reconstruction, abandoned projects and unemployment. In South Africa for instance, some public transportation facilities utilization is opposed by stakeholders as attributed to perceived inconvenience from its operations. The rejection in various reports is associated with poor planning and design parameters. For example, a non-acceptance of facilities such as Intermodal Transport Facility in Mangaung Central Business District (CBD) is observed to have caused problems like congestion around the area.

Although various studies have sought to investigate the causes of societal acceptance or non-acceptance of infrastructure projects in respective domains, findings from these studies have indicated the non-consideration of the influence of geometric parameters on the acceptance of public transportation infrastructure projects [5, 6]. The gap in the study of geometric parameters as factors influencing stakeholders' acceptance of transportation infrastructure motivates this research study. The objectives of this paper are to study the geometric parameters of Mangaung Intermodal Transport Facility and to obtain stakeholders' perception on various geometric parameters of a transportation infrastructure in Mangaung Metropolitan Municipality. To achieve this objective, the rest of the paper is structured as follows: literature review, research methodology, presentation and discussion of findings and conclusion.

2 Literature Review

A good transportation system in a country is an engine to its increased gross domestic product (GDP) through expansion of its productive capacity and make existing production inputs more active [7]. Developed countries like United States of America, United Kingdom and Canada as well as developing nations like China have paid serious attention to investments in transportation infrastructure. This has resulted in improved international trade and employment opportunities to citizens [8]. Despite the importance of transportation infrastructure, a number of factors do challenge its delivery, operations and management. Some of such factors include dwindling finances and rapid population growth, geometric design, design standard documents, traffic parameters, and stakeholders' management [9]. However, some of the important factors that largely influence planning and design of transportation infrastructure and their higher acceptability include stakeholders' engagement and involvement, socio-economic factors, influencing stakeholders' perception, geometric parameters for transportation infrastructure, and traffic management and control factors.

2.1 Stakeholders' Engagement and Involvement in Transportation Projects

Recent studies have revealed poor stakeholders' engagement and involvement in transportation infrastructure projects as a factor that affects both its delivery and sustainability. According to Bourne [10], Tengan and Aigbavboa [11], the success of public transportation infrastructure depends on stakeholders' acceptance. Their acceptance or opposition to its existence is however, influenced by the level of information provided to them about project impacts, sense of ownership and the process of involving and engaging relevant stakeholders [5, 6]. Woltjer [12] pointed out that Dutch citizens are not always giving adequate information about infrastructure projects. He added that the engagement and participatory planning and construction process is challenged by the resistance of the public to be actively involved on the ground of lack of professional skills or knowledge. It is also common in some situations that the integrity of the person/agency communicating and/or managing an infrastructure project influence other stakeholders' perception about the project. Where such a project is seen as a manipulative attempt to take away some benefits or cause environmental hazards in the end, the infrastructure project may face non-acceptance by stakeholders based on trust [13].

2.2 Socio-Economic Factors Influencing Stakeholders' Perception

According to Kumar, Amandeep and Singh [14], the charges associated with the use of public transportation infrastructure influence stakeholders' behaviors towards a project. It is expected that public transportation is affordable by most people so as to enhance effective mobility for trips demands [15]. It is further noted by Cohen, Reichl and Schmidthaler [16] that a change in transportation technology or land use of a place is a positive or negative motivator to acceptability of a transportation infrastructure at a particular location. He added that some stakeholders tend to oppose the siting of a transport facility because it might change their existing transport system or land use benefits. In Indian transportation planning and design for instance, the roles and contributions of relevant stakeholders are disregarded because of administrative or political decisions about a transportation infrastructure projects. It is commonly found with public transportation infrastructure projects that once the concerns of stakeholders are neglected, it results to protest or the existing anger becomes amplified uncontrollably [17]. Some stakeholders instead of protest decide to refrain from being part of the project especially where quality is compromised. Besides, there may be built-up conflict among stakeholders of the transportation infrastructure [18].

2.3 Geometric Parameters for Transportation Infrastructure

Geometric parameters define a transportation infrastructure in terms of the shape bounded by points or lines. It therefore gives various shapes within its context. In transportation, such geometric parameters are widths, lengths, curvature (vertical and horizontal) as well as grade. Different parameters have different specifications for an infrastructure. The lane width of 2.70 m to 3.65 m is recommended at straight of carriageway [19]. This lane width can however be increased to 3.70 m at horizontal curve [20]. Vehicles' turning considered very important in transportation infrastructure planning and design. The horizontal curves on highway are designed with consideration of varying vehicles length. Council for Scientific and Industrial Research [21] recommends minimum internal radius of 6.2 m for passengers' car unit, 12.8 m for single unit vehicles and 13.1 m for buses on highways. Roess and Prassas [22] added some road geometric factors that are a function of design speed. Such factors include the minimum turning radius Rmin = v2/127(e + f) where e is the angle of superelevation, f is the coefficient of friction and v is the design speed. Council of Scientific and Industrial Research [20] also uses speed function for determination of stopping sight distances (SSD) of 30 m for 30 km/hr, 50 m for 40 km/hr, 115 m for 80 km/hr and 210 m for 120 km/hr.

2.4 Traffic Management and Control

Traffic control system is designed to ensure that it is easy to understand by the users. This is possible especially where road markings are done using a national standard. At a signalized intersection for instance, the right of way should be given to pedestrian and cyclists travelling in parallel direction to conflicting vehicles [23]. Timing of traffic movement at an intersection, bus terminal or taxi rank has potential effect on human behaviors. Due to value that human beings attach to time, the duration within which a traffic waits at an intersection influences acceptance or non-acceptance to its utilization. Additionally, passengers' waiting time for vehicle arrival at a terminal or vehicle waiting time for passengers before departure is critical design factor for stakeholders' buy-in of a transport facility [24]. However, waiting time at a bus station is a subjective parameter as regards events that surround it. Fan, Guthrie and Levinson [25] argues that waiting time at bus stop or terminal without basic amenities like seats and shelter is felt more than equivalent duration at a facility with these amenities.

3 Study Area

For the purpose of this investigation, the transportation infrastructure such as the 'Intermodal Transport Facility' (popularly known as the new Taxi rank) in the CBD of Bloemfontein (a medium sized city located in Mangaung Metropolitan Municipality) has been used as the case study. The Intermodal Transport Facility is located in CBD where it is surrounded by public offices, shopping centers and malls. These offices and shops motivate travel demand for people within Bloemfontein and neighboring towns. Adjacent to the facility is the Interstate Bus Line Terminal. Both Intermodal Transport Facility and Interstate Bus Line Terminal are fed with traffic from a three-lane street, Hanger Street. Harvey Street is a two-lane one-way carriageway which traffic from the facility are expected to exit into it. The non-utilization of Intermodal Transport Facility has caused some traffic related problems such as congestion on Hanger's Street, St Andrew's Street and St George's Street during peak hours around these public transportation nodes as shown in Figs. 1 and 2. The facility was completed in 2012 but its utility is only piloted. Following users' protest against its functionality, the facility is left abandoned and locked against unauthorized access. The transport facility has three floors of parking lots, which can accommodate about 480 vehicles. The intermodal transport facility has one-lane entry and one-lane exit carriageways as shown in Figs. 3 and 4 respectively. It is observed that the traffic congestion in the CBD is as the result of the temporary taxi rank along Peet Street, Harvey Street, St Andrew Street and Douglas Street. The intersections of roads with high traffic volume in the area have functional traffic signals and appropriate road markings to guide movement of vehicles and pedestrians.

Fig. 1. Temporary taxi rank on Douglas Strt

Fig. 2. Minibuses parked on Peet Strt.

Fig. 3. Entry point into the facility.

Fig. 4. Exit point of the facility.

4 Research Methodology

This research adopted the case study of the Mangaung Intermodal Transport Facility in Central Business District (CBD) in Bloemfontein. In order to obtain required data, the study used physical observation, measurement and semi-structured questionnaire. This public transportation facility is used as case study because of the users' protest against its functions as transportation services facility and it is left abandoned. The study therefore uses the case to study transportation infrastructure geometric parameters through physical measurements and observations. It also conducted semi-structure interview with stakeholders to get their perception on different planning and design parameters and factors. Different empirical models were used to determine geometric parameters while content analysis was used to understand stakeholders' perceptions. Different design and planning factors that influence stakeholders' acceptance were identified from their stakeholders' responses.

4.1 Data Collection

This study took physical measurements of various parking lots, turning curve lengths and driveways width within Intermodal Transport Facility. The lengths and the widths of each parking lot was measured, the tangent lengths and horizontal curve lengths at entry and exit points of each parking lot and driveway horizontal curves were measured for each floor using grip wheel tape. The driveway and lane widths were also measured. The various floors geometric structure was drawn to guide physical measurements and data entry as shown in Figs. 5, 6 and 7. The parking lots are numbered from A1 to A21 on the first floor, B1 to B23 on the second floor but the third floor parking lots were not numbered. Accessibility and traffic signs were observed in the facility to examine their various functions. Nine local community members coded LC1 to LC9 and eight taxi association members coded TA1 to TA8 were selected by convenience purposive sampling. The choice of the sampling technique is because; users of the facility carried out the protest to its utility. The semi-structured interview conducted sought the opinions of users about the planning and design factors that influence stakeholders' acceptance of public transportation infrastructure in the study area. The interview continued with local community and taxi association members until their responses were seemed repeated or similar with more interviews conducted. The content analysis of the interview reports as coded for each interview was used to understand and identify various factors that can influence stakeholders' acceptance.

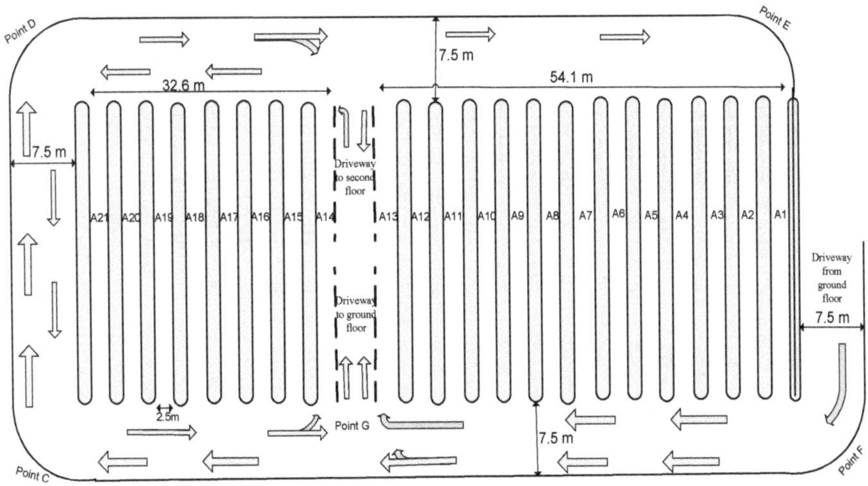

Fig. 5. First floor parking lots and driveway.

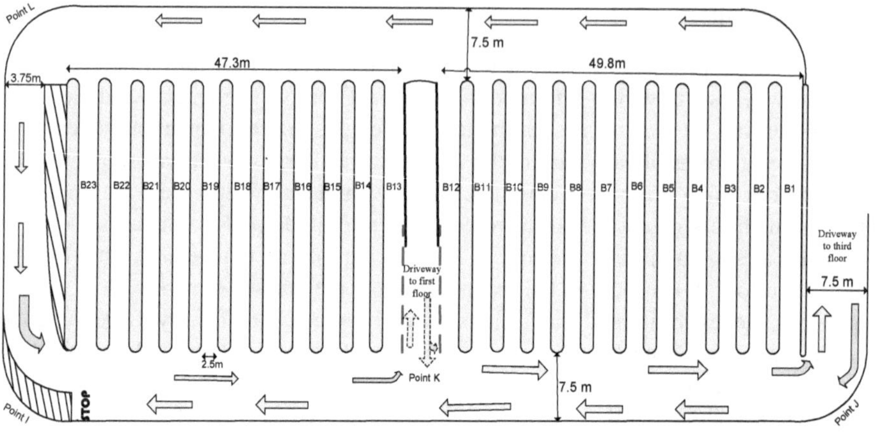

Fig. 6. Second floor parking lots and driveways.

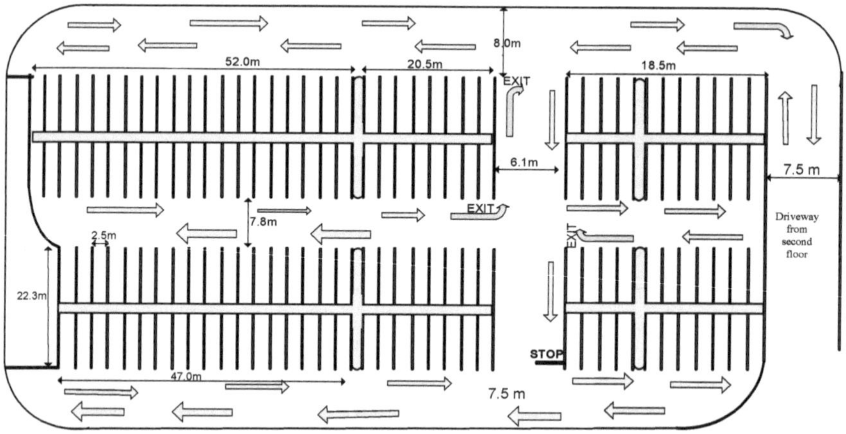

Fig. 7. Third floor parking lots and driveways.

4.2 Data Analysis

The physical measurement geometric parameters such as the lengths, widths and lengths of curves taken from Mangaung Intermodal Transport Facility were used to determine other required unknown geometric parameters. Different empirical models such as Eqs. 1 to 4 were used to calculate minimum turning radius, lane width, and number of cars in a parking lot.

$$Rmin = (180 X C)/\pi\theta \tag{1}$$

$$Rmin = 2C/\pi \tag{2}$$

$$\text{Rmin} = \text{T} \tag{3}$$

$$\text{N} = \text{L}/6.6 \tag{4}$$

Where Rmin = minimum turning radius; C = length of curve; θ = angle between two tangent lines; T = length of tangent line; N = the number of vehicles and L = parking lot length.

5 Results and Discussion

This section presents the results from the preliminary data collected for the research studies. Table 1 presents the geometric parameters of the first floor, second floor and third floor of the Intermodal Transport Facility in CBD of Mangaung Metropolitan Municipality in Free State of South Africa.

Table 1. Geometric parameters of intermodal transport facility floors.

			Entry point			Exit point		
Parking lot	Length (m)	Width (m)	Tangent length (m)	Angle between tangent lines (°)	Turning radius (m)	Tangent length (m)	Angle between tangent lines (°)	Turning radius (m)
First floor (21 parking lots)								
A1–A21	42.0	2.6	3.6	90	3.6	0.6	90	0.6
Second floor (23 parking lots)								
B1–B13	42.0	2.6	0.6	90	0.6	3.6	90	3.6
B14–B23	42.0	2.6	3.9	90	3.9	3.6	90	3.6
Third floor								
CA1–28	22.4	2.4	1.0	90	1.0	0.9	90	0.9
CB1–11	22.4	2.4	1.0	90	1.0	0.9	90	0.9
CC1–26	22.4	2.4	1.0	90	1.0	0.9	90	0.9
CD1–11	22.4	2.4	1.0	90	1.0	0.9	90	0.9

The first floor has 21 parking lots of 42.0 m long each, numbered A1 to A21. Each of these parking lot has a width of 2.6 m. This shows that the parking lot in the first floor can accommodate passenger's car unit (pcu) which measures 2.1 m wide without difficulties. This is similar to the second floor parking lots. The third floor parking lots are 2.4 m wide. The width of first and second floor is sufficient for passengers' car as recommended by South African Institute of Civil Engineers standard of 2.5 m as minimum. Roess and Prassas [22] also recommend a minimum parking bay width of 2.4 m. This means that the parking bay width is adequate for all parking lots in the transport facility. The parking lot of length 42.0 m can therefore accommodate 6 vehicles resulting to a total of 126 vehicles on first floor and 138 vehicles on the second floor (with 23 parking lots) as recommended by Roess and Prassas [22]. However, the South Africa Institute of Civil Engineers gives allowable space for each vehicle in the parking lot as 5.0 m. In this case, each parking lot can be occupied by 8 vehicles giving a total of 168 vehicles on first floor and 184 vehicles on the second floor. The third floor parking lots CA1 to CA28, CB1 to CB11, CC1 to CC26 and CD1 to CD11 are 20.0 m each with 2.3 m pedestrian walkways through them resulting to two of 10.0 m for each parking lot. The 5.0 m marked for vehicles on the floor is within the recommendation of South Africa Institute of Civil Engineers.

The angles between the parking lots and the driveways at the entry or exit point of parking lot are 90°. This makes the internal turning radius to be equivalent to the length of tangent lines. For instance, the turning radius at entry into parking lots is 3.6 m on the first floor for A1 to A21, 0.6 m for B1 to B13 parking lots, 3.9 m for B14 to B23 parking lots. The third floor parking lots at their entry points is 1.0 m. At the exit points of the parking lots, the turning radius at first floor for all parking lots is 0.6 m, all parking lots for second floor is 3.6 m while those for third floor is 0.9 m each. According to United States Institute of Transportation Engineering and South African Institute of Civil Engineers, the minimum turning radius for passenger cars is 2.3 m and 1.6 m respectively. This means that the minimum turning radii at the entry points are adequate for A1 to A21 and all B14 to B23 parking lots. The radius for B1 to B13 and all on the third floor are not adequate according to these standards. The turning radii of all the parking lots of second floor are above the minimum recommendation at exit points. The exit points radii for first and third floors are below the recommended standards. This implies that drivers of vehicles can conveniently maneuver at entry points of all parking lots of first floor, parking lots B14 to B23 and at exit points of all parking lots of second. However, maneuverability at all exit points of first and second floor parking lots and at entry points of B1 to B13 on the second floor is challenging to the vehicle drivers. The maneuverability on various lanes on the driveways in the Intermodal Transport Facility is presented in Table 2.

From Table 2, the geometric characteristics of various driveways in the facility are shown. The driveway from ground to first floor has entry point at first floor at point F. The width of driveway is 7.5 m with one lane of 3.75 m. The driveway that exits traffic from first floor to ground floor has two lanes in one way with 7.5 m total width. The driveway around the first floor is a two-lane two-way. This has internal and external lanes that have horizontal curves at points C and D. The table shows that all lanes on the facility have width of 3.75 m. The Council for Scientific and Industrial Research of South Africa states that the minimum lane width should be 3.7 m. Each width is therefore adequate for passenger cars. However, the minimum turning radii for internal lanes at points C, D, H and K and exit point G at first floor are below the recommended standard spelt out by South African Institute of Civil Engineers and US Institute of Transportation Engineering. The results from Table 2 shows that the lane width is sufficient for passenger cars in the facility as recommended by all standard of South African Council for Scientific and Industrial Research. The results are indication that the turning of vehicles on internal lanes of first floor at points C, D, G and H and internal lane at K on the second floor will be not be convenient for drivers of moving vehicles. This is also the same with the case of internal lanes on the third floor.

Table 2. Geometric characteristics of lanes in intermodal transport facility

Point	Lane position	Length of curve (m)	Tangent length (m)	Angle between tangent lines (°)	Minimum turning radius (m)	Width of lane (m)
First floor						
C	External lane	7.0	–	–	4.5	3.75
C	Internal lane	–	0.8	90	0.8	3.75
D	External lane	5.0	–	–	3.2	3.75
D	Internal lane	–	0.6	90	0.6	3.75
F		6.3	–	–	4.0	3.75
G		–	0.6	90	0.6	3.75
Second floor						
H	External lane	–	3.6	90	3.6	3.75
H	Internal lane	–	0.6	90	0.6	3.75
I		8.0	–	–	5.1	3.75
J	External lane	7.2	–	–	4.6	3.75
J	Internal lane	–	0.7	90	0.7	3.75

Table 3 presents the perception of the stakeholders who have stake in intermodal transport facility. 65% identified proximity of transport facility to residences and offices, 53% have considered the number of vehicles available to board passengers at public transport facility and time spent at a transport facility to travel as critical factors, 47% of interviewees said that transport fare, economic activities provided at the facility and community need for a transport facility are factors which influence stakeholders' participation in operations or utility of a transport infrastructure. On the impact of transport facility on environment, width of driveway and turning radius, 41% of respondents demonstrate agreement to the factors that they have influence on their perception of a transportation infrastructure. The size of parking space, passengers' accessibility to transport facility and the quality of services associated with a facility are motivators of their acceptance or non-acceptance as shown by 35% of them. Less attention is however, paid to other factors like land expropriation compensation, project ownership, project implementer and effect of weather changes. This shows that factors such as facility proximity to trip ends, availability of vehicles at transport facility, travel time, environmental impact as well as turning radius and driveway width should be taken serious during planning and design of transportation infrastructure for improved stakeholders' buy-in.

Table 3. Information collected from interviewees

No.	Information	No. of occurrence			
		LC	TA	Total	Percentage (%)
1	Transportation infrastructure's closeness to residence and public office	6	5	11	65
2	The fare for using transport facility	5	3	8	47
3	Number of vehicles to carry passengers at all time	4	5	9	53
4	The impact of transportation facility on the environment	4	3	7	41
5	Compensation on land expropriation	1	1	2	12
6	The transportation infrastructure project ownership	0	2	2	12
7	The quality of services of a transport facility to the public	2	4	6	35
8	The quality of public transport facility	3	2	5	29
9	Transport facility need to a community	5	3	8	47
10	Economic activities provided by a transport facility	2	6	8	47
11	Transport facility project implementation party/agency	1	0	1	6
12	The width of driveway in a transport facility	0	7	7	41
13	The effect of weather changes in transport facility	0	3	3	18
14	The size of parking lot	2	4	6	35
15	The length of parking bays	0	5	5	29
16	The nature of turning curves	1	6	7	41
17	The road markings and signs	2	4	6	35
18	Similarities or differences with other similar projects	0	2	2	12
19	Accessibility of a transport facility by passengers	3	3	6	35
20	Time spent on a transport facility to travel	5	4	9	53

6 Conclusion

Sustainable infrastructure development has been a challenge in developing countries. This challenge is commonly found with transportation, which is a critical infrastructure for social and economic development. One of the issues affecting sustainable transportation infrastructure is stakeholders' buy-in of transportation infrastructure projects or facilities. Different transportation infrastructures are abandoned as the result of public rejection to its operations. This paper has examined the influence of planning and design factors on stakeholders acceptance of this type of infrastructure. However, this study is carryout in the context of Mangaung Intermodal Transport Facility and Mangaung CBD. Further studies can be carried out on stakeholders rejected transport facilities in other areas. It can also be done by extending studies on other stakeholders of Mangaung Intermodal Transport Facilities like government transportation agencies and contractors.

References

1. Buhr, W.: What is Infrastructure? Volkswirtschaftliche Diskussionsbeiträge (107-03) (2003)
2. Janusova, L., Cicmancova, S.: Improving safety of transportation by using intelligent transport systems. Procedia Eng. **134**, 14–22 (2016)
3. Matebesi, S.Z.: Insurgent Citizenship and Sustained Resistance of a Local Taxi Association1. Narrating the Everyday: Windows on Life in Central South Africa, p. 420 (2018)
4. Nwanna, C.R.: Right to the city: lagos, an emerging revanchist city in Nigeria?. Adv. Soc. Sci. Res. J. **5**(4), 1–50 (2018)
5. Banerjee, A., Duflo, E., Qian, N.: On the Road: Access to Transportation Infra-structure and Economic Growth in China, No. w17897, National Bureau of Economic Research (2012)
6. Pradhan, R.P., Bagchi, T.P.: Effect of transportation infrastructure on economic growth in India: the VECM approach. Res. Transp. Econ. **38**(1), 139–148 (2013)
7. Glaeser, E.L., Ponzetto, G.A.: The political economy of transportation investment. Econ. Transp. **13**, 4–26 (2018)
8. Abuzeinab, A., Arif, M.: Stakeholder engagement: a green business model indicator. Procedia Econ. Finance. **18**, 505–512 (2014)
9. Oyedele, O.A.: The challenges of infrastructure development in democratic governance. Constr. Econ. Manage. **1**(6119), 1–15 (2012)
10. Bourne, L.: Targeted communication: the key to effective stakeholder engagement. Procedia-Soc. Behav. Sci. **226**, 431–438 (2016)
11. Tengan, C., Aigbavboa, C.: Level of stakeholder engagement and participation in monitoring and evaluation of construction projects in Ghana. Procedia Eng. **196**, 630–637 (2017)
12. Woltjer, J.: The' public support machine': notions of the function of participatory planning by Dutch infrastructure planners. Plann. Pract. Res. **17**(4), 437–453 (2002)
13. Beecher, N., Harrison, E., Goldstein, N., McDaniel, M., Field, P., Susskind, L.: Risk perception, risk communication, and stakeholder involvement for biosolids management and research. J. Environ. Qual. **34**(1), 122–128 (2005)
14. Kumar, G., Amandeep, K., Singh, K.K.: Public transport and urban mobility: perception of people on services of public transport in Bathinda City, Punjab, India. Int. Res. J. Soc. Sci. **3**(12), 8–14 (2014)
15. Huang, L., Duan, B., Bi, J., Yuan, Z., Ban, J.: Analysis of determining factors of the public's risk acceptance level in China. Hum. Ecol. Risk Assess.: Int. J. **16**(2), 365–379 (2010)
16. Cohen, J.J., Reichl, J., Schmidthaler, M.: Re-focusing research efforts on the public acceptance of energy infrastructure: a critical review. Energy **76**, 4–9 (2014)
17. Anderson, C.: The networked minority: how a small group prevailed in a local windfarm conflict. Energy Policy **58**, 97–108 (2013)
18. Das, D.K.: Influence of Administrative and Political Authorities' Decisions on the Construction of Community Development Projects in India (2017)
19. Belwal, R., Belwal, S.: Public transportation services in Oman: a study of public perceptions. J. Public Transp. **13**(4), 1 (2010)
20. South African National Road Agency Limited: Design Guidelines for Single Carriageway National Roads (2009)
21. Council for Scientific and Industrial Research: Geometric Design and Layout Planning (2009)
22. Roess, R.P., Prassas, E.S., McShane, W.R.: Traffic Engineering, 4th Edn. (2011)
23. Maurer, M., Gerdes, J.C., Lenz, B., Winner, H.: Autonomous Driving, vol. 10, pp. 978–1003. Springer, Heidelberg (2016)

24. Lu, H., Burge, P., Heywood, C., Sheldon, R., Lee, P., Barber, K., Phillips, A.: The impact of real-time information on passengers' value of bus waiting time. Transp. Res. Procedia **31**, 18–34 (2018)
25. Fan, Y., Guthrie, A., Levinson, D.: Waiting time perceptions at transit stops and stations: effects of basic amenities, gender, and security. Transp. Res. Part A: Policy Pract. **88**, 251–264 (2016)

Evaluating the Impact of Emerging Contractors' Failure to Secure Tenders from the Tender Market

Sikhumbuzo Lufele(✉)

Department of Construction Management and Quantity Surveying, Cape Peninsula University of Technology, P. O. Box 1906, Bellville 7535, South Africa
lufeles@cput.ac.za

Abstract. Purpose - The study evaluates the impact of emerging contractors' failure to secure contracts from the tender market as part of the Western Cape's Contractors Development programme (CDP). **Design** - The study adopted a quantitative research method, the study targeted emerging contractors in the Western Cape. The data was collected by using a questionnaire survey with closed-ended questions and was distributed to the population of 16 emerging contractors with CIDB grade 3 and 5. The descriptive and inferential statistics were used to analyse the study. **Findings** - The study has found that skills deficiency as the main reason for emerging contractors' failure to secure contracts from the tender market. The findings have further revealed among others emerging contractors are unable to tender, interpret construction drawings, plan, estimate and enter into negotiations. **Conclusion** - Emerging contractors on the Western Cape's CDP programme encounter difficulties to secure contracts from the tender market owing to their skills deficiency. This situation if further compounded by the fact that the Western Cape CDP has not set aside contracts for training purposes. **Practical implications** - The Western Cape CDP should review its policy with regards to training projects, contractors struggle to secure tenders from the tender market. The Western Cape CDP should not maintain the current status quo with regards to training projects otherwise it runs a risk of contractors not able to complete the mentorship programme.

Keywords: Emerging contractors · Tendering · Planning · Estimation · Construction drawings

1 Introduction

In South Africa, Small and Medium Enterprises (SMEs) play a pivotal role in term of economic growth. Moreover, SMEs also contribute in creation of jobs, income generation and output growth. The total contribution of SMEs in terms of employment is approximately 60% and 40% of output [4, 25]. This contribution is significant in terms of addressing socio-economic challenges such as unemployment, poverty and the shortage of skills. SMEs in the construction sector are sometimes referred to as emerging contractors. Emerging contractors demonstrate a potential but not yet fully

C. Aigbavboa and W. Thwala (Eds.): CIDB 2019, *The Construction Industry in the Fourth Industrial Revolution*, pp. 265–275, 2020.
https://doi.org/10.1007/978-3-030-26528-1_26

competent in all aspects of the business. Moreover, in many instances emerging contractors still require training and mentorship to enhance their technical and businesses skills [21]. It is common knowledge that throughout the processes of running their businesses, emerging contractors have been confronted with a number of challenges. Consequently, some of these emerging contractors are no longer in business and some kept their businesses were able to overcome all their challenges. These problems have been reported on by a number of researchers [11, 15, 22]. The latter have reported that the problems confronted by emerging contractors do not go away. Malongane [15] highlights that these challenges confronted by emerging contractors affect their businesses both internally and externally. Thwala and Phaladi [23]; Iruka and Shakantu [10] explains that emerging contractor businesses in south Africa do not remain in business for a very long time. Problems relating emerging contractors were reported on more than seven years ago but still appears to have an impact on the South African construction sector [6].

2 Literature Review

2.1 Tendering

Securing a contract from the tender market requires a contractor to put together a sound and competitive tender. It is therefore up to the contractor ensure compliance to tender requirements, if a contractor fails to adhere to the tender requirements then it could lead to disqualification or fail to secure a tender. Martin and Root [17] discuss problems confronted emerging contractors and highlights general lack of knowledge; deficiencies in the knowledge of pricing procedures, contractual rights and obligations, management techniques and principles, technology, as well as general law. Despite intense competition, that on its own does not prevent contractors from entering the market. According to Thwala and Phaladi [23] there is fierce competition at the lower grades of cidb who are competing against each other and that has a ripple effect on small contractors in terms of maintaining a sustainable workflow. Iruka and Shakantu [10] argue that it up to the emerging contractors to remain in business and still be competitive and economically sustainable by developing a dynamic business strategy that would embrace the changing trends and conditions in today's business world. Further Thwala and Phaladi [23] reveal that emerging contractors are confronted with increasing competition due to the decline in demand for construction. In response to the market conditions, contractors decide to lay off their workers.

2.2 Interpretation of Construction Drawings

Construction drawings are necessary for most spheres of the construction industry as being the best means of transmitting detailed and often complex information from the designer of the project to all those involved in the construction process. Ramaswamy [20] highlights that construction drawings are used to communicate the architectural and engineering design of a construction project. Babalola [2] states that construction drawing is one of several diagrammatic forms used in the building design process.

Babalola and Eastman [1] define construction drawings as graphically complex, depicting an assemblage of sub-assemblages and parts. Ordinarily, contractors would be issued with two types of construction drawings namely architectural drawings and engineering drawings. The architect together with other designers of the project should ensure that the message communicated via the construction drawings is clear and unambiguous so as to eliminate any possibility of misreading and misinterpretation of the drawings. Equally so, contractors and those who work with construction drawings should have an understanding of interpreting construction correctly. With regards to interpreting construction drawings as part of a planning process, emerging contractors on more occasions encounter difficulties and subsequently remain clueless. Thwala and Phaladi [23] show that emerging contractors are unable to read and interpret construction drawings due to lack of technical skills. Malongane [15] alludes that some of the emerging contractors start their construction businesses without possessing any technical skills. However, Chilipunde [5] maintains that one of the basic skills that would make a successful contractor is the ability to read and interpret constructions drawings. It is crystal clear that construction drawings in the construction industry are important and must be used by contractors as a guide in terms of steps to be followed in constructing a building.

2.3 Planning

De Marco [8] defines planning simply as mechanism to ascertain "What" is going to be done, "How" things are going to be done, "Who" will be doing activities and "How much" activities will cost. Planning is a very important step towards achieving successful project implementation, typically the construction industry values planning very highly as the project can never be undertaken unless planning has been done thoroughly. Cook and Williams [7] argue that it would be difficult to envisage a successful project without proper planning. If planning is done properly it would then highlight the potential of the project and not only from the clients' perspective but also from the contractors' perspective [7]. In most cases, planning is closely linked to improving construction project performance where it is possible to meet the project parameters which are cost, quality and time [14]. In the case of emerging contractors, planning is still a challenge as contractors often fail to undertake proper planning and implementation as the results projects are not completed on time [23, 24]. Van Vuuren et al. [25] argue that emerging contractors are unable to plan for projects due to inadequate management skills to deliver on the project. Thwala and Malongane [22] concur with the latter with regards to the inadequacy of management skills among emerging contracting firms. Thwala and Malongane [22] further argue that contractors need to ensure that the job is executed in a way to make a profit.

2.4 Estimation

The cost estimate can be defined as a process of putting together an estimate in terms of how much the project will cost. This is often done by considering various elements of the project. An estimate should only be treated as such and never be regarded as accurate [3]. According to Lester [13], estimation requires a structured approach and whatever method of estimation is chosen, the level of accuracy has to be indicated. Estimation, therefore, is a very important part of any project as it forms the basis for subsequent control. For this reason, estimation is crucial such that a contractor has to ensure that an accurate and competitive estimate is prepared that would lead to contract award. Chilipunde [5] argues the fact that contractors are unable to estimate or price, respond to the tender document and take into consideration inflation is evident enough to show how deep problems are in terms of emerging contractors. Thwala and Phaladi [23] also highlight poor estimation and lack of pricing of tenders as one of the challenges faced by emerging contractors and results in contractors' failure to secure projects from the tender market.

2.5 Negotiations

Negotiations in the construction industry are paramount to contractors because they are responsible in terms of procuring materials and other goods for the project from the suppliers. Perks and Oosthuizen [19] state that negotiations are a good and powerful tool used in concluding agreements with suppliers. Furthermore, negotiation is a first step in having a long-term relationship with the suppliers. A good relationship between the contractor and the supplier must exist for a successful relationship. In the absence of a cordial relationship, then contractors would be left vulnerable and find it difficult to carry out a construction project with a supplier as a business partner to the project. The main objective with regards to contractors is to obtain the best possible price from the right supplier in the right quantity, at the right time [9]. Emerging contractors are unable to negotiate in order to make money instead of contractors lose money due to poor preparation and poor negotiations [5].

3 Research Methodology

The study has adopted a quantitative research approach. A questionnaire survey with closed-ended questions were distributed in August 2017 to 16 CIDB grade 3 to 5 emerging contractors who have been selected to go through mentorship programme Western Cape CDP. The questionnaire comprised of 7 point Likert scale questions to establish the extent to which emerging contractors' failure to secure contracts from tender market has effected emerging contractors. The Likert scale was organised as follows 1 = Not affected, 2 = Slightly affected, 3 = Moderately affected, 4 = Affected, 5 = Highly affected, 6 = Extremely affected, 7 = Completely affected, U = Unsure. In addition, the questionnaire consists of two sections, namely Section A and Section B. Section A consisted of biographical data of emerging contractors, whilst Section B focused on questions pertaining to the emerging contractors' skills deficiency. Respondents were first informed of the focus and the purpose prior to the commencement of the survey.

Leddy and Armrod [12] highlights that to analyse quantitative data one would use descriptive or inferential statistics. Typically, the Statistical Package for Social Sciences (SPSS) software was used to capture and analysis of descriptive data. Descriptive analysis measures the central tendency and is divided into three categories namely mode, mean, and median. A reliability testing was conducted by making use of Cronbach's alpha. According to Welman et al. [26], reliability is concerned with the findings of the research and relates to the credibility of the results. Cronbach coefficient alpha is a reliability test for a single occasion data collection which is an estimate of internal consistency of responses to different scale items [24]. Reliability in terms Cronbach's alpha coefficient varies from 0 to 1; the closer the coefficient is to 1, the more reliable it is.

4 Findings

4.1 Profile of the Respondents

In terms of Table 1, the study sought to establish gender distribution of the 16 respondents. From the figure below it is shown that of 81.3% (13) of respondents are male and 18.8% (3) are female, and this suggests both genders were represented, although females were underrepresented in the study. The age distribution of respondents, where 37.5% were aged between 21 and 30 years, 12.5% were aged between 31 and 40 years, 50% were aged between 41 and 50 years. The race groups of the respondents, 6.3% of respondents are white, 31.3% black, and the 62.5% which is the majority comes from the coloured race group. From an educational point of view more than 56.3% of respondents have a matric certificate while 43.8% have not completed secondary education. It is a concern that none of the respondents has a tertiary or a post-graduate qualification. This could be problematic with regard to having at least at tertiary qualification that would give respondents an academic experience rather than site experience. Furthermore, respondents could also have a challenge in understanding the business and construction related modules on the advanced phase of the programme. The cidb grades for groups 1, 2 and 3. It is clear that 50% of respondents are registered for cidb grade 4, followed by 37.5% for cidb grade 5 and 12.5% for cidb grade 2.

4.2 Reliability Testing

Table 2 below shows the Statistical Package for Sciences Software (SPSS) was used to test the scaled questions for reliability. The testing in terms of Cronbach's alpha coefficient is interpreted as follows: Values that are lower than 0.60 degrees are considered unacceptable, values with 0.70 degrees are considered as having low reliability, 0.80 degrees are considered as having moderate reliability and 0.9 degrees are considered having high reliability [16]. Also, degree values tend to be low when items less than 10 are tested.

Table 3 displays the results of the test for the normality. A non-significant result (sig value of more than 0.05) indicates normality [18]. If the obtained significance value of 0.00 (equal or less than 0.05), it suggests the violation of the assumption of normality. However, the significant difference was 0.056 which is higher than 0.05 therefore, the normality was not violated, hence ANOVA was adopted.

Table 1. Profile of sample

Gender	No.	%
Female	3	18.8
Male	13	81.3
Total	16	100%
Age group	No.	%
21–30 years	6	37.5
31–40 years	2	12.5
41–50 years	8	50.0
Total	16	100%
Race group	No.	%
White	1	6.3
Black	5	31.3
Coloured	10	62.5
Total	16	100
Qualification	No.	%
Secondary - not completed	7	43.8
Matric certificate	9	56.3
Total	16	100%
CIDB grade	No.	%
Grade 4	8	50%
Grade 5	6	37.5%
Grade 6	2	12.5%
Total		100%

Table 2. Reliability testing

Variables	Number of times	Cronbach's alpha coefficient	Rank
Tendering	6	0.85	Moderate
Interpreting drawings	4	0.99	High
Planning	5	0.96	High
Estimation	5	0.97	High
Negotiations	3	0.99	High

Table 3. Tests of normality for skills deficiency

	Kolmogorov-Smirnov[a]			Shapiro-Wilk		
	Statistic	Df	Sig.	Statistic	Df	Sig.
Skills deficiency	.197	16	.097	.954	16	.560

Table 4 summarises the null hypothesis test on emerging contractor skills deficiencies. There was no statistically significant difference in gender (0.69), race (0.55) and cidb grading (0.51). The acceptable statistically significance level was based on standard value $p > 0.05$.

Table 4. Null hypothesis for skills deficiency

Skills deficiencies	Gender (Sig)	Race (Sig)	CIDB grading (Sig)
Tendering	0.15	0.98	0.18
Interpretation of drawings	0.12	0.96	0.36
Estimation	0.02	0.73	0.52
Planning	0.03	0.72	0.14
Negotiations	0.01	0.56	0.52
Average	0.07	0.79	0.34

4.3 Results

4.3.1 Tendering

With respect to tendering, Table 5 shows that comply with JBCC contracts in the building industry is ranked first with a mean score of (4.81), comply with the closing date for tenders (4.81) followed by complying with GCC2000 contracts in the civil engineering industry. It is therefore evident that contractors are struggling to secure contracts from the open tender market due to contractors' lack of understanding of the dynamics of a tendering system. The average mean score of (4.16) demonstrates the degree of being affected by the tendering system, this also shows that mentors need to expose contractors to all aspects of tendering including but not limited to types of tendering process, tender pre-qualification process and tendering procedures. Without all these tendering aspects contractors would find it difficult to tender or secure contracts from the market.

Table 5. EC's limitations on tendering system

Statement	No.	Mean	Std	Rank
Comply with JBCC contracts in the building industry	16	4.81	2.13	1
Comply with the closing date for tenders	16	4.18	1.60	2
Comply with GCC2000 contracts in the civil engineering industry	16	4.12	1.58	3
Understand the different types of tendering	16	4.12	1.58	3
Comply with the returnable documents for a tender	16	3.93	1.43	4
Understand the construction tendering system	16	3.81	1.47	5
Average	16	4.16	1.26	

4.3.2 Interpretation of Construction Drawings

Table 6 indicates that distinguish between a drawing for tendering purposes and a drawing for construction purposes is ranked first with a mean score of (4.06), followed by distinguishing between architectural and engineering drawings (4.06). This reveals that contractors on the Western Cape CDP mentorship programme are unable to read and interpret drawings. Moreover, the failure to read drawings is a limitation as contractors are expected to able to do as a requirement especially during the tendering process. The average mean of (4.03) shows that contractors are slightly affected by the lack of reading and interpreting construction drawings, this could subsequently exacerbate the situation unless Western Cape CDP address this aspect of mentorship.

Table 6. EC's interpretation of drawings

Statement	No.	Mean	Std	Rank
Distinguish between a drawing for tendering purposes and a drawing for construction purposes	16	4.06	1.28	1
Distinguish between architectural and engineering drawings	16	4.06	1.28	1
Interpret construction drawings	16	4.00	1.26	2
Read construction drawings	16	4.00	1.26	2
Average	16	4.03	1.27	

4.3.3 Planning

With respect to planning, Table 7 indicates that to prepare a Network Analysis is ranked first with a mean score of (4.25), followed by prepare a Gantt-chart (4.25). The findings show that contractors are negatively affected by the lack of planning for a project, consequently, contractors are unable to plan for a construction project. Furthermore, if a contractor fails to understand planning then it means that contractors do not undertake pre-tender, pre-contract and in-contract planning and can never deliver in a project within a given timeframe. The average mean score of (4.20) suggests that a gap exists in terms of planning and an intervention is warranted to emphasis on planning as it is a requirement to run any construction project successfully.

Table 7. EC's planning for a project

Statement	No.	Mean	Std	Rank
Prepare a Network Analysis	16	4.25	.85	1
Prepare a Ghant-chart	16	4.25	.93	1
Prepare for pre-tender planning	16	4.25	1.39	1
Prepare for in-contract planning	16	4.12	1.20	2
Prepare for pre-contract planning	16	4.12	1.20	2
Average	16	4.20	1.05	

4.3.4 Estimation

With respect to estimation, Table 8 indicates obtain rates from plant and equipment suppliers is ranked first with a mean score of (4.31), followed by obtaining prices from material suppliers (4.31). The respondents have shown they are struggling and not competent in terms of preparing a sound estimate for tenders as contractors are unable to obtain prices for both materials and plant from suppliers. The average mean score of (4.26) reveals that any tenders submitted by contractors participating on the Western Cape CDP mentorship programme may not be successful due to contractors not being able to put forward a financially sound estimate.

Table 8. EC's estimation of a project

Statement	No.	Mean	Std	Rank
Obtain rates from plant and equipment suppliers	16	4.31	1.25	1
Obtain prices from material suppliers	16	4.31	1.25	1
Estimate for construction projects	16	4.25	.85	2
Prepare a reasonable and acceptable estimate for a construction project	16	4.25	1.12	2
Obtain rates from subcontractors	16	4.18	1.04	3
Average	16	4.26	1.05	

4.3.5 Negotiations

With respect to negotiations, Table 9 indicates negotiate rates with suppliers of plant and equipment is ranked first with a mean score of (4.12), followed by negotiating rates with subcontractors (4.06). The findings reveal that contractors are not being able to negotiate material prices and rates with suppliers and subcontractors pose a threat in terms of making profits. The average mean score of (4.08) shows that contractors are slightly affected by this and require mentors to make changes in this regard so as to make more profits then loses.

Table 9. EC's negotiations of rates and prices with suppliers

Statement	No.	Mean	Std	Rank
Negotiate rates with suppliers of plant and equipment	16	4.12	1.25	1
Negotiate rates with subcontractors	16	4.06	1.18	2
Negotiate prices for materials with suppliers	16	4.06	1.18	2
Average	16	4.08	1.20	

5 Conclusion and Recommendations

The study has evaluated the impact of emerging contractors' failure to secure tenders from the tender market. Despite the fact that emerging contractors receive training to address their skills deficiencies, emerging contractors are not yet fully competent whilst on the mentorship programme to submit competitive bids. In addition, the emerging contractors' situation is further compounded by the lack of training projects from the Western cape CDP set aside for emerging contractors.

The Western Cape CDP should review its policy with regards to training projects, contractors struggle to secure tenders from the tender market. The Western Cape CDP should not maintain the current status quo with regards to training projects otherwise it runs a risk of contractors not able to complete the mentorship programme.

The following gaps were identified in the study and should be addressed in the following manner:

- To train contractors how to tender, tendering methods and tendering procedures in the public and private sector;
- To train contractors to interpret construction drawings and able to undertake construction projects according to the drawings provided to them by various consultants;
- To train contractors to undertake planning on various stages of the project including pre-tender, pre-contract and in-contract planning;
- To train contractors to build rates based on supplier's prices and sub- contractor's rates; and
- To train contractors to be able to negotiated prices and rates with material suppliers.

Further research should be done on evaluating the level training offered by service providers to emerging contractors on the mentorship programme. With regards to the training projects on the mentorship programme, further research is needed to establish whether lack of funds is a constraint faced by the Western Cape CDP to set aside such projects.

References

1. Babalola, O., Eastman, C.: Semantic Modelling of Architectural Drawings (2001)
2. Babalola, O.: A model based framework for semantic interpretation of architectural construction drawings. Ph.D. Thesis, Georgia Institute of Technology (2012)
3. Bennet, F.L.: The Management of Construction: A Project Life Cycle Approach, 1st edn. Butterworth-Heinemann, Burlington (2003)
4. Berry, A., Blottnitz, V., Cassim, M., Kesper, R., Rajaratnam, A., Seventer, D.E.: The economics of SMMEs in South Africa (2002). [http://www.tips.org.za/research-archive/trade-and-industry/item/204-the-economics-of-smmes-in-south-africa. Accessed 5 Aug 2016
5. Chilipunde, R.L.: Constraints and challenges faced by small, medium and micro enterprises contractors in Malawi. MSc Thesis, Nelson Mandela Metropolitan University (2010)
6. Construction Industry Development Board. 2004/2005. Annual Report 2004/2005. Pretoria
7. Cook, B., Williams, P.: Construction Planning, Programming & Control, 2nd edn. Blackwell, Oxford (2004)
8. De Marco, A.: Project Management for Facility Constructions: A Guide for Engineers and Architects (2011). http://www.springer.com/978-3-642-17091-1
9. Hugo, W.M.J., Badenhorst-Weiss, J.A.: Purchasing supply management, 6th edn. Van Schaik, Pretoria (2011)

10. Iruka, A., Shakantu, W.: A theoretical perspective of emerging contractors' economic sustainability: a strategic management approach. In: Proceedings: 9th Built Environment Conference, 2–4 August 2015, Durban, South Africa (2015)
11. Lazarus, S.: An integrated skills development model for emerging contractors in the Eastern Cape. Unpublished dissertation. Nelson Mandela Metropolitan University (2005)
12. Leedy, P.D., Ormrod, J.E.: Practical Research: Planning and Design. Pearson, Boston (2010)
13. Lester, A.: Project Planning and Control, 4th edn. Butterworth-Heinemann (2003)
14. Lines, B.C., Sullivan, K.C., Hurtado, K.C., Savicky, J.: Int. J. Constr. Educ. Res. (2015). https://doi.org/10.1080/15578771.2013.872733
15. Malongane, D.: Challenges facing emerging contractors in Gauteng. M.Tech. Thesis. Tshwane University of Technology (2014)
16. Maree, K., Pietersen, J.: Surveys and the Use of Questionnaires: First Steps in Research. Van Schaik, Pretoria (2007)
17. Martin, L., Root, D.: Engineering contractors in South Africa: Interactions and learning. J. Eng. Des. Technol. **8**(1), 64–79 (2010)
18. Pallant, J.: SPSS Survival Manual – A Step by Step Guide to Data Analysing Using the SPSS Program, 4th edn. McGraw Hill, Berkshire (2010)
19. Perks, P.S., Osthuizen, N.: Exploring supplier negotiations best practices and supplier relationships strategies in South Africa. Rev. Integr. Bus. Econ. Res. **2**(1), 333 (2013)
20. Ramaswamy, J.N.: Reading and interpreting construction drawings. Course notes #403 (2016). http://www.pdhsite.com/courses/Reading%20and%20Interpreting%20Construction%20drawings.pdf. Accessed 1 July 2016
21. South Africa. National small business amendment Act No. 102 of 1996 (1996)
22. Thwala, A.: Challenges facing black owned small and medium construction companies: a case study of Nelspruit - Mbombela municipality. J. Econ. Behav. Stud. **6**(10), 771–776 (2014)
23. Thwala, W., Phaladi, M.J.: An exploratory study of problems facing emerging contractors in the North West Province of South Africa. In: Proceedings 4th Built Environment Conference, 17–19 May 2009, Livingstone, Zambia (2009)
24. Tredoux, C., Durrheim, K.: Numbers, Hypotheses & Conclusions: A Course in Statistics for the Social Sciences. UCT Press, Cape Town (2002)
25. Van Heerden, H., Mashatole, A., Burger, M.: Constraints faced by small contractors in the Gauteng province of South Africa. In: Proceedings of Conference of Informatics and Management Sciences, 24–28 March 2014, Zilna, Slovak Republic (2014)
26. Welman, C., Kruger, F., Mitchell, B.: Research Methodology. Oxford University Press Southern Africa, Cape Town (2005)

Enhancing Financial Communication in Quantity Surveying Practice

Rolien Terblanche(✉), Obinna Ozumba, and David Root

School of Construction Economics and Management,
University of Witwatersrand, 1 Jan Smuts Avenue,
Johannesburg 2000, South Africa
{rolien.labuschagne,Obinna.Ozumba,
david.root}@wits.ac.za

Abstract. The construction industry is becoming increasingly complex, including the aspects that deal with communication. Subsequently, financial reports generated during the project delivery process are also becoming more complex and thus harder to comprehend. The purpose of this study to explore financial communication in quantity surveying practice, with a view to understanding their role, the financial communication emanating from them, guidelines for successful financial communication, and determining a reliable quality test for such financial communication. A systematic literature review was done on financial communication, readability tests for financial communication and financial communication in quantity surveying practice. The main limitation of the study is that it is based on a preliminary literature review. However results indicate certain requirements for accurate readability tests on financial communication including Plain English guidelines and the Gunning Fog Index test for textual components, while also taking standard construction terms into account, where there could be more than three syllables. The readability and legibility model has four main categories for optimisation of a document, each of which has elements that need perfecting in order to create transparent, efficient and effective financial communication. The implication of this research is that in the face of noted problems with financial communication from Quantity Surveyors, there is need to explore the quality of such information, and to determine the existence, if any, and the use of relevant quality assurance systems in their communication.

Keywords: Readability · Financial communication · Quantity surveying · Cost reports and feasibility studies

1 Introduction

The research interest in this paper relates to construction project financial communications emanating from quantity surveyors, and the impact of its comprehensibility on project outcomes. Laskin [1] argues that financial communication is much like other forms of communication. Financial communication, however, is situation-specific. In the context of the project delivery process, financial communication provides financial data regarding the performance of the project to investor/s, developer/s and the project team.

C. Aigbavboa and W. Thwala (Eds.): CIDB 2019, *The Construction Industry in the Fourth Industrial Revolution*, pp. 276–286, 2020.
https://doi.org/10.1007/978-3-030-26528-1_27

Xu, Fernando and Tam [7] found that investors expressed their concerns about financial communication that is becoming more complex. Financial communication are becoming less comprehensible to ordinary investors. Unreadable financial reports are the outcome of firms' desire to hide poor performance as well as the firms' complex operational communication systems. While between 70 and 90 percent of investors claim that they understand what they are reading from the financial communication provided, the comprehension tests proved otherwise [8]. Day [2] explains that construction projects are growing more complex and increase in scale and that teamwork is all the more important. It becomes more important for everyone to speak the same language, working with the same standards and embracing new technologies. Lu et al. [3] and He et al. [4] developed a measurement model for measuring the complexity of construction projects. According to their studies, as projects increase in complexity, so too does communication between project stakeholders. However, in the construction industry, different professions and organisations have developed standards and terminology, often independently of other project stakeholders. Increasing complexity of financial communication is a generic phenomenon but is reflected on construction projects as they also become more complex.

Complexity hinders both efficiency and productivity and reducing complexity as well as making quality enhancements are the most likely ways to overcome this (Clark [5]). The construction industry as a whole includes various forms of communication as well as stakeholders. This study, however, focusses on the financial communication from quantity surveyors. Among 33 causative factors of poor communication identified by Gamil and Rahman [6], five relate to poor financial communication, namely; lack of effective communication between construction parties; lack of effective communication system and platform; poor communication skills; lack of support for advanced communication technologies; and a lack of understanding among parties.

Thus, there is an emergent problem in that complex financial communication in the construction industry, is often ineffective, partly due to the confusing and incomprehensible communication which in some aspects, hinder the efficiency and productivity of construction projects and ultimately their sucess. The current study has a key objective of understanding the role of financial communication in quantity surveying practice. Another key objective is to determine how this financial communication can be measured in terms of readability and successfulness. The final objective is to gather guidelines for effective financial communication in quantity surveying practice.

2 Financial Communication

Financial communication is an important aspect of transparency and is a key feature for investor confidence and the quality and credibility as a whole (Brendin Prat, Cliff and Pricewaterhouse Coopers [9]). Brendin Prat, Cliff and Pricewaterhouse Coopers further explains that consistency in financial communication is vital as inconsistent financial communication could mislead investors. The information provided must be accurate, fair and true. EY [10] states that financial communication is a vital link between investors and the company. DVFA [11] compiled a structure of principles for effective financial communication. The three dimensions for effective financial communication are target group orientation, transparency and continuity. Thus, poor financial communication is inaccurate, inconsistent, confusing and misleading communication that creates a lack of confidence for investors. Almaşi, Gomoi, & Cuc [12] explains that financial communication is facing various obstacles concerning the homogenisation of communication forms. Due to globalisation, financial communication on global capital markets are visibly changing. People from different parts of the world still face language barriers in the financial communication field. Davenport et al. [13] identified eight key factors for the successful management of projects while studying 24 global companies with 31 knowledge management projects. Amongst the key factors are culture and processes, the development of a common purpose, and a mutual language for knowledge identification and selection as cited in Desouza and Evaristo [13]. According to the World Health Organisation [14], the use of standardised terminology will result in more efficient services. Standardised information could reinforce and improve cost analysis, statistical reporting, and expected performance reporting (financial communication). Walker [15] stated that the relationship between language use, typographic design, how a document is to be used and the effectiveness of its use is vital. Effective communication arise from the perfect combination of language use, typographic design while keeping the goal of communication in mind. Refer to Fig. 1 below.

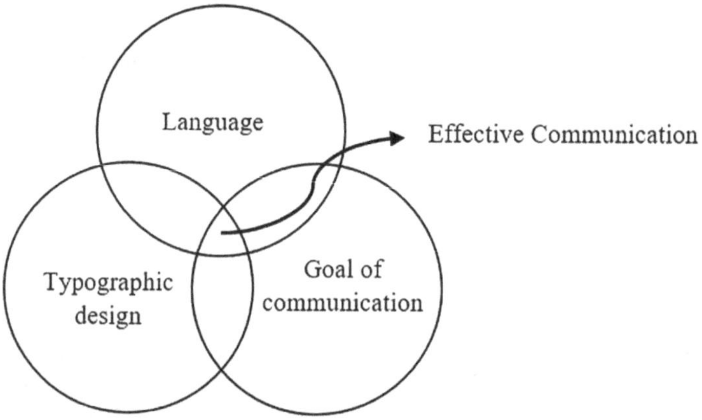

Fig. 1. The relationship between language, typographic design and the goal of communication (adopted from Walker [15]).

Palmieri, Perrin and Whitehouse [16] investigated the pragmatics of financial communication. They found that language skill, prior knowledge and text presentation plays a role in understanding financial documentation. In the empirical study they examined the extent to which contextual factors affected the ability of financial communication users to process information. The results indicated that the text presentation format dictates the level of skill needed to successfully comprehend financial documents. Successful financial communication calls for an integrated approach where obtaining and maintaining trust with investors is vitally important [17].

The success of financial communication thus depends on accuracy, truth, consistency, standardised terminology, text presentation, readability and a common language. The better the financial communication, the higher the investor confidence. The prior knowledge and language skill of the audience needs to be analysed in order to communication on the correct level of readability. This is equally applicable to the construction industry where similar concerns will apply. Financial communication needs to produce comprehensible information that contributes to successful investments in the fixed asset being delivered.

As the quantity surveyor is the consultant in the construction and property development industries whom advises clients on the optimum use of funds and uses financial communication extensively [18], it is vital to understand the how the quantity surveyor uses financial communication and to assess the quality of their financial communications. Quality in financial communications is an important aspect, such that 'readability tests' has been developed for quality assurance, which are discussed hereunder.

3 Readability Tests

Multiple readability tests have been developed in order to determine the level of success with regards to the understandability of financial communication and it is imperative to understand which test/s would be applicable for measuring the success of the communication. Hochhauser [19] explains that the effective communication of a document can be evaluated in terms of formatting, text factors, subheadings, headings and readability strategies. Readability refers to the ease of understanding a written document due to the style of writing, where legibility refers to the typeface and layout of a written document. Hochhauser [19] listed several guidelines for good documents design. These guidelines include: Work should be familiar to the reader; sentences must be short, simple and direct; readability analysis should be done, words containing more than three syllables should be avoided; the style of print must be easy to read; left margins needs to be justified and right margins needs to remain ragged; lower and upper case letters must be used; headings must be simple and close to the text; and plain English should be used. Gray and Leary [20] created four categories for the 228 identified elements that affect readability: Style, content, format and features of organization. They found that all four categories were deemed important to the readability of a document, however, content was deemed slightly more important, then style, format and lastly features of organization.

Palotti, Zuccon and Hanbury [21] investigated the readability of documents as presented to readers. They highlight that readability is an important aspect of documents. They named five of the most used readability measures: Automated Readability Index (ARI), Coleman-Liau Index (CLI), Flesch-Kincaid Grade Level (FKGL), Gunning Fog Index (GFI) and Simple Measure of Gobbledygook (SMOG). All these tools measure the readability of documents by analysing the length of the words and sentences in documents. The Gunning Fog Index is defined as a way to measure the readability of financial statements [22]. Loughran and McDonald [22], however found that the Gunning Fog Index lacks specifications when it is used to assess financial documents and produce errors. The Fog Index measures the average length of sentences in combination with the proportion of complex words. Complex words in the Fog Index is defined as words with more than three syllables. The longer the sentences and the greater the proportion of complex words, the harder the readability [23]. They argue that text used in business context has a high percentage of complex words which is easily comprehended by investors and recommend the use of the file size of 10K (annual report) as a substitute readability test for financial documents [23]. Bonsall, Leone and Miller [24] contradicted Loughran and McDonald and found that the file size 10K readability test has measurement errors. The file size 10K readability test does not separate textual and non-textual components and thus generated errors. It is suggested that the Plain English readability test captures the readability quality of financial documents better.

The problems that the Plain English Handbook [25] addresses includes: the passive voice, hidden verbs, unnecessary words, legal and financial jargon, terms defined numerously, abstract words, superfluous details, long sentences and an unreadable design and layout. Plain English serve thus as a guideline for readability and legibility. Matveeva, Moosally and Willerton [26] conducted a study on 'plain language' in the twenty-first century with the aim to reintroduce the use of plain language to the professionals. Plain language proposes an approach to language and design to produce comprehensible and readable documents. The 'plain language' started to trend in several countries in the 1970s. Recent legislation in combination with public and private sector initiatives in the United States of America encouraged the use of 'plain language' greatly. The principles and guidelines for plain language include: a careful audience analysis; logical organization of information; lists with bullets, active voice; words that are used every day; short paragraphs and sentences; and a usability test is required. Schriver [27] compiled the timeline of the development plain language from 1940–2015. According to this timeline, plain language only gained momentum from 2000. Schriver defines plain language as the following: "communication is in plain language if its structure, wording and design are so clear that the intended audience can easily find what they need, understand what they find, and use that information". Campbell et al. [28] did a study on the plain-style preferences of professionals in the United States of America. They found an overwhelming preference for plain language amongst the professionals, whose native tongue is English. The use of plain language helps combat systemic inequity and is therefore vital for ethical purposes (Cheung [29]).

Given the above, and the increasing complexity of information in the construction industry, the use of readability tests on financial communication could be used to evaluate the current status of the quality of financial communication generated by quantity surveyors, and the quality assurance processes for such communication, so as to improve such communication.

4 Quantity Surveying Practice and Financial Communication

Since the quantity surveyor is a professional that acts as the financial and development consultant to the construction and property development industries, advising clients on the best use of their funds as well as approaches to maximise the use of human and physical resources, the quantity surveyor is central to the generation of financial communication on construction projects [18]. The Association of South African Quantity Surveyors describe the responsibilities of a quantity surveyor as including preparing budgets and feasibility studies for potential projects and overseeing costs and reporting these to clients throughout the design and construction stages.

Feasibility studies are the evaluation of the viability of a development project. It is used as a tool for analysing if a proposed project can function under given assumptions, for example the technology used and the monetary aspects of the construction work (Hoffman [30]). Feasibility studies are an analysis of the viability of a proposed project (Hofstrand and Holtz-Clause [31]). The feasibility study helps answer the question: "should one proceed with the proposed idea?" Feasibility studies are thus a tool used to make decisions on capital investment. Thus, quantity surveyors fulfil a similar role to financial advisors and financial brokers. A feasibility study is a critical factor for the success of a project and projects can fail due to incorrect facts or assumptions (Mukherjee and Roy [32]). Feasibility studies must be reviewed and analysed by experts in order to enhance the quality thereof (Hyari and Kandil [33]).

The purpose of cost reporting is to advise a client in a construction development project of the expected closing cost thereof. The estimated final cost may be expressed as a variance against the budget or as an absolute value (Royal Institute of Chartered Surveyors [34]). Sullivan [35] explained that cost reporting could be conducted in a similar manner where everyone completes the same steps in a regulated and more sophisticated manner. Feasibility studies and cost reports are thus the means of financial communication to clients from quantity surveyors. Essentially the QS is responsible for cost reports and feasibility studies for construction projects, which constitute financial communication on which investment decisions are made. The readability of feasibility and cost reports are arguably imperative for successful investment decisions. It is important for investors and decision-makers to fully comprehend such financial information. It is therefore important to understand this critical area with regard to QS practice.

5 Research Design

This study aims to answer the following research question: How can financial communication in quantity surveying practice be improved? The research design followed in order to answer the research question is a systematic literature review. The databases used for this research includes Google Scholar as well as the Wits University library e-journal database that consists of various journal databases. The following key words were used consistently in all databases: financial communication, readability tests for financial communication and financial communication in quantity surveying practice. The criteria for the articles to be included in this study is the following: it must be relevant to the topic; it must be scientific and peer reviewed journals, conference papers, books, master and/or doctoral theses; most of the articles must not be older than five years; data must have been collected ethically.

6 Analysis of the Literature Review

There is an appreciable amount of research on financial communication, readability and readability tests. However, there is limited research on financial communication in the construction industry and specifically in quantity surveying practice.

Table 1. The findings of the readability tests.

Readability test	Findings
Automated Readability Index (ARI)	Top five most used test. Analyses the readability of documents by analysing the length of words and sentences in documents
Coleman-Liau Index (CLI)	Top five most used test. Analyses the readability of documents by analysing the length of words and sentences in documents
Flesch-Kincaid Grade Level (FKGL)	Top five most used test. Analyses the readability of documents by analysing the length of words and sentences in documents
Simple Measure of Gobbledygook (SMOG)	Top five most used test. Analyses the readability of documents by analysing the length of words and sentences in documents
Gunning Fog Index (GFI)	Top five most used test. Analyses the readability of documents by analysing the length of words and sentences in documents. This test is used for financial documents, however it produces errors as it does not take standard business words into consideration that is more than three syllables
File size of 10K	This test is used for analysing financial documents. It does not separate textual and non-textual components and thus generates errors
Plain English	Captures the readability quality of financial documents the best

Considering Table 1, the key result from the systematic review is that the most accurate readability tests for financial communication in quantity surveying practice, is the Plain English guidelines that should be followed together with the Gunning Fog Index test for the textual components, while taking standard construction terms into account that might be more than three syllables. The results of systematic literature review are further analysed in Fig. 2. As illustrated in the model (Fig. 2), there are four main categories of readability to take into consideration when aiming to achieve readable and legible financial document: Content, style, organisation and format. In order to get the **content** right, an audience analysis should be done and the goal of the communication should be well understood.

The **style** of the document can be tested by readability tests and should take the following guidelines in consideration: Use short paragraphs and sentences; use short common words (general construction terms are acceptable even though it might be

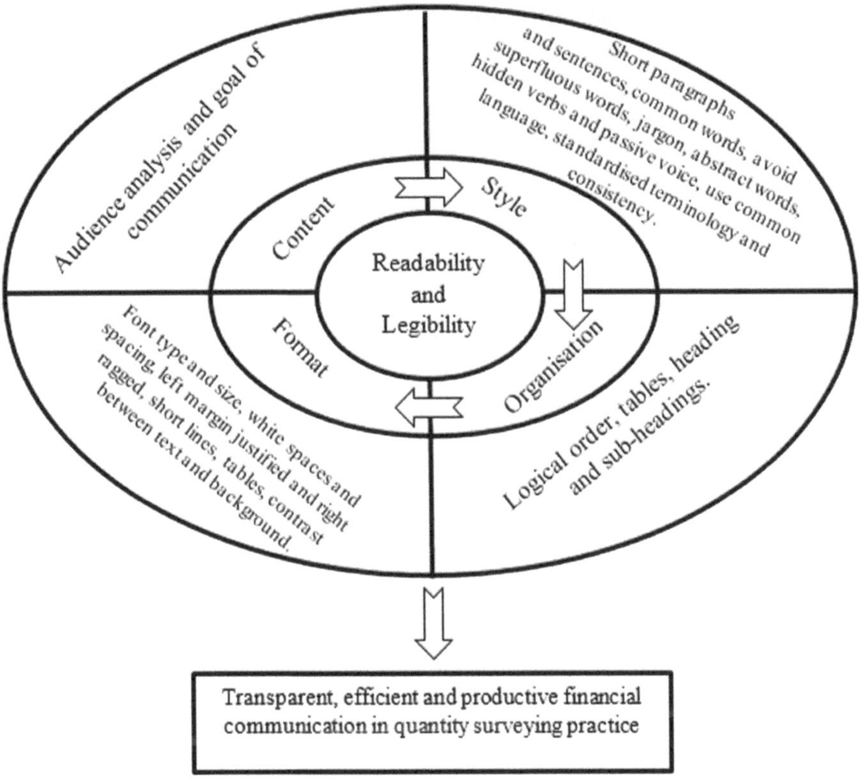

Fig. 2. Readability and legibility model (adopted and developed from [19, 20, 22, 23, 25, 26]).

more the three syllables); avoid superfluous words, jargon, abstract words, unnecessary details, passive voice and hidden verbs. Use a common language, standardised terminology and consistency. The **organisation** of the document should be done in a logical order with headings and subheadings to simplify the content. Lastly, the **format** of the document refers to the design and layout to the document and that contributes to the legibility thereof. The font size should be appropriate and the font type must be simple. White space and appropriate spacing should be used as too dense wording becomes difficult to read. Left margins needs to be justified and the right margins needs to remain ragged. Short lines should be used. Tables could be used to clarify information. The contrast of the text on the background should contribute to easy legibility. It is possible to derive a readability and legibility model from the information in Fig. 2. However note that each of the four categories in the conceptual model has elements that need perfecting in order to create a transparent, efficient and effective financial communication that could be adopted by the quantity surveying profession.

7 Conclusion, Limitations and Recommendations

The purpose of this study to explore financial communication in quantity surveying practice, with a set of objectives that would generate meaningful contributions which would contribute to better financial communication in quantity surveying practice. The following were explored; purpose of financial communication, the role of the quantity surveyor with regards to financial communication, and guidelines for producing readable and legible financial document that promote efficiency and successful financial and business investment decisions. Through a systematic literature review, analysis of findings, and synthesis of relevant elements, a conceptual model for readability and legibility, which is applicable to quantity surveying practice has been developed. Financial communication is understandably a key feature for investor confidence. Therefore quality assurance such as readability tests should be performed on relevant financial documents emanating from quantity surveyors. The current study has found out the relevance of the Plain English guidelines and the Gunning Fog Index test, in addition to developing the basis for a possible working readability and legibility model. Regardless, looking at the broader picture, a common language, standard terminology, and required consistency, would be the most probable future for financial communication, especially for quantity surveying and related fields.

Based on the main limitation, which is the lack of an empirical section, it is suggested that future studies include empirical data analysis based on relevant professionals in the construction industry. It would also be beneficial in future studies, to analyse Quantity surveyors' feasibility studies and cost reports.

References

1. Laskin, A.V.: The Handbook of Financial Communication and Investor Relations, 1st edn. Wiley, Hoboken (2018)
2. Day, B.: New direction. RICS Constr. J. September–October 2017, 7 (2017)
3. Lu, Y., Luo, L., Wang, H., Le, Y., Shi, Q.: Measurement model of project complexity for large-scale projects from task and organization perspective. Int. J. Project Manage. **33**(3), 610–622 (2015)
4. He, Q., Luo, L., Hu, Y., Chan, A.P.: Measuring the complexity of mega construction projects in China—a fuzzy analytic network process analysis. Int. J. Project Manage. **33**(3), 549–563 (2015)
5. Clark, A.: Digital revolution. RICS Constr. J. November–December 2017, 7 (2017)
6. Gamil, Y., Rahman, I.A.: Identification of causes and effects of poor communication in construction industry: a theoretical review. Emerg. Sci. J. **1**(4), 239–247 (2018)
7. Xu, Q., Fernando, G.D., Tam, K.: Executive age and the readability of financial reports. Adv. Account. **43**, 70–81 (2018)
8. Jack, L., Davison, J., Craig, R. (eds.): The Routledge Companion to Accounting Communication. Routledge (2013)
9. Cliff, B.P., Coopers, P.: Financial communication: framework and practices (2018). https://www.pwc.fr/fr/assets/files/pdf/2018/07/pwc-financial-communication-framework-and-practices-uk.pdf

10. EY: Excellence in financial communication (2015). https://www.ey.com/Publication/vwLUAssets/EY_IPO_-_Excellence_in_financial_communication/$FILE/EY-IPO-Excellence-in-financial-communication.pdf
11. DVFA: DVFA Principles for Effective Financial Communication (2007). http://www.dvfa.de/fileadmin/downloads/Publikationen/Standards/principles_effective_financial_communication.pdf
12. Almasi, R., Gomoi, B.C., Cuc, L.D.: The financial communication and the accounting-audit-valuation trinomial. J. Econ. Bus. Res. **21**(2), 153–158 (2015)
13. Desouza, K., Evaristo, R.: Global knowledge management strategies. Eur. Manag. J. **21**(1), 62–72 (2003)
14. Wold Health Organisation: eHealth: standardized terminology (2006). http://apps.who.int/gb/archive/pdf_files/EB118/B118_8-en.pdf
15. Walker, S.: Typography & Language in Everyday Life: Prescriptions and Practices. Routledge (2014)
16. Palmieri, R., Perrin, D., Whitehouse, M.: The Pragmatics of Financial Communication. Part 1: From Sources to the Public Sphere (2018)
17. Whitehouse, M.: Financial Analysts and Their Role in Financial Communication and Investor Relations. Handbook of Financial Communication and Investor Relations, pp. 117–126 (2017)
18. Association of South African Quantity Surveyors. http://www.asaqs.co.za/?page=Publications
19. Hochhauser, M.: The informed consent form: document development and evaluation. Drug Inf. J. **34**(4), 1309–1317 (2000)
20. DuBay, W.H.: The Principles of Readability (2004). https://files.eric.ed.gov/fulltext/ED490073.pdf
21. Palotti, J.R.D.M., Zuccon, G., Hanbury, A.: The influence of pre-processing on the estimation of readability of web documents. In: 24th ACM International Conference on Information and Knowledge Management, pp. 1763–1766. ACM (2015)
22. Loughran, T., McDonald, B.: Measuring readability in financial disclosures. J. Finance **69**(4), 1643–1671 (2014)
23. Loughran, T., McDonald, B.: Plain English, readability, and 10-K filings. University of Notre Dame (2009)
24. Bonsall, S.B., Leone, A.J., Miller, B.P., Rennekamp, K.: A plain English measure of financial reporting readability. J. Acc. Econ. **63**(2–3), 329–357 (2017)
25. United States: Securities and Exchange Commission. Office of Investor Education and Assistance. A plain English handbook: How to create clear SEC disclosure documents. The Office (1998)
26. Matveeva, N., Moosally, M., Willerton, R.: Plain language in the twenty-first century: introduction to the special issue on plain language. IEEE Trans. Prof. Commun. **60**(4), 336–342 (2017)
27. Schriver, K.A.: Plain language in the US gains momentum: 1940–2015. IEEE Trans. Prof. Commun. **60**(4), 343–383 (2017)
28. Campbell, K.S., Amare, N., Kane, E., Manning, A.D., Naidoo, J.S.: Plain-style preferences of US professionals. IEEE Trans. Prof. Commun. **60**(4), 401–411 (2017)
29. Cheung, I.W.: Plain language to minimize cognitive load: a social justice perspective. IEEE Trans. Prof. Commun. **60**(4), 448–457 (2017)
30. Hoffman, A.: http://www.godfreyhoffman.com/civil-engineering-blog/bid/298843/why-is-a-feasibility-study-important
31. Hofstrand, D., Holtz-Clause, M.: What is a Feasibility Study? (2009). https://www.extension.iastate.edu/agdm/wholefarm/html/c5-65.html

32. Mukherjee, M., Roy, S.: Feasibility studies and important aspect of project management. Int. J. Adv. Eng. Manage. **2**(4), 98–100 (2017)
33. Hyari, K., Kandil, A.: Validity of feasibility studies for infrastructure construction. Jordan J. Civ. Eng. **3**(1), 66–77 (2009)
34. Royal Institute of Chartered Surveyors: Cost Reporting, 1st edn. RICS, London (2015)
35. Sullivan, J.: Little and large. RICS Constr. J. November–December 2017, 18–19 (2017)

Empirical Evidence on the Measurement of Information Asymmetry on the Pricing of REITs on the Johannesburg Stock Exchange

Oluwaseun Damilola Ajayi(✉) and Omokolade Akinsomi

University of the Witwatersrand, Johannesburg, South Africa
1565694@students.wits.ac.za

Abstract. The no-consensus nature of literature on measuring information asymmetry as it affects the pricing of stocks is fast becoming topical. The paper examines the level of information asymmetry associated with the pricing of 37 real estate investment trusts (SAREITs) over the last ten (10) years (2008–2018). Based on an exhaustive search and yield of unbalanced data from the price section of the I-NET (BFA) McGreggor database, the daily closing and opening price data were of 17 listed REITs were retrieved. The ordinary least square regression (OLS) model was adopted in a bid to test the relationship between information asymmetry other variables. These variables including cumulative bid-ask spread, price volatility, firm size and trading volume of 17 SAREITs (due to availability of relevant data) were controlled and regressed. Findings revealed that SAREITs have a low level of information asymmetry. This suggests that investors are able to make optimum investment decisions.

Keywords: REITs · Information asymmetry · Bid-Ask spread

1 Introduction

The concerns about the performance of South African real estate investment trusts (SAREITs) following their evolution and conversion period (2013) is indeed a motivation in understanding how they are priced. Globally, there has been considerable documentation on the issues around the pricing and performance of REITs; one is of information asymmetry which appears to be on the front burner in the financial markets as documented by literature. Existing body of knowledge established that the unpredictable nature of trading volumes in the financial sector is fast becoming worrisome as it has been argued back and forth that information asymmetry impacts upon investment participation on the stock market (with particular reference to bonds, shares and other forms of investment portfolios). Ekumah and Essel [16]; Tumwebaze, Orobia and Kamukama [36]; and Afzal [2] who had documented that information asymmetry (information failure) was a case wherein a party to an economic transaction possesses greater knowledge than the other party, were amongst scholars who made this assertion. Findings revealed that shareholders' wealth maximization appear anchored on the need to provide potential and existing investors with the most appropriate proxy in measuring information asymmetry. Abdul-Baki [1] for instance, had documented that

C. Aigbavboa and W. Thwala (Eds.): CIDB 2019, *The Construction Industry in the Fourth Industrial Revolution*, pp. 287–296, 2020.
https://doi.org/10.1007/978-3-030-26528-1_28

little consensus had existed among scholars of finance literature on the most appropriate information asymmetry proxy globally. Furthermore, the validity of price formation by both parties is fast becoming unreliable; and by implication, naivety of investment decisions seems to be the order of the day and mainstay.

Within investment portfolios, Blazenko [7] had reported that managers in most cases act independently without shareholders participation; by implication, trade volumes are on a downward trend with optimum returns ultimately affected. The real estate investment trusts just like other stocks are a unique and preferred investment portfolio (perhaps because of their high percentage of contribution to their shareholders) despite the number of risks associated. Akinsomi, Kola, Ndlovu, and Motloung [4] had earlier provided evidence as to how both local and international investors beam investment porchlights towards the REITs market. The paper which focused on the REITs structure in South Africa documented it as the only African country represented in the FTSE EPRA/NAREIT index and the S & P Global REIT index; by implication, this strengthens its importance and significance in the global real estate market. Interestingly, the South African REITs market ranks as the 9th largest REIT globally. The introduction of REITs particularly in emerging markets like South Africa appears to have witnessed a litany of research focusing on the impact of information asymmetry on the pricing of REITs in emerging markets in which Wei, Hsieh and Sirmans [38] had opined that there is a greater degree of information asymmetry when the Bid-Ask spread is larger; by implication, such REIT firms are associated with higher levels of debt usage as against equity, all other things being equal and as corroborated in Qu, Wongchoti, Wu and Chen [33]; and a consensus on the most appropriate scientific proxy for information asymmetry measurement. A number of scholarly articles had argued about different proxies for information asymmetry measurement; (Glosten [18]; Easley, Kiefer, O'hara and Paperman [15]; Engle and Lange [17]; Matoussi, Karaa and Maghraoui [27] and Vayanos and Wang [37]) had documented evidences on lack of liquidity as an indication of information asymmetry; more so, because a potential buyer/investor would appear to be hesitant to participate in a transaction believed to be skewed towards a shareholder/seller who possesses better information about the stock. The Bid-Ask spread on the other hand is believed to be a more scientific proxy towards the information asymmetry measurement. Narayan, Mishra and Narayan [31] had in their paper opined that the Bid-Ask spread is a better model which treats the spread persistence as a predictor of spread.

Khemraj [21]; Mossman [28]; Buckle and Beccalli [10]; Roll [34]; and Krinsky and Lee [23] had equally argued that the Bid-Ask spread empirically documented as the "amount by which the ask price exceeds the bid price for an asset in the market". In other words, the Bid-Ask spread is essentially the difference between the highest price that a buyer is willing to pay for an asset and the lowest price that a seller is willing to accept. In as much as there has been a topical argument in favour of the Bid-Ask spread being a scientific information asymmetry measurement for the finance and stock sector in most developed markets, it appears the REITs market; and in emerging markets are yet to share in the effects of measuring information asymmetry. Without this, investors tend to make naïve investment decisions; hence, the purpose of the article is to seek to measure the degree of asymmetric information on the pricing of REITs stocks on the JSE with a view to providing insights on enhanced investment decision making. The research question is: Does asymmetric information have impacts on the pricing of REITs stocks on the JSE? It was therefore hypothesized that:

- H_0 - If there is a high value for the index, it indicates higher levels of information asymmetry.
- H_1 - If there is a low value for the index, it indicates lower levels of information asymmetry.

2 Literature Review

2.1 Empirical Review on Information Asymmetry

The theory of asymmetric information which was developed in the 1970s and 1980s had become a persuasive justification for common phenomena that mainstream general equilibrium economics appear not to elucidate; the theory merely opines it as information imbalance between buyers and sellers which can lead to inefficient outcomes in certain markets. Breeden and Viswanathan [8] presented an insight on why REITs firms hedge based on a theory of mana-gerial responses to asymmetric information. They hinged their reasons the fact that managers who have superior abilities with respect to some risks or uncertainties will try to ensure that their superior abilities are quickly discovered by the market. On the other hand, managers with inferior abilities have incentives to reduce the efficiency of the learning process. The scholars also noted that these managers undertake risky variance-increasing activities. However, given that superior managers undertake hedging activities, lower ability managers may or may not hedge. This study was carried out only in a developed market; and further demonstrated the inconclusiveness of the asymmetry theory. According to Cooper, Downs and Patterson [12], the impact of information asymmetry was tested on whether if affects the relationship between trading volume and expected returns of REITs firms; it was found that predictability of real estate returns is generally more indicative of portfolio rebalancing effects than an adverse selection problem. This suggests that information asymmetry has been a subject of inquiry for decades; this then presents an opportunity for further research as this would provide an empirical solution to investment timing and ultimately expected returns. Again, this study was carried out in developed markets and to the exclusion of emerging markets. A number of literature had established opinions about information asymmetry. In [16], Information asymmetry is a situation where relevant information is not known to all the parties particularly involved in a transaction. In [2], it is noted that Information asymmetry dates back to the lemons principle where one party knows more than the other party to the same contract. Information asymmetry exists when there is an imbalance in knowledge of relevant factors and details on a contract between two parties. The existence of information asymmetry is characterized by two major problems, which are adverse selection and moral-hazard problems [22]. By implication, what this suggests is that these parties in a transaction possess more information than the others. Its measurement has also become topical as some scholars have stated that Information asymmetry can be measured on the basis of information quality and quantity. In [36], it is opined that Information quality is inherent with some proxies including reliable, timely, complete, fair and consistent information presented in clear, simple terms, relevant and understandable to decision makers or users. On the other hand, [2] opined that Information quantity indicates the amount of information available to a decision maker or user. Meanwhile and few decades ago, Malhotra [25] also established that too little or too much information can lead to inferior decision making. Therefore, one can only deduce that the importance of an empirical solution to this issue is of urgency. Investors are fast becoming more sophisticated; and are always in a bid to make maximum returns off their investments per time. This has made them to seek information about the firm they are investing in. Lagoarde-Segot and Lucey [24] reported that stock market informational efficiency is crucial for economic growth and, as such it is vital to persistently measure and improve information asymmetry in capital markets.

2.2 Market-Microstructure Literature on Measuring the Impact of Information Asymmetry on Real Estate Asset Pricing

The empirical understanding of real estate asset pricing is fast becoming an appetite for researchers and scholars as literature on underpricing and overpricing reveal that asset pricing ultimately influences the performance of REITs. As much as there have been extant studies on the nature of pricing, many of these studies have only focused on developed markets and to the exclusion of emerging markets. It is now important to understand how specific trading mechanisms affect the price formation process of real estate investment trusts. The market-microstructure strategy is hereby adopted from Finance which ultimately would empirically understudy the role information asymmetry plays in the pricing of real estate. Managers who know more about the firm as they get private information (firm-specific information) before the market transactions appear better advantaged than investors (Deakins and Hussain [13]; Myers and Majluf [29]). In other words, the firm-specific information asymmetry corresponds only to a subset of total market information asymmetry about the firm as both managers and investors are likely to be equally well informed about market-wide determinants of firm value. Bagehot [5] argues that adverse selection due to the presence of better informed traders in the market affect the price-discovery process. It is only natural to believe that informed traders in the market are those closer to the firm's operations, such as the managers, employees, analysts and large and/or institutional shareholders (Bharath et al. [6]). Malkiel and Fama [26]) noted that literature on Market-microstructure acknowledges that asset prices may not fully reflect all available information at all times due to market frictions as hypothesized by the efficient-market hypothesis. This study had been carried out over few decades ago but only in developed markets. This suggests that there is need to understudy the emerging markets. Therefore the measurement of information asymmetry is a worthy trial for the nature of this research as it would seek to understand how managers and investors behave during transactions.

As identified in literature, two of the effects of information asymmetry has been adverse selection which presents a situation where the type or quality of an asset is unknown by one party in a transaction; while moral hazard presents a situation where there is a hidden action that results from the transaction. Of particular concern in this study is the adverse-selection problem. As an example, traders with better private information about the quality of a product will selectively participate in trades that benefit them, at the expense of the other traders. In a case where managers have inside information about the quality of the firm, they may offer or issue equity when they know the offer price exceeds their private assessments of the firm's value. O'Hara [32] extensively studied the problem of measuring information asymmetry about a firm's value and the returns of securities. The study noted that insiders of a firm such as managers have superior knowledge about the firm and are therefore better informed about the firm and its business. This study dates back as far as more than a decade ago; indeed, it understudied the issue of information asymmetry but neglected real estate investment. The inconsistency of the theories underpinning real estate asset price formation resulting in underpricing and overpricing appear fragmentary in emerging markets; hence, it would be important to examine the amount of price variability due to firm-specific information.

3 Data and Methodology

The REITs firms' quote and trade data from 2007 to 2017 were sourced from the price data section of the INET-BFA (McGreggor Database). In order to measure the degree of information asymmetry, an ordinary least square (OLS) regression was adopted with the bid-ask spread as the dependent variable with measures of inventory cost as the independent variables. Measures of inventory cost and information cost include price volatility, trading volume, Age, Leverage and firm size. The spread is a proxy for information asymmetry. This was calculated as the average rand bid-ask spread divided by the average transaction price of the REITs stock. The bid-ask spread was calculated as $Ask_{it} - Bid_{it}$; where (Table 1):

- Bid_{it} – The highest or best-buying price for firm i on day t
- Ask_{it} - The lowest or bestselling price for firm i on the day t.

$$\mathrm{LoGBA_{it}} = \alpha + \boldsymbol{\beta}_1\mathrm{LoGPV_{it}} + \boldsymbol{\beta}_2\mathrm{LoGVol_{it}} + \boldsymbol{\beta}_3\mathrm{LoGSize_{it}} + \boldsymbol{\beta}_4\mathrm{LoGAge_{it}} + \boldsymbol{\beta}_5\mathrm{LoGLeverage_{it}} + \boldsymbol{\beta}_6\mathrm{LoGIA} + \varepsilon_{it} \tag{1}$$

Table 1. Variable definition

Variable(s)	Definition
Dependent	Difference between the highest price that a buyer is willing to pay for an asset and the lowest price that a seller is willing to accept on day t
1. $\mathrm{LoGBA_{it}}$	
Independent	
1. $\mathrm{LoGPV_{it}}$	Volatility is the coefficient of variation of the daily cash flow for firm d on day t over the Year
2. $\mathrm{LoGVol_{it}}$	This is the average daily trading volume of firm d on day t over the year
3. $\mathrm{LoGSize_{it}}$	This is calculated as the logarithm of total assets
4. $\mathrm{LoGAge_{it}}$	The number of years since IPO
5. $\mathrm{LoGLeverage_{it}}$	Firm's debt to assets

This methodology is consistent with [38]; and Chipeta and Mbululu [11].

Hence, an OLS regression was applied in measuring the degree of information asymmetry which is as follows (Fig. 1):

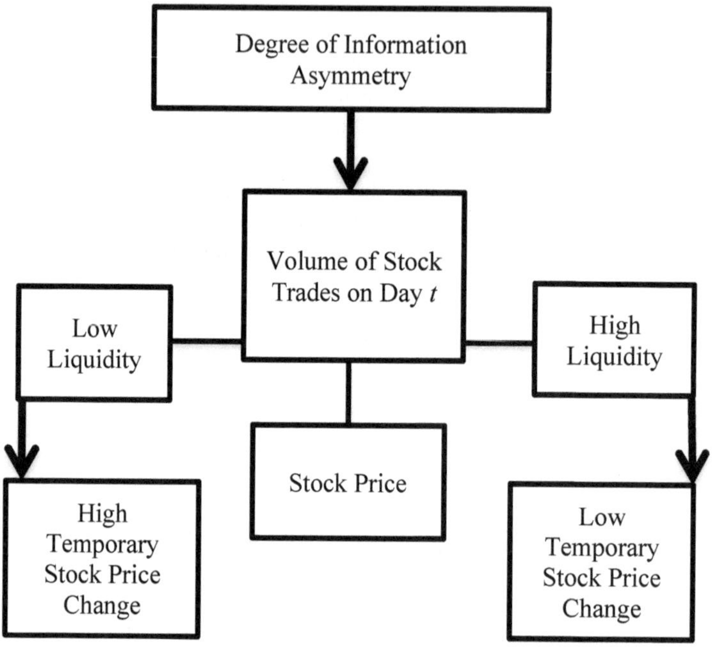

Fig. 1. Researcher's Conceptual Assumption on Information Asymmetry and Share Price Volatility

From the above conceptual framework, it sought to understand the effects of information asymmetry on the pricing of REITs. It was assumed that the degree to which investors will be involved in a trade will be subject to information available; and/or the nature of information; as such, it influenced the volume of trades. Again, it was envisaged that the nature of information available to an investor will ultimately influence how the stock of a REITs firm is priced. Volume moves prices.

4 Results

4.1 Introduction

First and chronologically, the section begins by providing some descriptive statistics (Table 2) of the sample. Next, it addresses the objective designed in the paper. Finally, a summary of key findings is presented, and then conclusion to the paper.

Table 2. Descriptive statistics

Statistics	Bid_Ask_Sp	Pr_Volty	Volume	Size	Leverage
Mean	2040.169	2018.625	282724.8	14447822	53.35604
Maximum	2549.050	2491.450	457527.5	26212848	67.57222
Minimum	1646.538	1600.231	176741.1	7362658.	35.22250
Std. Dev.	358.1098	363.9937	101612.8	6429519.	12.25100
Skewness	0.428704	0.199385	0.973906	0.799250	−0.524847
Kurtosis	1.600954	1.451652	2.345186	2.238681	1.630175
Jarque-Bera	1.121867	1.065167	1.759479	1.306171	1.240950
Probability	0.570676	0.587086	0.414891	0.520437	0.537689

Table 3 presents the result of the OLS regression on relationship between the bid-ask spread (dependent variable) and information asymmetry (predictors). The predictors comprises the log (LG) on average of price volatility (PV), volume of the share of SEO REIT traded (Vol.), the industry size (Siz) and leverage (Leg.) for the SEO REIT industry. The parameters extracted from the OLS regression model include the unstandardized beta (Un-Std. β_{it}), standardised beta (Std. β_{it}), Test statistics value (t-value), standard error (Std. Error) and significant level at 5% (Sig.). The analysis of effect of price volatility on the pricing of REITs showed a positive contribution as indicated by the beta (.521). This implies that frequent fluctuation in SAREIT prices especially when the volatility is resulted from increase in the stock price of the property stocks, could induce the enhancement in the performance of the REIT sector. Similarly, the result of the analyses on volume of share traded, size and leverage have positive beta coefficient of .091, .640, and .190 respectively. Thus, an improvement in the pricing of SAREITs can be explained by the healthy growth in the volume, size and leverage of the sector. For instance, higher volume of traded REIT shares can lead to higher dividend yield and returns. Similarly, as the total asset base of the REITs sector increase, there is a strong tendency that this may have positive impacts on the stock pricing and ultimately optimum returns. The interpretation therefore is supported by [38] wherein most SAREIT firms are subject to a greater level of information asymmetry and by implication, firm managers are subject to higher levels of debt usage; as such, this portends naivety of investment decision making by shareholders.

Table 3. Relationship between Bid-Ask spread and information asymmetry

Parameters	$LoGPV_{it}$	$LoGVol_{it}$	$LoGSiz_{it}$	$LoGLeg_{it}$
Un-Std. β_{it}	.493	.052	.282	.138
Std. β_{it}	.521	.091	.640	.190
t-value	2.484	.211	1.854	.519
Std. Error	.198	.248	.152	.267
Sig. @5%	.068	.844	.137	.631

Dependent variable: Bid-Ask spread

Table 4 shows the summary of the OLS regression model. The analysis of the predictors as a whole indicates that the explanatory variables contributed 87.5% to the prediction in the variance of the pricing of SAREITs as indicated by the adjusted R-square analysis. The implies that the collective contributions of price volatility, volume of the traded shares, size of the industries and leverage in explaining the variance in the pricing of SAREITs is 87.5%. However, the explanatory power of the collective contributions of the predictors is significant as indicated by the p-value (.011) which is less than 5% confidence level (0.05). By effects, it suggest that information asymmetry has a significant relationship with the pricing of the SAREITs; hence indicating that the level and the quality of information at the sector strongly determines their success or failure.

Table 4. Model summary of the OLS regression

Statistics	Result
Constant (a)	.047
Residue (ϵ_{it})	.250
R-Square	.968
Adjusted R-Square	.875
Std. Error of the Estimate (SEE)	.057
F-Stats	15.022
Sig.	.011

5 Discussion and Conclusion

The aim of this research was to measure the degree of Information Asymmetry on the pricing of SAREITs with a view to enhancing real estate investment decision-making. The extent of the relationship has also been measured to provide a coefficient/Beta value which shows that the relationship between information asymmetry and pricing are positively correlated. As against the findings in Corroborating [33], SAREIT firms are associated with a low degree of information asymmetry. With [10] in their findings, it was also documented that there was the presence of a high degree of asymmetric information; this study was carried out in a developed market. It may appear that the South African REITs market enjoys the benefit of information asymmetry due to its ranking on the global index of REITs; and with the fact that it is still in its incubation having evolved just about six (6) years ago. It is recommended that managers should continue to give more information including information in the REITs market, the acquired information and information that has not been reflected in the stock price. Also, investors are encouraged to buy real estate stocks as the performance is a reflection of the high number of volumes traded owing to the limpidity of transactions. Finally, new evidence has been provided with respect to the determination of the presence of information asymmetry. Researchers are now able to consolidate on this research by conducting a study on the appropriateness of proxy measurements in testing the presence of information asymmetry.

References

1. Abdul-Baki, R.: Do information asymmetry proxies measure information asymmetry? (2013)
2. Afzal, W.: Towards the general theory of information asymmetry. In: Information Seeking Behavior and Technology Adoption: Theories and Trends, pp. 124–135. IGI Global (2015)
3. Akerlof, G.: The market for "lemons": quality uncertainty and the market mechanism. Q. J. Econ. **84**, 488 (1970)
4. Akinsomi, O., Kola, K., Ndlovu, T., Motloung, M.: The performance of the Broad Based Black Economic Empowerment compliant listed property firms in South Africa. J. Property Investment Finance **34**, 3–26 (2016)
5. Bagehot, W.: The only game in town. Financ. Anal. J. **27**, 12–14 (1971)
6. Bharath, S., Dahiya, S., Saunders, A., Srinivasan, A.: Lending relationships and loan contract terms. Rev. Financ. Stud. **24**, 1141–1203 (2009)
7. Blazenko, G.: Managerial preference, asymmetric information, and financial structure. J. Finance **42**, 839–862 (1987)
8. Breeden, D.T., Viswanathan, S.: Why do firms hedge? An asymmetric information model. J. Fixed Income **25**(3), 7 (2016)
9. Breeden, D., Viswanathan, S.: Why do firms hedge? An asymmetric information model. J. Fixed Income **25**, 7–25 (2015)
10. Buckle, M., Beccalli, E.: Principles of banking and finance. University of London (2008)
11. Chipeta, C., Mbululu, D.: Firm heterogeneity, macroeconomic conditions and capital structure adjustment speeds: evidence from the JSE. Investment Anal. J. **42**, 69–80 (2013)
12. Cooper, M., Downs, D., Patterson, G.: Real estate securities and a filter-based, short-term trading strategy. Real Estate Res. **18**(2), 313–333 (1999)
13. Deakins, D., Hussain, G.: Risk assessment with asymmetric information. Int. J. Bank Mark. **12**(1), 24–31 (1994)
14. Downs, D., Nuray Güner, Z., Patterson, G.: Capital distribution policy and information asymmetry: a real estate market perspective. J. Real Estate Finance Econ. **21**, 235–250 (2000)
15. Easley, D., Kiefer, N., O'Hara, M., Paperman, J.: Liquidity, information, and infrequently traded stocks. J. Finance **51**, 1405–1436 (1996)
16. Ekumah, E.K., Essel, T.: Information is power: the problem with credit accessibility in rural banks in Ghana (2003). https://www.ilo.org/wcmsp5/groups/public/—ed_emp/documents/publication/wcm_041881.pdf
17. Engle, R., Lange, J.: Measuring, forecasting and explaining time varying liquidity in the stock market. SSRN Electro. J. (1998)
18. Glosten, L.: Insider trading, liquidity, and the role of the monopolist specialist. J. Bus. **62**, 211 (1989)
19. Jensen, M., Meckling, W.: Theory of the firm: managerial behavior, agency costs and ownership structure. J. Financ. Econ. **3**, 305–360 (1976)
20. Kerr, J., Ozel, N.: Earnings announcements, information asymmetry, and timing of debt offerings. Acc. Rev. **90**, 2375–2410 (2015)
21. Khemraj, T.: The local FX market and its relation to the real economy. In: Financial Development and Economic Growth. The Caribbean Centre for Money and Finance, Guyana, Georgetown (2015)
22. Knutsen, S.: Financial fragility or information asymmetry? The inter-war banking crisis in Norway. Presented at the EBHA Conference (2001)

23. Krinsky, I., Lee, J.: Earnings announcements and the components of the bid-ask spread. J. Finance **51**, 1523–1535 (1996)
24. Lagoarde-Segot, T., Lucey, B.: Efficiency in emerging markets—evidence from the MENA region. J. Int. Financ. Markets Institutions Money **18**, 94–105 (2008)
25. Malhotra, N.: Information load and consumer decision making. J. Consum. Res. **8**, 419 (1982)
26. Malkiel, B.G., Fama, E.F.: Efficient capital markets: a review of theory and empirical work. J. Finance **25**(2), 383–417 (1970)
27. Matoussi, H., Karaa, A., Maghraoui, R.: Information asymmetry, disclosure level and securities liquidity in the BVMT. Finance India **18**, 547–558 (2004)
28. Mossman, C.: The Small Firm Anomaly and the Bid-Ask Spread: Empirical Tests (1993)
29. Myers, S., Majluf, N.: Corporate financing and investment decisions when firms have information that investors do not have. J. Financ. Econ. **13**, 187–221 (1984)
30. Myers, S.: The capital structure puzzle. J. Finance **39**, 574–592 (1984)
31. Narayan, P., Mishra, S., Narayan, S.: New empirical evidence on the bid-ask spread. Appl. Econ. **47**, 4484–4500 (2015)
32. O'Hara, M.: Market Microstructure Theory (108). Blackwell Publishers, London (1995)
33. Qu, W., Wongchoti, U., Wu, F., Chen, Y.: Does information asymmetry lead to higher debt financing? Evidence from China during the NTS Reform period. J. Asian Bus. Econ. Stud. **25**, 109–121 (2018)
34. Roll, R.: A simple implicit measure of the effective bid-ask spread in an efficient market. J. Finance **39**, 1127–1139 (1984)
35. Stonier, T.: Towards a general theory of information II: information and entropy. In: Aslib Proceedings, vol. 41, pp. 41–55 (1989)
36. Tumwebaze, Z., Orobia, L., Kamukama, N.: Information asymmetry and stock market participation: evidence from the uganda stock exchange. ORSEA J. **4**, 2 (2017)
37. Vayanos, D., Wang, J.: Theories of liquidity. Found. Trends® Finance **6**, 221–317 (2011)
38. Wei, P., Hsieh, C., Sirmans, C.: Captive financing arrangements and information asymmetry: the case of REITs. Real Estate Econ. **23**, 385–394 (1995)
39. Welker, M.: Disclosure policy, information asymmetry, and liquidity in equity markets. Contemp. Account. Res. **11**, 801–827 (1995)
40. Worzala, E., Sirmans, C.: Investing in international real estate stocks: a review of the literature. Urban Stud. **40**, 1115–1149 (2003)

Barriers for the Adoption of Incorporating RFID with Mobile Technology for Improved Safety of Construction Professionals

Temidayo O. Osunsanmi(✉), Ayodeji E. Oke, and Clinton O. Aigbavboa

SARChI in Sustainable Construction Management and Leadership in the Built Environment, Faculty of Engineering and Built Environment, University of Johannesburg, Johannesburg, South Africa
osunsanmidayo@gmail.com, emayok@gmail.com, caigbavboa@uj.ac.za

Abstract. The frequent occupational hazards experienced within the construction industry is responsible for construction professionals having high exposure to occupational hazards than other occupations. The high incidence of hazards is related to the weak monitoring of construction activities on site. Towards improving the construction safety this study proposes the merging of RFID (Radio Frequency Identification) with mobile technology for monitoring construction professionals on site. Thus, this study appraises the barriers to the adoption of RFID and mobile technology for construction safety in South Africa. A random sampling method was used for administering the questionnaire to construction professionals in Gauteng Province South Africa. The data were analysed with SPSS V 24, using, mean score and frequency distribution. The findings from the study revealed that the adoption of RFID and mobile technology is hindered by the cost of procuring and low technical ability. The study contributes to establishing a proactive approach for construction safety management in South Africa.

Keywords: Fourth industrial revolution · Construction safety · Mobile technology · Occupational hazards · Radio Frequency Identification (RFID)

1 Introduction

The construction industry contributes significantly to the asset base of a nation's economy, as the final products of construction are recorded as investment goods [1]. Likewise, Baradan et al. [2] opined that the industry performs a significant function in the economy of a nation through the provision of employment opportunities, housing delivery, provision of infrastructure and many others. In support of the opinion as mentioned earlier [3] affirmed that the economy of every nation is dependent on the activities of the construction sector. The activities of the industry have a far-reaching effect on the national gross domestic profit and vice versa. It can be inferred from the above opinions that there is a correlation between the activities of the construction industry and national economic growth [4].

C. Aigbavboa and W. Thwala (Eds.): CIDB 2019, *The Construction Industry in the Fourth Industrial Revolution*, pp. 297–304, 2020.
https://doi.org/10.1007/978-3-030-26528-1_29

The activities of the construction industry are performed by the construction professionals responsible for the development of a construction project [5]. The professionals utilise a combination of their body and mind coupled with their skill to work on construction sites [6]. Windapo [7] asserted that construction professional skill is a crucial resource for the construction project. This is because the skill is used to combines all the other resources like; materials, plant equipment, and finance to produce the various construction projects. Ho [8] posited that the skill possesses by a construction professional is a determinant for achieving project performance. Liang and Leung [5] further noted that construction professionals in the construction industry are an essential asset to the construction sector. Similarly, Enshassi et al. [9] avowed that construction workers are critical to the industry, just as construction activities are significant to the economies of nations. The above opinions are an indication that construction professionals contribute significantly to the growth of the construction industry.

Despite the significant contribution of construction professionals to the industry they are often affected by hazards on construction sites [10]. Also, Idoro [11] affirmed that construction professionals have a higher exposure to occupational hazards than other occupations. Hinze et al. [4] indicated that construction professionals are prone to hazards due to the complex nature of construction sites, extensive use of sophisticated plants and machinery. Likewise, Muhammad et al. [12] asserted that construction sites are generally complex and unsafe therefore increasing the chances of a construction professional been injured on site. Charehzehi and Ahankoob [13] affirmed that the primary cause of hazards on construction sites is due to weak supervision and inadequate inspection of construction professionals. This implies that the safety of construction professionals could be improved through the effective inspection and monitoring of their activities.

Towards ensuring effective inspection and monitoring of construction professionals activities on sites this study proposes the incorporation of RFID with mobile technology for improved safety of construction professionals. Radio frequency identification (RFID) has been in existence for several decades ago, became popular after the advent of the fourth industrial revolution [14]. The use of RFID was initially developed for improving military warfare in Britain [15]. Ever since the adoption of RFID has been used in different sectors and industries including the construction industry. Authors such as; Lu et al. [16], Lindblad et al. [17] and Kasim et al. [18] have revealed the possibility for using RFID in the construction industry. They appraised its potential in the usage of construction project management, material handling process and construction material trackings. However, few studies have been conducted on the prospect of incorporating RFID with mobile technology. Although Osunsanmi conducted a Osunsanmi et al. [19] the study failed to evaluate the barriers to the adoption of RFID and mobile technology. Thus, this study aims to fill the gap in the literature and also contributes to improving the safety of construction professionals.

2 Construction Industry Safety and RFID

2.1 Safety in the South African Construction Industry

The construction industry in South African contributes enormously to the development of the country [20, 21]. It adds to the economy through the delivery of infrastructures and employment provision. Balogun et al. [22] opined that the country GDP rose by 6% due to the infrastructure provided by the construction industry. Concerning the employment capacity, the industry was responsible for employing over 3 million people in the formal and informal sector [23]. Therefore, it is reasonable to envisage that the construction industry in South Africa contributes to the growth of the economy.

Despite the industry contribution to the growth of the economy, it is confronted with numerous challenges. In support of this assertion [24] indicated that the construction industry is confronted with multiple problems. Windapo [7] concluded that one of the challenges facing the South African construction industry is the shortage of skilled construction professionals. The lack is experienced among construction professionals that require technical and formal training like electricians, welders, carpenters and many others. Windapo [7] discovered the shortage is due to the mandatory requirement that the construction worker must be certified before been employed. Erasmus and Breier [25] related the deficiency of skilled construction professional in South Africa to the state of the economy, ageing workforce and poor primary education for construction workers. Aside from construction professional shortage literature has shown that the South African construction industry is confronted with the health and safety of construction professionals [26, 27].

Towards reducing the occupational hazard experienced in the construction industry the Government of South Africa has established various act and policies [28]. The act includes; occupational health and safety act (OHSA) act no 85 of 1993 and compensation for occupational accidents, injuries, and diseases of 1993 that substituted the machinery and occupational safety act. The construction regulation act of 2003 as stipulated by section 43 of the occupational health and safety act of 1993. Among all this safety act the construction regulation was explicitly developed for the construction industry with the intention of reducing the occupational health and hazards experienced in the industry. Despite, the enactment of the construction regulation act the health and safety records in the industry has not improved significantly [29]. Likewise, Ayessaki and Smallwood [26] reported that the South African construction industry contributes to a large proportion of occupational injuries when compared to other sectors and is ranked the third highest death rate industry. This, therefore, calls for the need for a modern form of managing health and safety on the construction site.

2.2 Adopting RFID and Mobile Technology for Construction Monitoring

Xiao et al. [15], submitted that the RFID was initially designed for usage in improving warfare technology in Britain. Tulla, et al. [30] indicated Radio frequency identification (RFID) is a generic term for technologies that adopt radio waves for automatic identification of people and objects. Similarly, Lu et al. [16] asserted that RFID is a technology that provides automatic identification of an object using radio frequencies to capture and convey data from a tag, or transponder. It works through using radio waves to transmit data from a sender called a tag to a receiver or reader to perform a specific task or identify the location of an object depending on the reason the tag was installed [18]. It can be inferred that the major component of RFID is the tag and reader.

The RFID tags can be classified into four categories which are; passive, active, semi-passive and semi-active tags [31]. The passive tags operate without a battery; a battery powers active tags. Whereas the semi-passive and semi-active are in the middle because they both use a small battery. The semi-passive uses the small battery to power their chips and semi-active use the battery to power their antenna. Sardroud [32] reported that the most crucial component of the RFID is the tag which is followed by the reader also referred as the scanner or interrogator used for communicating with the tag made possible by the radio waves sent by the tag's antenna. The reader then converts the received radio waves into digital information that is passed to the backend system. The back-end system is referred to as the database for storing the information collected from the reader.

This study proposed the use of the Mobile phone as the RFID reader while the tag will be fixed on the protective equipment of the construction professionals. The use of mobile phones as RFID reader has been possible due to the growth of the industry 4.0 providing the opportunity to run various software's on the phone [33]. Venkataramani and Gopalan [34] affirmed that mobile technology has developed to the extent of attributing integrated RFID reader software on telephones. The use of Mobile phones as RFID tags will provide a fast and easy way for monitoring and construction workers activity remotely on site. Unfortunately, the construction industry has not tapped into the advantage of using RFID tags and mobile devices. This, therefore, leads to examining the barriers of incorporating mobile technology and RFID for monitoring construction safety.

3 Methodology

The study was conducted in Johannesburg Gauteng province, South Africa. Random sampling technique was adopted for selecting the respondents within the study area. Creswell and Creswell [35] reported that random sampling works on the concept of obtaining a comprehensive list of a larger population and select individuals randomly to be adopted for the sample. The idea behind random sampling makes it suitable for this study because they are numerous construction professionals within Johannesburg. Therefore, professionals with proximity to the researcher that are registered with their respective professional bodies are selected randomly. A total of 40 construction professionals were chosen randomly with the selection based on their involvement of health and safety on the construction site. Out of the selected construction professionals, 34 responded effectively, and their response was analysed. The study adopted a statistical package for social science (SPSS) version 24 using descriptive statistics such as; mean item score and frequencies.

4 Discussion of Findings

4.1 Background Information

The background information of the respondents is summarised in Table 1 below;

Table 1. Summary of background information

	Frequency	Percent (%)
Category of respondents		
Consulting base	17	50.0
Contracting organisation	17	50.0
Total	34	100
Highest academic qualification		
ND/Diploma	11	32.4
B.Sc/B.Tech	15	44.1
M.sc/MBA/MPM	5	14.7
PhD	3	8.8
Total	34	100

Table 1 shows that an equal proportion of response was received from respondents working with consulting and contracting organisation. Regarding the academic qualification of the respondents, Table showed that they are all educated thereby confirming their eligibility to answer the question posed by the research instrument. However, 44.1% of construction professionals are B.Sc/B. Tech degree holders, 32.4% of the respondents, are ND/Diploma holders, while 14.7% have obtained MSc/MBA/MPM degree and few (8.8%) have studied up to PhD level.

4.2 Barriers to the Use of RFID and Mobile Technology

The respondents were asked to rate their level of agreement to the factors that could hinder their adoption of RFID and mobile technology. Their response was acquired using a five-point Likert scale from not agree denoted by 1 to very agree represented by 5, and the outcome is presented by Table 2.

Table 2. Barriers to the use of RFID and mobile technology

	Mean	Rank
Cost	4.53	1
Low technical ability	4.50	2
Security of data	4.29	3
Communication range	4.24	4
Storage of data	4.12	5
The additional weight of the sensor on the PPE	4.06	6
Power availability	3.62	7

Table 2 showed that the major hindrance to the adoption of RFID and mobile technology for construction safety is cost and low technical ability of construction professionals with a mean score of 4.50 above. This finding coincides with studies conducted by Nasr et al. [33] and Venkataramani and Gopalan [34]. These studies recognised that the need to save cost by construction professionals and clients prevents the adoption of ICT for construction works. Also, Lu et al. [16] opined that most construction clients perceive that the use of ICT such as RFID will increase the overall cost of construction and thus avoid any form of modern construction technologies. Osunsanmi, et al. [19] reported that low research and poor awareness of construction professional's bodies reduced the technical ability of most construction professionals. Other factors that hinder construction professionals towards the adoption of RFID include; data security, communication range, data storage and adding weight to the PPE. Whereas, the respondents rated power availability has a less significant hindrance to the adoption of RFID. This finding did not correspond with a study conducted by Oladapo [36] in other developing countries like Nigeria. They conclude that the erratic power supply in the country affects the use of ICT on construction sites. This suggests that the location of a construction project determines the form of technology adopted.

5 Conclusion

The South African construction industry contributes to the nation's economy through the development of infrastructures employment opportunities. The construction professionals are responsible for developing, managing and maintaining the infrastructures within the country. Therefore, this makes the professionals crucial for industry growth and indirectly to the growth of the economy. Unfortunately, the construction professional in the South African construction industry is confronted with occupational hazards during the delivery of resources. Towards reducing the occupational hazards affecting the construction professionals, the Government of South Africa has enacted different safety policies and regulations. Regardless of the policy and rules promulgated by the Government on constructional health and safety, the professionals are still confronted with hazards on construction sites.

Towards reducing the occupational hazards confronting the construction workers this study propose incorporating RFID with mobile technology for improved safety of construction professionals. The study recognised that RFID comprises of a tag, reader and a backend system. The study proposed that the tag should be attached to the protective equipment of the construction professionals while the mobile phone will serve as the reader. It is expected that this will provide the opportunity for easy and effective monitoring of construction professionals on site. Thus, this study examined the barriers to the incorporation of RFID with mobile technology. It discovered that the cost of maintaining, procuring and low technical ability. The study contributes to establishing a proactive approach for construction safety management in South Africa. Thus, this study recommends that construction professional bodies should create more awareness regarding the use of RFID for construction safety. Also, RFID should be subsidised by the government for the construction industry.

Acknowledgements. The authors like to appreciate everybody that has contributed to this article directly or indirectly.

References

1. Olugboyega, O., Aina, O.O.: Analysis of building information modelling usage indices and facilitators in the nigerian construction industry. J. Logistics Inf. Serv. Sci. **3**, 1–36 (2016)
2. Baradan, S., Dikmen, S.U., Akboga Kale, O.: Impact of human development on safety consciousness in construction. Int. J. Occup. Saf. Ergon. 1–11 (2018)
3. Loosemore, M., Lim, B.T.H.: Linking corporate social responsibility and organizational performance in the construction industry. Constr. Manage. Econ. **35**, 90–105 (2017)
4. Hinze, J., Thurman, S., Wehle, A.: Leading indicators of construction safety performance. Saf. Sci. **51**, 23–28 (2013)
5. Liang, Q., Leung, M.Y.: Analysis of the relationships between mindfulness and stress for construction professionals. In: Proceedings of the 31st Annual Association of Researchers in Construction Management Conference, ARCOM 2015, pp. 469–478 (2015)
6. Makhene, D., Thwala, W.D.: Skilled labour shortages in construction contractors: a literature review (2009)
7. Windapo, A.O.: Skilled labour supply in the South African construction industry: the nexus between certification, quality of work output and shortages. SA J. Hum. Resour. Manage. **14**, 1–8 (2016)
8. Ho, P.H.: Labour and skill shortages in Hong Kong's construction industry. Eng. Constr. Architect. Manage. **23**, 533–550 (2016)
9. Enshassi, A., Al-Swaity, E., Abdul Aziz, A.R., Choudhry, R.: Coping behaviors to deal with stress and stressor consequences among construction professionals: a case study at the Gaza Strip. Palestine. J. Financ. Manage. Property Constr. **23**, 40–56 (2018)
10. Umeokafor, N.: An appraisal of the barriers to client involvement in health and safety in Nigeria's construction industry. J. Eng. Des. Technol. **15**, 471–487 (2017)
11. Idoro, G.I.: Comparing occupational health and safety (OHS) management efforts and performance of Nigerian construction contractors. J. Constr. Dev. Countries **16**, 151–173 (2011)
12. Muhammad, B.A., Abdulateef, I., Ladi, B.D.: Assessment of cost impact in health and safety on construction projects. Am. J. Eng. Res. **4**, 25–30 (2015)
13. Charehzehi, A., Ahankoob, A.: Enhancement of safety performance at construction site. Int. J. Adv. Eng. Technol. **5**, 303 (2012)
14. Lin, Y.-C., Cheung, W.-F., Siao, F.-C.: Developing mobile 2D barcode/RFID-based maintenance management system. Autom. Constr. **37**, 110–121 (2014)
15. Xiao, Q., Gibbons, T., Lebrun, H.: Rfid technology, security vulnerabilities, and countermeasures. In: Supply Chain the Way to Flat Organisation, ed: InTech (2009)
16. Lu, W., Huang, G.Q., Li, H.: Scenarios for applying RFID technology in construction project management. Autom. Constr. **20**, 101–106 (2011)
17. Lindblad, F., Bolmsvik, Å., Pettersson, J., Wiberg, S.: Efficiencies in the on-site material handling process by using radio frequency identification in the wood building construction industry. Int. J. Innov. Manage. Technol. **9**, 252–259 (2018)
18. Kasim, N., Sarpin, N., Mohd Noh, H., Zainal, R., Mohamed, S., Manap, N., et al.: Automatic materials tracking practices through RFID implementation in construction projects. In: MATEC Web of Conferences, p. 05001 (2019)

19. Osunsanmi, T.O., Oke, A.E., Aigbavboa, C.O.: Fusing RFID with mobile technology for enhanced safety of construction project team members. In: Proceedings of the International Conference on Industrial Engineering and Operations Management Pretoria, Johannesburg, South Arica, 29 October–1 November, 2018, p. 312 (2018)
20. Shakantu, W.M., Emuze, F.A.: Assessing reverse logistics in South African construction. In: IGLC 2012 - 20th Conference of the International Group for Lean Construction (2012)
21. Windapo, A.O., Cattell, K.: The south African construction industry: perceptions of key challenges facing its performance, development and growth. J. Constr. Dev. Countries **18**, 65 (2013)
22. Balogun, O.A., Agumba, N., Ansary, N.: Determinants of business overdraft accessibility within small and medium-sized enterprises in the South African construction industry: a case of gauteng province. In: Proceedings of the 21st International Symposium on Advancement of Construction Management and Real Estate, pp. 617–628 (2018)
23. CIDB: Construction employment monitor, 24 November 2018. http://www.cidb.org.za/publications/Documents/Construction%20Monitor%20-%20October%202018.pdf
24. Aigbavboa, C.O., Thwala, W.D.: Challenges facing black owned small and medium construction companies: a case study of Nelspruit–Mbombela municipality, South Africa (2014)
25. Erasmus, J., Breier, M.: Skills shortage in South Africa: case studies of key professions, 22 November 2018 (2009). https://www.hsrcpress.ac.za/product.php?productid=2257
26. Ayessaki, W., Smallwood, J.: Construction project manager health and safety interventions towards improving workers' performance. In: Proceedings of the 21st International Symposium on Advancement of Construction Management and Real Estate, 2016, pp. 513–521 (2018)
27. Smallwood, J.: The impact of a client contractor health and safety (H&S) programme on contractor H&S performance. Procedia Eng. **196**, 996–1002 (2007)
28. Emuze, F.: Foreseeing countermeasures for construction safety violations in South Africa. In: Proceeding of the 34th Annual ARCOM Conference, ARCOM 2018, pp. 587–596 (2018)
29. Goldswain, C., Smallwood, J.: Key inputs into a designing for construction health, safety, and ergonomics model in South Africa. In: Proceedings 29th Annual Association of Researchers in Construction Management Conference, ARCOM 2013, pp. 359–367 (2013)
30. Tulla, K., Vähä, P., Matinmikko, T., Tolman, A., Möttönen, V.: RFID technology changes FM services deliveries. Facilities **27**, 457–468 (2009)
31. Smith, S., Oberholzer, A., Land, K., Korvink, J.G., Mager, D.: Functional screen printed radio frequency identification tags on flexible substrates, facilitating low-cost and integrated point-of-care diagnostics. Flex. Printed Electron. **3**, 25–29 (2018)
32. Sardroud, J.M.: Influence of RFID technology on automated management of construction materials and components. Sci. Iranica **19**, 381–392 (2012)
33. Nasr, E., Shehab, T., Vlad, A.: Tracking systems in construction: Applications and comparisons. In: 49th ASC Annual International Conference Proceedings (2013)
34. Venkataramani, G., Gopalan, S.: Mobile phone based RFID architecture for secure electronic Payments using RFID credit cards. In: The Second International Conference on Availability, Reliability and Security (ARES 2007), pp. 610–620 (2007)
35. Creswell, J.W., Creswell, J.D.: Research Design: Qualitative, Quantitative, and Mixed Methods Approaches, Sage Publications (2017)
36. Oladapo, A.: An investigation into the use of ICT in the Nigerian construction industry. J. Inf. Technol. Constr. **12**, 261–277 (2007)

Organizational Culture for Construction Enterprises in the Fourth Industrial Revolution

Oluwayomi Babatunde[1(✉)] and Olalekan S. Oshodi[2]

[1] School of Construction Economics and Management, University of the Witwatersrand, Johannesburg, South Africa
oluwayomi.babatunde@wits.ac.za

[2] Department of Construction Management and Quantity Surveying, University of Johannesburg, Johannesburg, South Africa
oshodilekan2002@yahoo.com

Abstract. Construction industry is slow to innovate and adopt new technology. Construction innovation can be disruptive or enabling, depending on the organizational culture of construction enterprises. Since organizational culture shapes business practices, there is the need to understand the implications of organizational culture on innovation adoption. The aim of this conceptual study is to recommend the organizational culture type for construction innovation adoption in the Fourth Industrial Revolution (Industry 4.0). The methodology includes the interpretivism research philosophy, inductive research approach, and qualitative research methodological choice. Using the integrative literature review, qualitative/textual data were gathered on the dimensions of Industry 4.0, organizational culture types, and the dominant organizational culture types among construction enterprises in eleven countries. The findings include the need to manage change in the digitalization of processes and products involved in construction activities' value co-creation. The adhocracy culture is recommended as best supporting innovation adoption in the rapidly diffusing era of Industry 4.0. This could be subjective; hence, a limitation and theoretical implication for a future empirical study to validate.

Keywords: Construction enterprises · Industry 4.0 · Innovation · Interpretivism · Organizational culture

1 Introduction

Organizational culture and organizational climate are, often, confused. However, organizational culture adopts qualitative research methods and is contextualized while organizational climate uses quantitative methods and is generalized [1]. Organizational culture is about shared basic assumptions, values, and beliefs while organizational climate is about shared perceptions of policies, practices, and procedures [2]. Organizational culture is characterized by a nuanced view to truly understand and change organizations [3]. Resistance to change is rooted in the anxiety of uncertainty and stimulus overload surrounding a paradigm shift from the familiar to the unfamiliar practices [4]. This conundrum characterizes most construction enterprises to the extent that innovation is seen as a threat to their stability. However, innovation is inevitable

C. Aigbavboa and W. Thwala (Eds.): CIDB 2019, *The Construction Industry in the Fourth Industrial Revolution*, pp. 305–315, 2020.
https://doi.org/10.1007/978-3-030-26528-1_30

because, by nature, it is any idea, conduct, or mechanism that is novel. It is, often, uniquely different from the existing practices and, as a result, influences cultural change in people, processes, and products through diffusion [5].

This means diffusion and adoption of innovation can either disrupt an organizational culture or enables it to flourish. The Fourth Industry Revolution (Industry 4.0) is a smart factory characterized by decentralized decision making or social innovations for new practices to tackle social challenges [6]. Industry 4.0 is characterized by smart work geared at transforming the labor market through digitalization [7]. Industry 4.0 allows for sustainable manufacturing because it supports decentralization, virtualization, and interoperability [8]. Industry 4.0 allows for open innovation by facilitating combinations of technology, market and society [9]. With the unique nature of the construction industry, unlike the manufacturing industry, the primary research question is what are the implications of Industry 4.0 for construction enterprises in terms of organizational culture for innovation? Following, this study aims to investigate the organizational culture type for construction innovation adoption in Industry 4.0. Its three objectives include to: (1) establish the different dimensions of Industry 4.0 that can affect construction innovation; (2) describe the characteristics of the different organizational culture types and; and (3) propose the organizational culture type that best supports construction innovation adoption for Industry 4.0.

1.1 Research Question Mode

The mode of this study's research question (see the previous section) is gap-spotting; particularly, under-researched area [10]. This is because Industry 4 is yet to be fully-understood and fully-explored. It is an emerging concept that is yet to be investigated in relation to the other existing concepts. This leaves other areas such as implications for organizational culture under-researched.

1.2 Conceptual Paper Justification

A conceptual paper does not have numerical data due to its focus on integrating the existing relationships among concepts to advance alternative relationships [11]. As such, this study uses the integrative literature review to collect qualitative data to achieve its three objectives. The integrative literature distils representative literature on a topic to generate new perspectives [12].

1.3 Conceptual Framework

A conceptual framework situates a new study in the pertinent existing body of knowledge and is made up of the relevant theoretical and empirical works aimed at developing new knowledge about the existing associated concepts [13]. This study adopts the conceptual framework of the system view of construction innovation, which argues for rigorous interactions and partnerships [14]. Organizational culture affects innovation [15], which, in turn, influences adopter categories [16].

2 Methodology

Research methodology encompasses methods and techniques to systematically answer research question [17]. The research onion model offers an operative sequence to systematically design a research methodology [18]. This study is based on the interpretive phenomenology research philosophy. Phenomenology allows for subjective interpretations as a starting point in understanding a social phenomenon [19]. The research approach is inductive, which is often operationalized by collecting data to study a phenomenon under consideration to develop a theory [18]. This separates it from the deductive approach that is about testing a hypothesis.

Stemming from the research philosophy and approach, the methodological choice is qualitative, which centers around texts and observations to depict reality by studying and describing people in their actual contexts [20]. Integrative literature review was used following the checklist provided by [21].

The literature reviewed and discussed subsequently were selected via the Google Scholar search engine for a wider spread across the main and the alternative publication routes. The following five keywords that are central to this study were used: (i) Construction Enterprises, (ii) Fourth Industrial Revolution, (iii) Innovation, (iv) Interpretivism, and (v) Organizational Culture. Peer-reviewed authoritative/seminal conceptual and empirical journal articles and conference proceedings were retained for further analyses. The abstract, methodology, and conclusion sections of the papers retained were critically analyzed to judge significance using the seven criteria provided by [22]. More importantly, judging the validity and reliability of the papers retained the nine using the nine strategies recommended by [23].

2.1 Dimensions of Industry 4.0

The first objective was about the dimensions of Industry 4.0. The study by [24] was based on a brainstorming research method technique among eleven experts drawn from across the world. [24] analyzed Industry 4.0 in terms of institutions, technology as well as firm innovation and start-up strategy. Subject expertise boosts creativity [25] and, increasingly, experts' brainstorming session is being digitalized for ease of data management during the ideation process [26]. The strengths of subject expert brainstorming technique makes the analysis provided by [24], arguably, more authoritative and reliable in understanding the different dimensions of Industry 4.0. The four main dimensions of Industry 4.0 are discussed, in brief, subsequently.

The emerging *definitions*, in the forms of common phrases/terminologies used to refer to Industry 4.0, underscore that continuous and future transformational change in all industries is premised on digitalization of processes and products.

The transformational change necessitates *institutional response* at the macro and micro levels to nurture and boost creativity of the different organizations and departments that are involved in the co-creation of processes and products.

Consequently, since "smart work" characterizes Industry 4.0, *technological response* is inevitable. The design and fabrication of the different technologies must allow for flexibility in the forms of inter-connectedness and -operability.

Similarly, digitalization of processes and products does not have to negatively impact on the organizations. Equal attention should be paid to *organizational start-up strategy and innovation* to sustain value co-creation in the supply chain.

From the preceding discussions, it can be reasoned that there are benefits and, admittedly, challenges associated with Industry 4.0 Revolution. As a result, smart organizations are being strategically-positioned through digitalized and transformational changes for the people, processes and products involved in their value co-creation activities. Industry 4.0 is about integration, which could be disruptive to some construction organizations depending on their organizational culture types.

2.2 Organizational Culture Types

The Competing Values Framework (CVF) is an influential and extensively used model in the study of organizational culture [27]. CVF was initially used to define predictors of organizational effectiveness [28]. CVF is based on two dimensions of: (i) flexibility and discretion versus stability and control, and (ii) external focus and differentiation versus internal focus and integration [29]. The two dimensions describe four models of CVF including: (i) human relations model, (ii) open system model, (iii) rational goal model, and (iv) internal process model [29]. These four CVF models morphed into the four organizational culture types including: (i) Clan, (ii) Adhocracy, (iii) Market, and (iv) Hierarchy [30] as shown in Fig. 1 and discussed subsequently.

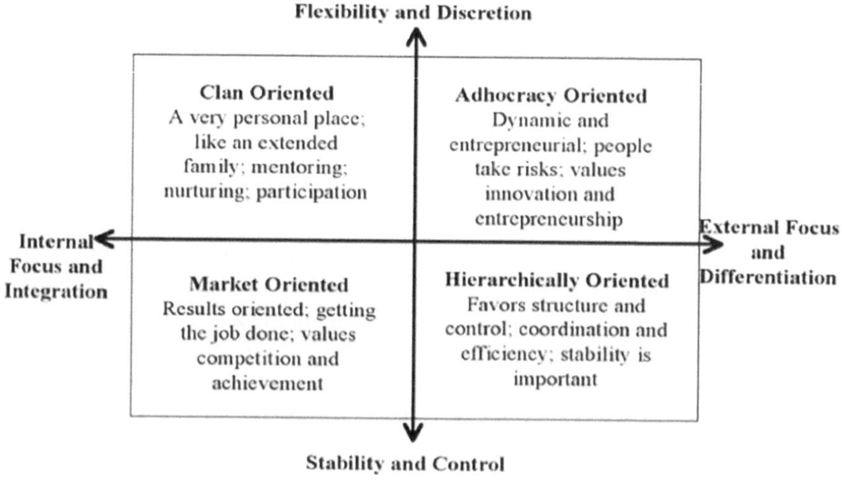

Fig. 1. The four culture-archetypes of CVF, Source [31]

Clan culture stresses flexibility plus discretion/change and concentrates on the internal organization. Clan-type organizations are characterized by teamwork, employee involvement, and corporate commitment. Clan-type organizations have shared values and common goals developed over a long period of time, steady association, non-existence of rigid options, and closely-guarded member interaction.

Adhocracy culture accentuates flexibility and change but it is externally oriented. Adhocracy-type organizations are characterized by creativity, entrepreneurship and risk taking. Consequently, adhocracy culture typifies a temporary institution that is inevitably terminated once the demand-driven organizational tasks have been completed and, later, reconstituted as soon as new tasks arise.

Market culture is externally focused and control oriented. Market culture-type organizations are characterized by productivity and competitiveness. The term market has been used figuratively to depict the organizational management's focus on the fundamental principle of maximizing profit and optimizing cost of production. The organizational goal is profit maximization through market competitiveness.

Hierarchy culture also favors control but focuses on the internal organization. Hierarchy-type organizations are characterized by productivity, synchronization, and close observance of rules and regulations. Rooted in the management concept of "bureaucracy", the hierarchy culture has a distinct organizational structure, uniform rules and procedures, stringent control, and clearly-defined responsibilities.

By and large, [32] distilled that while there are these four distinctive cultural categories, organizations are, in reality, unlikely to reflect only one cultural type because, to be effective, the adoption of some elements of each of the four organizational culture types is necessary. The common organizational culture types among construction enterprises are discussed in the subsequent section.

2.3 Organizational Culture Types Among Construction Enterprises

In relation to the second objective of this study, some empirical studies have established common organizational culture types among construction enterprises in different parts of the world using the organizational culture assessment instrument (OCAI). The OCAI diagnoses six aspects or dimensions of organizational cultures or "cultural subsystems" for organizations to assess their current and preferred culture types [31]. The six dimensions include: (i) dominant organizational characteristics, (ii) leadership style, (iii) employee management, (iv) organizational glue, (v) strategic emphasis, and (vi) success criteria. The subsequent OCAI-based empirical studies can, as a result, serve as a basis to identify the most prevalent organizational culture type among construction enterprises for suitability or otherwise for innovation adoption.

[33] conducted a time-constrained online survey to gain overall picture of the values in international and sub-units of international construction companies in Finland. 200 "white-collar workers" (33.3% response rate) were selected from construction sub-services and company size of less than 100 to 500 employees. Market culture was dominant with a preference for clan culture.

[34] steered a preliminary study on culture profiles of 159 senior managers (34.3% response rate) in five construction enterprises. The selection was based on (i) enrollment in the Chinese first-class construction enterprises, (ii) involvement in building construction and similar organization size, and (iii) regional diversity. Hierarchy culture was dominant with a growing preference for market culture.

[35] undertook an empirical study involving 826 managerial and non-managerial professionals from 134 contracting and architectural firms (38.2% response rate) in Turkey. The selection for participation was based on (i) origin of the firms as being

local, (ii) medium- and large employee-sized firms, and (iii) industry position measured by market share. Hierarchy and clan cultures were dominant.

[32] conducted an exploratory study among 56 managers of different levels across 56 contracting firms (38.6% response rate) in Singapore. The selection for participation was based on (i) local origin of the firms for cultural uniformity, and (ii) medium- to large-sized contractors measured by tendering limit and minimum paid-up capital and net worth. Hierarchy and clan cultures were found to be dominant.

[36] interviewed 139 local and expatriate project managers (71.6% response rate) in Hong Kong contracting firms. The selection was from a self-generated list of project managers. Clan culture was dominant at both project and organization levels while hierarchy culture was the least favored at both levels. The result for the hierarchy culture could be due to the expatriate project managers.

[37] conducted a pilot study among 15 high-, middle-, and low-managers across five private construction firms in Indonesia. Purposive selection criterion based on market position as leading contracting firms was used. The results showed an incongruent organizational culture profiles within the five Indonesian construction firms with the hierarchy culture still being be the most dominant.

[38] examined 71 small, medium, and large quantity surveying (QS) and building construction firms (19.3% response rate) in South Africa. Systematic sampling method based on company registrations with the relevant national bodies was used. Market culture was the dominant with a growing preference for clan culture, both among the QS and building construction firms.

[39] studied 59 project managers across four large project-based organizations in Australia, selected based on employee size, operating in heavy engineering, and a strong matrix structure of management. Hierarchy structure was dominant for knowledge sharing, combined with market culture or clan culture, depending on preference for competition or collaboration, respectively.

[40] assessed 61 senior managers (42% response rate) across three contracting firms in Botswana. Stratified sampling method based on registration with the Public Procurement and Asset Disposal Board was used. The market culture was found to be dominant across the three firms. The medium large firms preferred the clan culture while the small firms preferred the hierarchy culture.

[41] undertook a questionnaire-based descriptive study. It involved 74 construction companies (55.2% response rate) in Gaza Strip, selected based on classification by the Palestinian Contractors Union. The clan culture was the dominant current and desired organizational culture type. Small and medium organizations were predominantly market culture and clan culture, respectively.

Lastly, [42] investigated the links between organizational culture, innovation and performance. It involved 446 CEOs of Spanish organizations (27.9% response rate). Selection was based on having more than 15 employees and being located in southeast Spain. The results include: (i) adhocracy culture has a positive effect on innovation and performance, (ii) clan culture has no significant result on innovation but a positive effect on performance, (iii) hierarchy culture has a negative impact on innovation and performance, and (iv) market culture has no significant result on innovation and, paradoxically, has a negative effect on performance.

Most notably for construction innovation, [14] concluded that the configurations of innovation in the construction industry differ in diverse ways because the construction industry is project-oriented and, as a result, disjointed. This set-up makes construction innovation to remain unseen as it is jointly-developed at the project level.

2.4 Innovation Diffusion and Adopter Categories

As opposed to merely re-inventing the wheel, it is possible to plan innovation because it is "a process" whereby a novel idea, conduct, or mechanism is conceived of and brought into reality. This delineation of what truly qualifies as an innovation has influenced conceptualizing the innovation diffusion process as a normal curve of distribution as shown in Fig. 2. What this means is that innovation as a "finite product" cannot be planned or programmed. If the process or product of innovation is finite, it becomes what [43] has construed as "preventive innovations". Figure 2 should, as a result, be viewed as a way of classifying the active players in the social system based on their innovativeness [44], which also applies to construction enterprises.

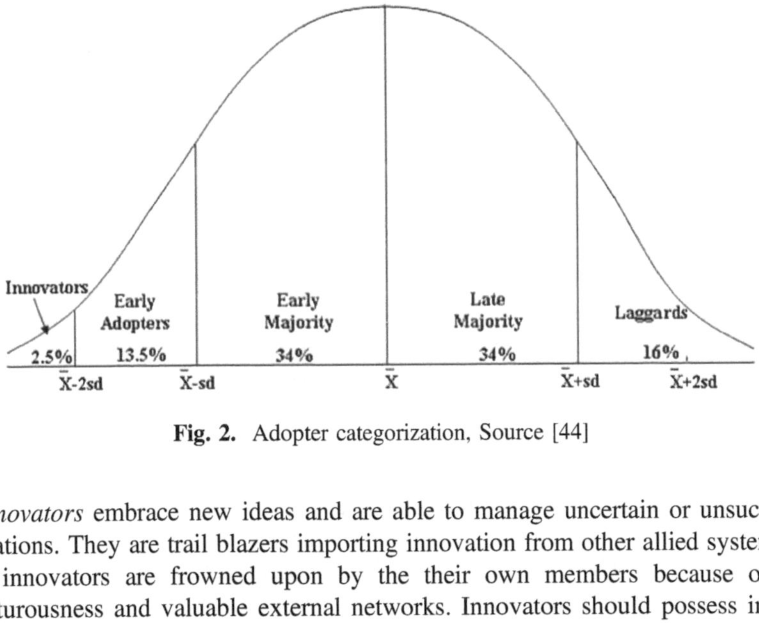

Fig. 2. Adopter categorization, Source [44]

Innovators embrace new ideas and are able to manage uncertain or unsuccessful innovations. They are trail blazers importing innovation from other allied systems. As such, innovators are frowned upon by the their own members because of their adventurousness and valuable external networks. Innovators should possess intricate technical know-how to be able to sustain their creativity and invention.

Early Adopters occupy leadership roles in the social system to distill an innovation for other members. Early adopters carry innovations forward. Their opinion leadership in adopting the innovation decreases other members' uncertainty about the innovation and this helps the innovation process to diffuse appreciably. They are, desirably, more discrete about their adoption choices than the innovators.

Early Majority also possess wholesome communication with external members; however, the absence of opinion leadership like the early adopters means that they are not as quick in adopting innovation. They adopt the innovation just before the other half because of their above average social status and contact with the early adopters. They are influenced by the competitive pressures from a "bandwagon" effect.

Late Majority are different from the early majority because they wait until most of their peers adopt the innovation. They are considered late because they adopt an innovation after the average member of the society. They adopt an innovation only after it "snowballs" from the early majority. This is due to their skepticism, below average social status, little to no financial lucidity, and opinion leadership.

Laggards are the most localized group of the social system. Their long decision-making process, due to their need for practical and safe innovation, means they are pathologically late to adopt. This is also due to their weak networks that mainly consists other members from the same category, absence of opinion leadership, limited resources and, lastly, the lack of awareness-knowledge innovations.

3 Results and Discussion

Related to the research question, this study aimed to ascertain the most appropriate organizational culture type for construction innovation adoption in Industry 4.0. Following from the integrative review of the literature as a methodological choice, for Objective 1, the multifaceted dimensions of Industry 4.0 necessitate diverse institutional and technological responses for the necessary innovation and start-up strategy by construction enterprises. This calls for a paradigm shift towards an open innovation culture to maximize the gains from Industry 4.0's collaboration, integration, and interoperability. For Objective 2, the hierarchy culture is the most dominant organizational culture type among construction enterprises, which can be attributed to the need to maintain control and focus on the internal organization. Moreover, since most construction enterprises have more than one organizational culture type [32], it was unsurprising to also find that the clan culture type and the market culture type were common among construction enterprises. The clan culture type emphasizes flexibility while the market culture is externally focused. For Objective 3, since the adhocracy organizational culture has been found to have a positive effect on innovation and performance [42], it is, therefore, proposed as the organizational culture type that best supports construction innovation adoption for Industry 4.0. This is a deviant proposition and a foundation for a future empirical study. More so, since the dominant culture type among construction enterprises is the hierarchy culture, combined with either clan culture, where collaboration is favored at project and organization levels or market culture, where competition is favored at project and organization levels.

4 Conclusion

Industry 4.0 is an emerging concept with opportunities and challenges. Construction enterprises should ensure that the people, processes and products involved in their value co-creation activities are open to the accompanying changes. As a managerial implication, the adhocracy culture best supports change and should be adopted by construction enterprises. It should be noted that this recommendation has stemmed from a conceptual study. A theoretical implication is the need for a future empirical study to test if this claim in favor of the adhocracy culture is valid or otherwise.

References

1. Denisen, D.R.: What is the difference between organizational culture and organizational climate? A native's point of view on a decade of paradigm wars. Acad. Manage. Rev. **21**(3), 619–654 (1996)
2. Schneider, B., Ehrhart, M.G., Macey, W.: Organizational climate and culture. Ann. Rev. Psychol. **64**, 361–388 (2013)
3. Watkins, M.: What is organizational culture? And why should we care. Harvard Bus. Rev. 15 May 2013. https://hbr.org/2013/05/what-is-organizational-culture
4. Schein, E.H.: What you need to know about organizational culture. Training Dev. J. **40**(1), 30–33 (1986)
5. Robertson, T.S.: The process of innovation and the diffusion of innovation. J. Mark. **31**(1), 14–19 (1967)
6. Morrar, R., Arman, H., Mousa, S.: The fourth industrial revolution (industry 4.0): a social innovation perspective. Technol. Innov. Manage. Rev. **7**(11), 12–20 (2017)
7. Eberhard, B., Podio, M., Alonso, A.P., Radovica, E., Avotina, L., Peiseniece, L., Caamaño Sendon, M., Gonzales Lozano, A., Solé-Pla, J.: Smart work: the transformation of the labor market due to the fourth industrial revolution (I4.0). Int. J. Bus. Econ. Sci. Appl. Res. **10**(3), 47–66 (2017)
8. Carvalho, N., Chaim, O., Cazarini, E., Gerolamo, M.: Manufacturing in the fourth industrial revolution: a prospect in sustainable manufacturing. Procedia Manufac. **21**, 671–678 (2018)
9. Lee, M.H., Yun, J.J., Pyka, A., Won, D.K., Kodama, F., Schiuma, G., Park, H.S., Jeon, J., Park, K.B., Jung, K.H., Yan, M-R, Lee, S.Y., Zhao, X.: How to respond to the fourth industrial revolution, or the second information technology revolution? Dynamic new combinations between technology, market, and society through open innovation. J. Open Innov. Technol. Market Complexity. **4**(21). https://www.mdpi.com/2199-8531/4/3/21
10. Hällgren, M.: The construction of research questions in project management. Int. J. Project Manage. **30**(7), 804–816 (2012)
11. Gilson, L.L., Goldberg, C.B.: Editors' comment: so, what is a conceptual paper? Group Organ. Manage. **40**(2), 127–130 (2015)
12. Torraco, R.J.: Writing integrative literature reviews: using the past and present to explore the future. Hum. Resour. Dev. Rev. **15**(4), 404–428 (2016)
13. Rocco, T.S., Plakhotnik, M.S.: Literature reviews, conceptual frameworks, and theoretical frameworks: terms, functions, and distinctions. Hum. Resour. Dev. Rev. **8**(1), 120–130 (2009)
14. Aouad, G., Ozorhon, B., Abbott, C.: Facilitating innovation in construction: directions and implications for research and policy. Constr. Innov. **10**(4), 374–394 (2010)
15. Hogan, S.J., Coote, L.V.: Organizational culture, innovation, and performance: a test of schein's model. J. Bus. Res. **67**(8), 1609–1621 (2014)
16. Rogers, E.M.: Diffusions of Innovations. The Free Press of Glencoe, New York (1962)
17. Kothari, C.R.: Research Methodology: Methods and Techniques. New Age International, Chennai (2004)
18. Saunders, M., Lewis, P., Thornhill, A.: Research Methods for Business Students. Pearson Education Limited, Harlow (2016)

19. Mark, L.: The philosophical underpinnings of educational research. Polyglossia **19**(1), 5–11 (2010)
20. Amaratunga, D., Baldry, D., Sarshar, M., Netwon, R.: Quantitative and qualitative research in the built environment: application of "mixed" research approach. Work Study **51**(1), 17–31 (2002)
21. Torraco, R.J.: Writing integrative literature reviews: guidelines and examples. Hum. Resour. Dev. Rev. **4**(3), 356–367 (2005)
22. Whetten, D.A.: What constitutes a theoretical contribution? Acad. Manage. Rev. **14**(4), 490–495 (1989)
23. Noble, H., Smith, J.: Issues of validity and reliability in qualitative research. Evid. Based Nurs. **18**(2), 34–35 (2015)
24. Lee, M.H., Yun, J.J., Pyka, A., Won, D.K., Kodama, F., Schiuma, G., Park, H.S., Jeon, J., Park, K.B., Jung, K.H., Yan, M.-R., Lee, S.Y., Zhao, X.: How to respond to the fourth industrial revolution, or the second information technology revolution? Dynamic new combinations between technology, market, and society through open innovation. J. Open Innov. Technol. Market Complexity **4**(21), 1–24 (2018)
25. Chan, J., Dang, S., Dow, S.P.: Improving innovation with expert facilitation. In: 19th ACM Conference on CSCW, pp. 1223–1235. ACM, San Francisco (2016a)
26. Chan, J., Dang, S., Dow, S.P.: IdeaGens: enabling expert facilitation of crowd brainstorming. In: 19th ACM Conference on CSCW, pp. 13–16. ACM, San Francisco (2016b)
27. Yu, T., Wu, N.: A review of study on the competing values framework. Int. J. Bus. Manage. **4**(7), 37–42 (2009)
28. Quinn, R.E., Rohrbaugh, J.: A spatial model of effectiveness criteria: towards a competing values approach to organizational analysis. Manage. Sci. **29**(3), 363–377 (1983)
29. Naranjo-Valencia, J.C., Jimenez-Jimenez, D., Sanz-valle, R.: Innovation or imitation? The role of organizational culture. Manage. Decis. **49**(1), 55–72 (2011)
30. Cameron, K., Quinn, R.E.: Diagnosing and changing Organizational Culture: Based on the Competing Values Framework. China Renmin University Press, Beijing (2006)
31. Kaarst-Brown, M.L., Nicholson, S., Von Dran, G.M., Stanton, J.M.: Organizational cultures of libraries as a strategic resource. Libr. Trends **53**(1), 33–53 (2004)
32. Koh, T.Y., Low, S.P.: Organizational culture and TQM implementation in construction firms in Singapore. Eng. Constr. Manage. Econ. **26**(3), 237–248 (2008)
33. Nummelin, J.: Measuring organizational culture in construction sector – finnish sample. In: CCIM Conference Proceedings, pp. 57–68. BUiD/CIB/CICE, Dubai (2006)
34. Liu, A.M.M., Zhang, S., Leung, M.: A framework for assessing organizational culture of chinese construction enterprises. Eng. Constr. Architectural Manage. **13**(4), 327–342 (2006)
35. Oney-Yazici, E., Giritli, H., Topcu-Oraz, G., Acar, E.: Oragnizational culture: the case of turkish construction industry. Eng. Constr. Architectural Manage. **14**(6), 519–531 (2007)
36. Fong, P.S.W., Kwok, C.W.C.: Organizational culture and knowledge management success at project and organizational levels in contracting firms. J. Constr. Eng. Manage. **135**(12), 1348–1356 (2009)
37. Coffey, V., Willar, D., Trigunarsyah, B.: Profiles of organizational culture in indonesian construction companies. In: 6th ISEC Conference Proceedings. ISEC, Zürich (2011)
38. Harinarian, N., Bornman, C.-L., Botha, M.: Organizational culture of the south african construction industry. Acta Structilia **20**(1), 22–43 (2013)
39. Wiewiora, A., Trigunarsyah, B., Murphy, G., Coffey, V.: Organizational culture and willingness to share knowledge: a competing values perspective in australian context. Int. J. Project Manage. **31**(8), 1163–1174 (2013)

40. Mufanebadza, J.M.: The Relationship Between Organizational Culture and Employee Job Satisfaction within the Botswana Construction Industry. MSc thesis, School of Construction Economics and Management, University of the Witwatersrand, Johannesburg (2017)
41. Tayeh, O.A., El-Hallaq, K., Tayeh, B.A.: The organizational culture of gaza strip construction companies. Int. J. Eng. Manage. Res. **8**(1), 40–64 (2018)
42. Naranjo-Valencia, J.C., Jimenez-Jimenez, D., Sanz-Valle, R.: Studying the links between organizational culture, innovation, and performance in Spanish companies. Revista Latinoamericana de Psicologia **48**(1), 30–41 (2016)
43. Rogers, E.M.: Diffusion of preventive innovations. Addict. Behav. **27**(6), 989–993 (2002)
44. Sahin, I.: Detailed review of Rogers' diffusion of innovations theory and educational technology-related studies based on Rogers' theory. Turkish Online J. Educ. Technol. **5**(2), 14–23 (2006)

Construction Education's Simulation Study in the Fourth Industrial Revolution

Oluwayomi Babatunde(✉)

School of Construction Economics and Management,
University of the Witwatersrand, Johannesburg, South Africa
Oluwayomi.Babatunde@wits.ac.za

Abstract. The increasing gap between academia and industry is of concern. In engineering, universities are introducing simulation studies into their construction management (CM) curricula for context-based simulated experience for graduating candidates. Consequently, this study investigated the important CM graduate skills and attributes as a basis for CM simulation course design. The methodology adopted a longitudinal study of two cohorts of CM graduates using semi-structured online questionnaire, consisting of 30 literature-informed CM graduate skills and attributes. The average response rate was 78%. Descriptive data analysis was used to categorize the CM graduate skills and attributes into criticality zones based on mean scores (minor = 0 to 2.50; moderate = > 2.50 to 3.75; and major = > 3.75 to 5.00). Wilcoxon rank sum test proved that the two cohorts were equal. Planning and controlling, time management, communication and leadership skills were ranked highest while environmental awareness, research and statistical analysis as well as marketing and entrepreneurship skills were ranked lowest by both cohorts. While the increasing need for soft or non-technical skills is supported, implications for CM education include the need for more problem-oriented nested learning activities, creating the opportunities to test solutions much more practically, and industry-academia collaboration in the design and assessment of simulated tasks.

Keywords: Capstone · Competence · Graduates · Simulation · Longitudinal study

1 Introduction

A gap exists between the skills acquired through university education and the graduate skills required to tackle real-world practical problems in general [1] and specifically for construction graduates [2] including in South Africa [3]. One of the interventions is simulation-based learning in construction engineering and management capstone courses [4]. Capstone is the word used as an allegory to refer to a milestone accomplishment from some preceding progressive or cumulative projects [5]. Consequently, engineering capstone course design is an aspect of undergraduate engineering program receiving considerable scrutiny [6]. Capstone course design is based on simulation exercises to bring situated learning experiences to the classroom [7], with the purpose of the capstone course being to integrate the prior disparate knowledge and skills from the preceding learning activities [8].

C. Aigbavboa and W. Thwala (Eds.): CIDB 2019, *The Construction Industry in the Fourth Industrial Revolution*, pp. 316–326, 2020.
https://doi.org/10.1007/978-3-030-26528-1_31

A well-designed engineering capstone course facilitates authentic learning by being able to simulate the real working environment to develop the necessary construction management (CM) competence in the graduating CM students [9] to be more competent graduate CM students. Simulation-based exercises are rooted in the constructivist approach [10], which is aimed at creating high challenge and support for the students [11]. A research question arises of what are the most important entry-level CM graduate skills and attributes to be simulated for students? This study aims to investigate the important CM graduate skills and attributes for the simulation course design. Its three objectives include: (1) identifying the important CM graduate skills and attributes (2) establishing if there is a significant difference in the CM graduate skills and attributes; and (3) recommending the process for a CM simulation course design.

2 Literature Review

The mode of the preceding research question is empirical need gap-spotting [12]. There are studies theoretically supporting the design of a CM capstone course. However, there is an empirical lack of support for the actual design of a CM capstone. Since theories lay the foundation for further progress [13], this empirical study uses integrative literature review to identify the different CM graduate skills and attributes. Integrative literature review critically analyses representative literature to generate new perspectives [14]. It is unique in how it studies the existing body of knowledge for a deeper understanding of the phenomenon under investigation [15].

2.1 Conceptual Framework

The conceptual framework situates a study in the existing empirical and theoretical bodies of knowledge to offer alternative insights on the concepts and associated thoughts [16]. The conceptual framework adopted in this study is the reflective practice of CM graduate students based on their real-life working experience vis-à-vis the simulated experience on the CM capstone course in their final year.

Students are central to education reform because they are co-creators of academic development initiatives [17], and are, as a result, a viable guide for curriculum improvements in culturally responsive contexts [18].

2.2 Construction Management Capstone Course Design

Engineering projects are, inherently, problem-based [19]. This unique feature makes engineering projects well-suited for a capstone course design. A capstone course is geared towards simulating problem-based learning experience to teach self-efficacy to be able to navigate in the dynamic real world [20].

The five shared features of engineering capstone projects among the world's top-ranked universities include: problem-oriented-pre-requisite courses, strong emphasis on group-based projects, develop and test program to gauge practicality, dynamic industry-academia collaboration, and, lastly, nested or progressive tasks for better sense-making [5]. These features support the idea of outcomes-based education model to curriculum design, based on the five philosophical orientations including: discipline-based, professional and academic, personal relevance, social relevance, and, lastly, systems design [21].

2.3 Construction Management Skills and Attributes

Different aspects of engineering capstone course have been investigated, including the teaching and learning activities [7] and the assessment thereof [6]. This present study takes a step back to map the requisite CM graduate skills and attributes that should be embedded in the design of a CM capstone course. CM graduate skills and attributes have been established both in an emerging country like South Africa [3] and [22] and beyond in the more advanced countries [23] and [24].

Following a mapping exercise of the CM graduate skills and attributes from the four immediate preceding studies, to eliminate overlaps and gaps, Appendix A presents 30 integrative-literature-review-informed CM graduate skills and attributes, arranged in an alphabetical order and coded for subsequent reference.

3 Methodology

This study adopted the positivism research philosophy, which is a philosophical standpoint of the physical scientist that entails working with a perceptible shared certainty to construct generalizations [25]. The research approach is deductive, which has the advantage of theory- or hypothesis- testing to confirm validity [26]. The research methodological choice used in this study is the quantitative method, based on numeric data collection as social facts for human behavior [27].

3.1 Research Design

Longitudinal survey design was adopted to produce data that represent quantitative descriptions about pre-selected characteristics of the population under investigation [28]. Objective 2 is premised on detecting a trend in the form of either a change or development [29] in the CM graduate skills and attributes. The advantages of the survey research strategy include: ample data within the limited time and cost, collecting data from a more representative sample, and producing empirical data based on real-world observations [30].

3.2 Research Method

A closed-ended structured online questionnaire was adopted to retrieve quantitative ordinal data from the target population. A researcher has the most control using the questionnaire as compared to other types of research instrument [31]. A closed-ended questionnaire has the advantage of providing the inquirer with quantitate or numerical data [32] while the online mode of administering the questionnaire allows for anonymity, flexibility, and ease of data management [33].

The questionnaire design has two main sections. Section A focused on the profile of the respondents while Section B focused on 5-point Likert scale rating of the level of importance (1 = not at all important and 5 = extremely important) of the 30 CM graduate skills and attributes.

3.3 Population and Sample

A population is an entire group about which some information is required to be ascertained. The population for this study was the CM graduate students in South Africa, which was, understandably, large and impractical to reach. Following, the target population or subset of the population used were two cohorts of CM honors graduate students from a university in South Africa. The sampling frame was the list of the consented private email addresses of the CM graduate students.

Purposive sampling of the 29 graduate students in Cohort A (Class of 2016) and 30 graduate students in Cohort B (Class of 2014) was adopted. Both cohorts have undertaken the simulated CM capstone course in the final semester of their CM honors program. Hence, additional stratified sampling technique used with the purposive sampling technique. The response rate for Cohort A was 72.41% and 83.33% for Cohort B; thus, giving an average response rate of 78%.

4 Data Analysis

Data collected through Section A of the questionnaire were analyzed descriptively. The analysis revealed that the post-graduation average work experience of both Cohorts A and B was 19 months each. Two graduate students from Cohort A were in non-construction sectors. More than 80% of the graduate students in both cohorts were working with private employers. Majority (>40%) in both cohorts were office- and site-based. Cohort B had a higher percentage (40%) of graduate students that were solely site-based. The job positions of Cohort A extended beyond CM to include one Senior Specialist Claims and a four quantity surveyors. Respectively, about 50% and 40% of Cohort A and Cohort B had changed jobs within their first two years.

A three-level data analysis was performed on the quantitative data collected through Section B of the questionnaire for Objectives 1 and 2. The first-level analysis, which served as the basis for achieving Objectives 1 and 2, entailed computing the mean scores of the level of importance of each of the 30 CM graduate skills and attributes. The values of the standard deviation (SD) were used as a tie breaker but the tie was maintained where the values of the mean scores and SD are, nonetheless, the same as presented in Appendix B.

The second-level analysis, for Objective 1, entailed using the mean scores to categorize the CM graduate skills and attributes into three criticality zones including minor (0 to 2.50), moderate (>2.50 to 3.75), and major (>3.75 to 5.00) based on the study by [31] as also shown in Appendix B.

Lastly, the third-level analysis, for Objective 2, entailed performing the non-parametric or distribution-free test Wilcoxon rank sum test on the re-ranked mean ratings/scores across the two cohorts still using the SD as the tie breaker as presented in

Appendix C then adding the total rank of all the items under each cohort. The Wilcoxon rank sum statistic W is defined as the total rank of the smaller sample. In this case, the samples have the same size ($n_A = n_B = 30$), so $W = 871$.

The five common steps involved in a non-parametric test includes: (i) setting up a hypothesis and deciding on the significance level, (ii) choosing a suitable statistical test, (iii) establishing the decision criterion, (iv) performing the statistical test, and (v) concluding based on the decision criterion [34]. The null and alternative hypotheses for this two-sided/tailed nonparametric test at 95% confidence interval include:

$$H_0 : \text{The two populations are different}$$
$$H_1 : \text{The two populations are not different}$$

From [35], for n_A, $n_B \geq 8$, it can be shown that, under H_0 that the two populations are equal, $W \approx N(n_A(n_A + n_B + 1)/2, s^2)$, where

$$s = \sqrt{[n_A n_B (n_A + n_B + 1)]/12}. \tag{1}$$

Using this normal approximation for the sample size of 30, W has mean $30(61)/2 = 915$ and $s = \sqrt{[900(61)/12} = 67.64$. Hence, the test statistic, $Z = 871 - 915/67.64 = -0.65$ which is lower than the critical value of 1.96 or −1.96 at 95% or 0.05 confidence interval. Hence, the null hypothesis is not rejected.

5 Results and Discussion

As shown in Appendix B for Objective 1, both cohorts ranked the following four CM graduate skills and attributes above the third quartile (i.e., among the top 25%): *Planning, scheduling and controlling*; *Time management*; *Active listening and verbal communication skills*; and, lastly, *Leadership skills*. Three of these four top-ranked CM graduate skills and attributes are mostly non-technical in nature (i.e., soft or generic skills). This result validates the increasing need for generic skills [36] or, more aptly, engineering soft skills [37]. Both cohorts ranked the following six CM graduate skills and attributes below the first quartile (i.e., among the bottom 25%): *Environmental awareness*; *Measurement, costing and estimating*; *Knowledge of the complex nature of the industry*; *Aptitude for conducting research (including performing statistical analysis*; *Know-how of surveying and levelling (or any other similar filed) apparatus*; and, lastly, *Marketing skills and entrepreneurship*. In terms of criticality, Cohort A identified 27 of the 30 CM graduate skills and attributes as "major" while Cohort B identified 23 as "major". None of the 30 skills and attributes was considered to be of minor criticality.

As shown in the preceding section's Wilcoxon rank sum test, based on Appendix C for Objective 2, there is a statistically significant evidence at the 95% confidence level to show that the two populations of CM graduate cohorts are not different in their rankings of the level of importance of the CM graduate skills and attributes. This being the case and all the 30 CM skills and attributes being either of moderate or major criticality validates the contextual-relevance [18] of the result of the mapping exercise presented as Appendix A. In the same vein, the mapping exercise is supportive of the idea of the outcomes-based education model to curriculum design [21]. More so, during the Fourth Industrial Revolution that self-efficacy based on a rich curriculum design for construction education is important.

Lastly, for Objective 3, the preceding results of Objectives 1 and 2 suggest that the successful design of a simulated CM capstone course is premised on practicality, authenticity and field experience of the teachers to truly minimize the cultural shock experienced by graduates new to the real world of construction. As supported by [5], CM simulation course design process should involve (i) problem-oriented-pre-requisite courses, (ii) strong emphasis on group-based projects, (iii) develop and test program to gauge practicality, (iv) dynamic industry-academia collaboration, and, (v) nested or progressive tasks for better sense-making.

6 Conclusion

Simulating a real-life experience is difficult but pedagogically-rewarding. The design process of the CM simulation course must be driven by the five common elements of engineering capstone projects emphasized in this study. The practical implication is a more unconventional pedagogical approach that nonetheless achieves the prescribed learning objectives. The theoretical implication is a program-level-constructive alignment design approach to make construction education's simulation study a worthwhile endeavor in the Fourth Industrial Revolution. The limitation of this study has been restricted to the minimal working experience of the respondents.

Appendix A. CM Skills and Attributes

CM skill and attribute	Code	CM skill and attribute	Code
Ability to conduct research	SKA1	Leadership capability	SKA16
Ability to exercise professional judgment	SKA2	Managerial knowledge	SKA17
Conflict and dispute resolution skills	SKA3	Marketing skills and entrepreneurship	SKA18
Know-how of surveying and leveling apparatus	SKA4	Measurement, costing and estimating	SKA19
Ability to work autonomously	SKA5	Numeracy/quantitative analytics	SKA20
Academic achievement	SKA6	Planning, scheduling and controlling	SKA21
Acceptance of responsibility	SKA7	Practical building knowledge	SKA22
Active listening and verbal communication	SKA8	Problem solving skills, creativity and innovation	SKA23

(*continued*)

(*continued*)

CM skill and attribute	Code	CM skill and attribute	Code
Malleability to dynamic work situation	SKA9	Supervisory skills and ability to train others	SKA24
Computer literacy	SKA10	Systems development ability	SKA25
Environmental knowledge	SKA11	Team building capability, trust and honesty	SKA26
Familiarity with construction quality management	SKA12	Time management	SKA27
Knowledge of the complex nature of the industry	SKA13	Up-to-date professional knowledge	SKA28
Financial management	SKA14	Work study	SKA29
Interpersonal skills	SKA15	Worker safety and health awareness	SKA30

Appendix B. CM Graduate Skills and Attributes: Ranking and Criticality

CM Graduate Skills[a]	Cohort A				Cohort B			
	Mean	SD	Rank	Criticality	Mean	SD	Rank	Criticality
SKA1	3.65	0.91	28	Moderate	3.52	1.20	28	Moderate
SKA2	4.42	0.82	7	Major	4.00	1.13	19	Major
SKA3	4.42	0.75	6	Major	4.16	1.16	16	Major
SKA4	3.45	1.02	29	Moderate	3.44	1.24	29	Moderate
SKA5	4.42	0.67	5	Major	4.12	0.82	17	Major
SKA6	3.95	0.94	22	Major	3.92	0.89	20.5	Major
SKA7	4.35	0.79	9	Major	4.54	0.76	4	Major
SKA8	4.55	0.67	3	Major	4.60	0.57	2	Major
SKA9	4.21	0.83	14	Major	4.44	0.80	5	Major
SKA10	4.26	0.71	12	Major	4.04	1.08	18	Major
SKA11	3.94	1.08	24	Major	3.64	0.93	25	Moderate
SKA12	4.35	0.65	8	Major	4.32	0.84	10	Major
SKA13	3.84	0.87	27	Major	3.52	1.06	27	Moderate
SKA14	4.35	0.85	10	Major	3.64	1.20	26	Moderate
SKA15	4.15	0.85	18	Major	4.60	0.75	3	Major
SKA16	4.47	0.60	4	Major	4.40	0.57	6	Major
SKA17	4.00	0.92	20	Major	4.16	0.97	15	Major
SKA18	3.39	1.16	30	Moderate	3.20	1.33	30	Moderate
SKA19	3.85	1.01	26	Major	3.68	1.26	24	Moderate

(*continued*)

(*continued*)

CM Graduate Skills[a]	Cohort A				Cohort B			
	Mean	SD	Rank	Criticality	Mean	SD	Rank	Criticality
SKA20	4.20	1.03	15	Major	3.76	1.42	23	Major
SKA21	4.60	0.73	1	Major	4.40	0.85	7	Major
SKA22	4.20	1.08	16	Major	4.24	0.76	11.5	Major
SKA23	4.30	0.71	11	Major	4.32	0.55	8	Major
SKA24	3.95	1.10	23	Major	4.32	0.68	9	Major
SKA25	4.00	0.84	19	Major	4.24	0.76	11.5	Major
SKA26	4.16	0.81	17	Major	4.16	0.83	14	Major
SKA27	4.56	0.68	2	Major	4.76	0.43	1	Major
SKA28	3.95	0.92	21	Major	3.92	1.16	22	Major
SKA29	3.90	0.99	25	Major	3.92	0.89	20.5	Major
SKA30	4.25	0.99	13	Major	4.20	0.85	13	Major

[a]Please refer to Appendix A for the full description of the CM graduate skills and attributes.

Appendix C. Ordered Data for Null Hypothesis

Code	Cohort A			Cohort B		
	Mean	SD	Rank	Mean	SD	Rank
SKA1	3.65	0.91	9	3.52	1.20	6
SKA2	4.42	0.82	51	4.00	1.13	24
SKA3	4.42	0.75	50	4.16	1.16	31
SKA4	3.45	1.02	4	3.44	1.24	3
SKA5	4.42	0.67	49	4.12	0.82	26
SKA6	3.95	0.94	20	3.92	0.89	15.5
SKA7	4.35	0.79	45	4.54	0.76	54
SKA8	4.55	0.67	55	4.60	0.57	57
SKA9	4.21	0.83	35	4.44	0.80	52
SKA10	4.26	0.71	39	4.04	1.08	25
SKA11	3.94	1.08	18	3.64	0.93	7
SKA12	4.35	0.65	44	4.32	0.84	43
SKA13	3.84	0.87	12	3.52	1.06	5
SKA14	4.35	0.85	46	3.64	1.20	8
SKA15	4.15	0.85	27	4.60	0.75	59
SKA16	4.47	0.60	53	4.40	0.57	47
SKA17	4.00	0.92	23	4.16	0.97	30
SKA18	3.39	1.16	2	3.20	1.33	1
SKA19	3.85	1.01	13	3.68	1.26	10

(*continued*)

(*continued*)

Code	Cohort A			Cohort B		
	Mean	SD	Rank	Mean	SD	Rank
SKA20	4.20	1.03	33	3.76	1.42	11
SKA21	4.60	0.73	58	4.40	0.85	48
SKA22	4.20	1.08	34	4.24	0.76	36.5
SKA23	4.30	0.71	40	4.32	0.55	41
SKA24	3.95	1.10	21	4.32	0.68	42
SKA25	4.00	0.84	22	4.24	0.76	36.5
SKA26	4.16	0.81	28	4.16	0.83	28
SKA27	4.56	0.68	56	4.76	0.43	60
SKA28	3.95	0.92	19	3.92	1.16	17
SKA29	3.90	0.99	14	3.92	0.89	15.5
SKA30	4.25	0.99	38	4.20	0.85	32
Total			**958**			**871**

References

1. Jackson, D.: An international profile of industry-relevant competencies and skills gaps in modern graduates. Int. J. Manage. Educ. **8**(3), 29–58 (2010)
2. Mahasneh, J.K., Thabet, W.: Rethinking construction curriculum: a descriptive cause analysis for the soft skills gap among construction graduates. In: 51st ASC Annual International Conference Proceedings, ASC, Texas (2015)
3. Chilesche, N., Haupt, T.C.: Industry and academic perceptions of construction management education: the case of South Africa. J. Educ. Built Environ. **2**(2), 85–115 (2007)
4. AbouRizk, S.: Role of simulation in construction engineering and management. J. Constr. Eng. Manage. **136**(10), 1140–1153 (2010)
5. Ward, T.A.: Common elements of capstone projects in the world's top-ranked engineering universities. Eur. J. Eng. Educ. **38**(2), 211–218 (2013)
6. McKenzie, L.J., Trevisan, M.S., Davis, D.C., Beyerlein, S.W.: Capstone design courses and assessment: a national study. In: ASEE Annual Conference & Exposition, pp. 1–14. ASEE, Utah (2004)
7. Lunce, M.M.: Simulations: bringing the benefits of situated learning to the classroom. J. Appl. Educ. Technol. **3**(1), 37–45 (2006)
8. Labossière, P., Roy, N.: Original concept for a civil engineering capstone project. J. Prof. Issues Eng. Educ. Pract. **141**(1). https://doi.org/10.1061/(ASCE)EI.1943-5541.0000199
9. Panas, A., Pantouvakis, J.P., Lambropoulos, S.: A simulation environment for construction project manager competence development in construction management. Procedia Soc. Behav. Sci. **119**(1), 739–747 (2014)
10. Shapira-Lishchinsky, O.: Simulation-based constructivist approach for education leaders. Educ. Manage. Adm. Leadersh. **43**(6), 972–988 (2015)
11. Larkin, H., Richardson, B.: Creating high challenge/high support academic environments through constructive alignment: student outcomes. Teach. High. Educ. **18**(2), 192–204 (2013)

12. Hällgren, M.: The construction of research questions in project management. Int. J. Project Manage. **30**(7), 804–816 (2012)
13. Koskela, L., Howell, G.A.: The underlying theory of project management is obsolete. In: PMI Research Conference. PMI, Washington (2002)
14. Torraco, R.J.: Writing integrative literature reviews: using the past and present to explore the future. Hum. Resour. Dev. Rev. **15**(4), 404–428 (2016)
15. Callahan, J.L.: Writing literature reviews: a reprise and update. Hum. Resour. Dev. Rev. **13**(3), 271–275 (2014)
16. Rocco, T.S., Plakhotnik, M.S.: Literature reviews, conceptual frameworks, and theoretical frameworks: terms, functions, and distinctions. Hum. Resour. Dev. Rev. **8**(1), 120–130 (2009)
17. Bovill, C., Cook-Sather, A., Felten, P.: Students as co-creators of teaching approaches, course design and curricula: implications for academic developers. Int. J. Acad. Dev. **16**(2), 133–145 (2011)
18. Jett, C.C., Curry, K.M., Vernon-Jackson, S.: Let our students be our guides: McNair scholars "guide" three urban teacher educators on meeting the needs of culturally diverse learners. Urban Educ. **51**(5), 514–533 (2016)
19. Zhou, C.: Integrating creativity training into problem- and project-based learning curriculum in engineering education. Eur. J. Eng. Educ. **37**(5), 488–499 (2012)
20. Dunlap, J.: Problem-based learning and self-efficacy: how a capstone course prepares students for a profession. Educ. Technol. Res. Dev. **53**(1), 65–85 (2005)
21. Roberts, P.: higher education curriculum orientations and the implications for institutional curriculum change. Teaching High. Educ. **20**(5), 542–555 (2015)
22. Chilesche, N., Haupt, T.C., Fester, F.: Assessing the readiness of building diplomates for the south african construction industry. J. Educ. Built Environ. **2**(1), 31–59 (2007)
23. Ahmed, S.M., Yaris, C., Farooqui, R.U., Saqib, M.: Key attributes and skills for curriculum improvement for undergraduate construction management programs. Int. J. Constr. Educ. Res. **10**(4), 240–254 (2014)
24. Benhart, B.L., Saurette, M.: Establishing new graduate competencies: purdue university's construction management curriculum restructuring. J. Constr. Educ. Res. **10**(1), 19–38 (2014)
25. Mark, L.: The philosophical underpinnings of educational research. Polyglossia **19**(1), 5–11 (2010)
26. Jahn, D.: Conceptualizing left and right in comparative politics: towards a deductive approach. Party Polit. **17**(6), 745–765 (2011)
27. Amaratunga, D., Baldry, D., Sarshar, M., Netwon, R.: Quantitative and qualitative research in the built environment: application of "mixed" research approach. Work Study **51**(1), 17–31 (2002)
28. Fowler Jr., F.J.: Survey Research Methods. Sage Publications, Thousand Oaks (2013)
29. Saunders, M., Lewis, P., Thornhill, A.: Research Methods for Business Students. Pearson Education Limited, Harlow (2016)
30. Kelley, K., Clark, B., Brown, V., Sitzia, J.: Good practice in the conduct and reporting of survey research. Int. J. Qual. Health Care **15**(3), 261–266 (2003)
31. Wilkinson, D., Birmingham, P.: Using Research Instruments: A Guide for Researchers. Routledge, London (2003)
32. Zohrabi, M.: Mixed method research: instruments, validity, reliability and reporting findings. Theory Pract. Lang. Stud. **3**(2), 254–262 (2013)
33. Newman, J.C., Des Jarlais, D.C., Turner, C.F., Gribble, J., Cooley, P., Paone, D.: The differential effects of face-to-face and computer interview modes. Am. J. Public Health **92**(2), 294–297 (2002)

34. Woolson, R.F.: Wilcoxon Signed-Ranked Test. Wiley Encyclopedia of Clinical Trials, pp. 1–3, March 2007. https://doi.org/10.1002/9780471462422.eoct979
35. Tan, W.: Practical Research Methods. Pearson, Singapore (2012)
36. Crebert, G., Bates, M., Bell, B., Patrick, C.J., Cragnolini, V.: Developing generic skills at university, during work placement and in employment. High. Educ. Res. Dev. **23**(2), 147–165 (2004)
37. Kumar, S., Hsiao, J.K.: Engineers learn "soft skills the hard way": planting a seed of leadership in engineering classes. Leadersh. Manage. Eng. **7**(1), 18–23 (2007)

Sustainable Project Delivery of Public Procurement Projects: Contractor Selection Procedure

James Akanmu[1], Oluwaseyi Ajayi[2(✉)], and Olusegun Akinsiku[2]

[1] Department of Civil and Environmental Engineering, University of Lagos, Lagos, Nigeria
[2] Department of Quantity Surveying, University of Lagos, Lagos, Nigeria
seyiajayi71@yahoo.com

Abstract. Selecting a competent contractor is a yardstick for sustainable project delivery. It ensures the project is completed within time, cost and quality standard. This study assesses the procedure of Nigeria's public procurement Act of 2007 and identifies the frequency of use of prequalification criteria. The research is both exploratory and quantitative in nature. The population for the quantitative study was public clients and consulting organization. Stratified random sampling technique was used. The sample size was calculated using Cochran formula. A total of 300 questionnaires were distributed using Cochran formula to generate the sample size. From the sample size calculated, 196 were from consulting firms and 31 from public clients' organization. Statistical Package for Social Science (SPSS) 21st version was used to analyse the data using frequency, percentage, relative importance index (RII) and t-test. There was an agreement between public clients' and consultant organisation on the frequency of use of contractors' prequalification criteria. In conclusion, it is important to consider the method of procurement, size of the project and complexity of the project to ensure the optimum balance of experience, positive references, working chemistry and teamwork.

Keywords: Contractors · Prequalification · Project delivery · Public Procurement · Tendering

1 Introduction

Public Procurement Projects system entails organizing, processing and procedure of actualizing construction project [1]. The process begins from the design stage to the completion of the project. However, the decision taken at the design stage has a direct effect on the construction stage and the performance of the project. One of the unique decisions taken by the client and his representatives at the design stage is the selection of the appropriate contractor for the project. The selection of a competent contractor for the proposed project influences the performance of construction activities on site. However, if an incompetent contractor was selected, it will leads to time and cost overruns, poor workmanship, rework, disputes and litigation [2]. Thus, it becomes necessary for a coherent approach is use to ensure a competent and suitable contractor is chosen for a project.

C. Aigbavboa and W. Thwala (Eds.): CIDB 2019, *The Construction Industry in the Fourth Industrial Revolution*, pp. 327–336, 2020.
https://doi.org/10.1007/978-3-030-26528-1_32

The Federal Government of Nigeria (FGN) established the Public Procurement Act (PPA) of 2007 to ensure that all government projects go through "Due Process" in order to ensure fairness, transparency, and accountability [3] and proper value procurement system. Similar Procurement Act had been developed in many other countries to guarantee efficient procurement system [4, 5]. In Egypt, where a Procurement Act of 9 was used for public projects, the Act failed to mandate point system for contractor evaluation. There is no appropriate criteria given to the project managers. However, evaluation of contractor was on price criterion [5]. In Ghana, the Public Procurement Act, 2003 (Act 663) was also enacted with the same purpose with that of Nigeria PPA of 2007 [6]. The main objective of the Public Procurement Board in these countries were to synchronize public procurement process in the public service in order to secure a transparent and objective judgment in the management of the use of the country resources in public procurement. It will helps to prevent fraudulent and unhealthy practices in the discharged of public procurement projects. For instance, in Nigeria PPA of 2007, the criteria were stated in the Act but the procedures for evaluation were not given. Thus, the stakeholders used different yardstick for evaluating contractors data.

Due to the risk involved, several researchers, [7, 8] have developed models to aid decision makers in choosing the appropriate contractor. Such model includes Analytical Hierarchy Process (AHP), Artificial Neural Network (ANN), Cluster analysis, Multi-criteria decision model (MCDM), Data-Base Management system (DBMs) and Multi-attribute analysis (MAA). However, these models have their weaknesses. The DBMs fails to address existing problems of prequalification and it increases subjectivity by restricting the range of decision criteria, thus, it does not reflect clients' goals and objectives [11].

Contractor selection method entails competitive and non – competitive [6]. Competitive contractor selection methods comprise of open and selective methods. While, nominated/negotiated method is known as the non - competitive contractor selection methods. In adopting the competitive contractor selection method, the criterion for selecting the contractor is on the lowest bidder. However, lowest bid selection criteria could sometimes result in delay, cost overrun, poor quality and disputes [12]. This suggests the need for a critical evaluation of contractor selection criteria in construction project procurement [10–13].

Many researchers have proffered solution in form of the use of multi-criteria approach in the selection of contractors [7–10]. Multi- criteria selection includes the use of price related and non- price related criteria. The non- price related criteria used at the prequalification stage of the contractor selection. Contractor's financial ability, experience, managerial capability, health and safety compliance, experience as well as contractors' reputation and image have been included in the non-price related criteria [14].

This study intends to provide an insight into how the prequalification of contractors are carried out in accordance with the provision of the PPA of 2007 as required by the provision of section 16 (6) of PPA, 2007. In addition, to compare the criteria listed in the PPA with criteria in literature in order to identify the frequency of prequalification criteria use. The project entail the construction of a reinforced concrete framed structured with an aesthetically appealing façade modern high-rise building on pile foundation in Lagos State, Nigeria. The scope of work includes amongst others: excavation and filling; earthwork support; preparing concrete pile and installation of 450 mm diameter bored piles; 43 m depth below pile cut of level including steel reinforcement with grade 30 grout mix; developing minimum 30 N/mm^2 work strength; fittings and fixtures; electrical and mechanical installations among others.

For this study, to evaluate the contractor's suitability in order to realize the construction of a multi-floor building. Criteria stipulated in the advertisement based on the provisions of Public Procurement Act, 2007 (PPA, 2007) as stipulated in section 16 sub-section 6, Sections 31 and 32 respectively were used. This was coupled with the provisions of the public procurement procedure manual on evaluation published by the Bureau of Public Procurement (BPP). The criteria stipulated are mandatory in compliance with section 16 (6) of PPA, 2007. This study also intends to seek the opinion of the Decision Makers (DMs) on the frequency of prequalification criteria according to the PPA.

2 Procedure of Procurement Process for Construction Projects

Development of criteria for prequalification and bidding documents was done before the advertisement of the project. Contractors interested in the advertisement collect the tender document with non – refundable fee and submitted as specified in the advertisement. Advertisements are usually done in the National Newspapers, official website, Federal Tender Journal (FTJ) and on the notice board of the procuring entity in line with the provision of section 25 of PPA, 2007. Following the advertisements for the Multi Floor Project, these following eight bidders responded. They are: Ax Project Limited, Bx Company Limited, Fl Construction Limited, Mf Nigeria Limited, Nd Nigeria Limited, Sg Limited, Ui Construction Ltd, and Vi Construction Limited. The prequalification documents were opened in line with section 30 of PPA, 2007.

2.1 Criteria for Pre-qualification

In accordance with the provisions of Section 16 (6) of PPA, 2007 the criteria for the selection of bidders as advertised were based on the following:

i. Evidence of Certificate of Incorporation with Corporate Affairs Commission (CAC)
ii. Evidence of Company Audited Account (2012, 2013, 2014)
iii. Evidence of Tax Clearance Certificate (201, 2013, 2014)
iv. Pension Clearance Certificate PENCOM
v. Evidence of compliance with ITF
vi. Two (2) registered structural engineers with COREN not less than ten years post registration with COREN
vii. Three (3) registered civil engineering technologist with not less than five years post registration with COREN
viii. Two (2) registered builders with not less than five years post registration with CORBON

ix. Two (2) registered Quantity Surveyors with not less than five years post registration with QSRBN
x. Two registered Architects with not less than five years post registration with Architect Registration Council of Nigeria (ARCON)
xi. Financial capability & Banking support
xii. Company's Technical staff experience & qualification
xiii. Performance on previous construction work of not less than five-storey building
xiv. Equipment and Technology capacity
xv. Annual turnover not less than five hundred million naira (#500M).

2.2 Prequalification Evaluation

The evaluation was carried out based on the provisions of PPA, 2007 using the advertised criteria as stipulated in section 16 sub-sections 15, Sections 31 and 32 respectively coupled with the provisions of the public procurement procedure manual on evaluation published by the Bureau of Public Procurement. The criteria stipulated are mandatory in compliance with section 16 (6) of PPA, 2007. The Evaluation of compliance of bidders indicates the level of compliance of the bidders to the criteria specified in the solicitation document.

3 Research Method

The study was both exploratory and quantitative in nature. The exploratory aspect illustrated how contractor selection practice in Public Procurement Project was carried out. While the quantitative entails sampling the opinion of public clients' and construction professionals on the procedure based on the frequency of prequalification criteria use. Thus, the construction professional entails Architects, Builders, Engineers and Quantity Surveyors. The study area was Lagos State because Lagos is a state with population of over twenty-three (23) million [15], making it 12% of Nigeria population (195 million) by the National Population Commission. Lagos has a land area of 3,577 sqkm, thus it the sixth most populous city in the world and the second largest city in Nigeria. It is one of the most populous cities in Africa [16] and the new biggest city in Africa [17]. Being the industrial as well as commercial centre of the country, the city had experienced an increased in population and abundant economic opportunities. It had led to over utilization of available utilities with attractive resources. The increasing rate of urbanization had resulted in pressure on land use; hence, it involves a lot of construction activities [18] to meet the expectation of its populace. The sampling frame was obtained from the professional bodies of the construction professionals as shown in Table 1.

Stratified random sampling technique was used to select the respondents. The sample size was calculated using Cochran formula as displayed in Table 2. It gives a sample size of 227. From the sample size calculated, 196 were from construction professionals in consulting firms and 31 were from public clients' organization.

Table 1. Sampling frame of respondents in Lagos

Respondents	Lagos State
1. Architects	250
2. Quantity Surveyors	345
3. Builders	220
4. Engineers	294
5. Public clients	108
Total	1217

Source: Institute of the respective professional bodies (Consulting firms) and www.Lagosstategov.ng/ **pagement (Lagos public clients)**

Table 2. Sample size

Respondents	Research area (Lagos State)
1. Consulting Architect	44
2. Consulting Quantity Surveyors	61
3. Consulting Builders	39
4. Consulting Engineers	52
5. Public clients	31
Total	227

$$\text{Cochran formula}\,(n_o) = \frac{t^2 \times s^2}{d^2}$$

Where: n_0 = sample size; t = t value for the acceptable margin of error (t = 1.96); s = estimate of variance in the population distribution (standard deviation (SD^2); d = acceptable margin of error (0.05).

300 questionnaires were distributed and 210 were collected giving a response rate of 70%. Statistical Package for Social Sciences (SPSS) was used for the analysis. Frequency, percentage, t-test for proportion and Spearman Rank Correlation statistical tools were used to generate the result from the data. This study was subjected to a reliability test using pre - test method. The reliability of the scale for the questionnaires was tested using Cronbach's alpha method, which was found to be 0.79. The result suggested that the questionnaires are highly reliable and there was an internal consistency. This was judged from the fact that, 0.79 was greater than 0.70 minimum reliability level [19, 20].

4 Data Analysis and Result of Findings

4.1 List for Prequalification

Table 3 displays the characteristic of consulting firms and public clients' organisation on ways they obtained list of contractors for prequalification. 81% of Architects, 89% of Builders, 63% of Quantity surveyors, 89% of Engineers and 80% of public clients' obtained their list of contractors through standing list of certain types and sizes. 19% of Architects, 11% of Builders, 22% of Quantity surveyors, 11% of Engineers and 20% of public clients used an ad-hoc list for a particular project to obtain list for prequalification. It shows that all the respondents from public clients' and consulting organisations agreed that the list of prequalified contractors were obtained from standing list based on the type of projects. Thus, such contractors must have registered with that organisation.

Table 3. List for prequalification

Prequalification list	Consulting organisation (Freq.)	%	Clients' organisation (Freq.)	%
Standing list for project of certain types and sizes	151	80	65	76
An ad-hoc list for a particular project	39	20	21	24
Total	190	100	86	100

Freq. = Frequency; % = Percentage

4.2 Review of Contractors' Information on Standing List

Table 4 presents the review of contractors' information on standing list for consulting and public clients' organisations. Due to the technicality involved in prequalification and tender evaluation procedure for purpose of selecting appropriate contractor for a proposed project, it shows that the information submitted by the contractors needs to be reviewed. Hence from the table, 48% of Architects, 42% of Builders, 40% of Quantity surveyors, 38% of Engineers and 38% of public clients' review contractors' information annually.

Table 4. Review of contractors' information on standing list

Review of contractors information	Consulting organisation (Freq.)	%	Clients' organisation (Freq.)	%
Never	4	2	3	3
Annually	81	42	41	47
Once in a while	62	32	27	31
Bi-annually	25	13	5	6
Half-yearly	22	11	11	13
Total	194	100	87	100

Freq. = Frequency; % = Percentage

4.3 Frequency of Prequalification on Type of Projects

Analysis of descriptive data of frequency of prequalification on type of projects was presented in Table 5. From the table, it reveals that the most ranked project by Architect (MIS = 0.82), Builders (MIS = 0.85), Quantity Surveyors (MIS = 0.84), Engineers (MIS = 0.79) and Pubic clients' organisations (MIS = 0.86) were commercial projects in comparison with other type of projects such as residential, industrial, religion and transport projects. The least ranked project by Architects (MIS = 0.53), Engineers (MIS = 0.54) and Public clients' (MIS = 0.62) were transport projects. While for Quantity surveyors, the least ranked were religion (MIS = 0.72) and transport projects (MIS = 0.72). It could be as a result of the value of the project and their functionality. In the study of [5] in Egypt, they concluded that prequalification should be carried out no matter how complex the project. They however, prequalified electro - mechanical, industrial and utility projects than building projects.

Table 5. Frequency of prequalification on type of projects

Type of projects	Consulting organisation		Clients' organisation		Overall	
	MIS	R	MIS	R	MIS	R
Commercial	0.83	1	0.84	1	0.83	1
Residential	0.79	2	0.83	2	0.80	2
Industrial	0.72	3	0.75	3	0.73	3
Religion	0.64	4	0.69	4	0.64	4
Transport	0.61	5	0.66	5	0.63	5

Very often = 4; often = 3; rarely = 2; never = 1; MIS = Mean Item Score; R = Rank

4.4 Frequency of Use Prequalification Criteria

Table 6 shows the frequency of use prequalification criteria. The overall most frequently used financial criteria were current fixed asset (87%), subcontractor (78%), balance sheet statement (76%) and annual turnover (74%). The least overall frequently used financial criteria were short term borrowing (55%), medium term borrowing (54%) and profitability (24%). For overall experience criteria, the most frequently used were skill including professional technical expertise such as qualification with experience expertise (97%), technical skills (97%), type of projects (96%), ability to handle projects (96%), ability to meet target dates (95%) and size of past projects completed (94%). The least overall experience criteria were the ability to efficiently integrate contract (85%) and national or local catchment (56%).

Frequently used overall managerial capability criteria were past performance (97%), quality control programme and quality of works on past projects (95%), quality workmanship (88%) and possession of quality assurance certificate (77%). For overall health and safety criteria, the top frequently used were provision of health and safety regulation (90%), company safety policy (88%), and level of adherence to health and safety regulation (87%) and the least ranked were accident book (62%) and experience in noise control (43%).

Table 6. Agreement between public clients' and consulting organisation on frequency of use prequalification criteria

Prequalification criteria	Spearman rank correlation Public clients' (r_1)	Spearman rank correlation Consultants (r_2)	$t_{cal.}$	$t_{tab.}$	Sig.	Remark
Financial	0.95	0.95	9.13	2.262	S	H_1 accepted
Experience	0.97	0.97	13.98	2.179	S	H_1 accepted
Managerial capability	0.00	0.97	6.99	3.182	S	H_1 accepted
Health and safety	0.96	0.96	5.9	3.182	S	H_1 accepted
Contractors' reputation and image	0.98	0.98	9.80	2.776	S	H_1 accepted

$\alpha = 0.05$; $t_{cal.}$ = t-test calculated; $t_{tab.}$ = t-test tabulated; Sig. = Significant; H_1 = Alternate hypothesis

Contractors' reputation and image criteria overall top ranked frequently used were capacity of work handled presently (93%) the amount of projects executed in the past 5 years (89%), permanent place of business (78%), and financial penalties previously levied in respect of failures to perform the terms of contract (68%) and the least was litigation tendency (53%).

The result of this study shows that organisation prequalified contractors but no particular prequalification criteria been used because clients' goals and objectives differs on projects. Thus clients' organisations advertise for contractors using open method of tendering and attached an expression of interest, which stated the prequalification criteria for the projects. Different prequalification criteria were been used depending on the type of projects or procurement method for the project.

Hypothesis 1

The null hypothesis that there is no agreement between public clients' and consulting organisation on frequency of use of contractors' prequalification criteria was tested using t-test for proportion to compare two proportions.

There was an agreement between public clients' and consultant organisation on frequency of use of contractors' prequalification criteria because, t-test calculated ($t_{cal.}$) for financial, experience, managerial capability, health and safety and contractors' reputation and image ($t_{cal.}$ = 9.13, 13.98, 6.99, 5.93, 9.80) is greater than t-test tabulated ($t_{tab.}$). Thus, the alternate hypothesis (H_1) is accepted. The Spearman rank correlation (r) from public clients' and consulting organisations as indicated in Table 6 shows a positive correlation among the variables.

From this study, it shows that prequalification criteria requirements for any construction projects were according to the clients' objectives/goal. [20] study on decision criteria for periodic prequalification in the UK found that there was no significant difference between the frequency of use of periodic prequalification criteria among client and contractor using chi-square. This study opposed that of [20] probably because the methodology differs.

5 Conclusion and Recommendation

This study provides an insight into how the prequalification of contractors are carried out in accordance with the provision of the Public Procurement Acts (PPA) 2007 as required by the provision of section 16 (6) of PPA, 2007 of the construction of multi – floor building. A quantitative study was also done to identify the frequency of use prequalification criteria. It was realized that prequalification criteria requirements for any construction projects were according to the clients' objectives/goal. Contractor selection in public sector has been a much-debated issue over the past years. It could be deduced that the tender submitted by the tenderers should be evaluated to check for arithmetical, basic rate and consistency of rate to ensure the appropriate contractor is given the project. It is also in support of other researchers [7-10-17] that, the lowest tender price, should not be the main criterion for selecting the appropriate contractor, hence, the argument of multi-criteria is supported by this study. However, the study of [5] placed more value on choosing contractors on lowest bidder even against the bye-law (Act 1989/1998) of the study research area (Egypt) that combines the result of the technical and financial evaluation. This study default the used of the lowest bidder as proposed by the Public Procurement Act (2007) of Nigeria due to the shortcoming of the criterion. [21] argued the used of multi-criteria procedure rather than mono-criterion.

References

1. Idiake, J.E., Shittu, A.A., Anunobi, A.I., Akanmu, W.P.: A comparative analysis of Traditional and Design & Build method of procurement in the Nigerian construction Industry. Int. J. Constr. Eng. Manage. **4**(1), 1–12 (2015)
2. Khodadadi, S.A.T., Kumar, B.D.: Contractor selection with risk assessment by using AHP Fuzzy method. Int. J. Adv. Eng. Technol. **5**(2), 311–318 (2013)
3. Udensi, C.U.: Problems in the implementation of due process. In: Proceedings of the 35th Annual General Meeting/Conference of Nigeria Institute of Building (NIOB) at Abia State, 10th–14th August 2005 (2005)
4. Wong, C.H., Holt, G.D.: Developing a contractor classification model using a multivariate discriminant analysis approach. RICS Found. Res. Pap. Ser. **4**(20), 1–23 (2004)
5. Salama, M., El Aziz, H.A., El Sawah, H., El Samadony, A.: Investigating the criteria for contractors' selection and bid evaluation in Egypt. In: Boyd, D. (ed.) Proceeding 22nd Annual ARCOM Conference, 4–6 September 2006, Birmingham, UK, Association of Researchers in Construction Management, pp. 531–540 (2006)
6. Awal, S.M.: Contractor Selection in Ghana. Unplished postgraduate thesis at Kwame Nkrumah University, Department of Building, Accra, Ghana (2010)
7. Holt, G.D.: Which contractor selection methodology? Int. J. Project Manage. **16**(3), 153–164 (1998)

8. Wong, C.H., Nicholas, J., Holt, G.D.: Using multivariate techniques for developing contractor classification models. Eng. Constr. Architect. Manage. **10**(2), 99–116 (2003)
9. Topeu, Y.I.: A decision model proposal for construction contractor selection in Turkey. Build. Environ. **39**, 469–481 (2004)
10. Zavadskas, E.K., Vilutiene, T., Turskis, Z., Tamosaitiene, J.: Contractor selection for construction works by applying SAW – G and Topsis Grey techniques. J. Bus. Econ. Manage. **11**(1), 34–55 (2010)
11. Ng, S.T., Smith, N.J., Skitmore, Martin: A case based reasoning model for contractor prequalification. Int. J. Constr. Inf. Technol. **6**(10), 47–61 (1998)
12. Xiao, H.: An analysis of the selection of project contractor in the construction management process. Int. J. Bus. Manage. **6**(3), 184–189 (2011)
13. Cristobal, J.R.S.: Contractor selection using multi criteria decision making methods. J. Constr. Eng. Manage. **138**(6), 751–758 (2012)
14. Hatush, Z., Skitmore, M.: Contractor selection using multi - criteria utility theory: an additive model. Build. Environ. **33**(2–3), 105–115 (1998)
15. Lagos Population (2018). http://populationof2018.com. Accessed 28 Feb 2019
16. Aderemi, A.E., Oyeyemi, B.F.: Legal redress for losses sustained through building collapse in Nigeria, BF, pp. 1–17 (2006)
17. Campbell, J.: Africa's new biggest city: Lagos, Nigeria (2012). http://www.TheAtlantic.Com/international/. Accessed 9 Oct 2014
18. Adelekan, I.: Private sector investment decisions in building and construction: increasing, managing and transferring risks; case study of Lagos, Nigeria. Background paper prepared for the Global assessment report on disaster risk reduction, Geneva, Switzerland (2013)
19. Asika, N.: Research Methodology in the Behavioural Sciences. Longman Nigeria Plc, Lagos (2002)
20. Field, A.: Discovering Statistics Using SPSS. Sage Publication Ltd, London (2009)
21. Mangitung, D.M., Emsley, M.W.: Decision criteria for periodic prequalification in the UK construction industry. In: Proceedings of the RICS foundation construction and building research conference, Nottingham Trent University, 5–6 September, pp. 1–13 (2002)
22. Ajayi, O.M.: Multi-criteria decision making model for contractor selection for construction projects in Nigeria. In: West Africa Built Environment Research (WABER) Conference, 27–28 July, Accra, Ghana, pp. 74–81 (2010)

Challenges to Lean Construction Implementation in South Africa

Thabiso Monyane[1,2](✉), Fidelis Emuze[1], Bankole Awuzie[1], and Gerrit Crafford[2]

[1] Department of Built Environment, Central University of Technology, FS, Bloemfontein 9300, South Africa
[2] Department of Quantity Surveying, Nelson Mandela University, Port Elizabeth, South Africa
tmonyane@cut.ac.za, gerrit.crafford@mandela.ac.za

Abstract. Efforts of the lean construction could prove to be highly worthwhile for the South African construction sector. Evidently exposed continuous poor performance of projects exhibit that the concept of lean thinking would push the construction industry's pursuit to advance quality and efficiency. While numerous countries internationally gained large benefits by adopting Lean Construction (LC) concepts, there still seems to be inadequate application of lean in South African construction industry, over the last two decades. A number of structural and cultural barriers appear to be militating against the successful implementation of lean construction. By not recognizing the factors that contribute to the successful application of LC, organizations will not be able to know what improvement efforts need to be made, where these efforts should be focused, or which efforts could obtain best results. For this reason, this study sought to identify and assess the possible barriers to the successful implementation of LC in South Africa. Grounded on a systematic literature review, followed by semi-structured interviews with Lean experts in the country, a number of barriers were acknowledged as significant barriers. Further investigation discovered that policy changes and cultural change are some of the barriers exposed as significant. The results of this study could be employed to help researchers, practitioners and companies in South African construction industry to focus their attention and resources on the significant issues, crucial to support the successful implementation of LC.

Keywords: Lean · Challenges · Barriers · Implementation

1 Introduction

The construction industry is considered an important sector in an economy of any country; hence, 73% of the people across the world agree to investment in infrastructure for the country's future economic growth [36]. Moreover, [36] predict that around US $5.25 trillion dollars is likely to be spent a year on infrastructure globally by the year 2030. Lean construction (LC) has been hailed as one of the innovative construction method ever created since the 1990s [32]. Due to the poor performance recorded continuously on construction projects, lean concepts have been recommended as a way to

C. Aigbavboa and W. Thwala (Eds.): CIDB 2019, *The Construction Industry in the Fourth Industrial Revolution*, pp. 337–344, 2020.
https://doi.org/10.1007/978-3-030-26528-1_33

improve efficiency and effectiveness of project's execution. Several studies has extensively research the benefits of lean and the achievements for over 27 years since its inception of the International Group for Lean Construction. The benefits derived by numerous studies include, improved productivity and reliability, better quality and customer satisfaction, improved forecasting, shortened schedules, waste minimization, cost effective and improved safety with the application of lean principles [8–12, 25, 32, 36, 40]. However, despite all these achievements, South Africa has not adopted the lean concepts for the construction industry's performance improvement. [16] realized that to integrate LC completely in a construction organization, the recommendation is to comprehend and anticipate the barriers that might be opposed to a proper implementation. A barrier in this case is a step or an action that prevents the attainment of the desired objective. Comprehension of such barriers is vital to propose remedies, mitigate their impact, or forewarn their occurrence and strengthen the identified conditions that contribute to the successful implementation of lean construction [16]. Based on this argument the main objectives of this study is to identify the barriers that may hinder successful implementation of lean construction initiatives. The study will first look at the background of lean construction in Sect. 2 of the paper, while Sect. 3 of the paper illustrate the structure of the research approach adopted in this study and Sect. 4 pronounce the SLR examination conducted to identify the barriers that may hinder the successful implementation of lean construction. The study focused only on barriers applicable to construction to limit the amount of information out there related to barriers of lean.

2 Background to Lean Construction

To define the word "lean construction" has been a controversial issue due to the lack of consensus Mossman [29, 36] including the International Group for Lean construction (IGLC) and Lean Construction Institute (LCI) communities. The machine that changed the world [48] presents a sketch picture of lean production [36]. Moreover, [36] imply that the term "Lean construction" emerged two years later through the contribution of [32] owing to lean production ideas in construction; however, [39] contrast this idea to have emanated from the term being coined by the International Group for lean construction in 1993. Lean production was coined by [34] to differentiate the Toyota production system from the western mass production system and subsequently popularized through the book the machine that changed the world [48]. However, [24] assert to the rich literature in case studies unfolding the worthwhile implementation of LC on real projects, while [33, 36] recite numerous benefits such as reduction of construction cost and shortened construction period witnessed when implementing LC in construction projects. There are several studies undertaken in various countries of lean implementation in the construction industry. Studies found in the literature are from developing and developed countries such as Chile [2], Uganda [3], Ghana Ayarkwa [6], Nigeria [41], Saudi Arabia [4], Australia [45], Germany [27], Singapore [21], The Netherlands [27], the UK [36], the USA [38], Turkey [47]. It is notable that none of the studies are from South Africa, and despite the benefits already cited the level of adoption is still non-existent to very low not only in South Africa but worldwide owing to barriers that may hinder successful implementation [36].

3 Research Approach

Figure 1 display a conceptual map of the overall structure of the research followed in this study to address the research objective formulated and stated in Sect. 1, and the sections of the article where the components of the research (i.e. research objectives, method and stages) are presented. As demonstrated in Fig. 1, the barriers to the effective implementation of Lean Construction (i.e. research objective) were identified and established by following a research method based on a systematic review of the academic literature.

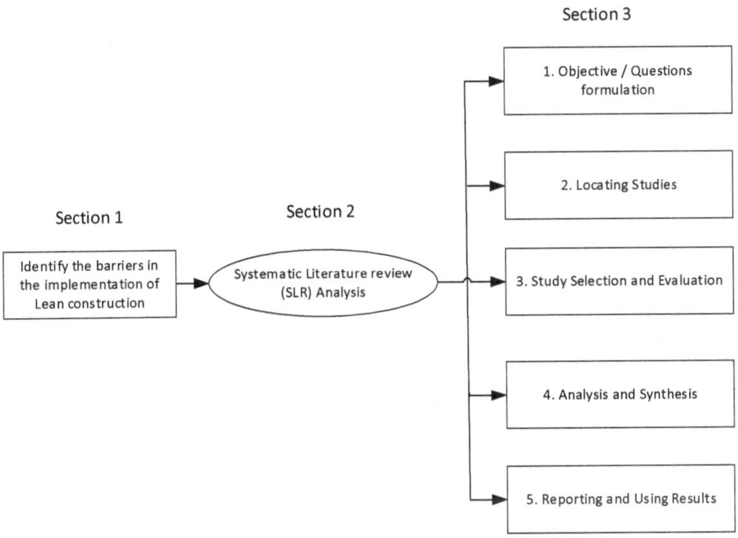

Fig. 1. Conceptual illustration structure of this study (Adapted from [17])

4 Identification of Barriers - SLR

As previously outlined a barriers in the context of this study is a challenge, a hurdle or obstacle, which does not assist, or restricts, progress to achieve a successful integration and implementation of lean [17]. To address the research questions and objectives [31] opines identification of barriers through an adoption of systematic review of the existing literature owing to its detailed, replicable, clear and meticulous approach [31]. Moreover, [20] suggests following the five sequential phases of SLR. These five phases entailed the following: (1) objective/question formulation, (2) locating studies, (3) study selection and evaluations, (4) analysis and synthesis and (5) reporting and using the results. Phase one is reported in Sect. 1 of this paper. The subsequent phases are addressed in the following sections beneath.

4.1 Locating and Selecting the Studies

The phenomenon under investigation guided the location of articles through usage of search strings linked to the main topic. According to [20], to facilitate an exclusion/inclusion criterion of the search strings a Context –intervention-mechanism-outcome (CIMO) can be followed, hence such was carried out in this study. Further relevant papers were identified using a combination of search strings such as Barriers, Challenges, failures, Obstacles, pitfalls, and Boolean operators (i.e. AND & OR). A specific search could be focused upon and this allowed the exclusion of irrelevant articles. When the same articles continued to re-appear a saturation point was considered to have been reached as a result. Input of search strings were done from various academic journals and proceedings in Emerald, ASCE, Scopus and Science Direct databases from 1990–2018. Academic books, google scholar and credible proceedings website of the IGLC on the topic are also included because of their impact on the advance of the field. Focus of the systematic literature included only lean implementation in construction and excluded implementation of lean in other sectors completely. The search criteria, a final sample of 177 articles that relates to lean implementation was identified. Conversely, only 86 of these discussed implementation barriers, and hence were further considered appropriate in this study. Figure 1 beneath is an illustration of the structure of this study.

4.2 Analysis and Synthesis

The qualitative data was synthesised using thematic analysis. There are various methods that can be employed to synthesize qualitative data, however, for this study thematic analysis proved most efficient and most appropriate due to its ability and structured way in recognizing important frequent themes [13]. Data obtained from the thematic analysis of the articles reviewed provided 30 barriers to lean implementation in construction. However, the authors would like to admit that due to length of a conference paper data could only be gathered from 46 articles at this stage. Data will be analysed further perhaps for a Journal article publication. Input from industry expert outlined the following barriers applicable to a South African context: Cultural barriers, financial issues, extensive use of subcontractors, lengthy approval process, and lack of training, lack of support by management as barriers that could hinder successful implementation of lean in the public sector. Table 1 beneath display a list of barriers identified from SLR.

Findings form the semi-structured interviews were similar to the ones identified from the literature. Policy changes was one of the critical barrier identified by six of the seven interviewers conducted. Although lean construction principles had to be explained to some of the interviewees, after comprehension from the respondents more barriers were identified as a result. Cultural change was also one of the key barrier outlined as a big factor in transforming the construction industry as a whole. Since most of the interviewees were consultants with more than 25 years' experience it would explain the culture change as a barrier. It would appear that senior consultant are comfortable with the experience they have gained all these years, and would not want to commence with new ways. However, they believed that lean stands a chance if

Table 1. Identification of barriers to lean construction implementation.

No	Barriers	References
1	Lack of technical capabilities	[3, 5, 6, 14, 23, 41, 44]
2	Heavy reliance on poorly skilled foreign workers	[21]
3	Non related work background to construction	[21]
4	Language and education barrier	[5, 14, 15, 21, 26, 36, 37, 41, 43]
5	Extensive use of subcontractors	[1, 7, 18, 21–23, 27, 30, 35–37, 41–43, 46]
6	Lack of long term commitment to change and Innovation	[1, 3, 21, 29, 37, 41, 43]
7	Price oriented tendering system	[21, 23]
9	Long term relationships with suppliers for fear of complacency	[2, 3, 23, 29, 30]
10	Limited communication via contract	[30]
11	Lack of quality management	[30]
12	Too many meetings and too much information	[19]
13	Lean beneficial to large projects not small projects	[30]
14	Extra resources required specific to deal with lean issues	[6, 41]
15	Difficult to tracks tasks from all planning forms	[21, 27, 43]
16	Lack of understanding of the concepts	[1, 2, 6, 7, 14, 15, 18, 20, 22, 28, 29, 36, 37, 41, 46, 47]
17	Lack of training	[2, 6, 14, 36, 37, 41, 43]
18	Lack of support from Management	[36]
19	Inflexible Legislation	[36]
20	Lean not part of University curriculum	[1, 36]
21	Inability to deal with change management	[26, 28, 36]
22	Lengthy approval process by client	[36, 43]
23	Culture and social issues	[1, 4, 14, 17, 29, 36, 37, 41, 43, 45]
24	Transparency	[2, 29]
25	Active client, user stakeholder involvement	[2, 5, 6, 29, 43]
26	Financial Issues	[2, 29, 36, 41, 43]
27	Lack of time for implementing new practices in the Projects	[2]
28	Challenge to create organizational elements	[2]
29	Lack of self-criticism to learn from errors	[2]
30	Resistance to change	[7, 10]

younger entrants begin with such ideas early on. Resistance to change would also be identified as a barrier as all seven of the interviewees identified culture change and not willing to commence with a new way for doing construction and a new mindset.

5 Conclusion

The study investigates the barriers that may hinder successful implementation of lean in the construction sector. This was attained by employing a systematic literature review [20, 31] was conducted as a methodology to identify the barriers that may hinder effectively adopting Lean in the construction sector. Employment of this method, 30 barriers were identified and analysed. Lean construction is still a novel approach in South Africa and the knowledge gained from identification of barriers will assist in promoting the drivers of implementing lean in future. Moreover, the contribution emanating from this study is advancing the knowledge of assisting the construction industry and forewarn on the possible pitfalls when lean gains momentum in the country. Knowledge attained from other industry will go a long way in guiding the South African construction sector not to make the same mistake already identified in other construction context. Moreover, lack of training about the new concept was highlighted by all seven of the interviewees as a key barrier. This was explained as time is always not on the side of execution of projects. This means most consultant are not willing to spend time learning and training to execute project using lean thinking. However, as the author asserts to the interviewees how they can embed lean in their daily lives, they showed interest and thought is an easier way of introducing lean construction.

References

1. Abdullah, S., Abdul-Razak, A., Abubakar, A., Mohammad, I.: Towards producing best practice in the Malaysian construction industry: the barriers in implementing the Lean Construction Approach. Faulty of Engineering and Geoinformation science, Universiti Teknologi, Malaysia (2009)
2. Alarcón, L.F., Diethelm, S., Rojo, O.: Collaborative implementation of lean planning systems in Chilean construction companies. In: Tenth Annual Conference of the International Group for Lean Construction (IGLC-10), August, Brazil, pp. 1–11 (2002)
3. Alinaitwe, H.M.: Prioritising lean construction barriers in Uganda's construction industry. J. Constr. Dev. Countries **14**, 15–30 (2009)
4. Alsehaimi, A., Tzortzopoulos Fazenda, P., Koskela, L.: Improving construction management practice with the Last Planner System: a case study. Eng. Constr. Architect. Manage. **21**, 51–64 (2014)
5. Asri, M., Naim, M.A., Nawi, M., Nasrun, M.: Actualizing lean construction: barriers toward the implementation. Adv. Environ. Biol. **9**, 172–174 (2015)
6. Ayarkwa, J., Agyekum, K., Adinyira, E., Osei-Asibey, D.: Barriers to successful implementation of lean construction in the Ghanaian building industry (2012)
7. Aziz, R.F., Hafez, S.M.: Applying lean thinking in construction and performance improvement. Alexandria Eng. J. **52**, 679–695 (2013)

8. Babalola, O., Ibem, E.O., Ezema, I.C.: Implementation of lean practices in the construction industry: a systematic review. Build. Environ. **148**, 34–43 (2019)
9. Bajjou, M., Chafi, A., Ennadi, A., El Hammoumi, M.: The practical relationships between lean construction tools and sustainable development: a literature review. J. Eng. Sci. Technol. Rev. **10**, 170–177 (2017)
10. Ballard, G.: The lean project delivery system: an update. Lean Constr. J. 1–19 (2008)
11. Ballard, G., Howell, G.: Lean project management. Build. Res. Inf. **31**, 119–133 (2003)
12. Ballard, G., Reiser, P.: The St. Olaf College Fieldhouse Project: a case study in designing to target cost. In: 12th Annual Conference of the International Group for Lean Construction, pp. 234–249 (2004)
13. Barnett-Page, E., Thomas, J.: Methods for the synthesis of qualitative research: a critical review. BMC Med. Res. Methodol. **9**, 59 (2009)
14. Bashir, A.M., Suresh, S., Proverbs, D.G., Gameson, R.: Barriers towards the sustainable implementation of lean construction in the United Kingdom construction organisations. In: ARCOM Doctoral Workshop, p. 1 (2010)
15. Brady, D.A., Tzortzopoulos, P., Rooke, J.: An examination of the barriers to last planner implementation (2011)
16. Cano, S., Delgado, J., Botero, L., Rubiano, O.: Barriers and success factors in lean construction implementation: survey in pilot context. In: Annual Conference of the International Group for Lean Construction (2015)
17. Cherrafi, A., Elfezazi, S., Garza-Reyes, J.A., Benhida, K., Mokhlis, A.: Barriers in green lean implementation: a combined systematic literature review and interpretive structural modelling approach. Prod. Plan. Control **28**, 829–842 (2017)
18. Choudhry, R.M., Hinze, J.W., Arshad, M., Gabriel, H.F.: Subcontracting practices in the construction industry of Pakistan. J. Constr. Eng. Manage. **138**, 1353–1359 (2012)
19. Cudney, E., Elrod, C.: Incorporating lean concepts into supply chain management. Int. J. Six Sigma Competitive Advantage **6**, 12–30 (2010)
20. Denyer, D., Tranfield, D.: Producing a Systematic Review. The Sage Handbook of Organizational Research Methods, pp. 671–689 (2009)
21. Dulaimi, M.F., Tanamas, C.: The principles and applications of lean construction in Singapore. In: 9th International Group for Lean Construction (2001)
22. Elazouni, A.M., Metwally, F.G.: D-SUB: decision support system for subcontracting construction works. J. Constr. Eng. Manage. **126**, 191–200 (2000)
23. Forbes, L.H., Ahmed, S.: Adapting lean construction methods for developing nations. In: Second LACCEI International Latin America and Carribean Conference for Engineering and Technology (LACCEI 2004), Challenges and Opportunities for Engineering Education, Research and Development, pp. 2–4 (2004)
24. Hamzeh, F., Kallassy, J., Lahoud, M., Azar, R.: The first extensive implementation of lean and LPS in Lebanon: results and reflections. In: Proceedings of the 24th Annual Conference of the International Group for Lean Construction, Boston, USA (2016)
25. Howell, G., Ballard, G.: Implementing lean construction: understanding and action. Proceedings of 6th Annual Conference of the International Group for Lean Construction (1998)
26. Jara, C., Alarcón, L.F., Mourgues, C.: Accelerating interactions in project design through extreme collaboration and commitment management–a case study. Proceedings of 17th Annual Conference of the International Group for Lean Construction (IGLC-17) (2009)
27. Johansen, E., Glimmerveen, H., Vrijhoef, R.: understanding lean construction and how it penetrates the industry: a comparison of the dissemination of lean within the UK and the Netherlands. In: Proceedings of 10th Annual Conference of the International Group for Lean Construction, Gramado, Brazil, pp. 6–8 (2002)

28. Johansen, E., Walter, L.: Lean construction: prospects for the German construction industry. Lean Constr. J. **3**, 19–32 (2007)
29. Jørgensen, B., Emmitt, S.: Investigating the integration of design and construction from a "lean" perspective. Constr. Innov. **9**, 225–240 (2009)
30. Kim, D., Park, H.-S.: Innovative construction management method: Assessment of lean construction implementation. KSCE J. Civ. Eng. **10**, 381–388 (2006)
31. Kitchenham, B., Charters, S.: Guidelines for performing systematic literature reviews in software engineering (2007)
32. Koskela, L.: Application of the New Production Philosophy to Construction, Stanford University Stanford (1992)
33. Koskela, L.: An exploration towards a production theory and its application to construction, VTT Technical Research Centre of Finland (2000)
34. Krafcik, J.F.: Triumph of the lean production system. MIT Sloan Manage. Rev. **30**, 41 (1988)
35. Miller, C.J., Packham, G.A., Thomas, B.C.: Harmonization between main contractors and subcontractors: a prerequisite for lean construction? J. Constr. Res. **3**, 67–82 (2002)
36. Millet, J.: A blueprint for modern infrastructure delivery: why things go wrong and how to put it right. Mace Insights Report (2019)
37. Mossman, A.: Why isn't the UK construction industry going lean with gusto? Lean Constr. J. 1–19 (2009)
38. Movaghar, E.M.: Identifying the Barriers of Implementing Lean Construction Principals in Developing Countries. Middle East Technical University (2016)
39. Nahmens, I., Ikuma, L.H.: An empirical examination of the relationship between lean construction and safety in the industrialized housing industry. Lean Constr. J. 1–19 (2009)
40. Nguyen, V.-H., Chang, L.-M.: Work structuring and the feasibility of application to construction projects in Vietnam. World Acad. Sci. Eng. Technol. Int. Sci. Index **62**, 662–670 (2012)
41. Nicolini, D., Tomkins, C., Holti, R., Oldman, A., Smalley, M.: Can target costing and whole life costing be applied in the construction industry?: evidence from two case studies. Br. J. Manage. **11**, 303–324 (2000)
42. Olatunji, J.: Lean-in-Nigerian construction: state, barriers, strategies and "Go to-Gemba" approach. In: Proceedings 16th Annual Conference of the International Group for Lean Construction, Manchester, UK (2008)
43. Oviedo-Haito, R.J., Jiménez, J., Cardoso, F.F., Pellicer, E.: Survival factors for subcontractors in economic downturns. J. Constr. Eng. Manage. **140**, 04013056 (2013)
44. Sarhan, S., Fox, A.: Barriers to implementing lean construction in the UK construction industry. Built Hum. Environ. Rev. 1–19 (2013)
45. Shang, G., Sui Pheng, L.: Barriers to lean implementation in the construction industry in China. J. Technol. Manage. China **9**, 155–173 (2014)
46. Stewart, R.A., Mohamed, S., Marosszeky, M.: An empirical investigation into the link between information technology implementation barriers and coping strategies in the Australian construction industry. Constr. Innov. **4**, 155–171 (2004)
47. Tam, V.W., Shen, L., Kong, J.S.: Impacts of multi-layer chain subcontracting on project management performance. Int. J. Project Manage. **29**, 108–116 (2011)
48. Tezel, A., Nielsen, Y.: Lean construction conformance among construction contractors in Turkey. J. Manage. Eng. **29**, 236–250 (2012)

Barriers to Construction Procurement Change in Higher Education Institutions

Xebiso Kamudyariwa(✉) and David Root

School of Construction Economics and Management,
University of the Witwatersrand,
1 Jan Smuts Avenue, Braamfontein 2000, South Africa
{Xebiso.Kamudyariwa, David.Root}@wits.ac.za

Abstract. This research looked at the barriers Higher Education Institutions face in changing their procurement practices. Since the South African Construction Industry tends to be laggard in terms of any change, the research assumed that barriers to procurement change would be linked to the barriers faced in the adoption of I4.0. These barriers would help determine the preparedness for I4.0 adoption. Qualitative research using semi-structured interviews was conducted with senior management in Facilities Management Departments in HEIs. Using thematic analysis, findings suggest that the identified barriers to I4.0 preparedness include lack of management buy in, slow pace of change, big data and human capital. It was concluded that a best value approach is a necessary prerequisite to any possible changes. A more proactive approach to change and a willingness to put learning and effort into understanding what change means/entails will result in a smoother change process for HEIs.

Keywords: Higher Education Institutions · Collaboration · Procurement · Culture · Technology

1 Introduction

The concept of change in procurement has been widely discussed within the construction industry. The Latham report puts this discussion at almost 25 years [1]. Change has been slow in its adoption, especially in an industry that tends to comfortable with its state regardless of its shortcomings [2].

There is global anticipation of changes expected with the 4th Industrial Revolution or Industry 4.0 (I4.0) as it is commonly known. These changes are not expected to be tied to a single industry, rather they "relate to all fields of life" [3].

Change tends to be inevitable, yet the global construction industry managed to remain embedded in century old practices [4]; South Africa was not exempt. Countries like the UK have adopted current best practices by a shift to more collaborative practices implemented through public procurement legislation [5, 6] which constituted clients driving change. The South African Construction Industry (SACI) on the other hand has largely ignored any shift to more collaborative procurement preferring instead to remain steeped in the industry-wide acceptance of the predominant use of more traditional procurement. Clients have also largely left all procurement to professionals

C. Aigbavboa and W. Thwala (Eds.): CIDB 2019, *The Construction Industry in the Fourth Industrial Revolution*, pp. 345–354, 2020.
https://doi.org/10.1007/978-3-030-26528-1_34

in the project team. Discussions around this tend to emphasize the fact that clients understand and accept that there will be problems; in essence challenges in terms of delays, cost overruns and quality, adversarial relationships and fragmentation [6–10] are bywords of the industry.

This condition of the industry is worrying especially considering the rapid changes occurring in various industries due to I4.0. It is expected that at some point, all industries will bow to the pressure of using technology to accomplish their work. This is of great concern in a construction industry like that of South Africa where change even to just more collaborative working has met resistance. One major question becomes whether the industry can adjust to applying I4.0.

Adjusting to I4.0 is about more than just using technology, the workforce needs to understand its purpose and how to actually utilize it in a manner that benefits their work. As pointed out, the SACI already struggles with the concept of working cordially together, it utilizes not so modern technology and has a focus on lowest cost procurement. These challenges are a red flag concerning the preparedness of the industry to apply I4.0.

The industry needs to adapt and in a timely manner. As with all change, a change leader/champion is needed to drive the process especially when it comes to technology [11]. Higher Education Institutions (HEIs) are expected by their very nature to be centers and creators of knowledge. With the South African government prioritizing funding for Science and Technology research and also major capital projects in HEIs [12, 13], it is to be expected that they (HEIs) would be disseminators of best practice and as an influential client, assist in implementing procurement change as a precursor to I4.0.

Unfortunately, public HEIs are not in a strong position to be drivers of change. While they espouse collaborative values, these have fallen short especially in construction procurement. It is of importance to determine what barriers HEIs experience as they try to adopt more collaborative ways of working which would in turn better prepare them for I4.0 application. The purpose of this paper is to investigate the barriers to more collaborative working and link these to the lack of preparedness the CI faces in adopting I4.0.

2 Literature Review

The concept of I4.0 was first introduced in Germany in 2011 and referred to as Industre 4.0. The term has varied definitions, with the common thread encapsulated in this definition by Hozdic who termed I4.0 as referring to the integration of physical objects, human actors, intelligent machines, production lines and processes across organizational boundaries, forming a system in which all processes are integrated and share information in real time [14]. I4.0 embraces the concept of all systems collaborating in real time in an integrated manner. While this concept is mostly referred to in terms of manufacturing processes [3, 15], it as a concept that will see those with modern ICT able to apply its principles for the benefit of their specific industries [3].

2.1 Various Revolutions

The fourth industrial revolution was preceded by revolutions that laid the work for (1) industrialization then (2) the foundation for the emergence of hard automation followed by (3) groundwork for computers and fluid automation finally culminating in (4) the ground work for I4.0 through modern information and communication technology [14]. Each revolution laid the ground work for evolution into more efficient systems. I4.0 is built around the collection and use of vast amounts of data [3, 14–18] to ensure more efficient systems, processes and human interactions.

I4.0 achieves cost reductions and improvements by the automation of production through the use and exchange of data in real time and use artificial intelligence [3]; these are termed also Big Data and Modern Technology [3, 14–17].

2.2 Barriers to I4.0 Adoption

Various authors have delved into the challenges involved in applying I4.0. Of note are the barriers noted by Slusarczyk who compiled a list of challenges South Africa faces in comparison to its global counterparts [3]. The table below adapted from Slusarczyk details the barriers to digitization and whether South Africa's position in relation to global averages (Table 1).

Table 1. Barriers to I4.0 adoption

	Loss of intellectual property control	Partners unable to digitally collaborate	Slow expansion of basic tech infrastructure	Lack of standards, norms & certification	Insufficient talent	Data security and privacy issues	High investment requirements	Unclear economic benefit	Lack of clear vision and leadership	Lack of digital culture and training
SA	7%	14 %	35%	12%	40%	14%	37%	35%	35%	58%

As can be noted in the table above, South Africa's greatest barriers to I4.0 are the slow expansion of basic technology infrastructure, insufficient talent in terms of skills for using technology, the high investment needs necessary to set up and the lack of digital culture and training. While unclear economic benefit and lack of clear vision and leadership scored lower than global trends, it was by a margin of at most 3%; therefore, these two can also be considered barriers to 14.0 application.

2.3 Procurement Change

Change in the SACI has been a protracted process. The current drive for change has been driven by National Treasury who have realized the benefit of utilizing procurement with a focus on Value for Money as opposed to Lowest Price [19]. This has resulted in a newly legislated process for procurement for public projects. The Department of Education and Training has also released draft copies of policies for procurement in HEIs [20]. The necessity for change has been prioritized due to the current economic pressures on government funding and the need for better accountability of public funds.

Change to a more collaborative approach has been touted as best practice [6]. Focus has been on more collaborative relationships through the type of contracting used. The table below details the envisioned changes. These changes all involve the concept of working together, decentralizing operations and sharing information (Table 2).

Table 2. Arms-length contracting vs. relational/obligational contracting

Arms-length contracting Adapted from [20]	Relational/obligational contracting Adapted from [6]
Fragmentation as each stakeholder has their part to do and it ends there	Stakeholders work together, e.g. contractor with designer
Silos as no one is interested in sharing or learning together	There is information sharing and transparency
Adversarial relationships	Cordial relationships built on trust
Constant overruns	Stakeholders work together to ensure schedule and budget are kept under control
Constant claims from the contractor	Reduced claims due to working together and preservation of relationships

3 Methodology

Qualitative research was conducted using semi-structured interviews with either Heads of Departments or knowledgeable senior management in the Facilities Management Departments of HEIs. Thematic analysis was used on the collected data.

There are currently twenty-six (26) public universities in South Africa made up of three (3) categories; traditional, comprehensive and universities of technology. These made up the population of the research. It should be noted that one of the participating Institutions (U5) was in charge of oversight for 2 of the newest Institutions. For this study, the population then reduces to twenty-four (24) HEI as U5 was managing procurement for the two Institutions.

To conduct the research, the researcher had to obtain permission from each university to be able to interview staff in their Facilities Management Departments (FMDs). When this was done, further permission needed to be sought from participants themselves.

While permission was obtained from eighteen (18) institutions, only five (5) participants agreed to be part of the study, giving a response rate of just over 20%. Semi structured interviews were conducted with the participants, some over the telephone (due to distance) and others face to face. Questions for the interviews were derived from literature and pertained to procurement and collaboration in their departments.

Thematic analysis was then conducted on the data. Themes derived from the characteristics of collaboration identified in literature were used as an initial point of reference in identifying themes from the interviews, with subsequent themes emerging from the data itself.

4 Analysis of Results and Discussion

The Institutions that participated in the study are referred to as U1, U2, U3, U4, U5. U5 is the Institution with two other institutions linked to it.

The observed themes included the following; functioning, procurement, relationships and future. Barriers were extrapolated form the themes. Additional data was collected on the experiences of U5 in terms of their change to more collaborative working.

4.1 Functioning

The Institutions had various models for their FMDs. While U2 and U3 have a central office dealing with both capital projects and maintenance, U1, U4 and U5 have separate divisions with specialty in either maintenance or capital projects. There is no "best practice" in terms of the models as they are dependent on the particular context (Table 3).

Table 3. Structures and policies of FMDs

Characteristic	U1	U2	U3	U4	U5
FMD structure	Separate	Central	Central	Separate	Separate
Procurement policies	One Policy Red Tape	One Policy Red Tape	One Policy Red Tape	One Policy Red Tape	Separate Policies

U1, U2, U3 and U4 also have stringent procurement policies in place which are not conducive to collaborative working. Total overhauls of their policies would be necessary for collaboration to take place as their current policies are geared towards goods and services. U5 on the other hand had a procurement policy for construction and a separate one for goods and services. This enables flexibility in a manner that was not possible when using a goods and services procurement policy. Working with one procurement policy is set to end though as National Treasury has legislated the separation of policies as done in U5. This is a huge step towards more collaborative working.

4.2 Procurement

As discussed in the literature review, traditional procurement is the "go to choice" for most projects in the SACI, it is therefore no surprise that Institutions would "prefer" this type of procurement. For most institutions, lowest price is a major consideration when assessing tenders. JBCC is a common choice of contract as it is well known and established; it tends to be most dominantly used contract. U5 uses the NEC as it is considered a flexible contract that enables strict control of the budget. It would be expected that this quality of strict budget control would make the NEC a preferred contract especially considering the funding problems most HEIs are facing (Table 4).

Table 4. Procurement strategy

Characteristic	U1	U2	U3	U4	U5
Tenders	Open 2 stage	Open 2 stage	Open 2 stage	Open 2 stage	Open 3 stage
Contracts	JBCC, FIDIC BOQ, Lump Sum	JBCC Lump Sum	JBCC, GCC	JBCC, GCC	NEC Target Cost, BOQ
Contractors	Concerns Building	Building	Building	Concerns Building	Design (ECI), Equal Partner

While stakeholder engagement is evolving globally, most HEIs are still treating the contractor as merely a builder. While contractor concerns could be considered, his job is clearly that of constructing the required infrastructure. In its collaboration drive, U5 brings in the contractor early to consult on projects and ensures that he is treated as an equal partner in the team. The logic here is that the contractor is actually a professional in his own right and should be treated as such.

4.3 Relationships

Institutions had varied relationships with their stakeholders. All except U5 have a 2-step conflict resolution process and have resorted to litigation in some cases. U5 has not had any litigation in 10 years and they attribute this to cordial relationships between stakeholders. Again, U5 was different in its approach to risk. While others allocated it upfront with possible assessments conducted over the project duration, U5 stood out as an Institution allocating risk with a consideration of which stakeholder was best suited to handle a particular risk. They also share trust and have long term relationships with other stakeholders. The other institutions had a problem with trusting especially contractors and their procurement policies did not cater for long term relationships (Table 5).

The Institutional approach to relationships infers its attitude to collaboration. Collaboration involves working together to ensure mutual benefit. Most Institutions are focused on solely the Client's benefit. This does not bode well in a relationship built on compromise and sharing. In addition to this, I4.0 involves big data sharing; if parties do not share trust, this acts as a solid barrier to adapting to change.

Table 5. Stakeholder relationships

Characteristic	U1	U2	U3	U4	U5
Conflict	Arbitration Litigation	Arbitration Litigation	Arbitration Litigation	Arbitration Litigation	Remediation Arbitration Litigation
Risk	Upfront Tender Doc	Upfront Tender Doc	Upfront Tender Doc	Upfront Tender Doc	Best suited allocation
Trust	Limited	None	Limited	Limited	Shared
Relationships	No long term	No long term	No long term	No long term	Long term

4.4 Future

Table 6 reflects the prospects or possibilities for each HEI. HEIs like U1, U2 and U4 saw the potential for change though current systems would create difficulties in implementing these changes. U3 was defeatist, envisioning no likelihood of change anytime soon. U5 on the other hand had problems mostly associated with what happens after handovers take place as opposed to working in a project like the other HEIs.

Table 6. Potential future changes

U1	ECI good idea, current structures do not allow Need to shift from goods procurement to construction procurement Challenging to move to NEC because of lack of capacity for administration
U2	ECI good idea but current policies do not allow
U3	Changes not really viable at the moment
U4	Incentives would be interesting to consider
U5	Need more integration between Capital Projects division and Maintenance

One common refrain seems to be that concerning policies in institutions. The most detrimental has been that of having a single procurement policy for both goods and construction. Besides limiting what should be a more flexible process, it also forces a focus on lowest price tendering. This is problematic especially as Institutions are being encouraged to prioritize value for money when procuring.

The results from the four themes show that U5 procures in a manner that is different from other Institutions; it has succeeded in adapting to and adopting collaborative practices. The comparison of U5 to the other institutions was therefore helpful in determining what barriers exist when changing from traditional procurement to collaborative procurement.

4.5 The Process of Change

Further information furnished by U5 showed that change is not an easy process. Before the introduction of any changes, there are prerequisites for the process (Table 7):

Table 7. Prerequisites to change

Management buy in	If management is not supporting the idea, change is unlikely to take place successfully. They also need to be prepared for a few mistakes along the way
Team to shape change	Change cannot happen without a someone driving the process, articulating what the process is and will achieve, encouraging stakeholders in the change, enforcing agreed upon rules
Large enough capital programme	A programme is needed in order to appreciate change. It also allows the flexibility to fine tune the process, discard what does not work and improve on what does

U5's experience showed that change does not have to be a sluggish process; they managed to implement all their changes at once. This is process of change would fit in with I4.0 considering that technology changes so rapidly, meaning organizations need to be prepared to implement flash changes.

Human capital also proved an important factor. Processes cannot run themselves, people need to be skilled and understand what change is and its purpose/benefit before committing. This applies to both management and general staff. Trust also needs to exist between stakeholders otherwise no information sharing will occur, there will be no transparency, and everyone will be working in silos.

U5 stressed the importance of the administrative process. Change takes time, energy, and investment in resources. One major resource is that is constantly overlooked is data. Data provides information for predicting patterns, receiving early warnings and making savings just to name a few.

From the issues raised, the barriers most relevant to preparedness for I4.0 are summarized in Table 8. Though barriers are linked to the difficulties faced in changing to more collaborative working, they are also a reflection of the problems likely to be faced with I4.0 application. It is important to take time to deal with these barriers in terms of collaboration as they open the door to working in a more flexible manner able to adapt to rapid changes. While it is unlikely that complete automation of the industry can occur, it would still be helpful to utilize technology as a means of improving the performance of the industry.

These barriers can be overcome if various stakeholders do their part in opening the SACI to change. Clients can make a huge difference in changing the way professionals in the industry work. In this case, HEIs through research can determine and insist on the utilization of best practice procurement in their capital projects.

Table 8. Barriers

Barrier	Relation to I4.0
Management buy in	When management does not understand benefits, this limits the change that can occur. This has been noted as an issue in the SACI
Slow change	I4.0 works on premise of constantly changing technology that works more efficiently and produces better processes. Change cannot be slow
Big data	There were concerns about the amount of data and work needed to implement e.g. changes to type of contract. If organizations unable to handle data, that limits benefits of I4.0
Human capital	Those working with technology need to be skilled and tech savvy, keeping up with trends. This is lacking in the SACI, which has been having limited success with change due to mostly archaic policies and comfort in the current status quo

5 Conclusion

In order for the SACI to be prepared for I4.0, there is a need for stakeholders to value best practice. The starting point of this would be in reviewing procurement policies and ensuring they have the necessary flexibility to accommodate change. All stakeholders also need to be able to work collaboratively together, sharing necessary information and making good use of available data as this will work towards improving the processes within the industry. They would also need to up skill as a means of ensuring they are able to acclimatize to a more technological environment. BIM is a byword in industry currently because of its "ability" to bring stakeholders together while supplying all necessary data. Utilizing such processes would be a great step towards preparedness for I4.0. Change is not impossible to accomplish, stakeholders need to be proactive about preparation instead of waiting to be forced into it when their processes are made obsolete by technology.

If HEIs take the initiative to change their manner of procurement, they have the ability to influence the SACI. Starting with their procurement policies and working on more collaborative relationships will make a difference in the route to preparedness for I4.0.

References

1. Latham, M.: Constructing the team (1994)
2. Gardiner, J.: Latham's report: Did it change us? (2014). http://www.building.co.uk/analysis/lathams-report-did-it-change-us?/5069333.article
3. Slusarczyk, B.: Industry 4.0 – Are we ready? Pol. J. Manag. Stud. **17**, 232–248 (2018)
4. Babic, N.C., Rebolj, D.: Culture change in construction industry: from 2D toward bim based construction. J. Inf. Technol. Constr. **21**, 86–99 (2016)
5. NBS: National BIM Report (Industry Report). Newcastle (2016)
6. Watermeyer, R.: Changing the construction procurement culture to improve project outcomes. In: Facilities Management, Procurement Systems and Public Private Partnerships. Presented at the Joint CIB W070, W092 and TG72 International Conference, Cape Town (2012)

7. Bowen, P., Edwards, P., Cattell, K.: Corruption in the South african construction industry: a mixed methods study. Presented at the ARCOM 2012, Association of Researchers in Construction Management (ARCOM), Reading UK, pp. 521–531 (2012)
8. Doree, A.G.: Collusion in the Dutch construction industry: an industrial organization perspective. Build. Res. Inf. **32**, 146–156 (2004)
9. Rust, F.C., Koen, R.: Positioning technology development in the South African construction industry: a technology foresight study. J. South Afr. Inst. Civ. Eng. **53**, 2–8 (2011)
10. Thwala, W., Mofokeng, G.: An exploratory study of problems facing small and medium sized contractors in the Free State Province of South Africa. In: Quah, C.H. (ed.) Business Dynamics in the 21st Century. InTech (2012)
11. Nam, C.H., Tatum, C.B.: Leaders and champions for construction innovation. Constr. Manag. Econ. **15**, 259–270 (1997)
12. DHET: Ministerial Statement on University Funding: 2017/18 and 2018/19. Department of Higher Education and Training South Africa (2016)
13. DHET: Green Paper for Post-School Education and Training (Green Paper). Department of Higher Education and Training South Africa (2012)
14. Agostini, L., Filippini, R.: Organizational and managerial challenges in the path toward Industry 4.0. Eur. J. Innov. Manage. **22**, 406–421 (2019)
15. Bienhaus, F., Haddud, A.: Procurement 4.0: factors influencing the digitisation of procurement and supply chains. Bus. Process. Manage. J. **24**, 965–984 (2018)
16. Luthra, S., Mangla, S.K.: Evaluating challenges to Industry 4.0 initiatives for supply chain sustainability in emerging economies. Process Saf. Environ. Prot. **117**, 168–179 (2018)
17. Mohamed, M.: Challenges and benefits of Industry 4.0: an overview. Int. J. Supply Oper. Manage. **5**, 256–265 (2018)
18. Rajput, S., Singh, S.P.: Identifying Industry 4.0 IoT enablers by integrated PCA-ISM-DEMATEL approach. Manag. Decis (2018)
19. National Treasury: Focus on: National Treasury Standard for Infrastructure Procurement and Delivery Management (Standard). Department: National Treasury Republic of South Africa (2016)
20. DHET: Macro-Infrastructure Framework (MIF) - a generic procurement strategy for universities (Draft for discussion with universities). Department of Higher Education and Training South Africa (2017)

An Examination of Contextual Factors that Cause Variations in Labour Productivity

Mochelo Lefoka(✉) and Abimbola Windapo

Department of Construction Economics and Management,
University of Cape Town, Cape Town 7701, South Africa
mochelo.lefoka@uct.ac.za

Abstract. The performance of labour is one of the factors that contribute to the success of the construction project. Labour productivity alone has been ascribed the sole measure of poor project performance without an examination of factors giving rise to this claim. Furthermore, there is limited research that examines the primary factors affecting labour productivity in the context of South African construction industry. This study examines the contextual factors affecting labour productivity on construction projects and whether there are key factors that impact on labour productivity on construction projects. The aim of this study will be achieved through a review of literature pertinent to contextual factors impacting labour productivity on construction projects. Additionally, a distribution of a questionnaire survey to construction companies listed on the cidb Register of Contractors in the grades 2–6. It emerged from the study of extant literature that delays in delivery of material, inspection delays from engineers, change orders from designers/consultants, lack of experience amongst workers and mismanagement of resources on site, critically and negatively impact on labour productivity on construction projects. This examination of contextual factors that affect labour productivity provides necessary information for use by construction organisations in implementing measures to properly address issues pertaining to poor labour productivity on construction projects. Based on these findings, the study recommends that construction organisations create a database that clusters contextual factors per region/location of the project. This is to ensure that the critical factors such as inadequate supervision and remuneration disputes that impact on the performance of labour are addressed, while those that are out of the organisation's control such as shortage of materials and industrial action are planned for at the pre-construction phase of the project. Further studies that investigate the contextual factors impacting on labour productivity, using empirical data is recommended.

Keywords: Delays · Labour · Material shortage · Productivity · Supervision

1 Introduction

Most global economies rely on the construction industry as a significant contributor to the Gross Domestic Product (GDP) [1]. This assertion holds for South Africa as the sector contributes close to 9% to the country's GDP [11]. Additionally, the construction industry further forms part of one of the larger industries that significantly strives to

C. Aigbavboa and W. Thwala (Eds.): CIDB 2019, *The Construction Industry in the Fourth Industrial Revolution*, pp. 355–363, 2020.
https://doi.org/10.1007/978-3-030-26528-1_35

create employment in both the formal and informal sectors of the economy. According to Statistics South Africa (StatsSA), over 10,8 million people are employed in the South African formal sector and 11% of this number work in the construction industry [12]. Additionally, between 2015 and 2016, 2.6 million people were reported to be employed in the formal sector of which the construction industry contributed 17% of the number of those employed. It can, therefore, be deduced that the impact of the construction sector to both employment and GDP is significant.

One of the endemic issues in the construction industry is the recorded poor performance of labour [1], and this has sparked global concern for the sustainability of the construction industry [2]. Notwithstanding the reported poor productivity, but it is also the general perception and sometimes practices, that the project performance in the construction industry is measured using the productivity of labour as a benchmark [3, 4]. This further allows contention between the different parameters that are used to measure the productivity of labour on a construction project.

Labour productivity is defined as a ratio of the output obtained to the input utilised to achieve the production [7–9]. This definition is expressed in Eq. 1 below;

$$\text{Labour productity} = \frac{\text{Product output}}{\text{Labour input}} \tag{1}$$

A criticism faced by the construction industry is the reported poor productivity that is evidenced by poor workmanship on projects and the inability of projects to be completed both to cost and on time [4, 13, 14]. Globally, a fundamental analysis of the factors that give rise to this issue has been explored; however, the South African construction industry has done this analysis marginally in comparison [3, 15]. It is crucial to unpack and identify contextual factors that are perceived to disrupts labour performance on construction projects and inadvertently, impact on the project performance.

In conclusion, this study aims to examine the contextual factors that affect the level of labour productivity on site and rank them according to the level of significance both within the designated cluster of factors and overall from a building project perspective. Additionally, this study will conclude the way forward that the construction industry as a whole can adopt and implement in their planning of activities, measures that can eliminate the excessive impact of these factors and their significant effects on project success.

2 Overview of Factors Impacting Labour Productivity

Due to the complex nature of the construction industry and the interconnectedness of activities, a synthesis of factors that affect the productivity of labour from a project, organisational and environmental perspective is essential to contextualise elements that have a direct impact on the productivity of labour [16, 17]. Through a critical review of the literature, critical factors influencing productivity were clustered together and explained and this is represented in Fig. 1 below.

As can be seen in Fig. 1, these factors have been clustered according to the components that when viewed as a whole, compliment the construction project. These

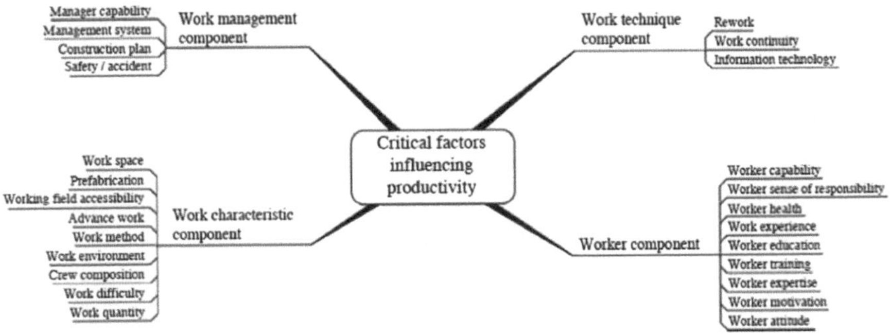

Fig. 1. Critical factors influencing labour productivity. Adapted from Jang *et al.* [16]

factors are adopted for this study and grouped into environmental, organisational, technical and social categories [17]. These categories are explained below.

2.1 Environmental Factors

The geographical environment or conditions under which a construction project is being executed, as well as the factors that are outside the control of the construction team, are termed the site environmental factors [18, 19]. These environmental factors comprise of the climatic conditions, ground conditions, as well as any other environmental considerations on the site that can impact the level of productivity. In critically assessing their impact and understanding what factors in this cluster are key to labour productivity, the environmental factors need further analysis in relation to the construction site.

2.2 Organisational Factors

Management and organisational dynamics have the potential to affect the level of labour productivity expended on a project, and as a result, the critical aspects of management are crucial to understanding the impact and the offset of the expected productivity output to that which is evidenced on site [18, 20]. It is postulated through literature and shown in Fig. 1 that the capacity of management and their skill, adopted management styles, economic changes that may destabilise the organisation and legislature and policy, are factors that indirectly affect onsite productivity [17, 18, 20, 21].

2.3 Technical Factors

Technical factors relate to both the capabilities and the skill set possessed by the labourers to execute the construction project [22]. Further arguments are that the nature of the construction projects is complex and closely linked from one trade or work, to the other, thus creating a relationship between activities which is sequential [23]. With this presented level of complexity, it is evident that the disruption of one activity can give rise to an important ripple effect of unproductivity in the following activities and as a whole, the construction endeavour becomes unproductive [16, 22]. Furthermore, the design of the project, method of construction employed, the innate complexity of the project as well as the inability of the engineers to give due feedback with regards to change orders can also impact labour productivity [23].

2.4 Social Factors

The construction industry is labour-intensive [24]. An often disregarded and not deeply explored level of research, is the relationship between the level of labour output and the social factors impacting the productivity of labour [25]. Socio-economic factors were evaluated in relation to their impacts on labour productivity by some scholars, and it was deduced that personal problems, substance abuse, commute between residential homes to construction sites, are some of the factors that impact the level of productivity [16, 24, 25]. It is thus essential to analyse the effect of social factors and their relationship to the level of expended productivity on construction projects.

3 Research Methodology

A thorough review of peer-reviewed journal articles was adopted for this study. Furthermore, keywords; labour, productivity, labour outputs were used to refine the search for literature pertinent to this study. A research questionnaire was developed and designed based on the synthesised literature and the intended objective of the study. This questionnaire was validated and approved through a consultative process. The questionnaire was administered using survey monkey web-based platform to contractors listed in Grades 2–6 of the cidb register of contractors. A total of 117 completed questionnaires was obtained after cleaning the data and used in the data analysis. However, for the objective of this paper, 58 contractors that operate on building projects were used in the analysis and the reporting of the data, and the other respondents that worked in the road and infrastructure projects were excluded from this analysis to be used in further studies. The Relative Importance Index (RII) method of analysis is employed in the analysis of data for this research to rank the factors that affect the productivity of labour on construction projects [26, 27]. Equation 2 shows the formula used in calculating the Relative Importance Index (RII) [28];

$$\mathrm{RII} = \frac{\sum_{i=1}^{n} W_i}{A \times N} \tag{2}$$

The numerator part of the equation computes the weight of the responses by multiplying the weight assigned to the variable with the number of respondents per variable. However, the highest factor that could be attributed to the variable is multiplied to the number of respondents to give us the denominator. The findings of the study are discussed in the following section. The RII method is used in order to identify what factors are perceived by the respondents to be critical to the productivity of labour on construction project. Further studies will draw correlation of the degree of impact these factors pose to the project performance.

4 Findings of the Study

The results of the study are presented in Table 1 below. The factors have been tabulated and ranked according to their respective RII values calculated using Eq. 2. As can be observed in Table 1, two different types of rankings are presented as Column A and Column B respectively. Column A represents the classification of factors across the four different categories of Site Environment, Organisation, Technical and Social factors affecting labour productivity. While Column B represents the ranking of factors within the specific group of factors impacting the productivity of labour.

Table 1. Perceptions of the respondents and the ranking of factors concerning IRR.

		Very low…Very high impact							
	Factors	1	2	3	4	5	RII	Group ranking [B]	Overall ranking [A]
Site environment	Inclement weather (rain/cold/heat)	6	19	4	16	13	0.638	5	12
	Unexpected site ground conditions	7	12	11	15	13	0.652	4	7
	Site location/environment	6	17	9	18	8	0.617	7	17
	Lack of experience amongst workers	5	15	6	19	13	0.669	2	4
	Delays in delivery of material	6	13	4	15	20	0.703	1	1
	Poor site supervision	9	19	3	12	15	0.617	7	17
	Lack of tools and equipment	10	17	5	11	15	0.614	9	21
	Mismanagement of resources on site	9	11	8	12	18	0.666	3	5
	Rework due to poor quality	10	15	6	12	15	0.624	6	15
Organisational factors	Non-payment of labour	17	5	5	11	20	0.641	3	11
	Delays in payment of labour	15	8	3	13	19	0.645	1	9
	Fluctuations in material/equipment prices	7	15	8	14	14	0.645	1	9
	Changes in management structure	14	13	10	12	9	0.562	8	30
	Shortage of labour/manpower	9	19	5	16	9	0.590	6	27
	Inexperienced supervisors	15	10	4	14	15	0.614	4	21
	Change in government legislature/policy	10	18	7	11	12	0.590	6	27
	Economic changes (change in VAT, inflation and foreign exchange rates)	11	15	6	11	15	0.614	4	21
Technical factors	Poorly designed project	9	11	10	13	15	0.648	4	8
	Incomplete drawings	8	13	5	18	14	0.659	3	6
	Methods of construction (prefabrication vs onsite)	9	16	10	13	10	0.597	8	26
	Poor coordination/planning of activities	11	11	7	18	11	0.624	6	15
	The complexity of the project	9	10	14	12	13	0.634	5	13
	Buildability of the structure	9	13	14	13	9	0.600	7	25
	Inspection delays from engineers	8	10	8	15	17	0.679	1	2
	Change orders from the designers/consultants	8	11	8	12	19	0.679	1	2

(*continued*)

Table 1. (*continued*)

	Factors	Very low…Very high impact 1	2	3	4	5	RII	Group ranking [B]	Overall ranking [A]
Social factors	Lack of experience of workers	12	13	6	13	14	0.614	4	21
	Personal problems	15	11	8	15	9	0.572	5	29
	Drug abuse	18	15	6	12	7	0.514	7	32
	Alcohol abuse	19	17	4	11	7	0.497	8	33
	Long commute periods to site	15	11	9	17	6	0.559	6	31
	The relationship between supervisors and labourers	9	13	7	17	12	0.634	1	13
	Labour unrest/rioting	16	9	6	8	19	0.617	2	17
	Uncertain job security	13	8	12	11	14	0.617	2	17

Further, the study sought to know what factors within the various clusters affect the level of labour productivity. It can be observed from Table 1 that under site environmental factors; Delays in delivery of material (RII = 0.703) is perceived by the respondents as the primary factor affecting labour productivity, followed by Lack of experience amongst workers (RII = 0.669), Mismanagement of resources on site (RII = 0.666), Unexpected site ground conditions (RII = 0.652) and Inclement weather (rain/cold/heat) (RII = 0.638) from a ranking perspective.

Moreover, under organisational factors; Delays in payment of labour (RII = 0.645) and Fluctuations in material/equipment prices (RII = 0.645) are ranked first. In the third rank, Non-payment of labour (RII = 0.641). The fourth-ranked factors are Inexperienced supervisors (RII = 0.614) and Economic changes (change in VAT, inflation and foreign exchange rates) (RII = 0.614).

The ranking of factors under the technical cluster of factors that impact labour productivity shows that: Inspection delays from engineers (RII = 0.617) and Change orders from the designers/consultants (RII = 0.617) are both considered by the respondents as primary factors affecting labour productivity, followed by Incomplete Drawings (RII = 0.659), Poorly designed project (RII = 0.648) and Complexity of the project (RII = 0.634) from a ranking perspective.

Lastly, under the social factors, respondents perceive that Relationship between supervisors and labourers (RII = 0.634) as the most critical factor affecting labour productivity from a ranking perspective, followed by Labour unrest/rioting (RII = 0.617), Uncertain job security (RII = 0.617), Lack of experience of workers (RII = 0.614) and Personal problems (RII = 0.572).

Overall, Delays in delivery of material (RII = 0.703) is perceived by respondents to be the critical factor affecting labour productivity, followed by inspection delays from engineers (RII = 0.679) and Change orders from the designers/consultants (0,679) both listed as the second factor, Lack of experience amongst workers (RII = 0.669) and Mismanagement of resources on site (RII = 0.666) from a ranking perspective.

5 Discussions of the Findings

The study sought to find out the factors affecting labour productivity on construction projects employing both a systematic review of extant literature and a survey of construction professionals. The outcome of this study resonate with this studies that conducted research in a similar research fiend, particularly on the subject of the productivity of labour on construction projects. It emerged from the study that Delays in delivery of materials is perceived as the key factor significantly affecting the productivity of labour and this finding is corroborated by literature sources [16]. The study also found that respondents see inspection delays from engineers to affects productivity as the factor ranks number one is technical factors and ranks second under a general ranking of factors. The inability of engineers to respond to inspections requested for work to commence, impact the productivity of labour as this contributes to idle time and inefficiency of labour.

Furthermore, this demoralises the productivity [24]. Additionally, change orders from designers also affect the productivity of labour as it is ranked second under general factors and again in first ranking under technical factors. The occurrence of change orders contributes to a change in work schedule which further impacts the productivity of labour as the disruption in the operational programme requires labour to work on tasks pre-planned for an additional period [4].

Lack of experience amongst workers is another factor that impacts the level of labour productivity. This factor is ranked fourth under general factors and in the second position under site environment factors. Lack of experience amongst workers gives rise to construction rework and additionally, cost and time overruns [4, 29]. Mismanagement of resources on site is another factor that impacts productivity on construction projects. This factor ranked fifth in the overall factors and third in Site environmental factors. The results are consistent with other studies as the results are consistent with the work that has been done before and in other countries [9, 19]. As part of the objective of the study, it was established that in factors that impact on the productivity of labour; site environment, organisational factors, technical factors and social factors, there are those specific factors that impact the level of productivity as evidenced from the analysed data. The degree of an impact amongst these factors is further evidenced by the level of ranking that these factors were assigned. Overall, it can be deduced from the study that Site Environmental factors are perceived to highly affect labour productivity more than the organisational, technical and social factors.

6 Conclusion

The findings of the study, show that delays in delivery of material, inspection delays from engineers, change orders from the designers/consultants, lack of experience amongst workers and mismanagement of resources on site are key factors perceived to impact the level of labour productivity in building projects. From a theoretical perspective, some of the factors that are found in literature have been tested for different project types and not specifically for building projects. This study has incorporated some of those factors to the research design and moreover, found that in the RII ranking matrix, some of the factors have a practical implication of affecting the performance of labour as per the respondents. Based on these findings, the study concludes that estimators and planners who do not consider the site factors during the project and planning stage will set the project up for failure during the construction stage - time and cost overruns. This must also elicit thorough planning and before a project is developed to prevent and eliminate the potential impacts that these factors affect the level of labour productivity.

Acknowledgements. This study was in part financed by the National Research Foundation (NRF). Additionally, the opinions expressed in this paper and the derived conclusions are those of the authors and not attributed to NRF.

References

1. Barbosa, F., Woetzel, J., Mischke, J., Ribeirinho, M.J., Sridhar, M., Parsons, M., et al.: Reinventing Construction: A Route to Higher Productivity. McKinsey Global Institute, New York (2017)
2. Dzadaza, B., Crafford, G.: Artisan's perceptions of factors affecting construction labour productivity on building sites. In: Fourth NMMU Construction Management Conference proceedings, Port Elizabeth, South Africa, pp. 335–344, 30th November–1st December 2015
3. Soekiman, A., Pribadi, K.S., Soemardi, B.W., Wirahadikusumah, R.D.: Factors relating to labour productivity affecting the project schedule performance in Indonesia. Procedia Eng. **14**, 865–873 (2011)
4. Baloyi, L., Bekker, M.: Causes of construction cost and time overruns: the 2010 FIFA world cup stadia in South Africa. Acta Structilia **18**(1), 51–67 (2011)
5. Alinaitwe, H., Mwakali, J., Hansson, B.: Labour productivity in the building industry. In: Understanding the Construction Business and Companies in the New Millennium, pp. 210–220 (2005)
6. Durdyev, S., Mbachu, J.: On-site labour productivity of New Zealand construction industry: key constraints and improvement measures. Constr. Econ. Build. **11**(3), 18–33 (2011)
7. OECD: Measuring Productivity - Measurement of Aggregate and Industry-Level Productivity Growth - OECD Manual Organisation of Economic Cooperation and Development, Paris (2001)
8. Tran, V., Tookey, J.: Labour productivity in the New Zealand construction industry: a thorough investigation. Australas. J. Constr. Econ. Build. **11**(1), 41–60 (2011)
9. Moselhi, O., Khan, Z.: Significance ranking of parameters impacting construction labour productivity. Constr. Innov. **12**(3), 272–296 (2012)
10. Yi, W., Chan, A.P.C.: Critical review of labour productivity research in construction journals. J. Manag. Eng. **30**(2), 214–225 (2014)
11. cidb: Labour & Work Conditions in the South African Construction Industry - Status and Recommendations (2015)
12. cidb: Construction Monitor, Employment Q3 (2017). http://www.cidb.org.za/publications/Documents/Construction%20Monitor%20-%20October%202015.pdf
13. Abdul-Rahman, H., Berawi, M.A., Berawi, A.R., Mohamed, O.: Delay mitigation in the Malaysian construction industry. J. Constr. Eng. Manag. **132**(2), 125–133 (2006)
14. Merrifield, A.: The Role of the Construction Industry in the Delivery of Infrastructure Mandate for Change 1994–1999. HSRC Publishers, Pretoria (1999)

15. Ofori, G., Hindle, R., Hugo, F.: Improving the construction industry of South Africa: a strategy. Habitat Int. **20**(2), 203–220 (1996)
16. Jang, H., Kim, K., Kim, J., Kim, J.: Labour productivity model for reinforced concrete construction projects. Constr. Innov. **1**(1), 92–113 (2011)
17. Juga, A.J., Hens, N., Osman, N., Aerts, M.: A flexible method to model HIV serodiscordance among couples in Mozambique. PLoS ONE **12**(3), e0172959 (2017)
18. Park, H.-S.: Conceptual framework of construction productivity estimate. J. Civ. Eng. **10**(5), 311–317 (2006)
19. Hughes, R., Thorpe, D.: A review of enabling factors in the construction industry in an Australian environment. Constr. Innov. **14**(2), 210–228 (2014)
20. Enshassi, A., Mohamed, S., Mustafa, Z.A., Mayer, P.E.: Factors affecting labour productivity in building projects in the Gaza Strip. J. Civ. Eng. Manag. **13**(4), 245–254 (2007)
21. Kadir, M.R.A., Lee, W.P., Jafaar, M.S., Sapuan, S.M., Ali, A.A.A.: Factors affecting construction labour productivity for Malaysian residential projects. Struct. Surv. **23**(1), 42–54 (2005)
22. Mawdesley, M.J., AL-Jibouri, S.: Modelling construction project productivity using systems dynamics approach. Int. J. Prod. Perform. Manag. **59**(1), 18–38 (2009)
23. Xia, B., Chan, A.B.C.: Measuring complexity for building projects: a Delphi study. Eng. Constr. Archit. Manag. **19**(1), 7–24 (2012)
24. Alinaitwe, H.M., Mwakali, J.A., Hansson, B.: Factors affecting the productivity of building craftsmen - studies of Uganda. J. Civ. Eng. Manag. **13**(3), 169–176 (2007)
25. Ulubeyli, S., Kazaz, A., Er, B.: Planning engineers' estimates on labor productivity: theory and practice. Procedia – Soc. Behav. Sci. **119**, 12–19 (2014)
26. Fugar, F.D., Agyakwah-Baah, A.B.: Delays in building construction projects in Ghana. Constr. Econ. Build. **10**(2), 103–116 (2010)
27. Jarkas, A.M.: Factors influencing labour productivity in Bahrain's construction industry. Int. J. Constr. Manag. **15**(1), 94–108 (2015)
28. Doloi, H.: Analysing the novated design and construct contract from client's, design team's and contractor's perspective. Constr. Manag. Econ. **26**, 1181–1196 (2008)
29. Kaming, P.F., Olomolaiye, P.O., Holt, G.D., Harris, F.C.: Factors influencing craftsmen's productivity in Indonesia. Int. J. Project Manage. **15**(1), 21–30 (1997)

Assessment of Political Risk Factors Associated with Public-Private Partnership Projects in Developing Countries: A Case Study of Lagos State Nigeria

Akintayo Opawole(✉) and Kahilu Kajimo-Shakantu

Department of Quantity Surveying & Construction Management, University of the Free State, Bloemfontein, South Africa
tayoappmail@gmail.com, kajimoshakantuk@ufs.ac.za

Abstract. Previous studies identify political risk factors as significant to the poor performance of public-private partnership (PPP) contracts in general. This study therefore aimed at developing a simplified approach to manage this category of risk factors. Quantitative data were obtained using a structured questionnaire which was self-administered to 5 architects, 7 quantity surveyors, 12 engineers, 1 lawyer, 1 economist/financial manager/banker and 8 builders involved at the design, procurement, financing, construction, and project management of PPP projects in Lagos State, Nigeria. The respondents were selected using a respondent driven sampling (RDS) approach. Data analysis was done using descriptive and inferential statistics. Key findings include that corruption and exchange rate policy were the most frequently occurring risk factors with indices of 0.806 and 0.748 respectively. Moreover, strict implementation of anti-corruption and transparency together with appropriate use of financial instruments were the appropriate mitigation measures. The study highlights implications for managing political risk factors associated with PPP projects that could guide investment decisions in infrastructure development by the private sector in developing countries.

Keywords: Infrastructure · Public-private partnership · Project · Political risk · Risk factor

1 Introduction

The success of public-private partnerships (PPP) is essentially influenced by sociopolitical and economic characteristics of a country. According to [1], maturities of legal regime and political stability have both been significant in different countries' contexts in the execution of PPP contracts. Unstable sociopolitical factors are significant and necessitate cautious assessments before arriving at any investment decisions in PPP transactions in developing countries [2]. Political risk factors such as; long rule of military, civil disturbance, change in law, political violence, expropriation, breach of contract, and currency convertibility have made PPP market immature and highly risky [3, 4].

C. Aigbavboa and W. Thwala (Eds.): CIDB 2019, *The Construction Industry in the Fourth Industrial Revolution*, pp. 364–372, 2020.
https://doi.org/10.1007/978-3-030-26528-1_36

Political risks have manifested in three main ways related to traditional, regulatory and non-compliance factors [5]. Traditional risk factors include opportunism, expropriation, currency convertibility and instability of policies. Regulatory risks are those related substantially to change in law and policies relating to PPP contracts as well as lack of mechanisms for enforcement of obligations and penalty. Non-compliance risks include default in buyer obligations, lack of mechanism for enforcement of buyer obligations and inability to comply resulting from forecast or managerial deficiency [5]. These types of risk factors have been identified as a critical success factors (CSF) PPPs in infrastructure delivery in almost all countries. For example, political factors were considered highly significant in the failure of transportation infrastructure by [sanni, 2012] in the case of the Ibadan-Lagos expressway project and delay in the delivery of Lekki-Epe road in the Nigerian context. Notwithstanding, the limited number of studies that have specifically focused on this category of risk have not enabled their proper management in PPPs within the developing countries' contexts. The purpose of this study is therefore to evaluate the political risk factors associated with PPP projects with a view to developing a simplified approach to their management. The study is expected to highlight implications for managing political risk factors that influence PPP projects and guide investment decision in infrastructure business by private sectors in developing counties.

2 Literature Review

Various limitations of the traditional delivery system of infrastructure delivery such as budget constraints [6, 7]; lack of political will by governments [8]; and competence of public sector organizations (PSOs) [9], had resulted in the adoption PPPs [10]. Although different variants of PPPs have emerged in recent times, the generic structure often comprise public sector organisations (PSOs) and private sector investors (concessionaire) [11].

A number of studies have identified political risk as highly significant to PPP transactions in developing countries [12, 13, 14]. Political risks such as approval risk and change in law were identified in [15]. Ameyaw and Chan [16] identified political interference, termination by government, change in government, political instability, corruption, weak regulatory and monitoring regime, uncertain tariff review, strained relationship, maturity of public party, competence of public sector organization, and public opposition. In [17], currency convertibility, expropriation, breach of contract, civil war/civil disturbance, legal and bureaucratic risks were identified. Lashinde et al. [18] identified currency risk, expropriation, taxation, change in law, approval bottlenecks and corruption. Moreover, [19] identified cancellation of project, construction permit delay, community risk such as protest by individual, formal/informal groups, outright confiscation or nationalization of asset, breach of contract, renewal of concession agreement, concession explicitly and ineffective judiciary system.

Political risk factors have affected PPP transactions most significantly through the withdrawal of concessionaire when the profitability on the investment is becoming uncertain [20]. The approach towards managing political risks include risk retention [3], and this has been effective in risk factors such as changes in regulation and delay in project approval. Excessive contract variations have been preferred to be shared by both parties [21]. Moreover, convertibility risk, expropriation, civil disturbance, change in law, war, political violence, expropriation, breach of contract, and currency convertibility risk had been mitigated using insurance mechanism [22]. Earlier studies [3, 17] and [23] have indicated varying frequency of occurrence, degree of impact and severity of risk factors on PPP contract performance in different countries' contexts. With limited studies that have specifically focused on political risk factors, it appears that the absence of framework for effective management of this category of risk factors in PPPs explains in part the poor performance of PPP contracts in Nigeria.

3 Research Method

The research design adopted was a quantitative descriptive approach. The study adopted the use of a structured questionnaire, which was designed to capture issues such as the identification of risk factors as well as their impact and mitigation measures. The respondents were selected using the respondent driven sampling (RDS) approach. RDS is an advanced snowball sampling technique that has been widely adopted in studies involving an undefined population [24]. This approach has been widely adopted in PPP studies due to the absence of comprehensive lists of PPP participants in the Nigerian construction industry [14, 25]. Li et al. Further, [26] defines the population of PPP participants in countries that are newly experimenting with PPPs as dynamic, which makes it difficult to compile a comprehensive population of participants. Respondents in this study were 5 architects, 7 quantity surveyors, 12 engineers, 1 lawyer, 1 economist/financial manager/banker and 8 builders, involved at the design, procurement, financing, construction, and project management of PPP projects in Lagos State. The respondents were considered suitable to provide information relating to political risk factors based on their experience and involvement at the various stages of PPP projects.

A structured questionnaire was utilised to enable a relatively simple and quick capture and assessment of the variables evaluated [27]. The first part of the questionnaire was designed to gather respondents' profile data such as the number of PPPs they have been involved on, type of PPP projects, academic qualifications, professional qualifications and years of working experience in the construction industry and on PPP contracts specifically. The second part of the questionnaire evaluated the specific objective of this study. This part was structured based on a 5 point Likert scale, where 1 and 5 represent the least and highest ranking respectively. Specific features relating to identification of political risk factors as well as their impact and mitigation measures were listed in tables for quicker evaluation by the respondents. The second part of the questionnaire was also designed in a closed ended format to ease the analysis of specific set of responses rated by the respondents. Data analysis was achieved using percentage, mean and relative importance index.

4 Results and Discussions

4.1 Profile of Respondents

The profiles of the respondents were evaluated based on type of organizations, coverage of organizations' operations since establishment, designation of respondents, highest academic qualifications, professional qualifications, type of projects undertaken, years of working experience and number of PPP contract executed. Thirty-four (34) copies of the questionnaires, which represent a response rate of 42.5% of the eighty (80) copies administered, were properly completed and analyzed to provide quantitative data for the assessment. The designations of respondents showed that engineers were the highest with 47.4% of the sample. Quantity surveyors were 22.8%, architects were 17.5%, and builders 1.8% of the respondents. Other professions/designations in the construction industry were 10.5%.

Twenty-six (76.5%) of the respondents were from public organizations and 23.5% represented private sector organizations. About 79.4% of the respondents have operated within their state, 2.9% within their geopolitical zone, 8.8% at national level and 8.8% at international level. About 14.7% of the sample were architects, while quantity surveyor, engineers and builder were represented at 35.3%, 20.6% and 23.5% respectively. Lawyers and financial administrators were represented at 2.9%. Analysis of the academic qualifications of the respondents showed that 29.4% held higher a national diploma (HND) and 8.8% were those with a postgraduate diploma (PGD). About 32.4% of the respondents obtained a first degree while those with masters degree and doctorate degree were represented at 26.5% and 2.9% respectively. The result of the professional qualifications showed that 17.6% of the respondents had professional affiliation to the Nigerian Institute of Quantity surveyors (NIQS) and 14.7% were members of the Nigerian Institute of Architects (NIA). Members who were affiliated to the Nigerian Society of Engineers (NSE) were 29.4%, 20.6% were members of the Nigerian Institute of Building (NIOB), and 2.9% were affiliated to Institute of Chartered Accountants Nigeria/Association of National Accountants of Nigeria (ACA/ANAN).

Respondents with some other professional affiliations represented 14.7% of the sample. About 19.1% of the respondents had been involved in transportation projects, 16.4% in power projects and 19.1% had been involved in housing projects. The respondents with experience in commercial projects, water, communication projects were represented at 27.9%, 0.2% and 0.4% respectively. The mean work experience of all the respondents was estimated at 7 years. The overall results of the analysis of the profile of the respondents justify their adequate education levels to supply reliable data for this study.

4.2 Evaluation of the Political Risk Factors Associated with PPP Contracts

Table 1 shows the frequency of occurrence (FoO) of the political risk factors evaluated. Corruption was ranked first with a RII of 0.806. In second place was currency exchange with an index of 0.748, while constraints on foreign currencies were ranked third with

(RII = 0.724). Community opposition, and variation to the contract scored indices of 0.700 and 0.706 to rank third and fourth respectively. Rethinking of nationalization of projects and failure to appropriate funds for proposed projects were equally rated with RII of 0.700, ranking fifth.

Table 1. Evaluation of the political risk factors

Factors	Frequency of occurrence		Degree of impact		Severity index
	RII	Rank	RII	Rank	
Corruption	0.806	1	0.853	1	0.688
Currency exchange	0.748	2	0.741	2	0.554
Political interference	0.694	8	0.712	3	0.494
Nationalization of project	0.700	5	0.682	4	0.477
Variation to contract	0.700	5	0.676	5	0.473
Government initiated change of project scope	0.664	12	0.676	5	0.449
Construction of competitive projects by government	0.670	11	0.671	7	0.450
Adjustment of construction duration	0.652	15	0.665	8	0.434
Community opposition	0.706	4	0.659	9	0.465
Change of output specification	0.658	13	0.653	10	0.430
Delay in land acquisition	0.606	25	0.647	11	0.392
Restriction on foreign currencies transfer	0.724	3	0.641	12	0.464
Restriction on foreign direct investment	0.658	13	0.641	12	0.422
Change in enabling law	0.648	19	0.635	14	0.411
Failure to appropriate necessary fund	0.700	5	0.629	15	0.440
Weak agreement between public and private parties	0.594	28	0.624	16	0.371
Breach of contract	0.652	15	0.618	17	0.403
Judicial and adjudication risk	0.652	15	0.618	17	0.403
Approval bottlenecks	0.682	9	0.606	19	0.413
Resistance to reveal vital information	0.600	26	0.606	19	0.364
Court injunctions	0.624	23	0.594	21	0.371
Nationalization	0.600	26	0.588	22	0.353
War/civil disturbance risk	0.624	23	0.576	23	0.359
Project/contract cancellation by government	0.682	9	0.571	24	0.389
Asset transfer	0.642	20	0.571	24	0.367
Volatile taxation policy	0.652	15	0.571	24	0.372
Outright cancellation of project by government agencies	0.636	22	0.565	27	0.359
Change in investment law	0.642	20	0.535	28	0.343
Expropriation	0.570	29	0.494	29	0.282

The risk factors with the least FoO were expropriation (RII = 0.570), weak agreement between parties (RII = 0.594), land acquisition (RII = 0.606), civil disturbance ranked (RII = 0.624), court injunction (RII = 0.624), and outright cancellation by public agencies (RII = 0.642). The findings that corruption, constraints on foreign currencies, community opposition, nationalization were rated high are consistent with earlier findings in [3] and [23].

An evaluation of the degree of impact of political risk factors showed that corruption is the highest with RII of 0.853 while currency exchange with RII of 0.741, and political interference with RII of 0.712, ranked first, second and third respectively. Government rethink for nationalization ranked fourth with RII = 0.682 and variation to the contract by public sector had the fifth degree of impact with RII of 0.676. This was closely followed by construction of competitive projects, which ranked sixth with RII of 0.671. Risk factors with the least degree of impact were; expropriation which ranked 29th with RII of 0.494, and change in law which ranked 28th with index of 0.535.

Outright cancellation, asset transfer and volatile taxation policies which were all ranked 24th with index of 0.571 each, civil disturbance which ranked 23rd with RII 0.576, and nationalization ranked 22nd with RII of 0.588.

The severity indices of the risk factors were calculated from the product of frequency of occurrence (FoO) and the degree of impact of the factors (RImI). The severity of the factors ranged as $0.282 \leq SI \leq 0.688$, with corruption, currency exchange, political interference and nationalization of the projects ranked first, second, third and fourth with indices of 0.688, 0.554, 0.494 and 0.477, respectively. Construction of competitive projects (SI = 0.449) and government initiated change of project scope (SI = 0.449) ranked sixth and seventh respectively. Corruption with SI of 0.668 ranked first as the political factors with the highest severity. The significance of corruption and currency exchange is far reaching as can be deduced from their rankings as factors with the highest FoO and RImI.

4.3 Evaluation of Mitigation Measure for Political Risk Factors

Table 2 shows the evaluation of mitigation measures against the political risk factors. The relative importance index (RII) of the factors ranged between $0.730 \leq SI$ 0.84, which is interpreted as significant or very significant on the likert scale adopted in the questionnaire used. The results show that all the identified measures were significant in mitigating political risks and agreed in part with [23]. The relative significance of the factors, however, shows strict implementation of anti-corruption and transparent standards ranked as first with RII of 0.841, while culture of open dialogue ranked second with RII of 0.835. Effective interaction between both parties, effective dispute resolution mechanisms, reliable and efficient administration and community engagement were equally significant at 0.824 to rank third. The least significant measures were integrated risk management system (0.729), establishment of export credit agency (RII = 0.730), multilateral investment guarantee agency (0.753) and transfer of political risk to public partner (0.759).

Table 2. Mitigation measure for political risk factors

Factors	RII	Rank
Strict implementation of anti-corruption and transparent standards	0.841	1
Culture of open dialogue	0.835	2
Effective interaction between both parties	0.824	3
Effective dispute resolution mechanisms	0.824	3
Reliable and efficient administration	0.824	3
Community engagement	0.824	3
Appropriate use of financial instruments	0.812	7
Government guarantee	0.812	7
Private partner measures	0.812	7
Public partner measures	0.812	7
Responsive business conduct	0.806	11
Joint public-private measures	0.806	11
Establishment of PPP agency	0.800	13
Efficient procurement and permit process	0.794	14
Political risk insurance	0.794	14
Stability of laws and regulations	0.788	16
World bank guarantee	0.782	17
Establishment of specific project governance agency	0.765	19
Establishment of multilateral organization provider of political risk insurance such as multilateral investment guarantee agency	0.759	20
Transfer of political risk to public partner	0.759	20
Multilateral investment guarantee agency	0.753	22
Integrated risk management system	0.729	23
Establishment of export credit agency	0.730	24

5 Conclusion

This study evaluates the political risk associated with PPP projects and thereby developed a simplified approach to understanding the management of the risk factors. Corruption and currency exchange were the political risk factors with the highest frequency of occurrence and degree of impact. However, expropriation was the least factor considering its frequency of occurrence and degree of impact. This high severity indices of the impact of the risk factors suggests that PPP contracts in Nigeria are likely to be subjected to multiple risks occurrences and high impact. The simplified approach to management of political risk factors provided in this study is therefore expected to guide parties involved in PPP contracts in making investment decisions and in preparing robust contact packages. Although the study provides highlights of the implications for management of political risk factors associated with PPP contracts, further assessment of the concept through a mixed methodology approach is recommended to provide findings that are more robust. Moreover, the approach adopted especially, the sampling method, could at best enable the generalisation of the finding in countries with similar PPP maturity markets as the study area. This study could also be extended to develop a multi-factor approach for the management of political risk factors in PPP contracts.

References

1. Eaton, D., Akbiyikli, R., Lemos, T., Gunnigan, L., Kutanis, R.O., Casensky, M., Ladra, J., El Sawalhi, N.: An examination of the suitability of a UK PFI model within the Czech Republic, the Republic of Ireland, Palestine (Gaza-West Bank), Portugal and Turkey. Constr. Innov. Info. Process Manag. **7**(1), 122–142 (2007)
2. Opawole, A., Jagboro, G.O., Kajimo-Shakantu, K., Olojede, B.O.: Critical performance factors of public sector organization in concession-based PPP projects in Nigeria. Prop. Manag. **37**(1), 17–37 (2019)
3. Ke, Y., Wang, S., Chan, A.P.C.: Equitable risks allocation of projects inside China: analyses from delphi survey studies. Chinese Manag. Stud. **5**(3), 298–310 (2010)
4. Boudet, H.S., Jayasundera, D.C., Davis, J.: Drivers of conflict in developing country infrastructure projects: experience from the water and pipeline sectors. J. Constr. Eng. Manag. **137**(7), 498–511 (2011)
5. Ferro, G.: Political Risk and regulatory Risk: Issues in Emerging Markets Infrastructure Concessions. Universidad, Argentina (2001)
6. Olufidipe, O.: Government Budgeting in Nigeria: Principle, Policies, and Practices. Obafemi Awolowo University Press Ltd., Ile-Ife (2003)
7. Opawole, A., Jagboro, G.O., Babatunde, S.O.: Factors influencing the implementation of public infrastructure in Nigeria. Global J. Res. Eng. **11**(6), 29–38 (2011)
8. United State Agency for International Development (2005)
9. Jones, D.: Policy development in Australia for public private partnerships – what more is there to do?. In: A Seminar on Providing Value for Money Through Public Private Partnerships: The Lessons Learnt So Far from Economic and Social Infrastructure Projects, vol. 26, June 2002. www.claytonutz.com/downloads/PPP%20paper%2026%20June%202002.pdf. Accessed 2 Jul 2010
10. Akinsiku, O.E., Babatunde, S.O., Opawole, A.: Adoption of public-private partnerships (PPP) procurement method on infrastructure delivery in Nigeria. J. Niger. Inst. Quant. Surveyors **58**, 3–12 (2014)
11. Kerf, M., Gray, R.D., Taylor, R.R.: Concession for infrastructure: a guide to the design and award. World Bank Technical paper, no. 399, pp. 1–194, Washington DC (1998)
12. Mouravie, N.: Risk management in public-private partnership and research agenda for transitional countries. Int. J. Manag. Bus. Stud. **2**(2), 37–41 (2012)
13. Grimsey, D., Lewis, M.K.: Evaluating the risks of public private partnerships for infrastructure projects. Int. J. Proj. Manag. **20**, 107–118 (2002)
14. Opawole, A., Jagboro, G.O.: Factors affecting the performance of private party in concession-based PPP projects in Nigeria. J Eng. Des. Technol. **15**(1), 44–57 (2017)
15. Fischer, K., Leidel, K., Riemann, A., Alfen, H.W.: An integrated risk management system (IRMS) for PPP projects. J Financ. Manag. Prop. Constr. **15**(3), 260–282 (2010)
16. Ameyaw, E.E., Chan, A.P.C.: Identifying public-private partnership (PPP) risks in managing water supply in Ghana. J. Facil. Manag. **11**(920), 152–182 (2013)
17. Voelker, C., Permana, A., Sachs, T., Tiong, R.: Political risk perception in Indonesian power projects. J. Financ. Manag. Prop. Constr. **13**(1), 18–34 (2008)

18. Lashinde, A.T., Ogunsemi, D.R., Awodele, O.A.: Financial and political risks influencing investors' decisions in concession infrastructure delivery in Nigeria. In: Ogunsemi, D.R., Oke, A.E. (eds.) Proceedings of 2nd Research Conference-NIQS Recon2, pp. 743–746. Nigerian Institute of Quantity surveyors, Akure, Nigeria (2015)
19. Bouston Consultin Group (2015)
20. United States Department of Transportation: Innovative Program Delivery, Risk Assessment for Public-Private Partnership: A primer (2012)
21. Patrick, X.W., Wang, S.W., Fang, D.: A life-cycle risk management framework for PPP infrastructure projects. J. Financ. Manag. Prop. Constr. **13**(2), 123–142 (2008)
22. Jensen, N.: Measuring Risk: Insurance Premiums and Domestic Political Institutions. Washington University, USA (2005)
23. Jagboro, G.O., Ojo, G.K., Opawole, A., Akindoyin, G.S.: Risks associated with concession-based contracts in Lagos state, Nigeria. American Association of Cost Engineering International Technical paper, RISK.1544.1, New Orleans (2014)
24. Heckathorn, D.D.: Respondent-driven sampling II: deriving valid estimates from chain-referral samples of hidden populations. Soc. Probl. **49**(1), 11–34 (2002)
25. Babatunde, S.O., Perera, S., Zhou, L., Udeaja, C.: Barriers to public private partnership projects in developing countries. Eng. Constr. Archit. Manag. **22**(6), 669–691 (2015)
26. Li, B., Akintoye, A., Edwards, P.J., Hardcastle, C.: Critical success factors for PPP/PFI projects in the UK construction industry. Constr. Manag. Econ. **23**, 459–471 (2005)
27. Kothari, C.R.: Research Methodology Methods and Techniques, 2nd edn. New Age International (P) Limited Publishers, New Delhi (2004)

Delphi Survey on the Influence of Succession Planning in Leadership Development

Murendeni Liphadzi(✉), Clinton Aigbavboa, Didibhuku Thwala, and Titus Kwofie

SARChI in Sustainable Construction Management and Leadership in the Built Environment, Faculty of Engineering and the Built Environment, University of Johannesburg, Johannesburg 2006, South Africa
mliphadzi@uj.ac.za

Abstract. Leadership succession is vital for the construction industry, and now, with the generation z and baby boomers entering the age of retirement, the industry must place greater endeavour in planning for succession amongst the current workforce. The study aimed to establish the influence of different succession planning factors on leadership development in the construction industry. The study adopted the Delphi survey system of data collection to examine the study objective. Delphi experts (Construction specialists and researchers) were engendered from peer-reviewed conference proceedings and the South African construction industry professionals' database. The evaluation of different succession planning factors was done by identifying the influence of each succession factor on leadership development; these factors were measured between no influence and very high influence. Data collected were analysed using mean item score and interquartile deviation. Of the different succession planning attributes evaluated, ability to understand organisational long term vision had a high influence on developing succession planning attributes for leadership development in the construction industry. The article contributes to the frame of knowledge on leadership development and succession planning in the construction industry.

Keywords: Human resources · Leadership development · Succession planning

1 Introduction

Humans are the most essential resource for many industries, and the construction industry is no exemption. The construction industry depends on the construction workforce of the organization; however, executives that establish and direct the companies are critical to the development of any construction organization [1]. Besides, potential leaders are even limited within Small Medium companies in the South African construction industry, and these companies are the majority of construction firms in the country [2]. Also, most of the experienced companies in the South African construction industry have an ageing generation on the verge of retirement, and this generation has various overtime carried out leadership and management responsibilities [3]. The succession of these leaders is an inevitable change that will effectively

C. Aigbavboa and W. Thwala (Eds.): CIDB 2019, *The Construction Industry in the Fourth Industrial Revolution*, pp. 373–379, 2020.
https://doi.org/10.1007/978-3-030-26528-1_37

influence many construction firms in the coming future [4]. Moreover, proper succession planning is essential for any construction firm. Xe et al. were of the view that if companies aspire to attain project success, they should make leadership transition a priority for their organizations [5].

Organizational professionals prominently discuss leadership succession planning. Yet, notwithstanding the interest from these professionals, the undertaking often falls with human resource managers of different firms to organize and coordinate efforts to establish and operate planned succession programs and avert the succession crises [6]. Without these programs, organizations will have difficulty maintaining leadership continuity or detecting appropriate leaders when a change is necessary. Planning for succession has been credited for reducing the impacts that come with construction project failures [7]. Moreover, construction firms within the industry should actively prepare for leadership transitions and minimize the negative influences that come with abrupt managerial changes. The research article appraised the Delphi experts' views on the influence of different succession planning attributes on leadership development within the construction industry.

2 Succession Planning

Studies in succession planning dates back to the early 20th century with Henry Fayol's 14 principles of management. Fayol's principles concedes the significance of developing and retaining key employees; he emphasized these principles by asserting the stability of tenure personnel. Nonetheless, it was not until the early 1960s that research in succession planning developed significantly [8]. Preferably, succession planning is a continuing process, a practice that is intertwined into the inclusive strategic plan of an organisation. Succession planning embraces developing talent from within the organisation, moreover, talent externally [9].

To succeed in the ever changing industries, it is vital for industry key personnel to understand the succession processes and to be able to deal with transformations and changes within different positions. It is to this end that succession planning is required to make sure that these competencies are not misplaced if and when people leave firms [17]. SFGate indicated that not having a succession plan sets the future of any organisation at risk [10]. Moreover, in this extremely competitive and fast-paced world, one needs every technical resource you can get, and in-relation to human resources; you need the best possible talent to keep up with the changing world. Rothwell revealed the importance of evaluating, selecting and developing the right candidate for successful succession planning [11].

Construction firms should not disregard the challenges and problems of planning for succession if they aspire for the firm's continuous success [5]. Previous studies established the benefits of planning for succession, but there is minimal research available to understand the unique attributes that influence succession planning for leadership development in the construction industry. Though human resource is vital for the construction industry [1], research that speaks to maintaining and replacing the construction workforce has afforded little support towards the South African construction industry. Moreover, these succession planning fundamentals are barely found in small construction companies [2]. Besides, it is essential for individuals selected for succession within a firm to be trained for them to become leaders within the construction industry and further align with the firms' strategic goals [6].

Dahlke indicated that different generations are collectively working together than ever before, and each generation brings its own set of principles and ways of working [12]. It is vital to effectively manage these different generations, though it is a challenge that has been noted by a few authors. Thus, effective succession planning processes are required to make sure that firms can deal with the above mentioned eminent challenge. Moreover, when selecting replacements for managers who leave firms, firms need to choose people who are proficient in dealing with the above discussed generational disparities.

It is also noteworthy to note that Baby Boomers are approaching retirement age. Although statistics indicate that many Baby Boomers are planning to work into their 70s, the reality is that employers will soon be facing a tremendous increase in the number of retirements and, therefore, open positions. The mass retirements of Baby Boomers will lead to a severe loss of skilled, experienced, knowledgeable employees, a situation that might be compounded and occur across all industries [12].

To reduce the impact of this significant shift in the workforce, it is even more critical to identify, train, and develop employees to fill these impending vacancies. Besides, globalisation has brought us a more culturally diverse workforce. As a result, it is vital for the succession-planning team to recognise the diversity of the workforce and take careful note of the skill levels of people from different parts of the world, while also paying extra attention to their customs, values, expectations, attitudes, preferences, and work styles [13]. Given, the emerging multigenerational, culturally diverse workplace, new sets of competencies are needed by people at every level from the CEOs to front-line employees.

2.1 Importance of Succession Planning

Succession planning will help construction firms clarify the short term and long-term staffing needs of any organization, it is also important to set up a tailor-made plan to identify, attract, and deploy the right people for the right positions and leadership transitions. It additionally helps with creating a road map for the future of the firm [12]. Succession planning and management can improve employee morale by reassuring promotion from within the firm. Indeed, promotions from within permit an organisation to develop the skills and abilities of individuals more successfully, and the opportunity to gain a promotion can serve as an incentive [14].

Moreover, Succession planning is needed to make sure that organizations keep up with technological developments. Not only do organizations need to train and develop personnel to be more tech-savvy, but they need a systematic plan to make sure that when people leave, new candidates can be equally technological [12]. Lastly, the construction industry that seeks to plan for the future will need ongoing analysis to prepare the workforce for changing roles, based on the changing business conditions and internal strategic initiatives.

3 Methodology

The methodology segment gives detail on the method used to determine the influence of different succession planning factors on leadership development in the South African construction industry. A Delphi survey was conducted amongst 13 experts (academics, built environment professionals and business owners) in the South African construction industry. A Delphi study is a group evaluation instrument compelling qualified experts who have an understanding of the subjects at hand [15]. Each expert was required to meet some of the following criteria; Knowledge: Knows construction management and project management; knowledgeable in leadership research, knowledgeable in the field of management theory. Academic Qualification: A degree related to any field and has had occupation and experience focusing on infrastructural development issues, psychology, construction management, project management and social sciences. Experience: Previously or currently performing consulting and contracting services within the construction industry or any related infrastructure projects. The experts should demonstrate a high degree in the subject matter and an also have an extensive theoretical understanding thereof. Research: Has done research on leadership and management issues. Membership: Member of professional organization or represent such organization. Finally, Enthusiasm: Panel members must be eager to participate entirely in the Delphi study.

The recommendations of Rowe et al. [16, 17] were assumed for the current study, whereby they stated that respondents should be enough to successfully consolidate the expert views, but not so large as to analyse the results in an uncontrollable manner. Five of these criteria items were measured than the customarily advocated two. Experts were required to be in managerial positions and also have an understanding of leadership development. The initial Delphi survey was made up of 20 experts comprising academics and construction professionals who were randomly selected based on the criteria. The experts applied their knowledge on the leadership concepts raised through the developed questionnaire. From the 20 experts requested to participate in the Delphi survey, 13 responded and completed all the three rounds. These amount of experts was considered sufficient grounded on the literature recommendations from scholars who have previously employed the Delphi technique, whereby they cited that, it viable to have between 6 and 16 experts, this makes 13 experts as seen in the current study reasonable [17]. Likewise, experts were asked to rate the impact factors influencing succession planning for leadership development in the construction industry as shown in Table 1. Data attained from the survey instrument were analysed from Microsoft Excel. The output from the analysis was a set of descriptive statistics whereby the mean items score and interquartile deviation (IQD) were determined through a scale that reflected no influence, low influence, medium influence, high influence and very high influence.

4 Findings

Results from the Delphi survey shows the (14) listed succession planning measurement variables, that were recognized by the experts as influencing succession planning for leadership development. When assessed, findings revealed all 14 measurement variables to have achieved consensus with IQD cut off (IQD $\leq$ 1) score (See Table 1). 14 variables managed to reach the median score of 7, which implied a high influence (HI: 7–8.99).

Table 1. Succession planning attributes

Table 1: Succession planning attributes	Median	Mean	SD	IQD
Knowledge management features	8	8.00	0.55	0.00
Future focused features	8	7.86	0.66	0.75
Succession time management features	8	7.64	0.50	1.00
Ability to understand organisational long-term vision	8	8.00	0.55	0.00
Leadership continuity features	7	7.29	0.47	0.75
Talent identification features	7	7.14	0.77	0.75
Talent management features	7	7.14	0.53	0.00
Transition management features	7	7.29	0.61	1.00
Change management features	7	7.36	0.74	1.00
Skills transferring features	7	7.29	0.61	1.00
Employee retention features	7	7.07	0.62	0.00
Leadership ability to document the succession plan	7	7.00	0.78	0.00
Ability to communicate the succession plan with immediate managers	7	7.36	0.63	0.75
Ability to review the succession plan continually	7	7.14	0.36	0.00

SD = Standard deviation; IQD = Interquartile deviation

Findings from the survey divulged that the following 14 succession planning dimension variables were contemplated by the experts to have varying influence on leadership development in the construction industry.

- Knowledge management features (HI)
- Future focused features (HI)
- Succession time management features (HI)
- Ability to understand organisational long-term vision (HI)
- Leadership continuity features (HI)
- Talent identification features (HI)
- Talent management features (HI)
- Transition management features (HI)
- Change management features (HI)
- Skills transferring features (HI)
- Employee retention features (HI)
- Leadership ability to document the succession plan (HI)
- Ability to communicate the succession plan with immediate managers (HI)
- Ability to review the succession plan continually (HI)

From the impact assessments of the factors; findings discovered that all 14 of the factors had high influence towards leadership development.

5 Discussion of Findings

Human capital in a firm is the most vital resource in the construction industry. Moreover, the absolute succession need for leadership creates undetected impacts to construction firms. These succession implementation challenges can cause entities to perform financially and operationally poor. This section bestows the discussions of the findings from the Delphi survey on the impact of succession planning variables on leadership development in the construction industry. Results reveal that knowledge management and succession management features are fundamental for leadership development and this corresponds with the work by [12, 14]. Further findings disclosed that it is essential for construction mangers and personnel to have the ability to understand the long-term organisational vision of their firms and this notion supports the effort by [7, 8, 13, 14].

6 Conclusion and Recommendation

The objective of this article was to evaluate several succession planning factors that influence leadership development in the construction industry. This was done using the Delphi survey technique. The study shows that succession planning has high level of impact on leadership change and development. This study offers a strong base for leadership succession research in the South African construction industry.

References

1. Yankov, L., Kleiner, B.H.: Human resources issues in the construction industry. Manage. Res. News **24**(3/4), 101–105 (2001)
2. Schrader, C.R.: Keys for effective succession planning. Qualified Remodeler **32**(2), 19 (2006)
3. Chavez, J.: The case for succession planning. Strategic Finance, 15–16, February 2011
4. Miller, D.: Some organizational consequences CEO succession. Acad. Manag. J. **36**(3), 644–659 (1993)
5. He, X., Wang, Z., Mei, L., Lian, Y.: The impact of founder turnover on firm performance: an empirical study in China. J. Chin. Entrepreneurship **2**(2), 148–164 (2010)
6. Toor, S., Ofori, G.: Leadership for future construction industry: Agenda for authentic leadership. Int. J. Project Manage. **26**, 620–630 (2008)
7. Behn, B.K., Riley, R.A., Yang, Y.: The value of an heir apparent in succession planning. Corp. Gov. **13**(2), 168–177 (2005)
8. Kesner, I.F., Sebora, T.C.: Executive succession: past, present & future. J. Manag. **20**(2), 327–372 (1994)
9. Perrenoud, A.: Effective succession planning in construction companies. Master's thesis, Arizona state university (2012)

10. SFGate: Succession planning few, executives say (2011). http://articles.sfgate.com/business/27104573_1_bloomberg-briefing-succession-releases-quarterly-results
11. Rothwell, W.: Effective Succession Planning: Ensuring Leadership Continuity and Building Talent from Within, 4th edn. American Management Association, New York (2010)
12. Dahlke, A.: Succession Planning for Dummies, Independent Organizational Consultant. Wiley, New York (2016)
13. Thompson, M.W.: Integrity and vision of firm's founders ease leadership transitions. Leadersh. Manag. Eng. **2**(2), 38–41 (2002)
14. Haroski, S.: The importance of succession planning, SPHR, FGP international (2009)
15. Okoli, C., Pawlowski, S.: The Delphi method as a research tool: an example, design considerations and applications. Inf. Manag. **42**, 15–29 (2004)
16. Rowe, G., Wright, G., Bolger, F.: Delphi - a re-evaluation of research and theory. Technol. Forecast. Soc. Chang. **39**, 238–251 (1991)
17. Creswell, J.W.: Research Design: Qualitative, Quantitative and Mixed Methods Approaches, 4th edn. Sage Publications, Thousand Oaks (2014)

Impact of Construction Health and Safety in Economic Growth

Sitsabo Dlamini[1] and Gcinithemba Masimula[2](✉)

[1] School of Construction Economics and Management, Faculty of Engineering and the Built Environment, University of the Witwatersrand, Johannesburg, South Africa
[2] Alakhiwe Project Management, Midrand, South Africa
Sitsabo.dlamini@wits.ac.za

Abstract. There is a great deal of interest in occupational health and safety (OHS) in the infrastructure sector of most developing countries. A lot has been said about how dangerous the infrastructure sector is on OHS. Given the poor record of OHS in the infrastructure sector, it is of interest to understand how this affects economic growth. This paper employs a descriptive analysis of data collected using a structured questionnaire to examine how OHS impacts on economic growth. The impact has been measured in terms of the degree of satisfaction of customers' needs in terms of budget. The use of the SERVQUAL model in the analysis of the results reveal that OHS has a positive impact on project delivery. This can be construed to mean that the implementation of OHS in the infrastructure sector has a potential to positively influence GDP growth in developing countries.

Keywords: Health · Safety · Economic growth · Project delivery

1 Introduction

The construction of physical infrastructure to provide society's needs such as water supply, flood control and transport is characterized by its large, elaborately designed and enduring features. It has been technology-driven to find better and efficient ways of meeting these needs. Networked delivery systems of infrastructure delivery have become common in recent years.

Construction is considered a major sector of the economy throughout the world. It influences investment. According to Hillebrandt, construction produces infrastructure that provides goods and services [8]. Construction products such as roads and factory buildings are considered as investment goods because they are used to create other commodities. Good transport infrastructure and the availability of factories attract investment into other sectors of the economy, such as manufacturing.

The infrastructure sector provides commercial facilities in which business operate. It builds dwelling facilities where people reside. The sector also builds education and health infrastructure. Increased demand for construction inputs such as cement and steel influences investment in the manufacturing sector. The provision of infrastructure is therefore vital for any economy to prosper.

C. Aigbavboa and W. Thwala (Eds.): CIDB 2019, *The Construction Industry in the Fourth Industrial Revolution*, pp. 380–389, 2020.
https://doi.org/10.1007/978-3-030-26528-1_38

In a case study of the infrastructure sector of Kenya, Muiruri and Mulinge ascertained that despite the importance associated with the sector, infrastructure is regarded as an accident-prone sector of the economy [13]. They asserted that the infrastructure sector was characterized with frequent accidents and problems of ill health. The fundamental challenge faced by most developing countries, especially in Africa is in effectively implementing occupational health and safety (OHS) procedures in construction sites.

What is OHS? The International Labour Organisation defines OHS not only as a means of avoiding accidents and preventing disease but also as all aspects of the worker's well-being [9]. In the same vein, Monk described OHS as a method through which the employees' safety, health and welfare is protected [11].

Monk went on to suggest that the maintenance of OHS issues in high regard in construction should be viewed as a fundamental human right. The nature of the construction work should not limit the manner in which workers carry out their work. Care must be taken to protect workers from any potential hazards. In most economies, OHS matters are regulated through appropriate legislation. Efforts are made to ensure that all employers are conversant with the risks involved and their obligations. It is not clear how the negative OHS picture of the infrastructure sector affects the sector's contribution to economic growth.

How does the poor OHS record of the infrastructure sector affect economic growth? It is not clear if this question has been explored in any depth. However, a vast amount of literature is available that points to the poor OHS record of the infrastructure sector. A descriptive survey undertaken by Smallwood revealed that inadequate enforcement of OHS legislation contributes to high project risk [16]. This negatively affects overall project performance, particularly project costs. The next chapter examines the established views on OHS as it relates to the infrastructure sector. The importance of OHS with respect to economic growth is then discussed.

2 Literature Review

Understanding of the link between OHS and economic growth in developing countries is poor. Using a multi-methods strategy design, Kheni, attempted to explain this link [10]. He found that the Ghanaian construction small and medium enterprises (SMEs) were major players in the economy of Ghana. He also observed though that failure by the construction SMEs to employ OHS in construction sites compromised the infrastructure sector's contribution to economic growth. He gave insight into the difficulties posed by inadequate skills, lack of support from government and weak institutions to facilitate the implementation of OHS in infrastructure projects.

What is the link between OHS and economic growth? Bloom, *et al.* tracked this link from the mid-nineteenth century [4]. They explored the evidence showing that the implementation of OHS contributes to growth in GDP. How? Bloom, *et al.* found that OHS on construction sites increased productivity. Another strand of supporting evidence that they found was that investment in OHS led to better employee wages, which raises the per capita income.

The study done by Bloom, *et al.* included a number of emerging economies in Africa and East Asia. Better wages in these areas motivated workers to save. Bloom, *et al.* asserted that behind the success of the economies studied was capital accumulation, which was driven by high levels of saving. This substantially boosted investment and economic growth. Boom, *et al.* also found that as developing countries implement OHS in their construction sites, this has a potential of attracting foreign direct investment. They observed that investors generally shunned areas where the infrastructure sector had a bad OHS record.

Why did the East Asia's exceptionally high rates of capital accumulation matter? In a time series statistical analysis of total construction output and GDP, Dlamini propounded that the infrastructure sector is part of a unified system of production, consumption and distribution [6]. It provides essential infrastructure to support other sectors of the economy, hence it is considered an investment sector. It plays a role in influencing investment through capital accumulation.

Dlamini went on to assert that apart from savings, capital and labour, the Harrod-Domar and Solow growth theories viewed growth as being influenced by technological factors, which were exogenous to the theories. Dlamini postulated that lately, the endogenous growth theory provided that over and above savings, capital and labour, economic growth was influenced by technological advance through deliberate actions of individuals within the economy. It is not just given as the Harrod-Domar and Solow theories purported.

Do all countries benefit equally from the link between OHS and economic growth? Apparently not! Bhargava, *et al.*, in a model they developed to investigate the effects of health on economic growth, ascertained that the issue of health was more important in the context of wages in developing economies than it was in developed economies [3]. Allied to this finding, Bhargava, *et al.* also argued that OHS was more important in economies that are transparent in their economic policies and governance structures. Bloom, *et al.* concurred with this view [4]. A major finding of their work was that those East Asian economies that achieved high economic growth served as a strong case to support the view that improvements in OHS can positively influence economic growth.

How does OHS affect project performance? In most developing countries, the project environment presents unique challenges arising from inherent risks such as bureaucratic contract procedures. These make it difficult to effectively implement OHS in construction sites. According to Akanni *et al.* such challenges have the potential to influence cost and time overruns on construction projects [2]. Failure to complete a project on time and within cost undermines project performance.

Faniran, *et al.* suggested that there was need to develop appropriate management tools and techniques. Such tools and techniques cannot be directly copied from developed countries [7]. They must be tailored to suit the specific project environment of developing countries. The implementation of OHS in construction sites demands that certain factors must be taken into consideration.

Walker identified a number of fundamental project environmental factors that need to be taken into consideration when implementing OHS in infrastructure projects [18]. These included political, legal, institutional, cultural and economic factors. In the same vein, some external environmental factors also need consideration. According to Ajayi, *et al.* these include community issues, weather conditions and government policy [1].

Does the implementation of OHS have a positive contribution to the economy? A study conducted by Thwala and Monese investigated the influences for productivity in building sites. The study suggested that the infrastructure sector has for some time suffered from poor productivity, which negatively affects project costs, resulting in unsatisfactory quality performance and time overruns [17]. Mthalane *et al.* asserted that health and safety issues in construction have major economic consequences in construction [12]. They contended that apart from the human cost of suffering an accident, the economic effect can be costly. This therefore suggests that OHS has both positive and negative impacts.

Given the poor record of OHS in the infrastructure sector, there is need to assess whether the implementation of OHS helps or hinders the sector's contribution to economic growth. Therefore, the research question for this study is: What is the impact of OHS on economic growth?

3 Methodology

This investigation seeks to assess the influences of OHS on economic growth in the infrastructure sector. Secondary data, adapted from Dlamini was used in the analysis [5]. The instrument used in collecting data was a structured questionnaire. The target population was composed of construction clients, design team members and contractors. The tool used to analyze collected data was the SERVQUAL model.

What is SERVQUAL? Developed by Parasuraman, *et al.* SERVQUAL is a model for measuring consumers' perceptions of service quality [14]. Shahin defined SERVQUAL as a multi-dimensional research model that can be used to measure service quality in relation to customer expectations [15]. Four variables were identified as factors having an impact on project delivery, namely, budget, time, quality and utility requirements. In the light of the fact that this paper is seeking to understand if OHS implementation hinders or enhances economic growth, the budget variable was chosen for the analysis.

4 Research Results

Survey findings on OHS and delivery of construction projects.

4.1 Data Collection

From a chosen sample of seventy-four individuals, fifty-eight people returned the completed questionnaire, which worked out to an 80% rate of response. The sample was selected randomly to include major disciplines, civil engineering and general building contractors. Professional councils such as the SACPCMP, SACAP and CIDB were also included. Other stakeholders included property developers, major clients and government agencies. Efforts were made to stratify the sample so that it was representative of major players in the construction sector. It was considered prudent that experienced professionals and line managers would be able to share valuable data, considering the complexity of the construction environment.

4.2 OHS Influence

4.2.1 Project Rollout

The total number of questionnaire received was 58. 14 out of 58 (24%) indicated that they agreed strongly with the assumption that the implementation of OHS was likely to improve the way in which a construction project is rolled out. 27 respondents out of the 58 (47%) shared the same opinion. Only 12 respondents out of the 58 (21%) were of the view that the implementation of OHS was not likely to improve the way in which a construction project is rolled out. 5 respondents out of the 58 (only 8%) felt very strongly that the implementation of OHS was unlikely to improve the way in which a construction project is rolled out. Overall, a total of 41 respondents suggested that the implementation of OHS was likely to improve the way in which a construction project is rolled out.

The collected data was further analyzed to establish if there were any strong descending views. The respondents' familiarity with OHS was assessed and 22 respondents (88%) who had used OHS confirmed that OHS enhanced the project rollout process. Even those respondents who had employed OHS in part also concurred with the view that OHS does positively influence the project rollout process. 8 respondents fell in this category, which represented 47% of the total respondents. It is also of interest to note that even respondents that had never really employed OHS in their projects also felt very strongly that OHS was likely to result in an improved project rollout process. 11 respondents (69%) were in this category while only 5 respondents (31%) had opposite views.

4.2.2 Client Satisfaction

The study also sort to establish the respondents' views on how they would rate the general level of project delivery where OHS was implemented. 50 respondents (86%) perceived that the delivery process ranked between reasonable and excellent. 30 respondents (52%) were of the view that the delivery process was reasonable. 14 and 6 respondents ranked the delivery process as good and excellent, respectively. Only 8 respondents (14%) were of the view that the process was poor or awful.

20 respondents (80%) who indicated that they had used OHS in some of their projects ranked OHS between reasonable and excellent. A majority of these respondents (13 to be exact, which represented 52%) suggested that implementing OHS in a construction project yielded a reasonable project outcome. 14 respondents (88%) that had never been exposed to OHS viewed the implementation of the system as ranging from reasonable to excellent (Table 1).

4.3 Measuring the Rollout Process

Table 1. GAP scores, perceptions & expectations

Dimension	Used OHS			Used OHS in part			Never used OHS			Overall		
	Expect.	Percept.	GAP score	Expect.	Percept.	GAP score	Expect.	Percept.	GAP score	Expect.	Percept.	GAP score
P1	3.792	3.522	−0.270	3.875	3.467	−0.408	3.250	3.467	0.217	3.639	3.485	−0.154
P2	3.458	3.182	−0.276	3.188	3.125	−0.063	3.625	3.385	−0.240	3.424	3.231	−0.193
P3	3.391	3.130	−0.261	2.938	2.733	−0.205	3.438	3.357	−0.081	3.256	3.073	−0.182
P4	3.409	3.136	−0.273	3.563	3.125	−0.438	3.375	3.429	0.054	3.449	3.230	−0.219
Tangibles	3.513	3.243	−0.270	3.391	3.113	−0.279	3.422	3.410	−0.013	3.442	3.255	−0.187
P5	3.870	3.000	−0.870	3.938	3.000	−0.938	3.563	4.071	0.508	3.790	3.357	−0.433
P6	3.696	3.217	−0.479	3.813	3.188	−0.625	4.125	3.929	−0.196	3.878	3.445	−0.433
P7	3.739	2.783	−0.956	4.063	3.125	−0.938	3.688	3.714	0.026	3.830	3.207	−0.623
P8	3.783	2.913	−0.870	4.125	3.000	−1.125	3.500	3.643	0.143	3.803	3.185	−0.617
P9	3.652	3.261	−0.391	3.750	2.813	−0.937	3.875	3.167	−0.690	3.753	3.080	−0.673
Reliability	3.748	3.035	−0.713	3.938	3.025	−0.913	3.747	3.705	−0.042	3.811	3.255	−0.556
P10	3.696	2.913	−0.783	3.938	3.313	−0.625	3.813	3.643	−0.170	3.816	3.290	−0.526
P11	3.696	3.042	−0.654	4.125	2.938	−0.187	4.067	4.071	0.004	3.963	3.350	−0.612
P12	4.261	3.696	−0.565	4.188	3.375	−0.813	4.400	4.333	−0.067	4.283	3.801	−0.482
P13	3.783	3.565	−0.218	3.938	4.063	0.125	3.133	4.188	1.055	3.618	3.939	−0.321
Responsiveness	3.859	3.304	−0.555	4.047	3.422	−0.625	3.853	4.059	0.206	3.920	3.595	−0.325
P14	3.565	3.087	−0.478	4.063	3.500	−0.563	4.188	4.143	−0.045	3.939	3.577	−0.362
P15	4.000	3.130	−0.870	3.688	3.133	−0.555	4.000	3.857	−0.143	3.896	3.373	−0.523
P16	3.087	3.087	0.000	3.250	3.250	0.000	3.929	3.929	0.000	3.422	3.422	−0.000
P17	3.304	3.304	0.000	3.563	3.563	0.000	3.667	3.667	0.000	3.511	3.511	−0.000
Assurance	3.489	3.152	−0.337	3.641	3.362	−0.280	3.946	3.899	−0.047	3.692	3.471	−0.221
P18	3.783	3.792	0.009	3.625	3.000	−0.625	3.733	3.533	−0.200	3.714	3.442	−0.272
P19	3.696	3.348	−0.348	3.500	3.813	−0.687	3.733	3.133	0.400	3.643	3.431	−0.212
P20	3.565	3.565	0.000	3.625	3.125	−0.500	3.563	3.500	−0.063	3.584	3.397	−0.188
P21	3.545	3.261	−0.284	3.500	2.750	−0.750	3.933	3.933	0.000	3.659	3.315	−0.345
P22	4.130	3.522	−0.608	3.875	3.180	−0.695	3.800	3.933	0.133	3.935	3.545	−0.390
Empathy	3.744	3.498	−0.246	3.625	2.974	−0.651	3.752	3.806	0.054	3.707	3.426	−0.281
Overall	3.670	3.246	−0.424	3.728	3.179	−0.549	3.744	3.776	0.032	3.714	3.400	−0.314

4.4 Statistical Analysis

Table 2 is an illustration of the sign confidence interval for the median worked out using the respondents' opinions on project delivery variables.

Table 2. Sign confidence interval

Variables	Median	Achieved confidence	Confidence interval
Improved project delivery	3.0	0.9363	(3.00, 3.00)

Table 3 presents the outcomes of the Kruskal-wallis tests on improved project delivery.

Table 3. Kruskal-wallis test

Variables	H	DF	P
Improved project delivery	6.50	2	0.390

The mood test was undertaken on the data supplied by the respondents and the outcome produced is illustrated in the following table (Table 4).

Table 4. Mood median test

Variables	Chi-square	DF	P	Overall median
Improved project delivery	7.80	2	0.020	3.00

One-way anova test analysis done produced the following table (Table 5).

Table 5. One-way anova test

Variables	F	P	Pooled SD
Improved project delivery	3.57	0.035	0.8443

The correlation analysis undertaken to ascertain the kind of relationships that exist between the variables produced the results outlined below (Table 6).

Table 6. Correlation between variables & structure

Variables	Correlation	P Value
Improve project delivery	−0.152	0.258

5 Discussion of Findings

Respondents drawn from professional disciplines in the construction sector constituted 45%. Those drawn from client bodies constituted 31%. It is important to note that client bodies are the ones who are affected by the service that professional consultants offer. Therefore, OHS matters have a direct impact on clients. Where construction is successfully undertaken, such outcome is often the result of efficiency in the implementation process. The investigation was thus focused on those contributors that have stake in the project implementation process.

Respondents in the prime age (middle age) group tended to provide more positive and mature responses. They constituted 73% of the respondents. A majority of respondents exhibited vast experience in the industry. It is to be expected that this increased the chances of getting more reasoned responses. 66% of the respondents had more than 10 years' experience in various roles in construction. The experience of such respondents was reflected in the depth and wealth of responses received.

Efficiency in the delivery of construction projects is crucial. The distinguishing characteristics of construction firms that that employ OHS in the implementation of their projects is but how efficient and effective the entire process is. OHS roles are often integrated into the core construction processes. However, it is not unusual to procure OHS services through subcontract arrangements. Improved project delivery has been shown to be highly likely where OHS is incorporated into the key project delivery mechanisms.

The most dominant responses received were clear that the implementation of OHS in construction results in more efficient project rollout sequences. Respondents who have utilized OHS were supported in their views by those never used the process in that the implementation of OHS brings lots of positives into the overall project rollout process. Common tendency and dispersion revealed a slightly different result from the one obtained when utilizing frequencies and percentages. Responses on improved project rollout process by familiarity with OHS was approximately one standard deviation for all three variables (used OHS, partially used OHS, never used OHS). The group that had used OHS was the only one that had a mean of 3.0. This was in agreement with the relationship between OHS and improved project delivery. The least common tendency was the group that partially used OHS, while the mean to the group that never used OHS was close to 3.0.

The median of 3.0 recorded by the group that used OHS and those that never used it indicate that the two groups were in agreement with the assertion that OHS improved project delivery. The median below 3.0 achieved by the group that partially used OHS suggest that they did not agree with the assertion. Again the group that partially used OHS seemed to agree with the assumption of inverse relationship between OHS and project delivery. 93–6% confidence was achieved in the median when using the sign confidence test for all variables. This suggests that there is a relationship between OHS and improved project delivery. The Mood test and Kruskal wallis test show that the relationships are not significantly associated given the high probability. Judging by the probability that was more than 0.5, it can be inferred that there was a weak negative correlation and no significant relationship.

The mean and median for overall project delivery was at 3.0 in all the different groups of respondents. This suggested the existence of a relationship between OHS and the manner in which the project was delivered. The statistical test suggested that there was no significant association.

Further analysis of data was undertaken using the SERVQUAL model. The aim was to ascertain understand some of the major perceptions held by respondents. The model revealed that the group that had never used OHS had the least gap score. This was followed by the group that had used OHS. The group that had partially used OHS exhibited the highest gap score.

To put project delivery into perspective, the SERVEQUAL model and its 22 questions were used. As far as tangibles are concerned, those companies that never used OHS had the least gaps. Those that had used OHS followed suit. The other aspect pertaining to reliability, responsiveness and empathy followed the same trend. The gap score for the assurance dimension was the highest with those companies that have used OHS and the lowest with those that had not.

In all the respondents, reliability had the highest gap score, while empathy had the least score for those companies that had used OHS and those that had not. Those respondents that partially used OHS gave the least gap score. These findings are important beyond the investigation of OHS. They may be used to in creating appropriate strategies that may be employed in improving the rollout processes of construction projects.

6 Conclusion

The implementation of construction health and safety has a positive impact on economic growth. This study has revealed the existence of a positive relationship between OHS and the manner in which construction projects are rolled out. Analyses of respondents' perceptions suggest that the implementation of OHS yields positive impacts on overall project delivery with respect to quality, budget, schedule and utility parameters.

Apart from the respondents who indicated that they had used OHS, the positive relationship of OHS to overall project delivery was ratified also by those who never used OHS and those that partially used it. Therefore, the identified positive relationship can be construed to suggest that OHS has the potential to positively impact on economic growth. The statistical analysis of data through frequency tables as well as percentages derived from collected data gave some insights on how OHS affects the rollout of construction projects.

The data analysis undertaken as part of this research was strengthened through descriptive statistics. These included employing common tendency and dispersion. The picture that emerged was that the relationship of OHS and project delivery was not necessarily significant. The utilization of quantitative statistical models can present limitations since the data in this research has been largely non-parametric (ordinal). It is thus imperative that the statistical tests undertaken should be used cautiously, with a view to avert asserting conclusions that might be misleading.

In the light of the fact that this study was exploratory in nature, it would be prudent that further research is undertaken to explore the relationships to some depth. Tracking the trends over time would yield more insights. In this vein, it would help to evaluate time series statistical analysis of construction output data. Allied to this, more studies should aim to fully explore the relationship at operational level over time, so as to establish objectivity as to how the implementation of OHS affects economic growth.

References

1. Ajayi, O.O., Smallwood, J.: The impact of Design and Construction Management on Construction Ergonomics, Paper Presented at Proceedings of the 3rd Built Environment Conference, Cape Town, pp. 6–8 (2008)
2. Akanni, P.O., Oke, A.E., Okpomiemie, O.A.: Impact of environmental factors on building project performance in delta state, Nigeria. Hous. Build. Nat. Res. Cent. J. **11**, 91–97 (2014)
3. Bhargava, A., Dean, T.J., Lawrence, L., Christopher, J.L.M.: Modelling the effects of health on economic growth. J. Health Econ. **20**, 423–440 (2001)
4. Bloom, D.E., David, C., Sevilla, J.: The effect of health on economic growth: a production function approach. World Dev. **32**, 1–13 (2004)
5. Dlamini, S.S.: The impact of design & build procurement system on project delivery. MSc, University of Bath (2010)
6. Dlamini, S.S.: The contribution of the infrastructure sector to economic growth. True Gospel Fortress, London (2016)
7. Faniran, O.O., Love, P.E.D., Smith, J.: Effective front-end project management – a key element in achieving project success in developing countries (2000). http://citeseerx.ist.psu.edu/viewdoc/download. Accessed 22 Feb 2019
8. Hillebrandt, P.M.: Economic Theory and the Construction Industry. Macmillan Press Ltd, London (2000)
9. International Labour Organization. Safety and health in construction, An ILO code of practice, Geneva (1991)
10. Kheni, N.A.. Impact of health and safety management on safety performance of small and medium-sized construction businesses in Ghana unpublished. PhD thesis, Department of Civil Engineering, Loughborough University, UK (2008)
11. Monk, V.: Occupational Health and Safety Management Systems and Safety Performance in the Building and Construction Industry. Worksafe Australia, Melbourne (1994)
12. Mthalane, D., Othman, A.A.E., Pearl, R.G.: The Economic and Social Impacts of Site Accidents on South African Society. Paper Presented at Proceedings of the 5th Post Graduate Conference on Construction Industry Development, Bloemfontein, 16–18 (2008)
13. Muiruri, G., Mulinge, C.: Health and safety management on construction project sites in Kenya. A case study of construction projects in Nairobi County. In: Proceedings Engaging the Challenges – Enhancing the Relevance, 16–21 (2014)
14. Parasuraman, A., Zeithaml, V.A., Berry, L.L.: Reassessment of expectations as a comparison standard in measuring service quality: implications for future research. J. Mark. **58**, 111–124 (1994)
15. Shahin, A.: Integration of FMEA and the kano model: an exploratory examination. Int. J. Qual. Reliab. Manage. **21**(7), 731–746 (2004)
16. Smallwood, J.J.: The Role of Health and Safety in Project Management, pp. 3–5. Regional African Project Management, Johannesburg (1999)
17. Thwala, W.D., Monese, L.N.: Motivation as a tool to improve productivity on the construction site. Paper Presented at Proceedings of the 5th Post Graduate Conference on Construction Industry Development, Bloemfontein, 16–18 (2008)
18. Walker, A.: Project Management in Construction, 2nd edn. BSP Professional Books, Oxford (1989)

Sustainable Water Supply in Buildings Through Rooftop Rainwater Harvesting

Chidozie Charles Nnaji[1,2(✉)]

[1] Department of Civil Engineering, University of Nigeria, Nsukka, Nigeria
Chidozie.nnaji@unn.edu.ng
[2] Faculty of Engineering and the Built Environment, University of Johannesburg, Johannesburg, South Africa

Abstract. This paper used a combination of theoretical concepts, literature review and real-life survey to establish the need for a global and aggressive adoption of rainwater harvesting (RWH) as sustainable means of water supply and flood control. The ideas presented in this paper were aimed at making RWH systems an integral part of building conception and design as well as enshrining compulsory RWH in national building codes. In areas of adequate rainfall, RWH can supply up to 100% of domestic water needs of bungalow dwellers. Non-reactive roofs such as aluminium roofing sheets were found to be more suitable for RHW than asbestos, galvanized roofing sheets and concrete roofs. RWH can be made more sustainable and cost-effective by using mass balance and optimization methods for providing adequate storage for harvested water. Simple point-of-use treatment methods such WaterGuard® and Purifier of Water® can be used to render harvested water potable.

Keywords: Rainwater harvesting · Water supply · Rainfall · Roof · Storage tank

1 Introduction

The coupled effects of population explosion and climate change are exerting a heavy toll on global water resources, leading to a high level of suffering in certain parts of the world. The economic development associated with urbanization increases the per capita use of water, as new technologies such as showers, washing machines and dishwashers increase residential water use [1]. Water scarcity has gone past an emerging global problem; it has been identified as one of the pressing global risks [2]. Sustainable development remains a mirage without sufficient water supply for domestic consumption, sanitation, hygiene, agriculture and industrial processes. Water is one of the most important needs of man and yet the most misused, especially among city dwellers. This could be because, in many parts of the world, water is so affordable that the burden of expenditure is hardly felt by anyone except, perhaps, the government. The culture of wasting water usually stems from the extreme perspective that water is a social good rather than an economic good.

Global water consumption rose form $200 \times 108\ m^3$ to $4400 \times 108\ m^3$ from 1900 to the year 2000. Arguments abound that despite the fact water is an economic good, it

C. Aigbavboa and W. Thwala (Eds.): CIDB 2019, *The Construction Industry in the Fourth Industrial Revolution*, pp. 390–400, 2020.
https://doi.org/10.1007/978-3-030-26528-1_39

is still comparatively and more importantly, a social good that guarantees several aspects of a dignified human life [3]. When defined as a social good, it may warrant a rights approach. However, the current global threat of water scarcity has proved intractable to governments. This has been the case for Cape Town which was predicted to be the first major city to run out of water in April 12, 2018 (day zero). Several other cities such as Cairo, Beijing, Mexico City, Moscow, Istanbul, Tokyo and London amongst others are facing the same fate. There is no doubt that if the current trend continues, many more cities will join the queue. Usually, government's inability to tackle a national threat forebodes helplessness for the citizenry. However, in the case of water scarcity, there remains a largely uncharted vista of relief – rainwater harvesting (RWH), and this can be explored without the assistance of the government. Hence this paper explores the potentiality of rainwater being a sole or supplemental source of water supply when incorporated into the building architecture.

2 Methodology

This research was accomplished by a combination of actual surveys, theoretical analyses and literature review. Part of the data used in this study came from a RWH survey carried out in Nsukka, Nigeria in 2017. The survey basically focused on RWH practices while the theoretical analyses were used to illustrate the geometric effect of land development (pavement) on urban runoff.

3 Discussion

3.1 Urban Flood Control Through Rainwater Harvesting

One of the numerous drawbacks of urbanization is indiscriminate paving of surfaces consisting of streets, parking lots, roads and driveways, which reduce the rate of infiltration and invariably aggravates urban flooding. In the United States alone, pavements and other forms of impervious surfaces cover over 43,000 square miles (111,369 km^2), which is nearly the size of the entire state of Ohio [4], one tenth of South Africa, one eighth of Nigeria or half the size of Ghana. Even at a moderate rainfall intensity of 10 mm/hr, this combined area of paved surfaces will produce a staggering 309,360 m^3/s of runoff which is eighteen times the discharge of River Mississippi or fifty-five times the discharge of River Niger. The foregoing paints a frightening image of the impending danger posed by impervious surfaces on the environment. In high density urban areas, rooftops constitute a substantial part of impervious surfaces that contribute to increased urban runoff and associated risk of flooding. Land development without corresponding efforts to equalize or attenuate the enormous amount of runoff generated represents a serious human and ecological risk resulting from flooding. RWH can be a cheap, unobtrusive and economical means of mitigating flood resulting from increased runoff engendered by land development.

3.2 Water Supply Prospects of Rooftop Rainwater Harvesting

RWH dates back to 3200–1100 BC down through the Hellenistic Roman, Byzantine, Venetian and Ottoman periods [5]. RWH is a relatively inexpensive means of small-scale water supply which basically requires a catchment area, a conveyance system of channels/gutters and a storage (Fig. 1). Depending on the nature of the catchment and conveyance systems, a simple treatment system may be required. However, the development of other forms of water resources in modern times which are able to provide water all year round almost obliterated the role of rainwater as source of water supply, until recently. Once again, RWH is being seriously considered in several countries as a major source of relief in the face of acute water shortages threating the existence of urban settlements. But even as the world grapples with dwindling fresh-water resources, there is still a substantial degree of inertia with respect to RWH despite its acknowledged great potential as a source of water supply.

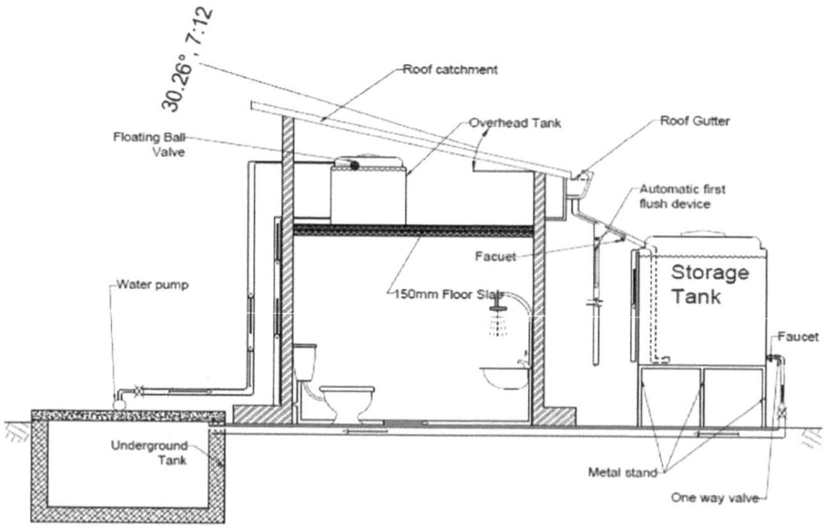

Fig. 1. Sectional view of RWH system

In recognition of RWH as a significant source of water supply, countries like Germany, Japan and Australia among others have made some policies in favour of RWH (Table 1). Other countries like UK and the USA have strong and active RWH associations namely: UK Rainwater Harvesting Association and American Rainwater Catchment Systems Association respectively. Though RWH has been gaining popularity in the United States, it is mainly used for irrigation [6, 7]. The United Kingdom (UK) had lagged behind other developed countries with respect to RWH practices with [8] reporting only 75 successful and operational RWH systems in 2001. However, by 2011, an estimated 7500 systems were identified in UK commercial and residential properties [9]. This is a far cry from the situation in Germany where about 50,000 systems were being installed every year [10]. RWH has always been practiced in the rural areas of Canada, but there has been a renewed interest in RWH for domestic purposes in urban areas [11]. However, uncertainty of costs and risks as well as lack of clear policies has impeded the rate of adoption of RWH as an alternative source of water in Canada [11]. France has a law that restricts the use of rainwater for washing clothes, bathing and drinking which is part of the reason for the limited uptake of RWH [12]. In Sweden RWH is motivated by ecological sustainable development in large community housing complexes [13]. In Japan where 50% of the total impervious area of the city is roof; and in China, reduction of urban flooding is the main reason for the adoption of RWH [14, 15]. However, in the semi-arid areas of China, the need to solve the problem of water scarcity is the main driver of RWH. In South Korea, urban flooding and combined sewer outflows during the wet seasons have sparked high interests in RWH and reuse systems [6]. In Australia, there was an increase in the number of people that practiced RWH from 24% in 2007 to 34% of households in 2013 [16]. This resurgence of interest in RWH was triggered by the millennium drought which occurred in Australia from years 2000 to about 2008, resulting in extensive investment in stormwater harvesting and rainwater tanks [6]. RWH is encouraged in Australia as a supplementary water source through financial incentives and regulation such as requirement in building codes and/or water conservation measures [6]. Brazil specifically stipulated compulsory RWH for buildings greater than 200 m^2. All these represent significant efforts towards RWH, but so much is yet to be done to maximize its potentials.

Table 1. Current status of RWH practices in various countries of the world

Country	Level of adoption of RWH	Motivation	Efforts in place
United Kingdom	Low – renewed interest	Non-potable domestic supply	Establishment of UK Rainwater harvesting Association (UKRHA) in 2004
United States of America	Low and in some cases prohibited	Irrigation and domestic supply	Some states offer tax exemption on rainwater treatment equipment, Establishment of ARCSA
China	Extensive in some areas such as Gansu Province	Potable water supply, irrigation, Urban flood mitigation	Workshops and conferences on RWH
Germany	Growing interest	Flood control, conservation of freshwater resources, toilet flushing, gardening	Provision of grants and incentives to households with RWH systems
Japan	Renewed attention	Flood mitigation, alternative source of drinking water, disaster response water supply, mitigation of urban heat island	Establishment of Rainwater Utilization Liaison Council for Local Governments, subsidy for RWH tanks, conferences

(continued)

Table 1. (*continued*)

Country	Level of adoption of RWH	Motivation	Efforts in place
Canada	Growing interest for sanitary uses, reluctance towards other uses	Toilet flushing, laundry, lawn and garden irrigation	Rebates for rain barrel
Australia	High rate of acceptance of RWH	Flood control, environmental sustainability, irrigation, high cost of piped water supply	Incentives for new buildings with RWH systems, stipulation of water conservation measures in the building codes
South Africa	Low, slow uptake	Acute water shortage	None identified
Nigeria	Scattered individual efforts	Poor perception of RWH, quality concerns, low annual rainfall in the arid and semi-arid region	Individual research efforts to promote RWH, lack of government support
Brazil	Rudimentary and low-level practice of RWH	Quality concerns, infeasible benefit-cost ratio, lengthy pay-back time	Establishment of standard for rainwater usage, compulsory RWH systems in buildings > 200 m^2
Malaysia	Low – declining interest from 1999 to 2006	Water quality concerns, mosquito breeding, cost	Introduction of RWH guideline in 1999 after the 1998 drought, compulsory installation of RWH systems in factories and schools with effect from 2006

3.3 Choice of Roofing Material for RWH

As rain drops glide through the atmosphere, they dissolve atmospheric impurities. More contaminants get dissolved as the water droplets reach the ground or roof of buildings, thereby making rainwater more contaminated. In RWH, the quality of water harvested is largely dependent on the ambient atmospheric conditions, the nature of catchment surface (roofing materials), type and sanitary condition of the conveyance system and the material and maintenance of storage tank. Unfortunately, RWH is not usually a consideration in the design of most residential building. The overriding consideration in the choice of roofing material and shape is aesthetics. However, there needs to be a paradigm shift in this trend if sustainable development is to be attained globally.

The nature, shape and age of roofing material are critical factors that largely determine the quality as well as quantity of harvested rainwater. More water can be harvested from smooth sloping roof with high runoff coefficient (C) than from flat rough roofs with much lower values of C. The range of runoff coefficients for various roof materials available in the literature are 0.9 for sloping concrete/asphalt roof, 0.81–

0.95 for sloping metal roofs, 0.7 for sloping aluminium roof and flat bituminous roofs, 0.8–0.85 for flat gravel roofs, tiles (0.8–0.9) and 0.81 for level cement roofs [17, 18]. In fact, [17] reported that smooth sloping roofs with C = 0.9 yield water of better quality and 50% more water than flat roof with C = 0.62. This can be explained by the fact that the residence time of water on sloping roofs is much lower than that of flat roofs. Low residence time of water on the roof reduces the risk of the dissolution and weathering of roof materials as well as other deposits on the roof by rainwater. The effect of type and age of roofing materials on rainwater quality has been the subject of several studies and there is a general consensus that water harvested from old roofing sheets are of much lower quality than those harvested from new roofing sheets. For instance, the total concentration of copper in old galvanized roofing sheets can be as much as 1.97 mg/l which is roughly twice that of new roofing sheets with a copper concentration of 1.0 mg/l [19]. Galvanized roofing sheets increase the concentrations of zinc and lead in rainwater. Aluminum roofing sheet yields harvested rainwater of higher quality than other common roofing sheets such as galvanized roofing sheets and asbestos. In a survey conducted in Nsukka, Nigeria as a part of this research, it was found that the most common types of roofing materials were concrete (4%) galvanized roofing sheet (40%) and aluminium sheet in (56%). The survey further revealed that long span aluminium roofing sheets are becoming popular as newer buildings are more likely to be roofed with them, resulting in a drastic decline in the use of other types of roof such as concrete and galvanized roofing sheets. This trend is by no means connected to a desire for improving the quality of harvested rainwater, rather it driven by aesthetics and class. The downside of this disposition is that in the future, there might be a shift to new roofing materials which may not favour harvesting of good quality rainwater.

3.4 Improving Reliability of RWH

Rainfall is seasonal and therefore necessitates adequate storage for water harvested in the rainy season for use in the dry season. With adequate storage system, rainwater can supply daily water demand all year round in areas with high rainfall amounts. The quantity of harvested water during each rainfall event is primarily a function of climatic factors such as rainfall intensity and duration; as well as building characteristics such as area and nature of roof. However, reliability is more intricately tied to the amount of water harvested per occupant of the building which is directly related to the roof area per capita. In this regard, high rise buildings present a severe disadvantage for rainwater harvesting since the roof size is most likely going to remain constant as the number of floors in the building increases. Increase in number of floors in a building invariably implies increased number of occupants for the same roof area. The concomitant outfall is drastic reduction of roof area available for rainwater harvesting per capita and impaired reliability of water supply (Fig. 2). The roof area per capita as shown in Fig. 2 is the portion of the entire roof plan area that supplies water to an individual occupant if the roof were divided equally among all occupants. For high rise buildings, as the number of floors or number of occupants increases, the roof area per capita reduces. Hence, high rise residentials buildings do not favour RWH and cannot justify the required investment or guarantee return on investment.

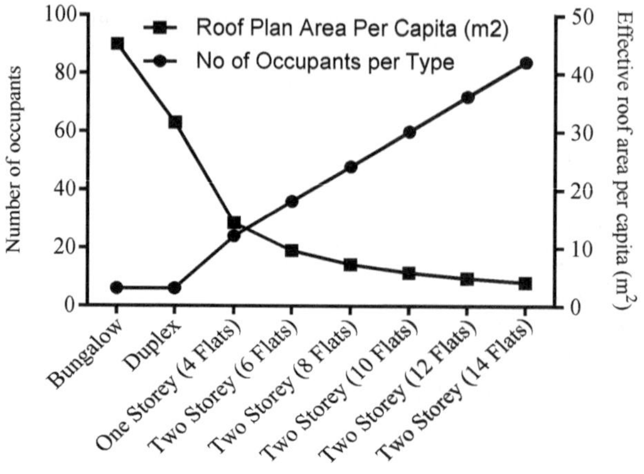

Fig. 2. Illustration of decline in per capita roof area for RWH for multi-storied buildings

One of the most important factors that determine the feasibility and sustainability of RHW is rainfall amount (depth) received in an area and their temporal distribution. It is not enough to have sufficient rainfall to meet water demand, it becomes almost infeasible to practice RWH if the entire annual rainfall occurs in a few months (short rainy season). This is because an enormous amount of storage volume, usually an underground tank, will be required for holding the water for use throughout the rest of the year. If it were possible to spread rainfall in a geographical location uniformly over the year, then there would be no need for an underground tank, and only nominal storage would be required, thus drastically reducing the cost associated with storage. Figure 3 illustrates a scenario where the annual rainfall (say 1000 mm) in a locality is spread uniformly over all the days of the year. It can be seen that the reliability of supply will decline asymptotically as the number of floors of the building increases, but at a point where the total annual demand exceeds the total annual rainfall, the required nominal storage remains constant. However, this is not so in real life because of the probabilistic and seasonal nature of rainfall. The constant storage capacity in Fig. 3 is because as the number of floors increases, the required storage capacity shifts from a demand-side to a supply-side system.

In an ongoing study on the lifecycle cost of RWH systems, results available suggest that storage (underground and overhead) can gulp between 62.9 and 89.1% of the capital cost of RWH systems depending on the level of water consumption or demand. However, if harvested rainwater is to be a supplemental source of water supply, the use of large underground storage tanks will not be justified. Hence, there is need for optimal sizing of RWH storage tanks to meet the level and degree of water demand. This implies that proper sizing of RWH storage tanks is necessary for optimal resource utilization. Determining a tank size depends upon a wide variety of factors including rainfall patterns in the given location, catchment area, runoff coefficient, the present and the future daily rainwater demands within the household. RWH storage tank sizing requires proper calculations and realistic assumptions, in order to optimize resource allocation. A number of methods have been proposed for sizing of rainwater tanks.

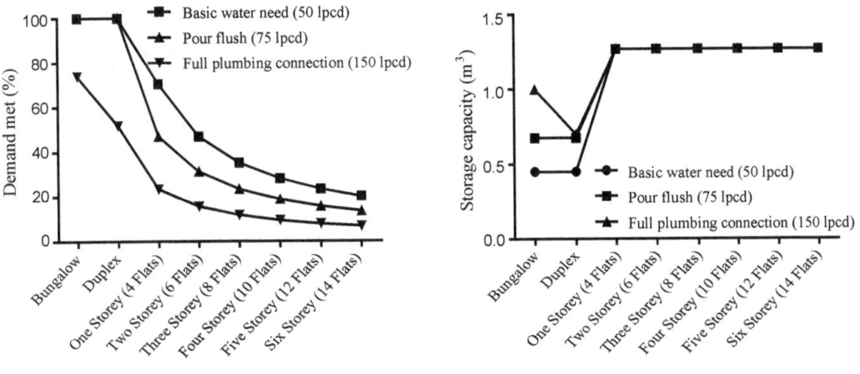

Fig. 3. Reliability and storage pattern for uniformly distributed annual rainfall of 1000 mm

One method suggests sizing storage tanks to cover the needs of R days of the year, based on the roof area, annual rainfall and runoff coefficient of the roof. Another method sizes rainwater storage tanks based on the longest average period without rainfall for the specific geographic location. The average monthly demand is multiplied by the period in months of the dry season and the resulting volume is the minimum storage capacity. Yet another method, uses the average annual water consumption and the dry season expressed in days and found as a ratio of the whole year (365 days). The ratio is multiplied by the annual consumption to find the minimum storage capacity of water. The major problem with these methods is the implicit assumption that rainwater will be allowed to accumulate in the tank without use until the dry season period. The reality is that rainwater is also utilized in the rainy season as it accumulates in the storage. Depending on the rate of water consumption and rainfall pattern, using these methods may yield oversized tanks which may be half-empty at the end of the rainy season or half full at the onset of the next rainy season. Hence, a water balance method which takes into cognizance continuous supply and consumption is more appropriate because it yields an optimized storage capacity.

Besides determining the appropriate RWH storage tank capacity, there is also the need to choose the type of tank to use. Choice of water storage tank is determined by three principal factors namely: cost, durability and size of tank. High density polyethylene (HDPE) tanks are usually cheaper than steel and ferro-cement tanks, are almost leak-proof, require very little maintenance (washing and changing of faulty faucets) and are easy to install. However, there is usually a limit to their size and they are prone to vandalism and accidental damage. HDPE tanks can undergo degradation under the influence of ultra violet radiation. Besides, research results have shown that black HDPE tanks can result in elevated water temperatures, thereby encouraging the proliferation of pathogens. Nevertheless, HDPE tanks have found very wide acceptance for RWH especially for above-ground small and medium scale storage. Steel and concrete tanks are the preferred options for underground storage because of their ability to withstand earth pressure when empty. However, concrete and ferro-cement tanks are susceptible to cracks due to shrinkage and therefore must be subjected to adequate analyses and design as well as proper construction both in terms of materials and workmanship. Ferro cement is the technology of choice for many rainwater harvesting programs because they are relatively inexpensive and with a little maintenance can last indefinitely.

3.5 Towards Improved Rainwater Quality

There is a prevailing perception that rainwater is contaminated, even when several studies have shown that rainwater can be consumed with minimal or no treatment. The first culprit of rainwater quality impairment is the atmosphere which contains acidic oxides that dissolve in rain to produce acid rain. This is responsible for the low pH of rainwater reported in the literature. Rainwater quality is further depleted as the raindrops come in contact with the roof where it weathers roof materials and dissolves pollutants deposited on the roof by wind and fecal matter deposited by birds, lizards and rats. Further deterioration occurs in the conveyance system and the storage tanks. Considering that one of the major objections raised against RWH is its questionable quality as perceived by many, there is therefore need to improve its quality by incorporating a treatment unit. The level of pollution of rainwater is almost infinitesimal compared to streams and rivers from which many cities and towns all over the world are served after treatment. Hence, given the low level of rainwater pollution, rainwater usually requires simple inexpensive treatment methods to render it potable. In most cases, the only treatment required is disinfection which can be achieved using the appropriate quantity of chlorine or other point use (POU) treatments such as Water-Guard Plus® and Purifier of Water®. Generally, however, the type of treatment adopted for rainwater will depend on the material of the catchment surface, atmospheric conditions and the intended use. The most practical approach to improving and maintaining rainwater quality is to adopt adequate measures for protecting, monitoring and treating the harvested water. Given the ease with which water is polluted, a multi-phased defence mechanism can be adopted by treating rainwater before and during storage, and equally at the point of use. The first line of defence in RWH system is to install a first flush diverter for removing pollutants that are deposited on the roof. The first flush diverter prevents the initial rain falling on the roof from entering the storage tank until the roof has been sufficiently cleaned by the abrasive impact of rain drops. The quantity of first flush water to be diverted depends on the amount and type of debris that accumulates, the number of dry days since the last rainfall, the properties of the catchment surfaces and the intended use of the water. However, [20] recommended that first flush water volume should range from 0.04–0.2 m^3 per 100 m^2 of roof. Each downspout will require its own first-flush diverter and the diverter should be sized according to the portion of the catchment that contributes to that downspout.

4 Conclusion

Integrating RWH into new buildings can form a veritable tool for sustainable global development. RHW can be employed to mitigate two of the foremost threats of our civilization viz: water scarcity and flooding. There is still a global widespread reluctance towards the adoption of RWH harvesting despite evidence of rapidly dwindling freshwater resources. However, compulsory requirement of RWH as integral part of new buildings in national building codes can be exploited to facilitate the adoption of this concept.

References

1. McDonald, R., Montgomery, M.: Water on an urban planet and the reach of urban infrastructure. Glob. Environ. Chang. **27**, 96–105 (2014)
2. Braumann, K.A., Ritcher, B.D., Postel, S., Malsy, M., Florke, M.: Water depletion: an improved metric for incorporating seasonal and dry-year water scarcity into water risks assessment. Elem. Sci. Atmos. **4**, 000083 (2016)
3. Gleick, P.H., Wolff, G., Chalecki, E.L., Reyes, R. The new economy of water: the risks and benefits of globalization and privatization of fresh water. Pacific Institute for Studies in Development, Environment, and Security, Oakland (2002)
4. Frazer, L.: Paving paradise: the peril of impervious surfaces. Environ. Health Perspect. **113**(7), A456–A462 (2005)
5. Antoniou, G., Kathijotes, N., Spyridakis, D.S., Angelakis, A.N.: Historical development of technologies for water resources management and rainwater harvesting in the hellenic civilizations. Int. J. Water Resour. Dev. **30**(4), 680–693 (2014)
6. Cook, S., Ashok, K.S., Ted, G.: Rain harvesting systems for urban developments. In: Ashok, K.S., Donald, B., Ted, G. (eds.) Rainwater Tank Systems for Urban Water Supply, pp. 1–14. IWA publishing, London, UK (2015)
7. Thomas, R.B., Kirisits, M.J., Lye, D.J., Kinney, K.A.: Rainwater harvesting in the united states: a survey of common system practices. J. Clean. Prod. **75**, 166–173 (2014)
8. Berndtsson, J.: Beneficial use of stormwater: a review of possibilities. Chalmers University of Technology, Gothenburg, Sweden (2004)
9. Diaper, C., Jefferson, B., Parsons, S.A., Judd, S.: Water recycling technologies in the UK. J. Chattered Inst. Water Environ. Manage. **15**(4), 282–286 (2001)
10. Nolde, E.: Possibilities of rainwater utilization in densely populated area including precipitation runoffs from traffic surfaces. Desalin **215**, 1–11 (2007)
11. Farahbakhsh, K., Despins, C., Leidl, C.: Developing capacity for large-scale rainwater harvesting in Canada. Water Qual. Resour. Can. **44**(1), 92–102 (2009)
12. Vialle, C., Sablayrolles, C., Lovera, M., Huau, M., Montréjaud-Vignoles, M.: Modelling of a roof runoff harvesting system: the use of rainwater for toilet flushing. Water Sci. Technol.: Water Supply **11**(2), 151–158 (2011)
13. Villarreal, E.L., Dixon, A.: Analysis of a rainwater collection system for domestic water supply in Ringdansen, Norrkoping, Sweden. Build. Environ. **40**, 1174–1184 (2005)
14. Ward, S.L.: Rainwater harvesting in the UK: a strategic framework to enable transition from novel to mainstream. https://ore.exeter.ac.uk/repository/handle/10036/106575
15. Zhang, X., Hu, M.: Effectiveness of rainwater harvesting in runoff volume reduction in a planned industrial park, China. Water Resour. Manag. **28**(3), 671–682 (2014)
16. ABS: Environmental issues: Water use and Conservation. Australian Bureau of Statistics, Canberra (2013)
17. Farreny, R., Morales-Pinzon, T., Guisasola, A., Taya, C., Rieradevall, J., Gabarrell, X.: Roof selection for rainwater harvesting: quantity and quality assessments in Spain. Water Res. **45**, 3245–3254 (2011)

18. Biswas, B.K., Mandal, B.H.: Construction and evaluation of rainwater harvesting system for domestic use in a remote and rural area of Khulna, Bangladesh. Int. Sch. Res. Not. Article ID 751952, 6 (2014)
19. Winter, N., Graunke, K.: Roofing materials assessment: investigation of toxic chemicals in roof. Washington State Department of Ecology, Washington (2014)
20. Waterfall, P.: Harvesting rainwater for landscape use. https://extension.arizona.edu/sites/extension.arizona.edu/files/pubs/az1344.pdf

Macro-BIM Adoption Study: Establishing Nigeria's BIM Maturity

Mansur Hamma-adama[1,2](✉) and Tahar Kouider[1]

[1] Scott Sutherland School of Architecture and Built Environment, Robert Gordon University, Aberdeen AB10 7GJ, UK
{m.hamma-adama, t.kouider}@rgu.ac.uk
[2] Department of Civil Engineering, Kaduna Polytechnic PMB 2021, Kaduna, Nigeria

Abstract. Construction Industry in Nigeria has since required a disruptive technology to change its construction business and improve its capabilities and productivity. As an on-going research (PhD work) to developing a strategy for an effective Building Information Modelling (BIM) adoption in Nigeria, a macro-BIM adoption study was carried out to establish BIM maturity within the Nigerian construction market. Online questionnaire was used as tool for data collection from the professional stakeholders in the industry. In the process to formulate a National BIM Roadmap, five conceptual macro-BIM maturity models were utilized. The models' findings act as a guide in developing a national BIM adoption policy. The five applied models helped classify the macro maturity components and the key policies' deliverables that must be addressed within both the initiation and consultation phases of proposing the Nigerian BIM roadmap. The results established positive progress in awareness and adoption level compared to the 2017 survey. Recommendations are made based on the study findings as to advance into policy development.

Keywords: Building Information Modelling · Macro-BIM · Adoption · Nigeria · Construction industry

1 Introduction

BIM is gradually becoming a norm in the built asset procurement, but its adoption around the world varies significantly. Effort by government is playing a significant role in facilitating BIM adoption around the world. For example, UK, USA, Finland, Russia, Denmark, Singapore etc. are some case study countries where government involvement played a significant role on BIM adoption [1, 2]. Moreover, more countries are keying into this strategy, to mention but a few, such as Canada, Germany, Japan, Ireland, Qatar and Spain. Some of these countries used the developed macro-BIM adoption models to streamline or develop BIM adoption roadmap and guidance for the development of their BIM adoption policy, and such countries include Ireland and Brazil [2, 3]. Other countries which are currently utilizing the macro-BIM adoption models include Spain, Canada, Russia, Brazil, Hong Kong etc.

C. Aigbavboa and W. Thwala (Eds.): CIDB 2019, *The Construction Industry in the Fourth Industrial Revolution*, pp. 401–411, 2020.
https://doi.org/10.1007/978-3-030-26528-1_40

BIM may be referred to as disruptive technological process that is changing the way construction work is being procured. The Nigerian construction sector is known to be a fragmented industry where professionals have monopolized construction information [4] and work more like a group than a team. BIM is not well established in the Nigerian construction market, but the level of awareness is rapidly progressing. The industry needs this disruptive process (BIM) to improve its productivity and capabilities through integrating the stakeholders' working process. To achieve this, BIM maturity must be established. This piece of work aims at determining the BIM maturity of the Nigerian construction market with the use of the BIM established adoption (macro) models.

2 Background of the Study

In 2015, Succar and Kassem developed five new conceptual constructs for assessing BIM adoption at macro (country) level. The developed models were subsequently refined as conceptual tools, developed additional assessment metrics to assist researchers and policy makers to analyze and improve or develop BIM diffusion policies within a market [5]. The developed macro-BIM adoption models include: Macro-Diffusion Responsibilities model; Macro-Diffusion Dynamics model; Diffusion Area model; Policy Actions model; and Macro-Maturity Components model.

Thus, the Nigerian macro-BIM adoption study aims to assist the policy makers in developing and or assessing the macro BIM diffusion policies, strategies and plans within the Nigerian construction market. Sequel to the completion of the assessment, the study aimed to achieve deliverables at Initiation Phase of Policy Development and specifically the development of a seed BIM policy framework and engagement with stakeholders. Finally, assessment and planning of diffusion roles are generated through mapping the macro player groups and the macro maturity components.

Having the macro BIM maturity models as one of the most cited and used maturity models [5, 6] and already applied in several countries like Peru, Russia, Ireland, Egypt, Spain, Hong Kong and Brazil (BIMexcellence); ultimately, the macro maturity model is considered as the most viable method to assess BIM adoption at macro scale.

As part of a process to develop a strategy for effective BIM adoption in the Nigerian construction market, the market maturity should be assessed ahead of policy development or adoption guide. There are several maturity models ranging from assessment of the derived benefits of BIM utilization [7] to the capability of National BIM Standards model that deals with BIM tools and maturity levels [8] and BIM proficiency matrix by Indiana University [9] etc. In spite of their derived benefits within their individual settings, the models do not offer full understanding of how BIM diffuses at macro level or comprehensive macro-BIM adoption [10].

This study is carried out mainly to assist the researcher in the development of a working strategy for an effective BIM adoption. Therefore, assessment of current market specific on BIM diffusion policies becomes necessary, and the developed macro-BIM maturity models by [11] or framework is thus adopted. The adopted framework consists of five conceptual models as illustrated in Succar & Kassem [11].

The precedence set by these models in their application of establishing BIM adoption at macro level ensured that the adopted framework is appropriate to achieve the researchers' objectives.

3 Nigerian Macro Maturity Model

Structured questionnaire was used as tool for data collection [6] hosted online using google forms. In addition, snowball method was adopted in targeting the survey respondents due to low level of BIM awareness and maturity in the country [12]. A few number of BIM experts volunteered to participate in the survey and subsequently more participants were recorded through them (initial respondents) – snowball. Thirty Seven (37) valid responses were recorded and analyzed quantitatively (see Table 1). This study was "market" specific; and the target was establishing the level of BIM "diffusion and adoption" in Nigeria.

Table 1. Profile of respondents *(field survey, 2018).*

Variable	Characteristics	Freq.	Percentage (%)	Total
Location of practice	North-Central	18	48.6	37
	North-East	2	5.4	
	North-West	8	21.6	
	South-East	2	5.4	
	South-South	3	8.1	
	South-West	4	10.8	
Years practice	<5 years	12	32.4	37
	5–10 years	13	35.1	
	11–15 years	5	13.5	
	>15 years	7	18.9	
Number of employees	<10 personnel (micro)	21	56.8	37
	10–50 personnel (small)	12	32.4	
	50–200 personnel (medium)	3	8.1	
	>200 personnel (large)	1	2.7	
Profession	Architecture	14	37.8	37
	Building engineering	1	2.7	
	Civil/structural engineering	14	37.8	
	Construction management	0	0.0	
	Electrical engineering	0	0.0	
	Mechanical engineering	1	2.7	
	Quantity surveying	6	16.2	
	Other	1	2.7	

(*continued*)

Table 1. (*continued*)

Variable	Characteristics	Freq.	Percentage (%)	Total
Specialization	Contractor/construction	8	21.6	37
	Designer or consultant	27	73.0	
	Client	1	2.7	
	Development authority	1	2.7	
Level of BIM utilisation	Modelling only - BIM stage 1	20	54.1	37
	Limited to collaboration - BIM stage 2	12	32.4	
	Up to integration - BIM stage 3	5	13.5	

There are two dominant BIM maturity classifications or capability stages; these are the Succar [13] descriptive BIM capability stages 1, 2 and 3 and the Bew-Richards' BIM maturity levels 0, 1, 2 and 3. The Succar's three-stage capabilities evaluate maturity from the first point of adoption (*POA*) just after the readiness ramp as BIM stage 1 (*modelling* only), to BIM stage 2 (limited to *collaboration*) and BIM stage 3 (up to *integration*). On the other hand, Bew-Richards' UK BIM maturity is prescribed based on levels, BIM level 0, BIM level 1, BIM level 2 and BIM level 3. The level 0 is an unmanaged CAD, predominantly two dimensional CAD system (2D) with paper or electronic paper as dominant information exchange mechanism [14]. Moreover, the level 0 appears to be of the same description of *POA* or pre-BIM in Succar [15], while the BIM level 1, 2 and 3 may be seen to be parallel to or matching with the Succar-Kassem's capability stages 1, 2 and 3 respectively. Going by a wide consideration of BIM capability stages in most of BIM studies, Succar-Kassem's maturity stages is specifically adopted for this section of work as it were in the Macro-BIM adoption conceptual models [11].

3.1 Model A: Diffusion Areas Model

Diffusion area model explains how BIM field types (process, policy and technology) relate with the BIM capability stages (integration, collaboration and modelling) to produce nine diffusion areas where BIM diffusion occurs; thus, such areas can be analysed and planned. Findings here demonstrated irregular distribution of rates (see Fig. 1). Nigeria and Ireland are reasonably mature in applying technology for modelling purpose with a little move at applying technology for collaboration as well as processes at modelling stage. There is very low level of inter-organisational collaboration and no model workflow at both fields not to talk of integration. On the other hand, there has been no policy or mandate by government [16].

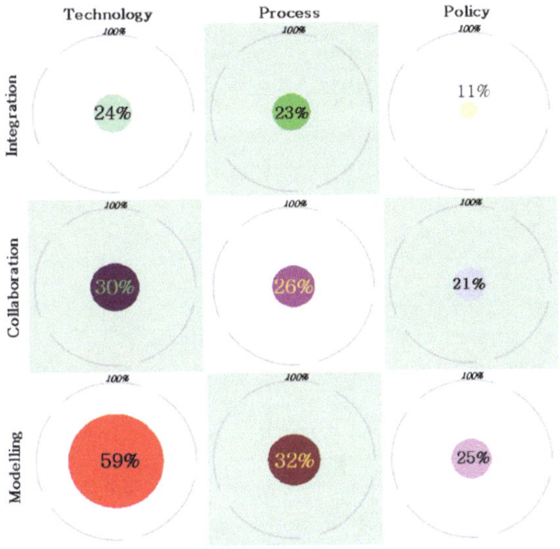

Fig. 1. Diffusion areas model for Nigeria

It is therefore established that by 2018, the diffusion levels of staged capability milestones in the Nigerian construction market are as follows:

25% diffusion rate of modelling capabilities,
21% diffusion rate of collaboration capabilities and
11% diffusion rate of integration capabilities.

3.2 Model B: Macro-maturity Components Model

There are eight complementary components within the Macro Maturity Components model used in measuring and establishing maturity of BIM at country level. The developed and refined components by Kassem [5] are as follows: Champions and drivers; Measurements and benchmarks; Noteworthy publications; Objectives, stages and milestones; Learning and education; Standardised parts and deliverables; Regulatory frameworks and Technology infrastructure.

Figure 2 below illustrates Nigerian maturity components that is, Nigeria's current maturity within each component. These components were assessed with BIM Maturity Index (BIMMI), which has different maturity levels (from the outer to the inner circle) as follows: ad-hoc – low maturity; defined – medium-low maturity; managed – medium maturity; integrated – medium-high maturity; and optimised – high maturity.

The components converge as they mature from *a* to *e* corresponding to *ad-hoc* to *optimised* or *low maturity* to *high maturity*. These components and their maturity index set a very clear description of all the eight components within a market. The closer these components are (converging), the mature they are. Assessments are made holistically based on granularly matrix as to compare relative maturity of one component over the

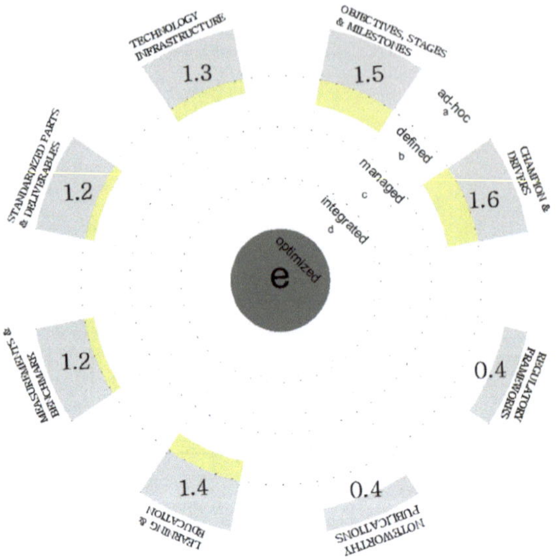

Fig. 2. Macro-maturity components model for Nigeria

other as prescribed in [11 Table 11]. Successively, each component is evaluated using component-specific metrics as described in [11 Table 3–10 pp. 70–72].

Nigerian construction market appears with a dominant '*medium-low*' maturity. Champions & drivers leading with 1.6 (between medium-low and medium maturity) on a Likert scale of 5 corresponding to maturity levels of *a, b, c, d* and *e* as prescribed above. These indicated that the components tangle between '*defined*' and '*managed*' levels (*b* and *c*), as such all the components needed a push. The evaluation suggests an early adopter with individuals as champions promoting the new concept.

Moreover, ranking regulatory framework lowest is an indication that government lacks policy consideration in this regard; and pending when regulatory requirement is considered, most of these components will not advance.

3.3 Model C: Macro-diffusion Dynamics Model

The macro-diffusion dynamic model was adopted [11 p.72 Fig. 7] primarily to assess the adoption trend within a market and compare with the directional pressures to how diffusion unfolds within a specific market. This model comprises three diffusion dynamics namely: Bottom-Up; Middle-Out and Top-Down [11]. Moreover, this model sets four directional pressure mechanisms who are laid over the three diffusion dynamics; these include Downwards, Horizontal Downwards, Upwards Horizontal and Upwards Horizontal pressures.

The study reveals Nigeria's diffusion dynamic as predominantly bottom-up, by '*majority*' response [11]; this result indicated smaller organisations are those pushing the adoption in the industry but not the bigger firms or the government. However, the bigger organisations seem to be picking up as the result suggests their suit.

The bottom-up diffusion dynamic assured the transmission by small organisations in an upward horizontal pressure mechanism with industry bodies, larger and other small organisations as pressure recipients and potential adopters. With current lack of policy in place [16] and unwillingness from most of the bigger companies to embrace the BIM concept, the bottom-up diffusion dynamic would possibly continue.

3.4 Model D: Policy Actions Model

The policy action model (Fig. 3) has nine policy actions generated from mapping the three implementation approaches (passive, active and assertive) and the three implementation activities (communicate, engage and monitor) [11]. Succar [11] developed this model as an assessment tool to generate activities/task, which are used in comparing policy actions across many countries for a structured policy intervention in achieving a market-wide BIM adoption.

The Nigerian policy action pattern recorded a full *active* with a partial *assertive* at engagement stage (see Fig. 3). This suggests government intervention at both engagement and monitoring stages. Moreover, incentivise and enforce (Fig. 3) are mostly prescribed by government/regulations. Therefore, the practitioners desired active government involvement approach.

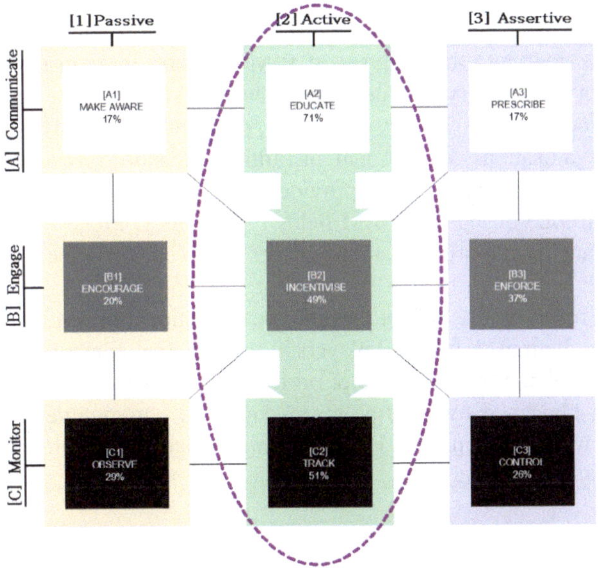

Fig. 3. Policy actions model of Nigeria

The evident result of diffusion of innovation within smaller organizations (bottom-up) has considerable influence in the behaviour of the bigger organizations or higher end of the supply chain [17].

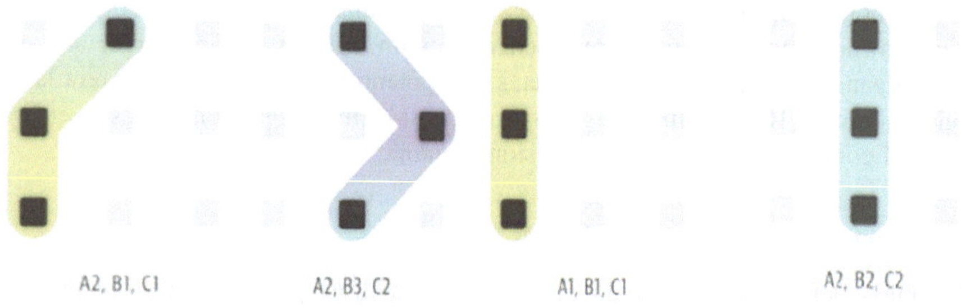

Fig. 4. Policy actions models of USA, UK Australia and Nigeria respectively

There are series of policy action model patterns at various country specifics that go along vertical stripe, alternating within passive, active and assertive action along the three implementation activities. For example, Fig. 4 above presents different sets of policy action models of USA (A2, B1, C1), UK (A2, B3, C2), Australia (A1, B1, C1) and Nigeria (A2, B2, C2).

3.5 Model E: Macro-Diffusion Responsibilities Model

The established BIM field types have their respective capability sets (that differ base on BIM stage) as group of players within construction industry and across the BIM field types [13]. This goes into the analyses of BIM diffusion through the players' (stakeholders) roles in the industry as a network of actors [11]. The nine player groups are: technology advocates, communities of practice, policy makers, individual practitioners, construction organisations, educational institutions, technology developers, industry associations and technology service providers (Fig. 5). Any of the player groups is either belongs to one of the three BIM fields type (Policy, Process & Technology) or intersection of any two; more to that, any player group has a number of player types as well.

The survey result reveals that at present, the educational institutions and individual practitioners are the most influential players in the Nigerian construction market. In the same vein, construction organisations & professional associations were acknowledged as key process players. However, policy makers and communities of practice were lowest players within this market. Figure 5 demonstrates the results of the model.

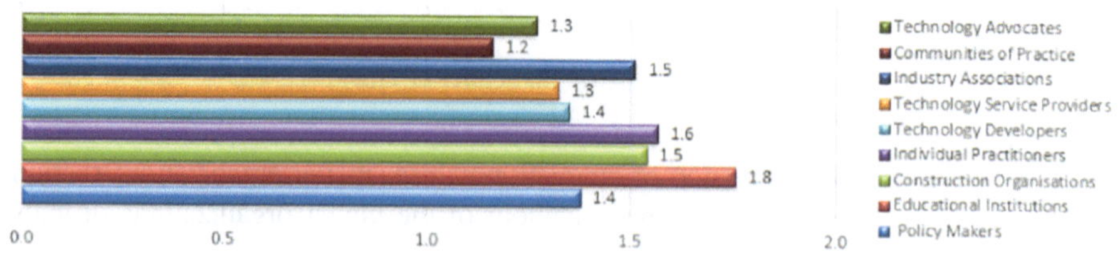

Fig. 5. Nigerian macro-diffusion responsibilities model

4 Development of BIM Policy Plans and Templates

The above models as equally explained in [10] have assisted in deeper understanding of BIM maturity in the Nigerian construction market and equally revealed grey areas where attention is needed. Succar and Kassem demonstrated how these models are utilised to provide basis for the BIM roadmap development at national level.

The policy plan is developed through three phases (Initiation, Consultation and Execution). However, due to a limitation to this study, only the first two (initiation and consultation) phases were dealt with for now.

4.1 Initiation Phase

The initiation phase is determined to institute "*task group*" (as a proposal) and the seed BIM Framework that will act as guidance to the National Framework. The following are set as applications of the three models (B, C and D) at initiation stage: model B is used in assessing BIM maturity or worldwide efforts, model C is used to identify the market specific diffusion dynamic and model D is used to establish a policy approach to be taken by policy makers.

The first part of the Initiation phase is the establishment of a task group; this includes the development of goals for the group and their corresponding objectives. There is currently no organisation taking similar responsibility in Nigeria. This research has been working to establish the BIM maturity within the Nigerian AEC industry for about 2 years. This has been attained through direct contact with Higher Education Institutions [18]; direct contact with the construction professionals (in the last 5 months of 2018); and direct discussions with some industry stakeholders [16].

As illuminated in [5 p. 294 Fig. 5], the task group targets the development of a seed BIM policy framework, where this section is considered achieved quantitatively in this study. The framework development involves investigation into similar efforts around the world and identify a suitable model approach to domesticate. Finding from the application of macro maturity components model on 21 countries suggests UK's framework as the strongest [10].

BIM has well-established guidance and workflows in the UK as such, those who adopted BIM concept in Nigeria considered UK BIM protocol as a source of guidance. The respondents largely agreed that the UK model provides substantial guide once adopted. Other potential countries that are worth learning from are USA and Australia, they have potentials in technologies and terminology, and their BIM participation at the world stage and availability of noteworthy BIM publications [19] are eminent. Any remodelled framework for the study context must certify acceptability to the country and its ecosystem.

The sequential input by model C and model D are explained based on the survey findings; as such, model C (diffusion dynamics) identified the Nigerian market diffusion dynamic as predominantly bottom-up. This will subsequently influence the next input (model D – policy approach). The policy approach as presented in model D [5 p. 294] is mostly active, hence putting further pressure on the proposed BIM framework whose smaller organisations are currently leading.

Although there is no mandate in place, there is still a substantial awareness mostly at lower or individual level. The awareness in the education sector is moving very fast since the launch of BIM Africa Student Advocacy Program mostly patronised by Nigerian students of AEC related courses. This program (initiated in 2018) is serving as a medium to create awareness and training to students of higher institutions. Although, the rudimentary training can increase awareness; however, some of the critical issues that will subsequently arrive are the availability of up to date software and BIM expert for training as multi-disciplinary class [18]. The organisational BIM adoption represents discrete approaches that need profound consultations with the professional stakeholders to confirm the level of execution, successes and challenges.

Primary website development as source of valuable information for the Nigerian AEC industry is the last stage of the initiation phase. This portal/website also serves as a medium for awareness, guidance and source of Noteworthy BIM Publications (NBPs). AEC related professional regulatory bodies and National Information Technology Development Agency (NITDA) are the key players in this aspect.

4.2 Consultation Phase

The consultation phase is explained as a stage where seed BIM framework is finally refined and transformed into a roadmap. The roadmap has a set of responsibilities that are assigned to selected stakeholders for action [5 p. 295 Fig. 6]. Model E is then deployed with performance indicators and timeframes. The initial stage involves identifying (from the survey undertaken) experienced stakeholders and conducting face-to-face interviews as a replacement to the round-table discussions and workshops [10]. As a result, this process aids in capturing of challenges and recommendation of the stakeholders as well as identifying champions at implementation stage.

The diffusion responsibility model helped in identifying sectors.and areas where the Nigerian construction industry is lacking the needed attention as priorities are also considered; adequate resources are to be provided all through as a recommendation. A roadmap is therefore designed with crucial dates and milestones labelled and connected to policy deliverables through a Macro Roadmap Template generated in 2017 as explained by Kassem [5 p. 296 Fig. 8].

5 Conclusions

The findings of this study provided the Nigerian construction industry's stand with regard to current BIM adoption and significant information where the country is lacking that has to be addressed in order to advance in macro adoption. As such, these include the following: the low diffusion level of 11% and low maturity components (especially in regulatory framework and NBPs) as suggested from the 'bottom-up' dynamic due to lack of regulations. Moreover, the active policy approach is also suggested as the government participation became paramount. This piece of work also demonstrated briefly how the findings could be used further to develop a roadmap for an effective BIM adoption in Nigeria. A proposed roadmap will reflect these findings and some other challenges that are not mentioned here through a series of recommendations based on other results from subsequent collected data. However, the execution phase remains out of this research scope and will require substantial resources to ensure its realization.

References

1. Hore, A., McAuley, B., West, R.: BICP global BIM study. CitA Ltd. (2017)
2. McAuley, B., Hore, A., West, R.: BICP global BIM study-lessons for Ireland's BIM programme (2017)
3. Kassem, M., Amorim, S.B.: Building information modeling no Brasil e na União Européia. Ministério do desenvolvimento, indústria e comércio exterior (MDIC), Brasília (2015)
4. Onungwa, I.O., Uduma-Olugu, N.N.E.Z.I., Igwe, J.M.: Building information modelling as a construction management tool in Nigeria. WIT Trans. Built Environ. **169**, 25–33 (2017)
5. Kassem, M., Succar, B.: Macro BIM adoption: comparative market analysis. Autom. Constr. **81**, 286–299 (2017)
6. Yilmaz, G., Akcamete-Gungor, A., Demirors, O.: A review on capability and maturity models of building information modelling. In: LC3 2017: Volume I – Proceedings of the Joint Conference on Computing in Construction (JC3), 4–7 July 2017, Heraklion, Greece, pp. 629–638 (2017)
7. Barlish, B., Sullivan, K.: How to measure the benefits of BIM—a case study approach. Autom. Constr. **24**, 149–159 (2011)
8. National Institute of Building Sciences: National BIM Standard: Verson 1 Part 1: Overview, principles and methodologies, BuildingSMART Alliance (2007)
9. Indiana University Architects Office: BIM Guidelines & Standards for Architects, Engineers, and Contractors, Indiana University (2009)
10. Hore, A., McAuley, B., West, R., Kassem, M., Kuang, S.: Ireland's BIM macro adoption study: establishing Ireland's BIM maturity (2017)
11. Succar, B., Kassem, M.: Macro-BIM adoption: conceptual structures. Autom. Constr. **57**, 64–79 (2015)
12. Hamma-adama, M., Salman, H.S., Kouider, T.: Diffusion of innovations: the status of building information modelling uptake in Nigeria. J. Sci. Res. Rep. **17**(4), 1–12 (2018). https://doi.org/10.9734/JSRR/2017/38711
13. Succar, B.: Building information modelling framework: a research and delivery foundation for industry stakeholders. Autom. Constr. **18**(3), 357–375 (2009)
14. BIM Industry Working Group: A report for the government construction client group building information modelling (BIM) working party strategy paper. Communications. London, UK (2011)
15. Succar, B., Kassem, M.: Building information modelling: point of adoption. In: CIB World Conference Proceedings, vol. 1, May 2016
16. Hamma-adama, M., Kouider, T., Salman, H.: State of building information modelling (BIM) adoption in Nigeria. Working paper presented at the 34th Association of Researchers in Construction Management (ARCOM) Conference, Belfast, United Kingdom, pp. 334–343 (2018)
17. Geroski, P.A.: Models of technology diffusion. Res. Policy **29**(4), 603–625 (2000)
18. Hamma-adama, M., Kouider, T., Salman, H.S.: Building information modelling uptake: tool training in Nigeria. Open Sci. J. **3**(3), 1–17 (2018). https://doi.org/10.23954/osj.v3i3.1728
19. Kassem, M., Succar, B., Dawood, N.: A proposed approach to comparing the BIM maturity of countries. In: Proceedings of the CIB W78 2013 30th International Conference, Beijing, China, 9–12 October 2013

Criteria to Be Considered in the Selection of Building Materials for Sustainable Housing Delivery

Bimpe Alabi(✉) and Julius Fapohunda

Department of Construction Management and Quantity Surveying, Faculty of Engineering, Cape Peninsula University of Technology, Cape Town, South Africa
alabim2010@yahoo.com

Abstract. Among the numerous factors that affect sustainable housing delivery is the inappropriate selection of building materials. It is significant to know the kind of building material to been used in a specific application, it is essential to know the measures to be taken for choosing such materials. Therefore, for a sustainable urban future and to avoid ecological loss, it is vital to choose materials wisely. Quantitative research approach was adopted to ascertain factor to consider in the selection of building materials for sustainable affordable housing. Construction stakeholders and site contractors are the selected sample to represent the population for the purpose of this study. The findings show the factor to be considered in the selection of sustainable building materials under the sustainability categorised are maintenance cost, health and safety, energy consumption, maintainability and expectancy of material. The identified factors to be considered in the selection of building materials if effectively considered, this would allow construction stakeholders to verify that the building production processes abide by the principles of sustainability from the early stage of building.

Keywords: Building materials · Selection of building materials · Sustainable housing

1 Introduction

According to [1] Sustainable housing means, "Housing which adds to community structure, towards community fairness and cost-effective at a confined level. Sustainability objectives may possibly be attained through considering factors such as environmental effect, economic impact, and market demand [2]. Hence, sustainable housing should bump into a certain number of objects in term of resource and energy efficiency, Cabondioxide and greenhouse gas emissions reduction, as well as pollution prevention, mitigation of noise, improved indoor air quality and harmonization with the environment [3]. Ideally, housing should be less expensive to construct, last endlessly with modest maintenance, and be able to return completely to the earth when abandoned [4]. Therefore, to actually develop sustainable housing, housing essential needs to be socially adequate, economically sustainable, environmentally approachable and technically

C. Aigbavboa and W. Thwala (Eds.): CIDB 2019, *The Construction Industry in the Fourth Industrial Revolution*, pp. 412–418, 2020.
https://doi.org/10.1007/978-3-030-26528-1_41

realistic [1]. This paper aims to investigate factors to be considered in the selection of building materials for sustainable housing.

Building materials remain the most substantial input in construction development because of this, play an undeniably significant role in the delivery of construction projects [5]. Notably, the building is identified by the materials used in its construction and its selection should be done along with sustainability principles to achieve cost-effective in construction [6]. Conversely, an unsuitable selection of materials may exclude the accomplishment of the anticipated sustainability goals [7]. In fact, an appropriate approach of selecting building materials in a sustainable way it's being noted to be the simpler means for construction stakeholders to begin integrating sustainable values in construction projects [3].

2 Criteria for Selection of Sustainable Building Materials

Sustainable development is the introductory principle towards ensuring an affordable life for the current and forthcoming generations. Construction can be said to be sustainable construction if it meets the goal of sustainable development [15]. Further elucidated that sustainable construction is attainable through management and organization of construction processes, material selection and construction approaches and resources consumption.

When choosing building materials to use for construction, it is essential to know the factors that need to be considered for selecting such materials [8]. With the emerging amount of building materials available, every single material has particular efficient demands and complex assembly necessities for building construction [9]. Construction projects are the main contributor to worldwide environmental degradation. Therefore, for a sustainable urban future and to avoid ecologically damages, it is vital to choose materials wisely The overall method for selecting building materials can be broadly divided into five categories [10]: resource efficiency; energy efficiency; water conservation; indoor air quality and affordability.

Resource Efficiency
Resource efficiency simply means the effective use of materials, energy and natural resources to produce materials with reduced resources and environmental impact. Notably, [6] emphasised that the selection of materials should be based on the environmental impact of the materials and this have to be integrated at the early stage of project and design stages. According to [9], resource efficiency can be made proficient by using a material that runs into one or more of the following criteria: recycled content, locally available, retrieved, remanufactured, and certainly durable [1]. Stated that the significance of reusing and recycling the building parts is the saving energy. He further verified using recycled materials to produce new building materials consumes less energy than using new materials.

Energy Efficiency
Energy efficiency is an imperative feature in enhancing the environmental sustainability of materials [11]. Thus, the main purpose of using materials which are energy efficient is to lessen the volume of generated energy in a structure. To achieve optimal energy

efficiency in building it is good to use materials or components that have less energy intake in a building [9].

Indoor Air Quality (IAQ)
In order to meet the wants of the existing a generation without conceding the ability of forthcoming generations to meet their own wants, taking cognizance of the quality of the air we breathe inside the building has gained considerable attention over recent decades [1]. Materials applied in the interior of a building can be a source of pollutants, significantly affecting the indoor air quality, which in turn is harmful to the health and comfort of building occupants [11]). According to [9] indoor air quality can be improved by using materials that adhere to these factors: little toxic, minimal organic releases, low-VOC assembly and dampness resilient.

Water Conservation
Materials that have water conservation features help in reducing the amount of water used on site. In addition, water consumption can be monitored and capped in buildings and landscaped areas by making use of materials that reduce water ingesting as well as increasing water recycling and reuse [9]. [1] highlighted examples of ways by which water can be conserved, by controlling the amount of water through a fixture and recycling water that has previously reached the site or been used. Thus, recycled gray water from sinks, showers, and laundry can be re-designated to flush toilets.

Affordability
Building materials are said to have cost-effective when less expensive than counterpart conventional building materials or within budgeted construction cost as well as low life-cycle cost [9]. Cost-related with energy, replacement, and maintenance costs are significantly reduced with the adoption of life cycle cost (LCC) and essential application of green construction practices. The LCC method is mainly applicable in knowing the cost of a building, which is economically realistic with reduced use cost when compared with another building that has a lower initial cost but greater use cost in terms of maintenance, operating, replacement and repair [11].

Cost Efficiency
The theory of sustainability concerning the construction of buildings relates to the effective cost and reducing investment costs [8]. Client demand on standard quality, lower cost and timely delivery of the project have been a considerable problem within the construction industry. In order to meet client demand, secure better contract and investment return, construction stakeholders should adopt cost efficiency strategies [12].

3 Methodology

The quantitative research approach was adopted to investigate criteria to be considered in the selection of building materials for sustainable housing. Quantitative research always includes the numerical inquiry of data collected by means of some kind of detailed questionnaire. Construction stakeholders (architects, quantity surveyors, engineers, construction managers, project managers, and materials suppliers) and contractors in Cape Town was selected as samples for the research population (Western

Cape Province). The samples were selected by the use of cluster sampling and purposive sampling technique. A total of sixty-nine (69) questionnaires out of (90) questionnaires were retrieved and used for analysis. Quantitative data were analysed with the use of Statistical Package for Social Sciences (SPSS). Respondents were asked to rate the level of importance of the derived criteria based on a four (4) point Likert scale: 4 = Extremely, 3 = important, 2 = Less important, and 1 = Not important.

3.1 Findings

Table 1 presents the positions held by survey participants. The largest groups of the respondents (20.3%) were project managers, followed by quantity surveyors which represented 17.4%; and site managers at 11.6%; construction managers at 13%; 11.6% were sales consultants; contractors at 7.2%, site managers at 5.8%; and architects at 13%. Table 1 present the work experience of respondents in the construction sector, 7.2% of respondents have worked more than five years in the construction industry. Respondents having six to ten years' construction work experience represented 10.1%. A substantial 36.2% of study participants had eleven to fifteen years' experience while 21.7% had sixteen to twenty years' experience, and 24.6% had construction experience of greater than twenty years.

Table 1. Profile of participants.

Profile of participant	Frequency	
Site engineer	8	11.6
Project manager	14	20.3
Construction manager	9	13.0
Sales consultant (building materials)	8	11.6
Contractor	5	7.2
Quantity surveyor	12	17.4
Site manager	4	5.8
Architect	9	13.0
Total	69	100
Work experience		
1–5 years	5	7.2
6–10 years	7	10.1
11–15 years	25	36.2
16–20 years	15	21.7
20 years above	17	24.6
Total	69	100

Criteria for Selection Sustainable Building Materials

Table 2 shows the outcomes for each factors group (e.g. socio-economic, environmental and technological). Respondents were requested to rate how important are the resulting benchmarks constructed on a four (4) point Likert scale: 4 = Extremely important, 3 = Important, 2 = Less important and 1 = Not important. Maintenance cost was rated as the foremost importance in the socio-economic group, with an (mv = 3.36). Energy consumption (mv = 3.34) was also ranked as having major significance under the environmental category, while maintainability (mv = 3.43) had the highest ranking in the technical category and interestingly was the highest among all the selection criteria.

Table 2. Criteria for selection of sustaiable building materials

Criteria	No.	Extremely important (%)	Important (%)	Not important (%)	Less important (%)	Mean value (mv)	Std.D	Rank (r)
Socio-economic criteria								
Maintenance cost	69	42.0	52.2	5.8	0.0	3.36	0.59	1
Health and safety	69	49.3	39.1	8.7	2.9	3.34	0.76	2
Initial cost	69	44.9	39.1	15.9	0.0	3.29	0.73	3
Disposal cost	69	34.8	56.5	8.7	0.0	3.26	0.61	4
Aesthetics	69	29.0	52.2	18.8	0.0	3.10	0.68	5
Labour availability	69	36.2	37.7	18.8	7.2	3.02	0.92	6
Recycle cost	69	29.0	42.0	21.7	7.2	2.93	0.89	7
Use of local materials	69	20.3	49.3	23.2	7.2	2.83	0.84	8
Transportation cost	69	20.3	36.2	40.6	2.9	2.74	0.82	9
Environment criteria								
Energy consumption	69	52.2	30.4	17.4	0.0	3.34	0.76	1
Environmental pollution	69	39.1	44.9	15.9	0.0	3.23	0.71	2
Air quality	69	42.0	39.1	17.4	1.4	3.22	0.76	3
Low or non-toxicity	69	43.5	36.1	18.8	1.4	3.21	0.80	4
Sound disposal options	69	34.8	44.9	18.8	1.4	3.13	0.77	5
Embodied energy in the material	69	34.8	43.5	21.7	0.0	3.13	0.75	6
Environmental statutory compliance	69	31.9	46.4	21.7	0.0	3.10	0.73	7
Green gas emission	69	29.0	50.7	20.3	0.0	3.08	0.70	8
Amount of likely wastage in the use of materials	69	27.5	53..6	17.4	1.4	3.07	0.71	9
Ozone depletion potential	69	27.5	47.8	20.3	4.3	2.99	0.81	10
Potential for recycling as well as re-use	69	24.6	43.5	27.5	4.3	2.88	0.83	11
Technical criteria								
Maintainability	69	49.3	44.9	5.5	0.0	3.43	0.61	1
life expectancy of material (strength, durability)	69	58.0	27.5	11.6	2.9	3.41	0.81	2
Fire resistance	69	42.0	46.4	11.6	0.0	3.30	0.67	3

(continued)

Table 2. (*continued*)

Criteria	No.	Extremely important (%)	Important (%)	Not important (%)	Less important (%)	Mean value (mv)	Std.D	Rank (r)
Resistance to decay	69	37.7	49.3	11.6	1.4	3.23	0.71	4
Energy saving and thermal insulation	69	46.4	29.0	23.2	1.4	3.20	0.71	5
Moisture resistance	69	39.1	40.6	17.4	2.9	3.16	0.82	6
Ease of construction	69	33.3	49.3	17.4	0.0	3.15	0.70	7
Material availability	69	29.0	59.4	8.7	2.9	3.14	0.69	8
Sound insulation	69	21.7	62.3	15.9	0.0	3.06	0.62	9

Table 2 reveals the perception of respondents on socio-economic criteria to be considered for selection of sustainable building materials. Maintenance cost was rated as the number one priority in the selection of sustainable building materials. High maintenance costs have negative effects on stakeholders and the construction industry as a whole [13]. However, building maintenance cost can be reduced through deep knowledge of building maintenance cost theories [14].

The finding also shows that energy consumption, environmental pollution, air quality and low or non- toxicity as the major criteria to be considered in the selection of building materials (Table 2). Effective uses of energy help to reduce energy intake in buildings and facilities [9]. [1] posit using low volatile organic compounds materials for construction can considerably decrease the release of explosive organic compounds. Thus, this is a vital factor to any effective project, also as a significant measure of sustainable construction.

Maintainability with a mean value of 3.43 under technical criteria as the most significant factor to be considered in the selection of sustainable materials. "Maintainability" has been a major fear among designer and engineers in the construction industry. Thus, preservation buildings are gradually to search by clients, to cut the running costs related to buildings [1]. In addition, the finding reveals life expectancy of material in term of strength and durability, fire resistance, resistance (Table 2) to decay as other notable factors to be considered in the selection of sustainable building materials.

4 Conclusion and Recommendation

The factors to be considered in the selection of building materials were further subdivided into a socio-economic criterion, environmental criteria and technical criteri based on sustainability. Effective considerations of building materials before selection in construction towards enhancing sustainable housing delivery based on socio-economic, environmental and technical criteria is a method to ensures that the sustainability supply chain requirements are met at the beginning of the phase of the project. Thus, this would allow the construction stakeholders to verify that the building production processes abide by the principles of sustainability from the early stages of the building.

Proficiently consideration of some factors when selecting materials will safeguard sustainable improvement in the building project. Maintenance cost, energy consumption, and maintainability as substantial factors to be considered in the selection of sustainable building materials this will enhance the delivery of affordable housing. Thus, Effective considerations of building materials before selection in construction towards enhancing sustainable housing delivery based on socio-economic, environmental and the technical criterion is a method to ensures that the sustainability supply chain requirements are met at the beginning phase of the project.

References

1. Akadiri, O.P.: Development of a multi-criteria approach for the selection of sustainable materials for building projects. Ph.D. Thesis - University of Wolverhampt, pp. 1–437 (2011)
2. Ljungberg, L.Y.: Materials selection and design for development of sustainable products. Mater. Des. **28**, 466–479 (2007). https://doi.org/10.1016/j.matdes.2005.09.006
3. John, G., Clements-Croome, D., Jeronimidis, G.: Sustainable building solutions: a review of lessons from the natural world. Build. Environ. **40**, 319–328 (2005). https://doi.org/10.1016/j.buildenv.2004.05.011
4. Farid, M.A.: Saving energy consumption in the phase of building saving energy consumption in the phase of building construction, a step forward (2018). https://doi.org/10.3217/978-3-85125-389-4
5. Akanni, P.O., Oke, A.E., Omotilewa, O.J.: Implications of rising cost of building materials in Lagos state Nigeria. SAGE Open. **4** (2014). https://doi.org/10.1177/2158244014561213
6. Abeysundara, U.G.Y., Babel, S., Gheewala, S.: A matrix in life cycle perspective for selecting sustainable materials for buildings in Sri Lanka. Build. Environ. **44**, 997–1004 (2009). https://doi.org/10.1016/j.buildenv.2008.07.005
7. Flórez, L., Castro-Lacouture, D., Irizarry, J.: Impact of sustainability perceptions on the purchasability of materials in construction projects, pp. 1447–1456 (2010). https://doi.org/10.1061/41109(373)145
8. Akadiri, P.O., Olomolaiye, P.O.: Development of sustainable assessment criteria for building materials selection. Eng. Constr. Archit. Manag. **19**, 666–687 (2012). https://doi.org/10.1108/09699981211277568
9. Čuláková, M., Vilčeková, S., Katunská, J., Krídlová Burdová, E.: Multicriteria decision analysis of material selection of high energy performance residential building. Sel. Sci. Pap. - J. Civ. Eng. **8**, 103–114 (2013). https://doi.org/10.2478/sspjce-2013-0023
10. Froeschle, L.: Criteria for green building materials.pdf
11. Das, S.E.: Thinking beyond …the box. In: Dynamic Modelling Approach for Designing Sustainable Green Buildings. Department of Built Environment Central University of Technology, Free State Private Bag X20539 Bloemfontein 9300, South Africa (2013)
12. Dzeng, R.-J.: The cost efficiency of construction industry in Taiwan. Open Constr. Build. Technol. J. **6**, 8–16 (2012). https://doi.org/10.2174/1874836801206010008
13. Akinyede, I.J.: Framework for effective management of cost constraint on building project delivery in South Africa (2014). http://etd.cput.ac.za/bitstream/handle/20.500.11838/1063/213209667_Akinyede_IJ_MTech_const_eng_2014?sequence=1&isAllowed=y
14. Ogbu, C.P.: Building quality condition and maintenance cost : the case of public low income housing in Abia State, Nigeria. J. Constr. Bus. Manag. **1**, 24–34 (2017)
15. John, G., Clements-Croome, D., Jeronimidis, G.: Sustainable building solutions: a review of lessons from the natural world. Build. Environ. **40**(3), 319–328 (2005)

Contracting Firms Compliance to Health and Safety Measures on Construction Sites in Nigeria

Olusola Festus Akinradewo(✉) and Bamidele Temitope Arijeloye

Department of Quantity Surveying,
Federal University of Technology, Akure, Ondo State, Nigeria
{foakinradewo,btarijeloye}@futa.edu.ng

Abstract. Every year, millions of people around the world die due to construction works related accidents. For this reason, this study evaluates adherence of construction organisations to health and safety practices while executing construction projects on site with a view to improving compliance with existing policies. The study was done by distributing well-structured questionnaires to external and in-house construction professionals who have been involved in ongoing construction projects at the Federal University of Technology, Akure. Percentile and mean item score were used to analyse the retrieved data. Results indicates that the usage of health and safety tools is very low by the construction firms because 12 out of 20 identified health and safety tools have mean scores below average. The study concluded that construction professionals are highly aware of health and safety regulations in the study area. The study recommended that a regulatory body to oversee enforcement of adherence to health and safety policies should be commissioned and empowered by the government.

Keywords: Compliance · Construction firms · Health and safety · Measures and tools

1 Introduction

The construction industry, as noted by so many researchers, is one whose level of risk is very high. Gharibi [1] noted that construction sites pose a lot of risks to workers' life and therefore, it is part of the highly ranked risky workplaces in the world. Similarly, [2] noted that the construction industry frequently record high rate of hazards to practitioners, construction workers and eventually the end users with accidents. Muiruri [2] further stressed that all over the world, construction workers have twice or thrice tendency to die or sustain dangerous injury while working than workers in other sectors. In addition, [3] stated that the industry experience double of accidents and fatality rates of other sectors on average, with an increase in the rate of minor accidents. Furthermore, [4] noted that the construction industry is widely acclaimed to be dangerous and highly risky due to the out-of-proportion incidence of accidents that takes place on construction sites.

There are wide ranges and probable hazards during construction works which include trips, slides, dips, hand arm vibrating syndrome, noise, collapse, respiratory

C. Aigbavboa and W. Thwala (Eds.): CIDB 2019, *The Construction Industry in the Fourth Industrial Revolution*, pp. 419–427, 2020.
https://doi.org/10.1007/978-3-030-26528-1_42

diseases, electrocution among others [5]. The reasons for high-risk level in construction, according to [6] arises from the nature of work operations in construction, construction methods, physical environment, heavy equipment in use, materials used for construction, coupled with the construction project physical properties. Hassan [7] in addition noted that the modus operandi of construction work is very dynamical and composite. Hassan [7] further noted that this is due to the different phases of construction such as the planning and viability phase, design phase, demolition and clearance phase, execution phase, commissioning phase among others. Putting these facts into view, there is a necessity to adopt construction site health and safety (H&S) measures. It is worthy of note that due to the facts mentioned above, the construction sector is perceived to be the hardest in implementing Health, Safety and Environment (HSE) measures [8].

Considering the construction sector's contributions to the economy of the country, there is a need for stakeholders to seriously look into the issue of construction H&S. If it is left unchecked, it will not only bring about economic loss but also affect the growth rate of investment. It would thus be to the advantage of organisations, construction industry stakeholders and the Government to invest in any system and processes that can decrease incident rates. Thus, it is on this note that this study aims at evaluating the adherence of construction firms to H&S policies on construction project sites.

2 Literature Review

H&S in the construction industry deal with the psychological and physical health and well-being of labourers and professionals on the construction site together with other individuals who can be at health risk. It is, therefore, a necessity for clients, employees, stakeholders, governments, employers, and general project participants to embrace the implementation of H&S policies [2]. Construction H&S management covers the activities a project/construction manager can engage in to train workers and motivate them to practice safe and productive construction activities. Workers' H&S are critical components in work environment. According to [9], a work environment that is safe reduces the level of absenteeism, the cost spent on companies' insurance bills but increases productivity. Globally, people die more while at work than when at warfront. International Labour Organization [10] reported that on average, about 2.3 million workers die yearly due to diseases and accidents that are work-related with about 360,000 resulting from disastrous accidents while about 1.95 million results from work-related diseases. This can be explained to indicate that after each day at work, workers up to 1 million will account for accidents while about 5,500 workers die at their workplace. H&S have been an issue to study since the development of a construction site and other related work settings [11].

"Occupation Safety and Health (OSH) also known as Occupation H&S (OHS) or Workplace Health and Safety (WHS)", by [12], is an interdisciplinary field that focuses on guaranteeing the health, welfare, and safety of workers while in their workplace taking others who may be partakers directly or indirectly into consideration. This is why [13] submitted that OHS covers the social, mental, as well as the physical welfare of workers in the workplace.

2.1 H&S on Construction Sites in Developing Countries

Construction activities in developing countries are considered dangerous, and a quick inspection of a construction site will show various H&S risks. Dutton [14] noted that H&S is a worldwide phenomenon which requires serious attention most especially by small and medium organisations. Unfortunately, the importance of practicing construction H&S in developing countries has been undermined. Farooqui [15] rightly noted that developing countries are faced with significant H&S challenges. The fact that construction work environments are considered to be extremely hazardous does not equally indicate that its proneness to accidents cannot be controlled. According to [16], the risks and hazards are largely dependent on "work situation" or "working environment" which can be controlled humanly. This can be ascertained from the safety reports and records of construction industries from advanced countries. Thus, the grounds on which [17] and [18] expresses worries concerning developing countries' worse safety practices still continue to surface such as Nigeria. Muiruri [2] opined that factors standing as barriers to managing developing countries' H&S in construction sector include "poor infrastructure; problems of communication due to low literacy level; unregulated practices on construction sites; adherence to traditional methods of working; non-availability of equipment; extreme weather conditions; improper use of equipment and corruption". The prevalence of OHS issues in developing countries has been associated with not giving enough attention to OHS by stakeholders, practitioners and government [19]. Puplampu [19] further opined that the causes might be as a result of more emphasis being laid on profitability and productivity with H&S procedures, standards, and policies compromised.

A lot of international and non-governmental companies mostly ask the reason behind most African countries struggling to promote effective OHS in the workplace. The irony of the situation, as noted by [15] indicated that there is an inadequacy of OHS systematic procedures while business owners and construction industry stakeholders are not devoted to OHS measures. This has led to accidents being under-reported in some of these African countries such as Nigeria and Ghana [18, 20] as well as the allocation of little resources to OHS [18]. In addition, [21] noted that developing countries' construction environment experiences issues that serve as barriers to implementation of OHS on construction sites among which are insecurity, corruption, and poor safety culture. Furthermore, poor regulation of OHS in developing countries like Nigeria results in a lackadaisical and indifferent attitude of construction professionals and workers towards OHS [21]. Similarly, [2] noted that the culture of developing countries' construction industry has not given room for encouraging H&S practices. Procurement process that uses competitive tendering which leads to the lowest bidder being awarded most public contracts obliges contractors to beat their rates down by taking out some cost aspects of the project thereby affecting OHS incorporation in many developing countries [2].

According to [22], although Nigeria participated in the OSH Convention held in Geneva in 1981, the country still falls behind in implementing OHS on construction sites. Furthermore, [18] noted that even the most relevant and up-to-date safety records by contractors in Nigeria indicated considerably increase in number of casualties on construction sites. From a study of forty-two Nigerian contracting organisations, there was poor performance on OHS practices with an average of five injuries per 10 labourer and two accidents per One hundred labourers recorded [18]. This corroborated the submission of [23] that these figures most times are worse in practice as concealment and under-reporting culture is normally done by these contracting firms.

Presently, the Nigerian construction industry has no H&S regulation policy in use which is a major challenge to the adoption and effectiveness of H&S practices in the industry.

3 Research Methodology

Research methodology according to [24] is an organization of conditions, a blueprint or an outline for data collection, measurement, and analysis, in a way that conflates relevance to the purpose of the research with the economy in procedure. This study, therefore, adopted quantitative research method such that data were retrieved chiefly using a well-structured questionnaire. The questionnaire adopted a five-point Likert scale to measure a range of opinions from the respondents. The research focused on Federal University of Technology, Akure, Ondo State, Nigeria as a case study while the target population included Mechanical and Electrical Engineers, Architects, Quantity Surveyors and Civil Engineers who have undertaken various construction works for the University. It also included the consulting team of professionals who are in the employment of the University at the Physical Planning Unit. Census sampling technique was employed by this study to retrieve data from the respondents. This was used simply because the target population falls between the reasonable size that can be reached with ease. 39 questionnaires were distributed and 27 were retrieved which were found suitable for analysis indicating a 69% success rate according to the submission of [25] that about 20–30% return rate for questionnaire survey is acceptable. Retrieved data was analysed using percentile and Mean Item Score for ranking of the variables based on the 5-point Likert scale.

4 Findings and Discussions

4.1 Respondent's Background Information

From the analysis of data retrieved, respondents' demographic information is presented in Table 1. The table showed that respondents' type of organization ranges between contracting firms, consulting firms and government firms. The table also shows that all respondents work in indigenous organizations. Results from the table also revealed that most of the respondents' year of experience ranges between 6 years to 15 years and they have handled an average of 6–15 projects. This shows that the respondents possess the required experience and expertise needed to provide professional answers to the questions of this research study.

Table 1. Characteristics of respondents

Factor	Variable	Frequency	Percentage
Type of organization	Consulting	7	25.9
	Contracting	11	40.7
	Government	9	33.3
	Total	27	100
Category of organization	Indigenous firm	27	100
Professional background	Building	5	18.5
	Engineering	11	40.7
	Architecture	5	18.5
	Quantity surveying	6	22.2
	Total	27	100
Academic qualification	HND	1	3.7
	PGD	2	7.4
	B-degree	7	25.9
	M-degree	17	63
	Total	27	100
Years of experience	1–5 years	2	7.4
	6–10 years	18	66.7
	11–15 years	6	22.2
	16–20 years	1	3.7
	Total	27	100
Public project handled	1–5	7	25.9
	6–10	10	37
	11–15	5	18.5
	16–20	5	18.5
	Total	27	100

4.2 H&S Regulations Awareness

Table 2 reveals a high awareness level of H&S regulations by the respondents. This is revealed as most of the regulations identified in the study has a mean score of over 4.0. It can, therefore, be inferred that construction workers know about H&S regulations. However, just as it is opined by [26], knowing the H&S regulations does not automatically translate to compliance. Several other factors including management commitment, involvement of workers, enforceable regulatory framework among others must be considered also for project H&S performance to be improved.

4.3 Compliance to H&S Measures in Construction

Table 3 showed how construction professionals comply with H&S measures. Most of the measures identified had a mean score a little bit above average. It is worthy of note that the use of first aid facility scored the highest with a mean score of 4.22. In contrast, results indicated a low level of the report of incidents and near misses as the measure

Table 2. Awareness level of H&S regulations

H&S regulation	Mean	Rank
Ensuring the availability, maintenance, usage and correct storage of safety equipment	4.74	1
Logout and tag of faulty and idle electrical devices	4.63	2
Supervision of employees to ensure the use of protective equipment and safety devices	4.52	3
Restriction of access to electrical equipment and installations to qualified and certified electricians	4.48	4
Having programs to develop and maintain health incentives for workers	4.22	5
Maintenance of emergency phone number, H&S poster, and others	4.19	6
Proper storage and restriction of access to hazardous materials	4.19	7
Regular H&S inspection	4.11	8
Use of safety signs and signals	3.85	9
Use of guardrails and scaffold fall protection for heights above 1.8 m	3.82	10
Report of accidents resulting in fatality or more than 3 hospitalization within 8 h	3.67	11

had a mean score of 2.93. The study also showed that provision for regular training regarding H&S is a little above average having a mean score of 3.26. This is similar to the findings of [26] which opined that the compliance level of professionals to construction H&S is below expectations and lower than the acceptable standard. This implies that there is a need for construction firms to show more commitment to upgrading their knowledge and to the training of workers concerning H&S. Considering the fact that inadequate training of workers particularly in recognition and avoidance of job hazards is one major cause of accidents on site [27], it is necessary that construction firms infuse a convinced behaviour towards H&S. Improvement of safety awareness in their workers through personal adherence to construction H&S regulations, personal contact with workers, training on H&S from time to time, and regularly scheduled safety meetings should be focused on also [28]. Furthermore, it was opined in [2] that a Personal Protection Equipment (PPE) program can be implemented on construction sites. It is believed that "this program would address the hazards present; the selection, maintenance, and use of PPE; the training of employees; and monitoring of the program to ensure its ongoing effectiveness". Failure to do so, according to [21], does not only imply a loss of confidence in those responsible for the training of workers, but also negligence of social obligations and responsibilities.

4.4 Level of Usage of H&S Tools

Considering the usage of various H&S tools by construction workers during construction, Table 4 revealed a wide disparity between the usage of these tools. The level of usage of some tools like Hard Hat and First Aid are very high with mean scores of 4.67 and 4.33 respectively, compared to other tools like restroom, foot and leg protection, eye protection, hand and arm protection with mean scores of 1.96, 1.93, and

Table 3. Compliance of contracting firms in the study area to the various H&S measures to be provided during construction

H&S measure	Mean	Rank
Provision of first aiders and first aid facility	4.22	1
Issuance personal protecting equipment	3.74	2
Compliance with up to date h & s policy	3.67	3
System for keeping up to date with OSH act and its regulation	2.37	4
Enforcement of use of personal protecting instruments	2.36	5
Safety signs and symbols	2.32	6
Program for regular training regarding H&S	2.26	7
Availability system for reporting and recording incidents and near misses	1.93	8

1.78 and 1.74 respectively. This indicates a low level of usage of H&S tools which is contrary to the suggestions given by [2] that the "Personal Protective Equipment (PPE) used in the construction sites include; eye protection and face protection, hearing protection, respiratory protection, hand and arm protection, foot and leg protection, head protection and body and fall protection mechanisms".

Table 4. Level of usage of various H&S tools.

H&S tool	Mean	Rank
Hard hat	4.67	1
Scaffold	4.63	2
Ladders	4.56	3
First aid box	4.33	4
Visibility jacket	4.15	5
Storage facility for hazardous materials	4.04	6
Safety boots	2.89	7
Healthy drinking water	2.67	8
Gloves	2.48	9
Safety and warning signs	2.44	10
Scaffold fall protection and guard rails	2.44	10
Lightning protection	2.44	10
Site fencing	2.33	13
Dust mask	2.22	14
First aid training	2.19	15
Fire safety training	2.00	16
Restroom	1.96	17
Foot and leg protection	1.93	18
Eye protection	1.78	19
Hand and arm protection	1.74	20

5 Recommendations and Conclusions

With the research study findings, conclusion can be drawn that construction professionals are highly aware of H&S regulations in the study area. Construction professionals know that they are to ensure the availability, maintenance, usage and correct storage of safety equipment; log out and tag faulty and idle electrical devices; ensure the use of safety equipment among other regulations. It is therefore recommended that a regulatory body to oversee enforcement of adherence to H&S policies be commissioned and empowered by the government. This regulatory body will be able to effectively monitor and ensure the provision of H&S tools by contracting firms on construction sites. It is also recommended that contracting firms should create avenues for regular training of their workers concerning various issues of H&S while they make provisions for H&S tools for their workers on site during construction. Further study can be carried out to evaluate causative factors of low adherence to H&S measures in the Nigerian construction industry.

References

1. Gharibi, V., Mortazavi, S.B., Jafari, A.J., Malakouti, J., Bagheri, M., Abadi, H.: The relationship between workers' attitude towards safety and occupational accidents experience. Int. J. Occup. Hyg. Occup. Heal. Assoc. IJOH **543514**, 63–145 (2008)
2. Muiruri, G., Mulinge, C.: Health and safety management on construction projects sites in Kenya a case study of construction projects in Nairobi County. In: FIG Congress, pp. 140–146 (2014)
3. Scholes, B.: Construction: the 10 most common health and safety risks. http://blog.papertrail.io/construction-safety-risks
4. Smallwood, J., Haupt, T., Shakantu, W.: Construction health & amp; Safety in South Africa (2008)
5. Mohammed, A.R.: Health and safety management on construction sites (2014)
6. Menzel, N.N., Gutierrez, A.P.: Latino worker perceptions of construction risks. Am. J. Ind. Med. **53**, 179–187 (2010). https://doi.org/10.1002/ajim.20735
7. Hassan, S.A.: Health, safety and environmental practices in the construction sector of Pakistan (2012)
8. Eurostat: Health and Safety at Work Statistics. http://epp.eurostat.ec.europa.eu/statistics_explained/index.php/Health_and_safety_at_work_statistics
9. Moeti-Lysson, J., Boy, R.L.: A study of health and safety practices in small and medium sized enterprises: a case of Botswana. E3 J. Bus. Manag. Econ. **2**, 125–131 (2011)
10. Organization, I.L.: Facts on safety and health at work (2009)
11. Cooney, J.P.: Health and safety in the construction industry: a review of procurement, monitoring, cost effectiveness and strategy. J. Built Environ. **4**, 107–119 (2016)
12. Kalejaiye, P.: Occupational health and safety: issues, challenges and compensation in Nigeria. Peak J. Public Heal. Manag. **1**, 16–23 (2013)
13. Quansah, F.: Occupational health and safety. http://www.ghanabusinesscode.com
14. Dutton, J.E., Ragins, B.R.: Exploring Positive Relationships at Work: Building a Theoretical and Research Foundation. Lawrence Erlbaum Associates Publishers, Mahwah (2007)
15. Farooqui, R.U., Ahmed, S.M., Lodi, S.H.: Assessment of Pakistani construction industry-current performance and the way forward. J. Adv. Perform. Inf. Value **1**, 51–72 (2008)

16. Olutuase, S.O.: A study of safety management in the Nigerian construction industry. IOSR J. Bus. Manag. **16**, 01–10 (2014). https://doi.org/10.9790/487x-16350110
17. Koehn, E., Ahmed, S., Jayanti, S.: Variation in construction productivity. AACE Int. Trans. **14**(20), 54–57 (2000)
18. Idoro, G.I.: Health and safety management efforts as correlates of performance in the nigerian construction industry/Sveikatos Ir Saugos Darbe Valdymo Pastangos Nigerijos Statybų Pramonėje. J. Civ. Eng. Manag. **14**, 277–285 (2008). https://doi.org/10.3846/1392-3730.2008.14.27
19. Puplampu, B.B., Quartely, S.H.: Key Issues on occupational health and safety practices in Ghana: a review. Int. J. Bus. Soc. Sci. **3**, 151–156 (2012). https://doi.org/10.1016/j.shaw.2014.12.002
20. Kheni, N.A., Gibb, A.G.F., Dainty, A.R.J.: The management of construction site health and safety by small and medium- sized construction businesses in developing countries : a Ghana case study. In: Boyd, D. (ed.) Proceedings of the 22nd Annual ARCOM Conference, pp. 273–282 (2006)
21. Umeokafor, N., Umeadi, B., Jones, K., Igwegbe, O.: Compliance with occupational safety and health regulations in Nigeria's public regulatory entity: a call for attention. Int. J. Sci. Res. Publ. **4**, 2250–3153 (2014)
22. Adeogun, B., Okafor, C.: Occupational health, safety and environment (HSE) trend in Nigeria. Int. J. Environ. Sci. Manag. Eng. Res. **2**, 34–37 (2013)
23. Ezenwa, A.O.: A study of fatal injuries in Nigerian factories. Occup. Med. (Chic. Ill) **51**, 485–489 (2001). https://doi.org/10.1093/occmed/51.8.485
24. Kothari, C.R.: Research Methodology: Methods and Techniques. New Age International (P) Ltd., Publishers, New Delhi (2004)
25. Akintoye, A., Fitzgerald, E.: A survey of current cost estimating practices in the UK. Constr. Manag. Econ. **18**, 161–172 (2000). https://doi.org/10.1080/014461900370799
26. Okoye, P.U., Ezeokonkwo, J.U., Ezeokoli, F.O.: Building construction workers' health and safety knowledge and compliance on site. J. Saf. Eng. **5**, 17–26 (2016). https://doi.org/10.5923/j.safety.20160501.03
27. Williams, O.S., Adul Hamid, R., Misnan, M.S.: Accident causal factors on the building construction sites: a review. Int. J. Built Environ. Sustain. **5** (2018). https://doi.org/10.11113/ijbes.v5.n1.248
28. Department of Workforce Services OSHA Division Wyoming: Comprehensive Health and Safety Program for Construction (2014)

Supplementing Quantity Surveying Traditional Services with Arbitration as One of the Alternative Dispute Resolution (ADR) Methods

Xoliswa Mangese and Mart-Mari Els(✉)

Department Quantity Surveying and Construction Management, University of the Free State, Bloemfontein 9300, South Africa
Archerm@ufs.ac.za

Abstract. The quantity surveying profession has once been defined as a profession on the brink, prompting quantity surveyors to reinvent themselves by adopting services outside their professional domain. One of the services they adopted was Alternative Dispute Resolution Methods, one of them being Arbitration. However, it is of concern that most quantity surveyors do not provide Arbitration as an additional service. This gives rise to the importance of the study. The aim of this study was to examine the knowledge and competency level of quantity surveyors regarding arbitration as a plausible cause for the limited number of quantity surveyors providing arbitration as an additional professional service. By making use of questionnaires, the researcher investigated whether the knowledge and competency level were not the cause of the small number of quantity surveyors who are arbitrators.

Keywords: Arbitration · Arbitrator · Alternative dispute resolution · Quantity surveyor

1 Introduction

Quantity surveyors, also known as cost engineers, building economist and estimators, are specialists in the building industry; they oversee expenses and funds of an undertaking, as well as provide advice on procurement to assist with construction agreements and the documentation thereof. They are mostly involved in the following sectors: the government and its parastatals, private-sector building and civil engineering companies (including consulting firms), bank institutions and property developers as well as project management organisations [10].

Cavill [13] issued the article, *The QS: a profession on the brink* in their 1999 building magazine. The magazine directed its own particular review of more than 12,000 quantity surveyors, and the staggering inclination was one of concern about their future. The review demonstrates that QSs are worried about the quick disintegration of their traditional role, and their failure to convince clients that they can fundamentally contribute to their projects.

C. Aigbavboa and W. Thwala (Eds.): CIDB 2019, *The Construction Industry in the Fourth Industrial Revolution*, pp. 428–439, 2020.
https://doi.org/10.1007/978-3-030-26528-1_43

Harun and Torrance [4] suggest that quantity surveyors ought not to position themselves inside the traditional limits of cost management. One of the ways the profession has evolved was by supplementing traditional services with Alternative Dispute Resolution (ADR), in particular arbitration.

The danger of the Quantity Surveying profession becoming extinct prompted them to reinvent themselves by supplementing their traditional professional services with services outside their professional domain, one being Arbitration. However, it is of concern to note that only a few Quantity Surveyors do in fact provide Arbitration as an additional service. This gave rise to the problem of the research study, which states, *"the majority of the QS professionals do not exploit Arbitration as one of the ADR services that they can provide"*.

The research study then raised questions as to why the Quantity Surveying fraternity was not interested in providing arbitration as a service and questioned whether the small number of Quantity Surveyors as arbitrators was due to the competency level required for one to become an arbitrator. The purpose of this study was therefore to examine the knowledge and competency level of Quantity Surveyors regarding arbitration as a plausible cause for the limited number of quantity surveyors providing arbitration as an additional professional service. The objectives that the study aimed to achieve were to determine the cause of the low interest amongst Quantity Surveyors to offer arbitration as an additional service and to investigate the competency level of Quantity Surveyors in offering arbitration services.

2 Overview of the Quantity Surveying Profession

A quantity surveyor is a professional involved within the building environment; their responsibility is concerned with building costs [8]. The Quantity Surveyor obtains his/her qualification through one or a combination of the following routes: formal education, specific training and knowledge. These afford one with a universal set of abilities that are then applied to multiple problems. These primarily transmit to costs and contracts on construction developments.

The primary function of the Quantity Surveyor is to advise the client on project cost. Over the most recent couple of decades, the QS profession has faced many developments. The progressions were motivated by fear of the quantity surveying profession becoming obsolete as a formal profession [7].

A senior partner of Davis Langdon and Everest was cited as saying, *"traditional quantity surveying, especially the calculation of BoQs, is on the way out"*. His statement was confirmed by numerous senior QSs of large firms, who also believed the profession was heading into specialisation, in particular management consultancy in construction. The latter currently allude to themselves as construction economists. This article appeared to spell the end for the Quantity Surveyor as more articles that sang from a similar psalm trailed it. "Adapt or Die" (2000) prompted QSs to discover another specialty in the industry as their conventional role of measuring and producing BoQs was fast becoming obsolete. It alluded to QSs as an 'endangered species' and that it was an instance of survival of the fittest, which means that the fittest would be the surveyor that had acclimatised and differentiated to an evolving industry [13].

A year later (2001), the building magazine published a further article titled "*What is the use of QSs*", expressed over a survey that the broad public practically had no clue of what a QS did. The writer proposed that construction law was a suitable landing spot for expanding a QS's role, mentioning that their cost knowledge could be a valuable tool in dispute resolution. The article additionally took after the subject of others, suggesting that the traditional role was dying [13].

Disputes in the built environment are mostly associated with cost, claims and time. The Quantity Surveyor might be the best-suited professional to deal with these kinds of disputes involving extension of time, disrupt claims, etc., because they are thorough and would advise on the best resolution for the dispute based on the knowledge in terms of cost planning, control and advice. Quantity Surveyors can also take up the arbitrator role, should the dispute be resolved by arbitration [14].

Quantity Surveyors are more acquainted with a project's contract conditions and should always advise their clients to limit disputes arising from the contract [13]. Asaqs.co.za [1] supports these findings by stating that QSs have the knowledge and are equipped in the field of costs and contracts, which enables them to act as arbitrators for disputes arising in the built environment.

Arbitration as the last dispute resolution method is a process whereby an unbiased third party or a group of neutral third parties known as an Arbitrator/s, usually chosen by the parties in disagreement, make a final and binding judgement (award) with regard to a dispute between the parties [2].

Truter [12] further elaborates that the parties to the dispute select an arbitrator of their choosing. Failing to do so within a specified timeframe, they can request the Arbitration Foundation of Southern Africa (AFSA) or the Association of Arbitrators South Africa (AASA) to assign an appropriate and qualified arbitrator. The contract regularly decides the strategy for selecting an arbitrator. Arbitration in South Africa is administered by the Arbitration Act 42 of 1965; however, there must be a valid agreement recorded in writing for the Act to be enforceable. By doing so, this will enable the parties to arbitrate any current disputes or future disputes identifying with an issue determined in the agreement.

Education requirements to become an arbitrator are not standard; the requirements may vary by position (job function) and mostly focus on real-world experience. However, having a degree or work experience in law-related matters is common for arbitrators. Experience is important in that it gives an understanding of the legal influences for decision-making [4].

It was confirmed that as of the 26th of April 2018, only one arbitrator affiliated with AFSA is also a Quantity Surveyor by Profession [9]. It was also confirmed that as of the 24th of April 2018, there are 37 Quantity Surveyor arbitrators who are members of the AASA [6]. It is believed that these numbers do not include candidate QS Arbitrators who are underway to qualify as Arbitrators. The total of 38 Arbitrators is seen to be insignificant compared to the total number of Quantity Surveyors. This implies that there are only 1.73% of Quantity Surveyor Arbitrators, compared to a high number of 2 194 PrQS. This number (2194) excludes CanQS and the unregistered Quantity Surveyors.

3 Research Methodology

The study was conducted by undertaking a quantitative approach. This was achieved through analysing arbitration as well as quantity surveyor services in the literature review, followed by the development of questionnaires. Quantitative analysis is as an approach used to test theories by examining the relationship between variables [3]. Therefore, a quantitative approach was used, because the study attempted to examine the relationship between the lack of knowledge and competency and the small number of arbitrators who are QS.

Questionnaires were the data collection method employed for this study. Close-ended questions based on the literature review were modelled for the participants and administered via email and google forms, as the target group was assumed to be computer-literate, and also to erase any geographical constraints [11]. The questionnaire was divided into four sections. Section A titled "Demographics" consisted of the participants' demographics. Section B titled "Knowledge Level Regarding Arbitration as a Service" tested the knowledge level of the participants regarding arbitration. Section C titled "Interest in Becoming an Arbitrator" asked the participants if they were interested in becoming Arbitrators and what would motivate them to become Arbitrators. The concluding section, Section D, titled "Interest Level in Providing Arbitration as a Service" was a scaled section from 1 (*Strongly Disagree*) to 5 (*Strongly Agree*) with four questions (regarding quantity surveyors' knowledge, expertise and arbitration as a specialised field) wherein the participants were requested to scale their opinions. Research ethics were considered, and the study adhered to ethical rules.

Convenient Sampling (also called availability sampling) was employed for this study. It is one of the most common non-probability sampling methods. It entails the use of those participants who are available or easily/readily accessible for the study [11]. For this paper, the selected individuals were QSs who worked within the built environment.

The QS profession population is large and therefore the entire group of interest was not studied. Subsequently, 100 questionnaires were distributed to Quantity Surveyors.

The data analysis was completed using SPSS 23 (Statistical Package for the Social Sciences) software. The descriptive statistics was rendered with this software and presented illustrating percentiles, mean, standard deviation, and correlation, using the Kruskal Wallis test to show any relationships between the variables.

4 Research Results and Discussion

Section A

In section A respondents were ask about their professional environment, year experience and qualification. From the table below it is clear that the majority of the respondents resides in Gauteng and a total of 67% of the respondents are candidate quantity surveyors. 60% of the respondents has less that 5 years experience in the industry. The majority (67%) of the respondents are working for consulting firms or private companies. 64% of the respondents indicated that their highest qualification is honours degree (Table 1).

Table 1. Respondents' profile

Province of employment	GP 46%	EC 19%	WC 14%	FS 12%	None 10%
Professional registration	CanQS 67%		PrQS 17%		Other 15%
Years experience	<5 years 60%		5–10 years 23%		>10 years 17%
Employment sector	Consulting 67%		Public sector 17%		Contractor 15%
Qualification	Masters 4%	Honours 64%	Bachelor 12%	B.Tech Degree 15%	Diploma 6%

(Source: Mangese, 2018: Adopted from SPSS 23)

Section B

Question 1: Do you know what an Arbitrator is/does?
98.1% of the surveyed respondents knew what an arbitrator is or does, while only 1.9% specified that they were not sure. The finding indicated that all the respondents except one knew what an arbitrator is or does, indicating that they had knowledge about an arbitrator is and what he does. Thus, it can be concluded that the knowledge level was very high.

Question 2: Are you an Arbitrator?
None of the surveyed respondents were arbitrators as all (100%) of them picked "no" as response. This implied that even though they knew what it means to be an arbitrator, as well as what the responsibilities of an arbitrator are, and possibly knowing that a QS can be an arbitrator, none of them was yet one or planned to become an arbitrator. The reason could be because the surveyed respondents lacked the experience or required qualification to be a qualified and practising arbitrator. This implies that traditional quantity surveying services have not yet been fully supplemented with Arbitration as one of the alternative dispute resolution (ADR) methods within the South African context.

Question 3: Do you know any QS who is an Arbitrator?
51.9% of the surveyed respondents picked "Yes", while 48.1% picked "no" as response when they were asked whether they knew any QS who is an arbitrator. This indicated that about half of them knew QSs who were arbitrators, while almost half of them did not know any QS who was an arbitrator.

Question 4: Which of the following do you think qualifies a QS to be an Arbitrator?
Table 2 reveals the responses to what qualifies a QS as an arbitrator from the surveyed quantity surveyors in the study area. The results indicate that 53.8% agreed that only experience qualifies a QS to be an arbitrator; 13.5% agreed that only qualification qualifies a QS to be an arbitrator; while 30.8% of them agreed that both experience and qualification qualify a QS to be an arbitrator. This implies that experience is a major requirement for being an arbitrator, although qualification is also important for a QS to be qualified as an arbitrator.

Table 2. What qualifies a QS to be an arbitrator?

	Frequency	Percent	Valid percent	Cumulative percent
Experience	28	53.8	53.8	53.8
Experience qualification	16	30.8	30.8	84.6
None of the above	1	1.9	1.9	86.5
Qualification	7	13.5	13.5	100.0
Total	52	100.0	100.0	100.0

(Source: Mangese, 2018: Adopted from SPSS 23)

Section C

Question 1: Do you have any knowledge that QS can offer Arbitration as a service?
A high number (80.8%) of the respondents expressed their opinion as 'yes', which means they had knowledge that QS could offer arbitration as a service. Only 7.7% picked 'no' as response, which means that they had no knowledge that a QS could offer arbitration as a service, whereas 11.5% were not sure. From the result of this study, it can be inferred that the respondents surveyed have knowledge that QS can offer arbitration as a service.

Question 2: Having knowledge that QSs can become Arbitrators, would you consider becoming one?
55.8% of the respondents picked 'yes', which indicates that they were interested in becoming an arbitrator based on their knowledge that a QS can become an arbitrator. 26.9% picked 'unsure', which connotes that they were yet to decide, while 17.5% picked 'no', which suggests that they clearly specified their intention that they did not consider becoming an arbitrator, despite the fact that they knew that a QS could become an Arbitrator.

Question 3: What would motivate you to become an Arbitrator?
Table 3 indicates what would motivate the respondents to become an arbitrator. 32.7% believed that becoming an Arbitrator would broaden their portfolio; 19.2% specified that it was an opportunity for them to offer expertise in their field; while 11.5% agreed that becoming an arbitrator would be for financial reward only. Some of the respondents (13.5%) agreed that motivations to be an arbitrator included broadening one's portfolio and offering expertise in their field, while 15.4% of them upheld the fact that what motivated one to be an arbitrator should be more than broadening one's portfolio and offering expertise in one's field, but also include financial reward. It could be inferred that broadening one's portfolio, offering expertise in one's field and financial reward are good motivators for becoming an arbitrator. Some of the respondents also believed that becoming an arbitrator provided the opportunity to solve problems and ensure fair dispute resolution.

Table 3. What would motivate you to become an arbitrator?

	Frequency	Percent	Valid percent	Cumulative percent
Broadening your portfolio	17	32.7	32.7	32.7
Broadening your portfolio; Offer expertise in your field	7	13.5	13.5	46.2
Broadening your portfolio; Offer expertise in your field; Financial reward	8	15.4	15.4	61.5
Financial reward	6	11.5	11.5	73.1
N/A	2	3.8	3.8	76.9
Offer expertise in your field	10	19.2	19.2	96.2
Solving problems	1	1.9	1.9	98.1
To ensure fair dispute resolution	1	1.9	1.9	100.0
Total	52	100.0	100.0	100.0

(Source: Mangese, 2018: Adopted from SPSS 23)

Section D: Question 1-4

The study further identified the expected requirements for a QS to become an arbitrator. These requirements include having knowledge and being equipped in the field of costs, contracts administration, and being equipped to offer services outside their domain. They should not believe that Arbitration is a specialised field and the services should be provided by law-qualified professionals only (Table 4).

Table 4. QS Perception on QS competency to provide arbitration services

	Mean	Std. deviation	Rank
QSs are more knowledgeable and equipped in the field of costs, such that they should be the ones offering arbitration services on building disputes	4.27	0.888	1
QSs are more knowledgeable and equipped in the field of contracts, such that they should be the ones offering arbitration services on building disputes	4.10	0.823	2
QSs are equipped to offer services outside their traditional domain	4.04	0.907	3
Arbitration is a specialised field and the services should be provided by law qualified professionals	3.04	1.371	4

(Source: Mangese, 2018: Adopted from SPSS 23)

From the table above, all the expected requirements were embraced and given overwhelming support as revealed by the mean score, which is far above the average value of 3.5. This includes having knowledge and being equipped in the field of costs (mean score = 4.27), having knowledge and being equipped in the contracts administration (mean score = 4.10), being equipped to offer services outside their domain (mean score = 4.04), and low score for marking arbitration a specialised field, which should be limited to only law qualified professionals only (mean score = 3.04).

The results implied that for a QS to be qualified to become an arbitrator, he/she is expected to have knowledge and be equipped in the field of costs, have knowledge and be equipped in the contracts administration, and be equipped to offer services outside their domain. There should not be any room for an inferiority complex, believing that arbitration is a specialised field that should be limited to law-qualified professionals only.

This question is related to the problem. Determining how the QS profession views their competency level will have a direct effect on the number of QS offering arbitration services, because in most cases people's interests are hampered by their incompetency, which keeps them from exploring their interests.

To test the hypotheses, further analysis using the Kruskal Wallis test was carried out on the data collected to examine the respondents' views, based on the profession/job description (as previously mentioned, through their job function, their knowledge on arbitration can be determined) and the years of work experience (competence) as a quantity surveyor to determine whether there was any correlation between the knowledge and competence level and the low number of QS arbitrators. The results show that the Kruskal-Wallis Test sig. values are greater than 0.05 for the variable that measures what qualifies a quantity surveyor to be an arbitrator. This indicates that there was no significance difference in the views of the respondents based on the profession/job description and years' working experience as a quantity surveyor. This is revealed from the result of the analysis that gave Kruskal-Wallis Test significant values greater than 0.05 for all the variable tested.

Hypothesis 1: Due to the lack of knowledge by QS that they can provide arbitration as a service there is a low number of arbitrators who are QS

Hypothesis 1 test summary

	Null hypothesis	Test	Sig.	Decision
1	The distribution of 1 Do you believe that QS's are more knowledgeable and equipped in the field of costs such that they should be the ones offering arbitration services on building disputes? Is the same across categories of Professional/Job Function categorised	Independent samples Kruskal-Wallis test	.350	Retain the null hypothesis
2	The distribution of 2 Do you believe that QS's are more knowledgeable and equipped in the field of contracts such that they should be the ones offering arbitration services on building disputes? Is the same across categories of professional/job function categorised	Independent samples Kruskal-Wallis test	.289	Retain the null hypothesis

(*continued*)

(*continued*)

Hypothesis 1 test summary				
	Null hypothesis	Test	Sig.	Decision
3	The distribution of 3 Do you believe that QS's are equipped to offer services outside their traditional domain? Is the same across categories of professional/job function categorised	Independent samples Kruskal-Wallis test	.408	Retain the null hypothesis
4	The distribution of 4 Do you believe that arbitration is a specialized field and the services should be provided by law qualified professionals? Is the same across categories of professional/job function categorised	1	.329	Retain the null hypothesis

Asymptotic significances are displayed. The significance level is .05.

Hypothesis 2: The majority of the QS population do not meet the requirements for one to become an arbitrator, consequently, not qualifying to render arbitration services

Hypothesis 2 test summary				
	Null hypothesis	Test	Sig.	Decision
1	The distribution of 1 Do you believe that QS's are more knowledgeable and equipped in the field of costs such that they should be the ones offering arbitration services on building disputes? Is the same across categories of years of experience as a Quantity Surveyor (QS) categorised	Independent samples Kruskal-Wallis test	.446	Retain the null hypothesis
2	The distribution of 2. Do you believe that QS's are more knowledgeable and equipped in the field of contracts such that they should be the ones offering arbitration services on building disputes? Is the same across categories of years of experience as a Quantity Surveyor (QS) categorised.	Independent samples Kruskal-Wallis test	.792	Retain the null hypothesis
3	The distribution of 3. Do you believe that QS's are equipped to offer services outside their traditional domain? Is the same across categories of years of experience as a Quantity Surveyor (QS) categorised	Independent samples Kruskal-Wallis test	.229	Retain the null hypothesis

(*continued*)

(*continued*)

Hypothesis 2 test summary				
	Null hypothesis	Test	Sig.	Decision
4	The distribution of 4. Do you believe that Arbitration is a specialized field and the services should be provided by law qualified professionals? Is the same across categories of years of experience as a Quantity Surveyor (QS) categorised	Independent samples Kruskal-Wallis test	.578	Retain the null hypothesis

Asymptotic significances are displayed. The significance level is .05.

The implication of this is that neither knowledge gained from the profession/job function nor competency influences the perception of the surveyed quantity surveyors on what qualifies a quantity surveyor to be an arbitrator. The results are consistent with the findings, as it was revealed that 80.8% of the respondents had knowledge that QS provided arbitration services. However, this did not influence their interest in pursuing arbitration as a service. Likewise, the findings were also consistent with the literature review, for all categories QSs confirmed that they were competent to perform arbitration services, because their tasks equipped them to offer such services that are outside their professional domain; they had the experience and the qualification.

Therefore, the null hypotheses are both retained, and the alternative hypotheses rejected.

5 Conclusion and Recommendation

In the literature review it was discovered that most disputes in the built environment were due to cost and contract-related issues. One of the methods to resolve disputes is arbitration and the QS are therefore equipped to resolve such disputes. However, most of them were not arbitrators. The research study aimed to investigate the lack of knowledge and competency as the causes of the small number of QS arbitrators. The research study revealed that a high number of the QS population were aware that they could provide arbitration as a service and that their job function and years' experience, together with their qualification, render them competent to offer arbitration services. It has emerged that the lack of knowledge that QSs could offer arbitration services and QSs being competent to render the services were not the reasons for the small number of QSs who are arbitrators. As noted, there could be several reasons for the small number; however, QSs having knowledge had to be ruled out as the first cause, followed by the requirements for QSs to become arbitrators.

The primary objectives of this study was to determine the cause of the low interest amongst QSs to offer arbitration as an additional service and to investigate the competency level of QSs in offering arbitration services. To achieve the objectives, questionnaires were distributed amongst the participants. The questions tested whether the participants knew that they could offer arbitration services, having the knowledge that they could offer arbitration as a service, should they consider becoming arbitrators. Their years' experience and whether they believed that their qualification and job function render them competent to be arbitrators.

The 13 questions posed to the participants were comprised of the participants' demographics, their knowledge level regarding arbitration as a QS service, their interest in becoming arbitrators and providing the service. All the questions were formulated to answer the research questions and consequently the research study achieved its main objectives.

As it was revealed that the small number of QS arbitrators was not attributed to lack of knowledge nor to the competency level of QS, the researcher made the following recommendations:

1. In order to have a successful dispute resolution in the built environment, an expert in the subject matter of the dispute is required. The cost and contract administration experts are none other than QS and therefore there could be a need and a market for QS arbitrators.
2. Having knowledge that a QS can offer arbitration services does not necessarily mean that QSs know exactly what arbitration entails and how it can benefit the construction industry. Therefore, it is important that QSs enlighten themselves on the benefits that arbitration brings about.

Due to the time constraints and the degree level the research study was narrowed to investigate the "first degree" causes of the small number of QS arbitrators. The researcher therefore suggests the following for further research:

1. Investigation into the average cost and contract related disputes that arise in a year to determine if the number necessitates more QS arbitrators.
2. Investigation into the average of cost and contract-related disputes that actually end up in arbitration cases per year; this will also determine whether there is a need for more QS arbitrators. In parallel, investigate the number of contracts that stipulates arbitration as a resolution method because the parties cannot request for arbitration if it is not agreed in the contract.
3. An in-depth investigation on other reasons for the small number of QS arbitrators such as personality requirements, because arbitration is not for the fainthearted.
4. As an extension to the knowledge of arbitration, investigate the curricula of the universities in South Africa to determine the extent to which the universities relate to arbitration as a QS service.

References

1. Asaqs.co.za. (n.d.): What is a quantity surveyor? – Association of South African Quantity Surveyors (ASAQS). http://www.asaqs.co.za/?page=what_is_QS. Accessed 11 Mar 2018
2. Ashman, D.: The impact of alternative dispute resolution (ADR) in employment law. Unpublished Thesis (Master's Degree). West Lafayette, Indiana: Purdue University, p. 13 (2011)
3. Creswell, J.: Research Design, 4th edn, p. 32. SAGE Publications, Los Angeles (2014)

4. Frost, S.: Qualifications for becoming an arbitrator (2018). https://work.chron.com/qualifications-becoming-arbitrator-5992.html. Accessed 17 Oct 2018
5. Harun, H., Torrance, J.V.A.: Drivers of change: New challenges for the quantity surveyors. In: Proceedings of the International Conference on Construction Industry 2006: Toward Innovative Approach in Construction and Property Development, Kuala Lumpur, Malaysia, p. 38 (2006)
6. IBM Corp. IBM SPSS Statistics for Windows, Version 23.0. Released 2015. Armonk, NY: IBM Corp. (2015)
7. Mthembu, H.: (Office Coordinator: Association of Arbitrators South Africa). Correspondence on number of quantity surveyor arbitrators. Bloemfontein, 24 April 2018
8. Ofori, G., Toor, S.: Role of leadership in transforming the profession of quantity surveying. Aust. J. Constr. Econ. Build. **9**(1), 37–38 (2012). https://epress.lib.uts.edu.au/journals/index.php/AJCEB/article/download/3013/3195. Accessed 21 Mar 2018
9. Oyewobi, L.: Diversification and performance of quantity surveyors in Nigerian construction industry, p. 8. ResearchGate, Cape Town (2016). https://www.researchgate.net/publication/303565051_Diversification_and_performance_of_quantity_surveyors_in_nigerian_construction_industry. Accessed 24 Mar 2018
10. Terk, I.: (Registrar: Arbitration Foundation of Southern Africa). Correspondence on number of quantity surveyor arbitrators. Bloemfontein, 26 April 2018
11. Thanikal, J.: MBA in construction economics and quantity surveying (2018). http://ricssbe.org/qs/. Accessed 21 Mar 2018
12. Trochim, W., Donnelly, J.P., Arora, K.: Research Methods: The Essential Knowledge Base. Nelson Education, Scarborough (2015). pp. 65 & 118
13. Truter, A.: Arbitration in South Africa (2017). http://www.polity.org.za/article/arbitration-in-south-africa-2017-05-05. Accessed 12 May 2018
14. UKEssays: The changes to profession from its traditional roots construction essay (2015). http://www.ukessays.com/essays/construction/the-changes-to-profession-from-its-traditional-roots-construction-essay.php?vref=1. Accessed 03 Apr 2018
15. Yeshwanth, B.: Modern quantity surveying- a complete package adding values to projects, p. 3. Yeshwanth Babu Reddy, Australia (2017). https://www.researchgate.net/publication/314152911_Modern_Quantity_Surveying-_A_complete_package_adding_values_to_projects. Accessed 11 Mar 2018

Perceptions of Urban Stakeholders Concerning the Value of Urban Open Spaces in Bloemfontein

Lindelwa Toba(✉)

Department of Built Environment, Faculty of Engineering Built Environment and Information Technology, Central University of Technology, Bloemfontein, South Africa
tobaM@cut.ac.za

Abstract. Different perceptions concerning the value of urban open spaces have created value conflicts in terms of the quest for sustainable neighbourhoods. These value conflicts negate the ability of planners to achieve planning objectives. Addressing the increasing incidence of value conflict management has become imperative for urban open space sustenance. As such, this study seeks to contribute to filling this gap through an exploration of the diverse perceptions of different urban stakeholders as it pertains to the value of urban open spaces in Mangaung Township, Free State Province. A qualitative case study research design is utilized in the study. The accruing data was analyzed using Qualitative Content Analysis (QCA). Findings indicate that urban stakeholders have different perceptions concerning the value of urban open spaces like economic, environmental and social values respectively. These findings have considerable implications for practitioners within the urban planning and socio-economic development praxes in the Municipality and beyond.

Keywords: Participation · South Africa · Urban open spaces · Urban stakeholders · Value conflicts

1 Introduction

Urban open spaces play a critical part in sustainable development of communities. From a sustainability perspective, these spaces contribute to improved environmental conditions and quality of life. Urban open spaces are defined in terms of the space, location, development and function. An open space is "any open space in ownership of the municipality which is situated outside normally built-up areas and …but is not limited to nature reserves, game farms, riverine vegetation and private open space" [1]. Urban open spaces comprise of portions of the urban area that contribute to its ambience through a combination of green and civic spaces. From a town planning perspective, developments must comprise of open spaces even though no criteria are provided concerning the number, location and usage of these spaces. During the planning phase, different land uses are allocated to built-up and non-built up environments. Built-up environments comprise of residential, industrial, commercial areas and the non-built environments are made up of open spaces. Planning of urban open

C. Aigbavboa and W. Thwala (Eds.): CIDB 2019, *The Construction Industry in the Fourth Industrial Revolution*, pp. 440–449, 2020.
https://doi.org/10.1007/978-3-030-26528-1_44

spaces is carried out through, supply and demand approaches. According to Maruani and Amit-Cohen [2], the demand approach to open spaces focuses on the fulfilment of human needs, namely, the attributes of the targeted community in reference to its size, values and preferences, residential distribution and density. But the supply approach focuses on the conservation of natural environment. The main targets to the latter include the selection of the site, the size and amount of open spaces in that community, recreational activities and the design of the site. Urban open spaces provide a vast range of benefits, values, opportunities and challenges globally. They define the sense of space, image and attractiveness as well as the competitiveness of urban areas. Also, they add value to social, economic, environmental, ecological and planning dimensions of urban areas. Value within the urban open spaces' context is described based on its functionality and its impact on the quality of life in the urban context [3]. The perceptions of urban stakeholders regarding urban open spaces include recreational, economic, environmental and housing values. Since the value of urban open spaces is dependent on functionality, these different perceptions lead to the incidence of value conflicts; a conundrum that negates sustenance of urban open spaces.

However, the future of land use planning relies on an understanding of these conflicts and the solutions that will allow them to create sustainable communities [4]. In addition, all these urban stakeholders have roles to play in the planning practice, management of urban open spaces and creation of sustainable communities. Therefore, there is need for the perceptions of the urban stakeholders concerning value of urban open spaces to be gauged and considered during the planning and development of urban centres. This will inadvertently lead to effective design and management of these open spaces. Yet, quite a few studies have sought to explore this phenomenon in this regard. This is the gap which this study seeks to explore and, possibly fill.

2 Theoretical Perspective

In this section, reportage in extant literature concerning the perception of the value of urban open spaces in the urban context and contributions thereof to the concept of creating sustainable communities is reviewed.

2.1 Defining Urban Open Spaces in Sustainable Communities

Urban open spaces are often considered as an integral part of sustainable neighborhoods. Campbell [5] refers to an open space as any unbuilt land found within a designated boundary of a neighborhood which provides direct or indirect benefits to the community. These spaces assume salient roles, both within a democratic society and in social space. In the former, urban open spaces are regarded as "places where democracy is worked out, quite literally, on the ground, and therefore, the way such spaces are designed, managed and used demonstrates the realities of political rhetoric" [6]. This involves different cultural groups with differing perceptions interacting as a community for recreational purposes. Furthermore, the community gains access to open spaces especially those without monetary value. As a social space, urban open space focuses on achieving urban integration instead of isolated entities. A relationship must be established between the urban open spaces and its users [6]. Urban open spaces involve different urban stakeholders working together in planning, managing, sustaining and protecting urban open spaces. Key factors for effective urban open space planning include community participation and consumer education.

2.2 Understanding the Value of Urban Open Spaces

2.2.1 Urban Stakeholders in Planning

Scholars like Walker, Borne and Rowlinson [7] describe the term 'stakeholder as consisting of individuals or group of individuals who either have an interest (stake) in an endeavour or are influenced by an endeavour. Stakeholders can be classified into three categories; those who affect the project, those who are affected by the project and those with an interest in the project [7]. Within the urban planning domain, Campbell [8] describes urban stakeholders as a group of individuals who have different backgrounds, roles, expertise and represent different aspects of urban complexity being witnessed in contemporary times. Continuing, the author maintains that urban stakeholders can be classified according to two broad classifications. The first set comprises of individuals or organizations which are involved in the delivery of the project and play salient roles in determining the context of the project [8]. This group of urban stakeholders comprises of planners, project managers, developer, investor, environmentalists and human settlement experts.

A second category of urban stakeholders comprise of those who may be directly or indirectly affected by a project. Members of this second category consist of the users of urban land (urban open spaces), owners of properties (houses) and the community member encroaching urban open spaces [8]. Also, urban stakeholders can be described as individuals who may be interested in a project such as researchers and potential users of future projects [9]. All these persons share common interests in planning practice or in the creation of sustainable communities.

2.2.2 What Values Do Urban Stakeholders Place on Urban Land?

Urban stakeholders place different values on urban land/open spaces, and this depends on how they perceive the contributions of the urban environment to their performance/sustenance [9]. Their perceptions are majorly influenced by the distinctive nature of engagement with existing land use patterns within an urban ecosystem [9]. For instance, investors and developers in the urban context view urban land in terms of its economic value potential. Based on economic value, cities are viewed as economic entities as they assist in creation of sustainable environments. Part of urban land, that is, urban open spaces, has a positive impact on the entire environment of the city because they make cities attractive, enhance tourist spending, boost the economic development of the city [10]. Urban open space has a considerable property value because it also attracts other community members who value urban open space benefits to reside near these spaces [10]. As urban stakeholders, planners are concerned with the creation of sustainable environments. Whilst planning for urban open spaces, the planner focuses on the attainment of improved quality of life for human beings and this includes meeting demands for recreational space and environmental quality [8]. Yet, environmentalists concern themselves with open space conservation by protecting the existing natural values [11]. These refer to the "demand approach" and "supply approach" [2]. CABE Space [11] alludes to the direct relationship between investment in environmental infrastructure such as the use of parks as integral aspects of urban drainage systems and an increase in the urban area's ability to capture excess water runoffs thereby mitigating the incidence of flooding.

Community members are concerned with safety and security in urban open spaces. Also, their concerns include increased cultural and social vitality, better quality of life including better and improved health, more inclusive open spaces, sense of place and accessible environments [12]. Those who do not have shelter place a housing value on urban land hence an increasing incidence of urban open space encroachment [12]. The housing value leads to gradual disappearance of urban open spaces and this threatens sustainable planning of urban open spaces. Socially, urban open spaces contribute to community cohesion, visual amenities, social integration and interaction among different urban stakeholders [13]. The interaction of urban stakeholders in urban open spaces promote the sense of community and sense of belonging [11, 13]. Culturally, different strategies for planning, designing and managing of urban open spaces is used to encourage different ethnic groups to use these spaces. Cultural and personality traits of individuals also determine the manner in which urban stakeholders view urban open spaces [14]. In addition, urban open spaces can be used as a cultural focus hence the significance of understanding people's perception on the use and value of urban open spaces within their cultural context [15]. Planners have to understand why different cultural groups use urban open spaces. Urban open spaces are also perceived as having an educational value for some urban stakeholders [15]. According to Woolley [15], the National Conservancy Council encouraged schools to be developed within a few minutes' walk to an ecological area. This added value to their education component. Facilities such as nature trails, field study sites and information centres should be provided within the reach of the local neighbourhood [15].

2.2.3 How Do These Value Conflicts Influence Urban Open Space Encroachment?

Effective planning for the use for urban open spaces considering the different perceptions of value held by relevant stakeholders requires extensive interaction of all urban stakeholders. Such interaction is buttressed by the involvement of different urban stakeholders and community participation in planning for urban open spaces thus enabling the management of encroachment [16, 17]. However, urban stakeholders view the value of urban open spaces differently. This leads to value conflicts on the value of urban open space hence planners struggle to achieve predetermined levels of planning objectives for these spaces through effective management of these variance.

With lack of awareness of the different existing perceptions about the value of urban land or urban open space by different urban stakeholders, a conflict may arise. The different value conflicts influence urban open space encroachment. For instance, drivers for placing housing value on urban open spaces is homelessness or lack of shelter. Other factors resulting in the encroachment of urban open spaces include; failure to manage and implement land use regimes by the planners, use of outdated town planning schemes by planners, and delays in approval of land use change application [17]. The major cause of concern is non-participation of community members during the planning of projects for sustainable neighborhoods [19]. Available evidence indicates that relevant authorities have neglected the value of urban open spaces with a renewed focus on planning for other social amenities [18].

2.3 Urban Open Space Encroachment in Mangaung

Mangaung Metropolitan Municipality in Free State Province, South Africa has the population of approximately 747 431 and measures 9 887 km^2 in extent. It comprises of three prominent urban centres, which are surrounded by an extensive rural area. It is made up of Bloemfontein, Botshabelo, Thaba Nchu, Soutpan, Dewetsdorp, Wepener and Vanstadersrus [19]. Mangaung is faced with urban open space encroachment because of people who do not have shelter or cannot afford decent housing. It is challenged with housing back-log, which is mostly caused by, among other factors, rapid urbanization. Urbanization forms an integral part of poverty eradication when it is properly planned and managed [20]. In 1998, the Mangaung Metropolitan Municipality took a resolution in one of its Council meetings that families illegally residing on erven (properties) with the land uses other than "Residential" in areas for which the town planning has already been finalized as well as other existing or future townships areas, will not be accommodated in terms of town planning, surveying and the provision of services [19].

However, when land use applications on municipal land are lodged, especially for informal settlement upgrading on land earmarked for urban open spaces, an exception is made. Due to this exception, urban open space encroachment on municipal land has become a common practice in Mangaung. Values of urban open spaces are discarded, and all the planning processes are followed by means of rezoning from an urban open space to residential. In this instance, a housing value, which is a concern for some community members, is considered for residential purposes (Fig. 1).

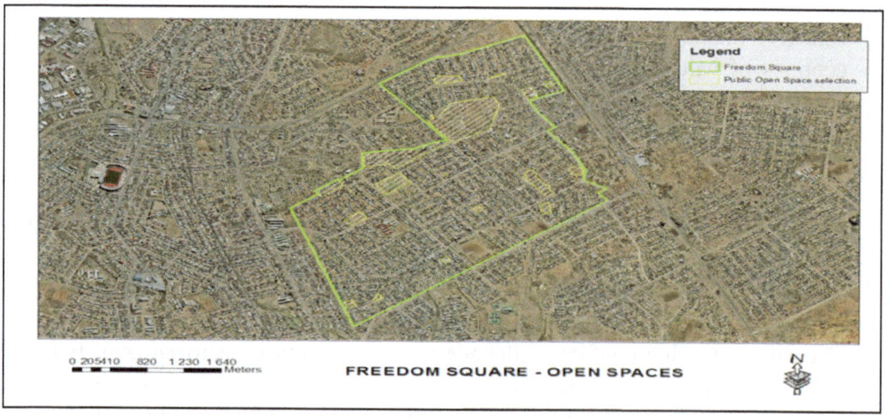

Fig. 1. Urban open space encroachment in mangaung source: [19], indicates the study area with urban open spaces that are encroached

3 Methodology

The study seeks to explore, in its entirety, perceptions of different urban stakeholders about the use and value of urban open spaces in Mangaung Townships, Free State Province of South Africa. This qualitative study adopted a case study research design. A variety of techniques such as focus group discussions, face-to-face semi-structured interviews and personal observation were employed for data elicitation at different intervals. A case study research design explores a contemporary bounded system through detailed, in-depth data collection [21]. The potency of this research design to engender in-depth data collection is enabled by the deployment of a plethora of techniques like observations, interviews, documents, etc. within the bounded area [21]. Semi-structured interviews focus on similar and not identical questions being asked to the same sample size. Personal observations as a form of data collection assists researchers to observe the behavior and lifestyle of participants in their natural setting [22]. Field notes were also taken during this process. Focus groups, as a method of data collection, refers to the interactive discussion with a group of people with the knowledge of the topic [23].

In this case, individual semi-structured interviews were conducted with purposively recruited town planning, human settlements, land invasion and environmental management professionals from the local government. The community members who had encroached upon urban open spaces and those residing on properties around urban open spaces formed part of the focus group discussions. The ward councilors and ward committee members governing the study area were selected to form part of these discussions. Personal observations were conducted around the study area to get first-hand information on the state of urban open spaces in Mangaung and notes (memos) were taken during this exercise. Ten (10) interviewees were interviewed for the study. This sample was drawn from the population mentioned previously. The accruing data was analyzed thematically relying on a set of pre-set themes which evolved from literature.

4 Presentation and Discussion of Findings

The study's findings are discussed according to pre-determined themes resulting from the study's guiding research question: How can the incidence of value conflicts among different urban stakeholders affect the burgeoning rate of urban open space encroachment? The thematic areas consist of the perceptions of urban stakeholders regarding the values of urban open spaces. The two selected themes for discussion include: (1) Evidence of varying perceptions of value among different urban stakeholders and, (2) Relationship between value conflicts witnessed and poor management of urban open spaces by planning authorities and communities.

4.1 Evidence of Varying Perceptions of Value Among Different Stakeholders

Different urban stakeholders display different perceptions on the value of urban open spaces. These perceptions range from economic, social and environmental value and influence planning choices. Investors and developers view urban land in general in terms of the economic value and their main aim is to secure investment [12]. The planner has a responsibility to create sustainable communities and to provide quality of life to human beings. This can be achieved by fulfilling the human demand of urban open spaces for recreational purposes and environmental quality. The demand approach in planning for urban open spaces refer to the satisfaction of the community needs [2]. Community members who are the rightful owners of the properties surrounding urban open spaces are concerned about the safety and security, increased cultural and social vitality, better quality of life that include better and improved health, more inclusive open spaces, sense of place and accessible environments [12]. Those who illegally occupy municipal land, that is, encroaching urban open spaces, view urban open spaces as unbuilt space or land without any function hence mushrooming of informal settlements. Most of the open spaces that are encroached are parks that are not properly managed. In order to avoid conflict, the community must be recognised as the key stakeholders in the management of open spaces.

Focus group discussants occupying the urban open spaces stated that they are in dire need for housing and these spaces are the only solution for them to access better houses. These open spaces have been left open for many years and are utilised for criminal activities or dumping sites. Some confirmed that they are aware of the values of urban open spaces, but their main concern is housing. What is more critical is non-involvement in planning processes and they are willing to be part of the planning and management of urban open spaces. This make stakeholder engagement to be encouraged. Participation is the distribution of power to society so that they can have influence in decision-making. Non-involvement of the community members in decision-making regarding policies affects the community negatively, they will not be empowered [24]. Residents occupying properties around the urban open spaces are willing to be educated on the sustainability of these spaces to create change in the quality of their lives. Richards, Carter and Sherlock [25] add that stakeholders must not only be given an opportunity to participate in decision-making, but they must be capacitated. Capacity building refers to the efforts taken in building relations and competency of the society in the participation process [20]. Despite the value conflicts by different urban stakeholders, they have to find a common goal which will assist the planner to achieve the objectives of planning.

4.2 Relationship Between Value Conflicts Witnessed and Poor Management of Urban Open Spaces by Planning Authorities and Communities

Abbasi, Alalouch and Bramley [26] highlight three main factors related to the effective use of open spaces and these include the needs of the users, quality of the physical features and the spatial structure of the space. These spaces have a positive impact on people's sense of quality of life, their physical and psychological well-being. Open spaces provide the community a space for interaction, relaxation, restoration, contact with nature and they offer many opportunities or leisure purposes. In Mangaung, urban open spaces are less prioritized as concentration is on other land uses like residential, educational and health facilities. This has led to mismanagement or negligence of urban open spaces. Similar cases are evident in some African countries and this result in poor implementation of planning projects [27]. According to Cilliers, Timmermans, Van Den Goorbergh and Slijkhuis [10] urban open spaces are not prioritized because they lack monetary value. This has caused value conflicts in terms of different land uses, conservation and development. Land uses like residential, commercial, and community facilities are highly prioritized because they have monetary value. Low prioritization of urban open spaces has led to under-provision and amendment for residential purposes. The municipal officials indicated that there are many urban open spaces in the township that have been rezoned for housing purposes. Discussants stated that they are residing in properties that were initially zoned as urban open spaces but are rezoned for residential purposes. This is referred to as "informal settlement upgrading". This leads to a challenge in balancing the importance of building sustainable future and addressing the need to prioritize urban open spaces [28].

In addition, Mashalaba [29] highlights that all the stakeholders must be part of the management of the urban open spaces. This will assist in achieving sustainable and meaningful contributions in decision making. Authorities who are responsible for urban open space management must consider open space audit. These audits must include the location, size, characteristics, quality and the purposes of open spaces. This will assist the authorities to know their space and manage it better [5]. Proper planning and management of urban open spaces, especially in townships, must be considered. In the case of Mangaung Metropolitan Municipality, management of urban open space in the township, city centre and the surrounding suburbs is different. The functions of the urban open spaces in the city centre and the suburbs are clearly stated on the municipality's land use schemes. Municipalities, especially in planning practice, need guidelines or framework on how to manage urban open spaces. This must be inclusive of community participation or involvement of the different stakeholders involved in management and sustaining urban open spaces.

5 Conclusion

Changes in the value and occupancy patterns on urban open spaces poses a major concern in urban planning and development. This study aimed at exploring perceptions of different urban stakeholders about the value of urban open spaces in Mangaung Townships, Free State Province. Focus of the study is in one of the townships in Mangaung, Free State because there is recurring gradual disappearance of urban open space due to encroachment for housing value by the community members who do not have shelter or cannot afford decent housing. For data elicitation, face-to-face semi-structured interviews, focus groups and personal observations were used. Findings indicated that urban stakeholders view urban open space values differently and this leads to the incidence of urban open space encroachment.

Mangaung townships has lost some urban open spaces due to value conflicts by different urban stakeholders and poor management urban open spaces. The different urban stakeholders involved in planning display different perceptions on the uses and values of urban open spaces. These different values include economic, recreational, environmental and housing. This permits different value conflicts among urban stakeholders and it makes it difficult for the planners and other professionals to achieve the planning objectives for urban open space and for other issues, such as housing and land use management. However, all urban stakeholders need to collaborate with each other to help in facilitating sustainable environments. Municipal planners are the key actors in promoting sustainable neighborhoods and they have to formulate policies that will curb the encroachment of urban open spaces.

References

1. Mangaung Local Municipality. By-laws regarding urban open spaces (2008)
2. Maruani, T., Amit-Cohen, I.: Open space planning models: a review of approaches and methods. Landsc. Urban Plann. **81**, 1–13 (2007). Depart of Geography and Environment, Israel
3. Baycan-Levent, T., Vreeker, R., Nijkamp, P.: Multi-dimensional evaluation of urban green spaces: a comparative study on european cities. Research Memorandum, Vrije Universiteit, Faculty of Economic Business and Administration, pp. 1–18. Vrije Universiteit, Amsterdam (2004)
4. Godschalk, D.R.: Land use planning challenges: coping with conflicts in visions of sustainable development and livable communities. J. Am. Plann. Assoc. **70**(1), 5–13 (2004)
5. Campbell, K.: Rethinking open space, open space provision and management: a way forward. Report Presented by Scottish Executive Central Research Unit, Edinburgh, Scotland, UK (2001)
6. Thompson, C.W.: Urban open space in the 21st century. Landsc. Urban Plann. **60**, 59–72 (2002)
7. Walker, D.H.T., Bourne, L., Rowlinson, S.M.: Stakeholders and the supply chain. In: Walker, D.H.T., Bourne, L., Rowlinson, S.M. (eds.) Procurement Systems: A Cross Industry Project Management Perspective. Taylor And Francis, Oxford/New York (2008)
8. Campbell, L.: Stepping back: understanding cites and their systems. In: ALNAP Working Paper. ALNAP/ODI, London (2016)
9. Mathur, V.N., Price, A.D.F, Austin, S.A., Moobela, C.: Defining, identifying and mapping stakeholders in the assessment of urban sustainability. In: International Conference on Whole Life Urban Sustainability and Its Assessment, Glasgow (2007)
10. Cillier, E.J., Diemont, E., Stobbelaar, D., Timmerman, W.: Sustainable green urban planning: the workbench spatial quality method. J. Place Manage. Dev. **4**(2), 214–224 (2011)
11. CABE SPACE: Does money grow on trees? (2005). www.cabe.org.uk/files/does-money-grow-on-trees.pdf. Accessed 25 Mar 2019
12. Carmona, M., De Magalhães, C., Edwards, M.: Stakeholder views on value and urban design. J. Urban Design. **7**(2), 145–169 (2002)
13. Wakaba, D.: An assessment of the quality of open spaces in Komarock Estate, Nairobi, Kenya. Unpublished dissertation. University of Nairobi (2016)

14. Özgüner, H.: Cultural differences in attitudes towards urban parks and green spaces. Landsc. Res. **36**(5), 599–620 (2011)
15. Woolley, H.: Urban Open Spaces. Routledge, Taylor & Francis Group (2003)
16. Haaland, C., van den Bosch, C.K.: Challenges and strategies for urban green-space planning in cities undergoing densification: a review. Urban Forest. Urban Green. **14**, 760–771 (2015)
17. World Health Organization (WHO): Urban green spaces: a brief for action, Europe (2017)
18. Mensah, C.A.: Urban green spaces in Africa: nature and challenges. Int. J. Ecosyst. **4**(1), 1–11 (2014)
19. Mangaung Metropolitan Municipality: Integrated Development Plan 2017/18. Bloemfontein (2018)
20. UN-Habitat: Capacity-Building-in-New-Urban-Agenda-HABITAT-III Nairobi (2015)
21. Creswell, J.W., Poth, C.N.: Qualitative Inquiry and Research Design: Choosing Among the Five Approaches, 4th edn. Sage, Thousand Oaks (2018)
22. Bernard, H.R., Ryan, G.W.: Analyzing Qualitative Data: Systematic Approaches California. SAGE Publications, Thousand Oaks (2010)
23. Merriam, S.B., Tisdell, E.J.: Qualitative Research: A Guide to Design and Implementation, 4th edn. Jossey-Bass, USA (2016)
24. Arnstein, S.R.: A ladder of citizen participation. J. Am. Inst. Plann. **35**(4), 216–224 (1969)
25. Richards, C., Carter, C., Sherlock, K.: Practical Approaches to Participation. Macaulay Institute, Aberdeen (2004)
26. Abbasi, A., Alalouch, C., Bramley, G.: Open space quality in deprived urban areas: user perspective and use pattern. Proc. - Soc. Behav. Sci. **216**, 194 (2015)
27. Mensah, C.A.: Destruction of urban green spaces: a problem beyond urbanization in Kumasi City (Ghana). Am. J. Environ. Protect. **3**(1), 1–9 (2014)
28. Cilliers, J., Cilliers, S.: Planning for Green Infrastructure: Option for South African Cities. South African Cities Network, Johannesburg (2016)
29. Mashalaba, Y.B.: Public open space planning and development in previously neglected townships. Unpublished dissertation. Department of Urban and Regional Planning. University of the Free State (2013)

Probable Challenges Facing Servitization Adoption in Construction: Lessons from the Manufacturing Sector

Rotondwa B. Ramafalo[1(✉)], Bankole O. Awuzie[1], and Clinton Aigbavboa[2]

[1] Department of Built Environment, Central University of Technology, Bloemfontein, Free State, South Africa
{rramafalo,bawuzie}@cut.ac.za

[2] Department of Construction Management and Quantity Surveying, Faculty of Engineering and the Built Environment, University of Johannesburg, Johannesburg, South Africa
caigbavboa@uj.ac.za

Abstract. The construction industry has acquired a reputation for product delivery without consideration for associated services. This dissatisfaction of clients due to the non-functionality of such products has led to a clamour for the adoption of servitization. Servitization entails a transition from selling products to offering products that are blended with services to add value to the end-user. This study explores probable challenges associated with the adoption of servitization within the construction industry based on lessons learnt from the manufacturing sector. A systematic review of the servitization literature was conducted. The literature selection was predicated on a set of keywords from various databases. These articles were analyzed using qualitative content analysis based on pre-selected themes. Results indicate that poor communication flow and nature of contractual arrangements which foster further fragmentation as well as lack of clear service-orientated strategies posed as probable barriers to the adoption of servitization in the construction industry.

Keywords: Clients · Product-Service System · Systematic literature review · Value-in-use

1 Introduction

Construction industry clients are increasingly committing to supporting societal aspirations through their operations. Of significance is their unwavering commitment to sustainability and sustainable development (SD) whilst also entrenching improved levels of efficiency in their operational activities [14]. Also, the impact of national commitments to global agenda has brought about policies which necessitate acute compliance on the part of these client organizations [19]. This has, inter alia, given verve to the increasing agitation for green and sustainable, eco-friendly, energy-friendly and low carbon buildings/products by clients, which will in turn enable an optimization of their core activities.

C. Aigbavboa and W. Thwala (Eds.): CIDB 2019, *The Construction Industry in the Fourth Industrial Revolution*, pp. 450–458, 2020.
https://doi.org/10.1007/978-3-030-26528-1_45

Yet, the penchant of the construction industry for under-delivery has continued to haunt it, negating attempts at improved service delivery to its clientele [6, 8].

Various initiatives like: sustainable construction, lean construction, and agile project management have been adopted by the industry over time to provide for these agitations. But such interventions appear not to be yielding the desired results clients still complain of the delivery of non-functional assets which do not provide desired levels of value-in-use. This is further buttressed by the seeming fixation of the industry on project management success as assessed by the ability of the project team to deliver projects according to time and budget with quality being undermined [13, 18]. Often, the client's business plan for initiating the project is not considered.

The manufacturing sector was plagued by similar circumstances prior to its adoption and implementation of the servitization praxis [3]. Although this transformation has not been without attendant challenges and increased risks [4], it has been heralded as delivering on the value-in-use proposition the clients [2]. Noticeably, a plethora of interventions towards improving construction industry standards as well as deliverables have been derived from other sectors especially the manufacturing sector. Obviously, this can be traced to the similarities in the prevailing modus operandi within these sectors. For instance, the interventions associated with lean production was adopted from the Toyota Production System (TPS) and subsequently adapted to the construction industry [15, 29]. As applied in the manufacturing sector, servitization has been identified as contributing towards the production of products and services that are customer-centric and not product-oriented services due to the attendant synergy in the creation of knowledge during the product conceptualization and development stages [2, 26]. The concept has been applauded for enabling a transition from transaction-based to relationship-based interactions between the manufacturer and the client-a transition that is far-fetched in construction despite the introduction of partnering, alliancing, etc. [2].

The construction industry as presently constituted has shown its inability to deliver in terms of client satisfaction. Therefore, it is trite to continue to explore vistas that have worked in other contexts within the industry in the attempt to get it right this time around. This notion endeared the authors to the concept of servitization and its probable utility in the construction industry. It has been proposed as a panacea for integrating various facets of the industry in manner that will prevent further fragmentation. But the wholesome adoption of this concept cannot be executed without understanding the probable challenges which may accompany the adoption. As such, this study set out to explore the challenges encountered in the manufacturing sector during the adoption of the concept therein, which may also occur in construction.

2 Literature Review

2.1 Servitization in the Manufacturing Sector

For the first time [23] evolved the term 'servitization', which is the process of adding services to products with the aim of creating value or meet the needs of the end user. Since then, the concept has been evident in the manufacturing sector for its use as a competitive manufacturing technique [20]. Since the invention of the term 'servitization', there is an increasing interest for this theme in the academics, communities, and business ventures and it has expanded beyond the manufacturing sector [11]. But a lot still believe that transiting towards servitization is a way to deliver added value capacities for conventional manufacturing companies.

The expanding interest by firms moving towards more creative and collaborative associations with their clients through 'servitization' are evident also in the automotive industry. For example, within the aerospace sector, engine suppliers, for example, Rolls-Royce do not simply fabricate engines, yet in addition give through-life support to their engines and rent out the utilization of their engines through what is known as "Power by the hour" [22]. While the applicability of cross-industry development must be noted [7], the construction of the Rolls-Royce Total Care Model, and additionally comparable innovations from other engine manufacturers within the aerospace division, raise doubt about whether the production network in building construction can play a more effective part in the operation and maintenance of building frameworks. Be that as it may, even before Rolls-Royce changed its plan of action and received "power by the hour", the firm still used to offer extras and offer repair and upgrade services [27].

Servitization includes the construction of the abilities of an association and procedures with the goal that it can better make common incentive that allows a transition from product-based offering to service-orientated products [3]. A few authors propose that manufacturing companies incorporate services among their principle offers [5]. As indicated by [21] manufacturers can discover new business openings by adding services to their chain, completing exercises previously undertaken by the client or presenting adjoining chains. This is supported by [5], who mentions that to increase the value of the primary chains, services can be joined with substantial products to separate them from contending products. On the other hand, [12] adds that the majority of products offered on the market today consist of both products and services, with a greater emphasis on the product offered.

Manufacturing firms implement services to increase their revenues and enhance customer satisfaction. Additional services may enable manufacturing firms to access new business opportunities, improve efficiency and enhance the offering differentiation and customer relationships [20]. In this manner, servitization evolved from the notion that clients stand to benefit from products when they are complemented by services, which ultimately deliver value-in-use [24, 25]. Endmost, the process of servitizing and building products and services into one interesting provision resounds quite evidently with the public–private partnership (PPP) or private finance initiative (PFI). Again, the importance is not only set upon the deliverance of the products yet in addition the servicing of these products over a 20 to 25-year time frame [1].

2.2 Servitization and Nature of the Construction

The paradigm shift by product manufacturers towards full service provision was first recognized in a study by [23], wherein the term "servitization" was first used. Developing patterns of servitization have increased and gained significance within the construction industry as expansion of private finance initiative (PFI) projects challenges contractors to adopt a more holistic strategy towards the design, construction, commissioning, operations, maintenance and post-occupancy of buildings [16].

The construction industry is a very complex and fragmented sector comprising of many team players at various stages of the construction project resulting in problems associated with fragmentation, such as lack of coordination between design and construction, and isolation of professionals, which has resulted in the development of a knowledge gap. The level of complexity and fragmentation in the construction industry becomes a barrier in adopting and implementing servitization. However, this brings the need for knowledge creation in the construction industry at the early stages of a project (design process) to reduce failure and improve functionality of the final product to the end users.

3 Research Method

A systematic review of the selected literature was conducted to explore the challenges of adopting servitization in the construction industry from 1994 till 2018 to ensure validity across two decades. The search approach was developed by first selecting keywords and identifying the relevant databases within the set timeframe in which the research papers where published. Finally, identification and analysis of the papers was done by reading through the abstracts of each identified paper. At a later stage, after filtering the relevant papers, the entire content of the papers were read to identify the patterns and themes to help prepare and present the findings (Fig. 1).

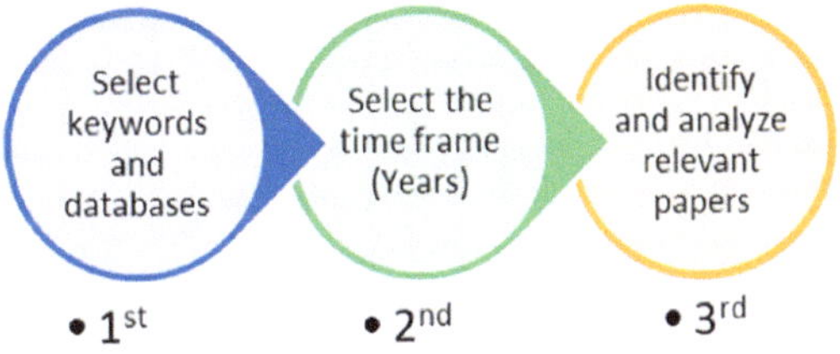

Fig. 1. Procedure for systematic review of literature

At first, a wide range of major databases was consulted with intentions to cover a variety published articles as per the researcher's choice. These databases included Compendex, Elsevier, Scopus, Springer link, ProQuest Central, ISI web of science, and EBSCOHost web (Table 1). The publications for the literature reviewed in this study were gathered in two parts. The first part gathered 237 full peer-reviewed scientific articles on the main keywords: such terms as Servitization and Product-Service Systems. The second part gathered 63 articles based upon the combinations of keywords, such terms as sustainability, service design, servitized products, value-in-use, service-centred, and service-oriented. The total reviewed papers of part one and part two amounted to 300, as shown in Table 1. However, not all the reviewed papers were relevant, only 102 of the reviewed papers were central to the study. Table 1 only shows the number of papers acquired from the databases by just reading through their abstracts before they were pruned down through reading the entire paper to check their relevance towards building up the paper. These databases represent the main academic journals related to the subject.

Table 1. Main keywords and combinations.

Part 1									Part 2	
Main keywords	ISI	SPR	PQC	CPX	SPS	EBS	EVR	No: of articles	Combination keywords	No: of articles
Servitization	45	19	17	14	60	5	12	172	Sustainability	29
Product - Service System (PSS)	15	7	8	9	16	3	7	65	Value-in-use	6
									Service design	5
									Servitized products	3
									Service-orientated	12
									Service-centred	8
Total from the databases	60	26	25	23	76	8	19	237	Total combination	63
Databases legend ISI = ISI web of science SPR = Springer PQC = ProQuest Central CPX = Compendex SPS = Scopus EBS = EBSCOHost web EVR = Elsevier										

Many of these keywords were combined with 'construction industry' in various search instances in order to ensure that they are relevant towards the goal of this study. Moreover, the set of discovered articles were then expanded and refined due to their appropriateness. All the remaining abstracts were considered, not unless they seemed to be inappropriate, then a need to read the entire paper was key. Later, the main topics and contributions were apprehended and gathered using techniques of a mind map to analyze the data. Thus, the findings of this paper were based on the analysis made from the main topics. More specifically, it is an addition to the work by [3] and identifies the existence of research gaps in the literature.

4 Presentation and Discussion of Findings

The findings deriving from the reviewed literature of servitization will be presented in this section in accordance to the designated themes.

Challenges to Servitization in the Manufacturing Industry
The following challenges were identified from the articles sourced through the literature review, which appear to be the barriers to the adoption and subsequent implementation of servitization (Fig. 2).

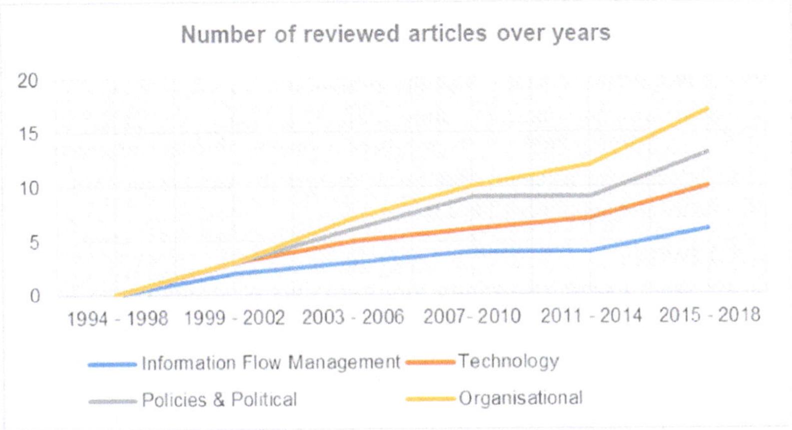

Fig. 2. Number of published articles reviewed over years in relation to challenges of achieving servitization

The presented data shows that there is a continuous increase in the barriers of achieving servitization due to the flow of information and factors resulting from the fragmentation of the construction industry. The studies showed that a significant number of servitizing firms were attempting to receive the financial related benefits publicized by past authors. This connects with a noticeable change in the recent tone of the servitization literature which tends to concentrate more vigorously on the difficulties related with making the progress towards more servitized provisions. The degree to which expanded innovation in building energy performance administrative structures will trigger servitized conduct from construction contractors will be reliant on the attractiveness of the agreed payment mechanisms. Without contractors having absolute control of energy supply and use, finding a quantifiable output from which contractor performance can be precisely judged remains a key challenge [9].

The following challenges appear to be the most common from the reviewed literature:

Information Flow Management
Information Flow Management is also another challenge since construction is fragmented, the level of fragmentation affects the flow of information which makes it

difficult to deliver products that are accompanied by services to convey value in use to clients through offering building with lower carbon and more energy efficient. Sustainable construction ought to deliver environmental, societal and economic execution, and value (as far as quality) to the end client and to society. Performance and value must be delivered if these destinations are shared through the value chain, i.e. if common intentions, targets and necessities are passed on undisrupted between ensuing players in the value chain [1].

Technology
Clearly, there is more intensity of issues talking about barriers from an industrial perspective that there is 'Technology Breakdown'. When technology is brought to construction, it should be made sure that there is adequate technology across the design, operational and management team. The differences in the design of products and services lies in the essence that the design for services are more difficult to delineate, hence they are vague in nature [23]. As a results, companies may be discouraged by intensifying their services offering due to the fact that need consider competition beyond the usual market form unexpected competitors which includes their trust clients and suppliers [20, 23].

Policies and Politics
While on the other hand [10] identify the constraints of legally contractual arrangements without the satisfactory help of relational governance. This can be seen within the more extensive challenge of endeavouring to embed a service-orientated culture within traditionally product-based associations [20]. [12] talk about the need for an internal alignment of performance metrics in achieving this cultural move in mind-set towards product-service delivery.

Organisational
At this point, difficulties are experienced relating to defining strategies of organizations that are necessary help deliver products that are complemented by services to the customers/clients [28]. Manufacturing firms have consistently delivered services, by providing spare parts, installing equipment, training staff, or providing client support. In competitive markets however, firms look for better approaches to separate their business, including an expanded effort on services and product-service systems, normally alluded to as servitization. Regardless of the focus, studies demonstrate that manufacturers struggle to servitize; numerous new service and PSS (product-service system) ideas are produced yet fail once they are taken to the market, and even manufacturers that discover how to dispatch new offerings fail to accomplish their vital targets of servitization [3, 23]. Adopting a downstream position, such as the provision of installed base services, organisations have to be service oriented and value services [20]. These organisations provide solutions through product-service combinations and tend to be client-centric by providing customised, desirable client outcomes organised around capabilities, competences, and client requirements. Finally, in the design for service provision, consideration should be given to clear and simple communication strategy towards describing the value proposition to clients/customers (Table 2).

Table 2. Challenges to servitization in the manufacturing industry

Categories	Challenges	No. of articles	Example of authors
Information flow Management	• Communication Flow	19	• [20]
Technology	• Design of products and services	12	• [20, 23]
Policies & Political	• Contractual arrangements	9	• [10, 17]
Organisational	• No clear service-orientated strategies	10	• [20]

5 Conclusions

This paper has used the servitization literature to comprehensively assess the probable challenges hindering the adoption of servitization in the construction industry. Furthermore, the paper seeks to explore the applicability of the concept of servitization in construction industry in assuring client value through delivering buildings that are accompanied by services to meet the clients purpose/use. This paper highlights the need for a better understanding of how servitized business models can leverage upon functionality and satisfaction through offering coordinated blend products and services that deliver value in use. It proposes that construction business models find ways of adding value by adding services to the product to satisfy client's needs, and create awareness with respect to concerns associated with servitization. one of the keys to success is creating an environment that is service-orientated and discovering individuals that are willing to deliver the required services. Thus, managers should take into cognizance that people are their main assets in order to transit from deliver of products only to delivery of products that are accompanied by services. To end with, companies gain an understanding of their clients' needs when they offer services to them, and they can start offering custom-made products to their clients.

References

1. Akintoye, A., Hardcastle, C., Beck, M., Chinyio, E., Asenova, D.: Achieving best value in private finance initiative project procurement. Constr. Manag. Econ. **21**, 461–470 (2003)
2. Baines, T., Ziaee Bigdeli, A., Bustinza, O.F., Shi, V.G., Baldwin, J., Ridgway, K.: Servitization: revisiting the state-of-the-art and research priorities. Int. J. Oper. Prod. Manag. **37**, 256–278 (2017)
3. Baines, T.S., et al.: State-of-the-art in product-service systems. In: Proceedings of the Institution of Mechanical Engineers, pp. 1543–1552 (2007)
4. Benedettini, O., Neely, A., Swink, M.: Why do servitized firms fail? A risk-based explanation. Int. J. Oper. Prod. Manag. **35**, 946–979 (2015)
5. Brax, S.: A manufacturer becoming service provider–challenges and a paradox. Managing Serv. Qual. Int. J. **15**, 142–155 (2005)
6. Egan, J.: Rethinking Construction Accelerating Change (2002)

7. Enkel, E., Gassmann, O.: Creative imitation: exploring the case of cross-industry innovation. R&D Manag. **40**, 256–270 (2010)
8. Farmer, M.: Collaborate or die. RIC Constr. J. **2017**, 15 (2017)
9. Gruneberg, S., Hughes, W.: A review of performance-based contracting. London Royal Institute of Chartered Surveyors (2011)
10. Guo, L., Ng, I.: Behaviour transformation: an examination of relational governance in complex engineering service. In: Complex Engineering Service Systems. Springer (2011)
11. Hewitt, P.: The government's manufacturing strategy. Secretary of State for Trade and Industry, Westminster (2002)
12. Hytönen, H.: A model for value-based pricing of industrial services. Master of Science in Technology, Helsinki University of Technology (2005)
13. Joslin, R., Müller, R.: Relationships between a project management methodology and project success in different project governance contexts. Int. J. Proj. Manag. **33**, 1377–1392 (2015)
14. Kibert, C.J.: Sustainable Construction: Green Building Design and Delivery. Wiley, Hoboken (2016)
15. Koskela, L., Howell, G., Ballard, G., Tommelein, I.: The foundations of lean construction. In: Design and Construction: Building in Value, pp. 211–226 (2002)
16. Leiringer, R., Bröchner, J.: Service-led construction projects. Constr. Manag. Econ. **28**, 1123–1129 (2010)
17. Martinez, V., Bastl, M., Kingston, J., Evans, S.: Challenges in transforming manufacturing organisations into product-service providers. J. Manuf. Technol. Manag. **21**, 449–469 (2010)
18. Mir, F.A., Pinnington, A.H.: Exploring the value of project management: linking project management performance and project success. Int. J. Project Manage. **32**, 202–217 (2014)
19. Ofori, G.: The construction industries in developing countries: strategic review of the book. In: Ofori, G. (ed.) New Perspectives on Construction in Developing Countries, Spon. Routledge, Abingdon (2012)
20. Oliva, R., Kallenberg, R.: Managing the transition from products to services. Int. J. Serv. Ind. Manag. **14**, 160–172 (2003)
21. Sawhney, M., Balasubramanian, S., Krishnan, V.V.: Creating growth with services. MIT Sloan Manag. Rev. **45**, 34–44 (2003)
22. Smith, D.J.: Power-by-the-hour: the role of technology in reshaping business strategy at Rolls-Royce. Technol. Anal. Strateg. Manag. **25**, 987–1007 (2013)
23. Vandermerwe, S., Rada, J.: Servitization of business: adding value by adding services. Eur. Manag. J. **6**, 314–324 (1988)
24. Vargo, S.L., Lusch, R.F.: Evolving to a new dominant logic for marketing. J. Mark. **68**, 1–17 (2004)
25. Vargo, S.L., Lusch, R.F.: Service-dominant logic: continuing the evolution. J. Acad. Mark. Sci. **36**, 1–10 (2008)
26. Vasantha, G.V.A., Roy, R., Lelah, A., Brissaud, D.: A review of product–service systems design methodologies. J. Eng. Des. **23**, 635–659 (2012)
27. Waters, N.: Engine health management. Ingenia **39**, 37–42 (2009)
28. Wise, R., Baumgartner, P.: Go downstream: the new profit imperative in manufacturing. IEEE Eng. Manage. Rev. **28**, 89–96 (2000)
29. Womack, J.P., Jones, D.T.: Beyond Toyota: how to root out waste and pursue perfection. Harvard Bus. Rev. **74**, 140–158 (1996)

Resolution of Management Issues with Mega Project

Ayodeji Olatunji Aiyetan(✉) and Mhlongo Scelo

Department of Construction Management and QS,
Faculty of Engineering and the Built Environment,
Durban University of Technology, Durban, South Africa
ayodejia@dut.ac.za, scelo@chsdevelopments.co.za

Abstract. South Africa is a developing country and there are many construction projects that are underway. This study aims to contribute to the resolutions on the management of mega projects, since previous studies indicated that there are many failures in the construction of mega projects. The study was based in 3 provinces in South Africa which are KwaZulu-Natal, Gauteng and Western Cape. A mixed method research approach incorporating both qualitative and quantitative methods were used in the study to collect empirical data from stakeholders working on mega projects. The sample frame consist professionals such as the project managers, architects, quantity surveyors, foremen and site agents. The data were collected using a structured questionnaire and focused on individual interviews. A total of ninety-two questionnaire were analysed for this study. The study established that there are management issues that contribute to the success of the mega projects such as the lack of stakeholder engagement, poor plant management, and poor public participation. The paper recommends that in order for mega projects to be successful, stakeholder decisions must mitigate turnaround time on taking decisions for a project. Also, sound decisions must be taken for mega project success. It is recommended that regular plant schedule checking is done on a daily basis and proper plant checking must be done daily, prior to plant operation.

Keywords: Stakeholder · Plant · Production · Supervision

1 Introduction

There is always a percentage of risk involved when erecting mega construction projects; however, there are many ways to improve the odds of escaping the negative consequences of such risks. Construction of mega projects is an essential aspect of development in cities, states and individual livelihoods. Substantial infrastructure projects are normally economic transformers, for example, the Dubai international airport is the world's most eventful, and responsible for 21% of Dubai's employment and 27% of its GDP [1].

The inability of the client and the project team to come to mutual agreements and understandings based on the comprehensive overview of the construction processes from inception to completion is a persistent reason for delays in delivery dates.

C. Aigbavboa and W. Thwala (Eds.): CIDB 2019, *The Construction Industry in the Fourth Industrial Revolution*, pp. 459–467, 2020.
https://doi.org/10.1007/978-3-030-26528-1_46

Either party may be unaware of the environmental effects that may hinder the construction process. This view is supported by [2] who emphasize the problem of delays in the global phenomenon. The construction undertaking involves a plethora of tasks, requiring task management personnel to drive the construction team for effective production. It was estimated in the year 2010 that there were more than 1500 large construction projects worldwide at dissimilar stages of financing or construction, in areas such as oil, power, conveyance and engineering. Furthermore, the number, the intricacy and the scope of the projects have been increasing hurriedly over the past few decades [3].

Owing to this, several guides have been drawn up to help participants understand the management of mega projects. Articles and books give guidelines of how to manage the mega project without failure; an example would be 'Management of Mega Construction Projects (National Economic Developments office of Construction)' where the author identifies the importance of the strength, caliber and leadership of the project/construction manager. Reports such as these will assist in increasing information and equipping participants for the successful implementation of many mega projects. The rapid rate of modernization and projected elevation of progress of the state's economy have created a colossal market for the construction business over the preceding decades. With the outline of a number of fiscal policies intended at increasing investment in fixed assets and to fast-track the infrastructure developments, an increase in extraordinary profile mega construction projects such as King Shaka International Airport have commenced in the preceding five years.

Mega construction projects are characterised by hefty investments, prodigious complexity, additional stakeholders and extensive influence. Despite these achievements of extensive influence, the construction business has been disparaged for producing inferior construction work, delays and budget overruns with truncated competence and value. Difficulties with dodging submission to the procurement system, untruthful contracts and unlawful sub-letting are some of the common challenges encountered in construction procurement. Regardless of a succession of government transformation in the construction trade, these adverse conditions persist. With this in mind, it is imperative to implement a rational and operative instrument to advance project management and governance of mega construction projects.

Intercontinental mega projects are probable to become progressively common in the engineering and construction trades. Participants at the Engineering Foundation Research session demarcated a mega-project as 'a high impact technically complex project which requires careful advanced planning, last three or more years, has significant impact on the public and industry, employs thousands of people and typically over billions of rand's'. The upsurge in the necessity for mega-projects are attributed to weakening and insufficiencies of present infrastructure, chiefly thoroughfares, conduits, water and sewage management plants; considerable infrastructure development required by third world nations to develop their economy; improved methods for the accrual and reduced bearing of perilous wastes on the environment, and the ongoing requisite for forceful private development and massive industrial ventures.

The very nature of these requirements infers that accomplishing them falls within the definition of a mega project. It is then indispensable that both advocates and implementers of mega-projects wholly understand the reasons involved in their preparation, implementation and overall management. [4] states that on one approximation, 65% of mega projects are botches. Project scoping and risk apportionment are fundamental to refining these consequences. Overall, the success rate of mega projects is less than the success rate of smaller projects.

2 Literature Review

This section discusses issues of management of mega projects. This is relative to the factors that influence the management aspects. Many mega projects face difficulties due to escalating management issues. Two main issues namely, stakeholder and plant management issues are discussed in this study.

2.1 Stakeholder Management

This concerns the relationship between an organization and the interested groups or stakeholders. These relationships affect the individuals and their organizations, and could be a positive or negative influence on any successful project []. In the past years, stakeholder management has been advocated to be the main component of increasing the delivery of construction projects [5]. Previous research focuses on addressing the need for a practical guide to carry out stakeholder management duties in the construction industry. The following factors are possibly the main causes of poor stakeholder management: stakeholder management decisions, meeting responsibilities through the project life cycle, internal stakeholder collaboration in carrying out stakeholder management tasks, stakeholder dynamics and the use of available techniques for stakeholder engagement [6]. The long process of design and implementation of construction projects involve interaction, collaboration and negotiations among many project stakeholders, which involve but are not limited to the client, designers, contractors, local authorities and general project environment [7].

The relationships created by project stakeholders both directly and indirectly influence the construction project and have an impact on the overall successful completion of the project [8]. The deference participants involvement both directly and indirectly on the project are also considered categories since the project stakeholders management is vital to the success of the project. The other concerning factors affecting the stakeholder management process is the hiring of a project manager with high competency and transparent evaluation of alternative solutions. This project manager should also ensure effective communication between the project and its stakeholders, set common goals and objectives for the project, and explore the stakeholders' needs and expectations.

2.2 Impact of Stakeholder and Plant Management

Plant management refer to the management of construction plant and equipment on construction site. The construction of mega projects has exacerbated the rate of change and witnessed widespread development of mega civil engineering and construction projects. As a result, the demand on machines became increased and seen as an economical alternative to costly labour resources [9]. In the earlier days, the powering system of the machines were steam. This was clumsy and inefficient, but in the 1950's new modern diesel engine machines were introduced, having more potential for work [10].

Modern construction project managers must therefore pursue the effective utilization of their equipment, and thus ensuring survival of company [9].

According to [11], three factors affect plant management. These are plant maintenance, downtime and health/safety. In the construction of mega projects, many believe that the construction plant has different costs, which require deferent utilizations in order to make it efficient in the construction site [12]. Other factors that affect plant management, such as employing expert mechanics to increase the quality of the plant and having enough fuel to avoid plant down time [13]. Within the broader subject of construction management, other principle themes are identified namely, productivity, optimization, robotics and automation, machine control and operators and competence [14]. The impact of stakeholder's decisions and plant management on a project are relative to time, cost, productivity, quality, company's image and competitive advantage. These factors might have adverse impact if not well attended to.

3 Research Methodology

This study was conducted within three provinces in South Africa, namely, KwaZulu Natal, Gauteng and Western Cape. The sampling frame consist of Engineers (24) Site Agent (23) QS's (24) Foremen (14) Social facilitator (9) and Safety officer (4).

A non-probability sampling approach was employed, namely, selective method of sampling. The research instruments for this study were a questionnaire survey and a structured interview. These were administered through e-mail and personal contacts. Responses for the questionnaires were received through the same means. Overall, ninety-two (92) questionnaires representing 90% response rate achievement recorded. Descriptive statistical tool was employed for data analysis. All respondents have been involved in the management of a mega project.

Respondents between the ages of thirty and forty predominate in the sample investigated. Regarding the qualification of respondents, (20%) have bachelor's degrees, and predominate in the sample. Followed by National Diploma's and B-tech Degree (15.5%) each; Masters Degrees at (7.5%) and Honors (7%). Majority of respondents (40%) were found to have fifteen years' experience. Respondents have involved in the construction of mega projects such as hospitals, malls, sport fields and residential units. The strategical means for data analysis used were means scores. The Cronbach alpha (>.60) and factor analysis (>.70) test were conducted and are satisfactory.

3.1 Data Presentation and Analysis

This section presents the data obtained of the study and analysis (Table 1).

The factor of that has the most significant influence under stakeholder management is: stake holder decision (MS = 3.13). The probable reason for the first factor are participants who take decision that will benefit themselves not the project. Decision that are not fair for all parties involved in the project.

Table 1. Stakeholder and plant management issues

Factor	Unsure	Response (%)						Mean score	Standard deviation	Rank
		DN	Minor............................Major							
			1	2	3	4	5			
Stakeholder related issues										
Stake holder decisions	0.0	0.0	1.3	20.0	43.8	35.0	0.0	3.13	0.77	1
Stake holder responsibility through project life cycle	0.0	0.0	1.3	12.5	62.5	23.8	0.0	3.09	0.64	2
Ensuring effective communication among the stake holder	0.0	0.0	0.0	17.5	56.3	26.3	0.0	3.09	0.66	2
Transparent evaluation of alternatives solutions	0.0	0.0	0.0	16.3	62.5	21.3	0.0	3.05	0.61	4
Stake holder collaboration	1.3	0.0	0.0	23.8	66.3	8.8	0.0	2.85	0.56	5
Techniques for stakeholder engagement	0.0	0.0	1.3	45.0	41.3	12.5	0.0	2.65	0.71	6
Stake holder dynamics	6.3	0.0	7.5	35.0	46.3	5.0	0.0	2.52	0.72	7
Common goals and objectives for project	1.3	0.0	6.3	51.3	37.5	3.8	0.0	2.39	0.67	8
Hiring competent project manager	3.8	2.5	41.3	38.8	12.5	1.3	0.0	1.72	0.75	9
Plant management issues										
Plant downtime	0.0	0.0	0.0	7.5	20.0	63.8	8.8	3.74	0.72	1
Plant maintenance	0.0	0.0	0.0	7.5	28.8	51.3	12.5	3.69	0.79	2
Health and Safety	0.0	1.3	5.0	22.5	58.8	12.5	0.0	2.80	0.72	3
Plant productivity	3.8	0.0	5.0	30.0	56.3	5.0	0.0	2.64	0.67	4
Robotics and automation of plant	1.3	2.5	10.0	32.5	48.8	5.0	0.0	2.51	0.75	5
Plant operators and competence	1.3	0.0	21.3	31.3	35.0	10.0	1.3	2.38	0.98	6
Plant optimization	25.0	0.0	2.5	45.0	26.3	1.3	0.0	2.35	0.58	7

The least significant factor in this category is hiring the competent project manager (Ms = 1.72). This implies that mega project requires and demand the competent project manager for the project to be successful. The inexperienced project manager who lacks management skills can take wrong decisions, which can fail the project.

The factor that has the most significant influence under plant management related issues is plant down time (Ms = 3.74). Most of the time in mega project heavy machines are used which some of them are imported and there not maintained regularly. Poor plant management and planning of work as well can also have an impact.

The least significant factor in this category is plant optimization (Ms = 2.35). The reason for this can be the lack of correct required plant on site. The failure to allocate plant on site can also have a negative impact on plant optimization.

Table 2 presents the impact of stake holder and plant management issues on key performance parameters, which include: cost, quality, time, production, company's image and competitive advantage, these are discussed below:

Table 2. Impact of stakeholder and plant management issues on mega project delivery

Factor	Response (%)					Mean score	Rank
	Minor.................................Major						
	1	2	3	4	5		
Cost							
Plant management	0.0	1.3	1.3	12.7	84.8	4.8	0.5
Stake holder management	0.0	0.0	13.8	31.3	55.0	4.4	0.7
Quality							
Stake holder management	0.0	3.8	6.3	16.3	73.8	4.6	0.8
Plant management	1.3	0.0	3.8	37.5	57.5	4.5	0.7
Time							
Stake holder management	0.0	0.0	1.3	5.0	93.8	4.93	0.31
Plant management	0.0	0.0	3.8	21.3	75.0	4.71	0.53
Production							
Stake holder management	0.0	1.3	1.3	5.0	92.5	4.8	0.5
Plant management	0.0	1.3	2.5	16.3	80.0	4.7	0.6
Company's image							
Plant management	1.3	3.8	23.8	50.0	21.3	3.8	0.8
Stake holder management	0.0	18.8	51.3	23.8	6.3	3.2	0.8
Competitive Advantage							
Plant management	1.3	26.3	32.5	27.5	12.5	3.2	1.02
Stake holder management	5.0	38.8	38.8	13.8	3.8	2.7	0.90

Relative to Cost

The most significant factor between stake holder and plant management that impact cost of the project is plant management (MS = 4.8). Plant management in the mega project need to manage by a very experienced personal since it can have a great impact on cost and once the cash flow of the mega project are affected, the entire project delivery also get affected.

Stake holder management (MS = 4.4) has less impact on cost compared to the plant management, this depends on the manner, experience and professionality of the stake holder.

Relative to Quality

The most significant factor that impact the quality on the management of mega project is plant management (MS = 4.6), faulty plant is likely to give wrong result. This is relative to the weighing of different materials for mixing, Also the adequacy of tools and plant to execute task neatly and more quickly. Therefore plant management have both positive and negative impact on the quality of production.

The least significant factor that influence plant management is stake holder management (MS = 4.5). Comparing to plant management it has less influence on quality of the mega project however it most important factor that may affect quality if it not management properly.

Relative to Time

On the table reveals the influence of management issues on time the factor that mostly affect time is plant management (MS = 4.93). Plant that are in good condition, given regular maintenance are likely to be used for a considerable long time without breaking down. A hitch free performance of plant engenders high productivity and early delivery of the project. The contrary is the case, if plant are not regularly maintained.

Stake holder management (MS = 4.71) has less influence compering to the plant management. The setup of the management team can have a positive or negative influence in time, therefore the way stake holder engaged each other can affect the project.

Relative to Productivity

Plant management (MS = 4.89). This corroborates the earlier analysis, stating that mega projects are driven by machines. This have great impact on production. While stake holder management (MS = 4.7) has less impact than plant management. There is minor difference on the influences between plant and stake holder management. This shows that stake holder management is also very critical on production management.

Relative to Company Image

The most significant factor that influence company image the most is stake holder management (MS = 3.86). Impact on company image. A good management will engender good operation and result. A bad management may also have negative impact or the operation of the company revealing the image of the company.

Plant management (MS = 3.2) has the less influence company image compered to stake holder management. However the company that fails to manage plant may fail to manage mega project which may lead to project failure and company image can be negatively influenced.

Relative to Competitive Advantage

Stakeholder management (MS = 3.24) has most influence in competitive advantage, there are knowledgeable and unknowledgeable clients in the construction industry, their contribution to the construction stages of a project could either make good of product or negatively affect. These contribution in either way having impact on the final product ultimately affect contractor's competitive advantage in both ways. A knowledgeable client will contribute positively in terms of idea on construction methods with positive impact on quality of production. While plant management (MS = 2.7) has a less influence on the competitive advantage.

4 Conclusion

From the analysis of the data, it is found that stakeholder's decisions and stakeholder's responsibility through the life cycle of projects are those that mostly influence the delivery of mega projects, from the stakeholder's point of view. While, on plant management issues, plant down time and maintenance are the factors that most influence the productivity of plant. Stakeholder and plant management issues adversely impact time, cost productivity and quality in this decreasing order. Based on the conclusions reached from the data analysis, the following recommendations are made:

Stakeholders should endeavor to make quick and qualitative decisions that will enhance progression on project.

A plant maintenance schedule should be produced and strictly compliance to plant maintenance be enforced, in order to mitigate against plant down time.

These two major decisions will mitigate against delay, cost increases, ensure high productivity and quality of products.

Acknowledgments. The authors herby acknowledge the financial contribution of Durban University of Technology relative to the conduct of this research.

References

1. Garamo, N., Matzenge, S., Palite, R.: Mega project: the good, the bad and better, McKensey Global Institute (2013). http://www.mckensey.com
2. Sambasivan, M., Soon, Y.W.: Courses and effect of delays in Malaysia construction industry. J. Proj. Manag. **25**(5), 517–526 (2007)
3. Miller, R., Lesserd, D.: Understanding and Management Risk in Large Engineering Projects. MIT Sloan School of Management paper 4214-01 2
4. Charrett, D., Loots, P.: Challenges in achieving successful mega project, Melbourn TEC Chambers (2000). http://mtecc.com.au
5. Pampliega, S.: Stakeholder management in the construction sector: stakeholder management in construction project. CEU Business School (2013)
6. Mowlus, J., Endogan, S., Gunlat, P.: Stakeholder Management in Construction Projects: A Life Cycle Based Framework. Doctor of Philosophy in Construction of Heriot Watt University (2014)
7. Winch, G.M.: Managing Construction Project: An Information Processing Approach, 1st edn. Blackwell Science, Hoboken (2010)
8. Takim, R.: Book of Innovation, Communication and Engineering, Islamic University of Malaysia (2009)
9. Hendrickson, C.: Intelligent Construction Plant Management. The Plant Management Research Group, The Built Environment Research Units, University of Wolverhampton (2000)
10. Edward, D.J., Aloke, D., Li, H.: Intelligent construction plant information management. In: 17th Annual ARCOM Conference, University of Salford, vol. 1, pp. 169–177, 5–7 September 2001
11. Edwards, D.J., Holt, G.D.: Construction plant and equipment management. J. Eng. Des. Technol. **7** (2009)

12. Aadal, H., Rad, K.G., Fard, A.B., Sabet, P.G.P., Morshedi, R.: Impact of plant and equipment in construction industry. Res. J. Appl. Sci. Eng. Technol. **7**(11), 2371–2375 (2014)
13. Bahru, J., Malaysia, T.: Major emergency assessor. University Manual of Occupational Safety and Health. University Teknologi Malaysia (2014)
14. Edward, D.J., Holt, H.G.: Construction workers health and safety knowledge: Initial observation on some test–result data. J. Eng. Des. Technol. **6**, 1 (2008)

An Assessment of the Challenges in the Delivery of Umgeni Water Project in South Africa

Ayodeji Olatunji Aiyetan(✉) and Naidoo Ashok

Department of Construction Management and QS,
Faculty of Engineering and the Built Environment,
Durban University of Technology, Durban, South Africa
ayodejia@dut.ac.za, ashok.naidoo@umgeni.co.za

Abstract. Implementing a suitable plan, design and management strategy is critical to the success of a project. Project delivery challenges encountered within the built-environment is a global phenomenon. Factors causing project delivery challenges in construction projects differ between countries, due to various fundamental reasons. The aim of the study is to assess lack of performance and its effects on project delivery for construction infrastructure projects with reference to Umgeni Water. The study was conducted at Umgeni Water, Pietermaritzburg in Kwazulu-Natal. Respondents for the study were Project Managers, Civil Engineers, Planning Engineers, Quantity Surveyors, Servitude Administrators and Environmental Project Managers. A non-probability sampling technique was employed in the selection of samples. A total of twenty three questionnaires were analysed for the study, Descriptive statistic was employed for the analysis. Findings revealed that insufficient planning and designing, and a weak project management approach endorsed and implemented by the project manager influence project success at Umgeni Water. This implies the need for the study to pursue strategies that will improve project and service delivery at Umgeni Water. Recommendations include that attention should be given to implementing a workable project delivery plan and management approach that will result in a better understanding and successful delivery of a project.

Keywords: Challenges · Design · Project delivery · Management approach

1 Introduction

Understanding the project delivery challenges and its associated influences are very important to achieve project success. [1] explains approaches endorsed and undertaken at different stages of the project may exert positive or negative influences on the project. Positive influences are such that it assists the project to reach its desired outcome and within the initial estimated time, cost and quality, while negative influences are such that will affect initial plans and exceed anticipated successful project delivery parameters. Strategic planning and management, which can be referred to as Process Model are expected to assist project managers including all stakeholders throughout the project life cycle. [2] defines a project life cycle as a set of sequential

C. Aigbavboa and W. Thwala (Eds.): CIDB 2019, *The Construction Industry in the Fourth Industrial Revolution*, pp. 468–476, 2020.
https://doi.org/10.1007/978-3-030-26528-1_47

phases that a project goes through from its beginning to its completion. [3] states that having an effective project delivery model increases the stakes of implementing and completing a successful project. [4] describes a project delivery model as a methodology within a framework. This means that a project delivery model must follow a certain process within the boundaries of an established setting. [3] states the quality and workability of the project delivery model affects the parameters. These could result in inadequate participation and contribution by the client, insufficient planning and designing, project risks, estimating errors, communication problems, project quality, and challenges, which hinder the implementation of a good project management approach. Therefore, this study aims at identifying leading challenges together with their associated influences and impact relative to the delivery of the project.

2 Literature Review

2.1 South African Water Infrastructure Projects

[5] reported that prior to 1994, water supply obligation was split with no singular government department accountable for its management. This caused inconsistent levels of service delivery within South Africa. After 1994, the government saw the critical need for new policies for the county, which included changes in the water sector. The department of water and sanitation is currently the third largest contributor to public sector infrastructure expenditure, with an estimated value of R132.1 billion between the period 2012/13 to 2018/19 [6]. This contribution by the department provides an indication of water infrastructures importance.

2.2 Project Delivery Challenges and Influencing Factors

2.2.1 Inadequate Participation and Contribution by the Client During the Project

[7] state that clients have to be more interactive with the project team and should be hands-on relating to contribution and participation. According to [8], these factors could be measured as client understanding of design drawings and specifications, awareness of the project processes, authoritative decision making, decisions made in a pressure situation, contribution of ideas, capabilities of understanding project constraints and productiveness of briefings to the project team.

[9] opined that client participation have high impact on project delivery, which in many cases may determine the project's success or failure; and that the clients' involvement during the project life cycle provides the fundamental link to the project.

2.2.2 Insufficient Planning and Designing Done During the Project

[10] investigates the effects of placing enough effort and cost into the early planning and design phases of capital expenditure projects in order to try to eliminate as many problems as possible when executing the project, labelling this process as front end loading (FEL). Factor that influence this are: the amount of time given to plan and design the works, procurement processes that favours the cheapest rather than experienced consultants and skill shortage experienced by consulting firms.

2.2.3 Project Risks That Are Poorly Identified, Assessed, Mitigated and Controlled During the Project Execution Stage

[2] defined the objectives of risk management as: to increase the probability and effect of positive events and decrease the probability and influence of negative events during the project. According to [11], risk management analysis should be performed in order to give the project manager significant, acceptable and valuable information on which future decisions will be based on. The findings by [12] suggest that there is a direct relationship associated with achieving project success and good risk management; and that the impact of not properly identifying, assessing, mitigating and controlling risk often results in project delays and cost overruns.

2.2.4 Unsatisfactory Cost Estimating Resulting in Errors and Oversights

[13] conclude that the initial cost estimates fail in considering all costs. Generally, costs are managed during the execution phase of a project and costs are not sufficiently managed during the planning, and design phases, when 80% of project costs are committed. [14] identify some of the main factors that measure cost estimating and cost control; Conditions of contract, market requirements, type of project and its complexity, project duration and contract period, error in judgement, lack of historical data, incomplete project information and estimating techniques.

2.2.5 Communication Problems Within the Project

It can be identified from [15] certain measurable components within communication in the construction industry. They are accuracy, procedures, barriers, understanding, timelines and completeness.

[16] explain that poor communication has grave consequences and effects in the construction environment, such as time overruns, cost overruns, disagreements, and finally project failure.

2.2.6 Oversight of Critical Quality Factors During the Project

[17], explain that project quality is often side-lined and insufficient attention paid on quality because project teams are more focused on the time and cost factors. [18] list some items that need to be addressed in order to improve on quality which positively affects the success of a project, these items are; designs, project management, materials, workmanship, quality management systems and methodologies.

2.2.7 Inadequate Attributes and Capabilities Endorsed by the Project Manager that Hinder the Implementation of a Good Project Management Approach

[19] validate that being a manager that leads is an important aspect in managing projects and performing at a high level as a leader is a vital manager's obligation to ensure the effort of people are aimed towards the same objective. [20] reinforces the perception that managers and leaders mould the environment and efficiency of the work surroundings. [21] classify the causes that are challenging the successful implementation of a good project management approach as; lack of leadership, failure to coordinate, inexperience, incompetence and lack of commitment.

2.3 Factors Impacting on the Delivery of Umgeni Water Project

There are many factors that can impact either negatively or positively on a project. These include: [22] insufficient planning and designing, [13] that reflecting incorrect costs at the beginning of a project has serious effects on the project, [19] poor project management adversely affects three fundamental aspects of a project, which are, cost, time and quality. [23] risks have the most influence during the construction phase if not managed correctly, client's level of participation and contribution on project delivery has enormous influence on the project, [17] state re-work is a product from poor quality both in designs and on the construction site. Further, he explain that redoing work due to poor quality is extremely costly and adversely impact on project time, and [15] identify that incomplete, inaccurate or misunderstood and these communication problems have a knock on effect on time, cost and quality.

3 Research Methodology

The focus of the study was to identify and assess challenges and its associated impact on project delivery in Pietermaritzburg, South Africa. The sampling frame consisted of Project Managers (10), Civil Engineers (5), Planning Engineers (3), Quantity Surveyors (2), Servitude Administrators (2) and Environmental Project Manager (1).

The sample size formula used: $n = \frac{N}{1+N(e)2}$

n = required sample size, N = the applicable population size, and e = precision or e-value

The study adopted a case study research approach; therefore, a non-probability sampling approach was employed due to limitation in number within the sample frame, namely, a selective method of sampling. The research instrument for this study was a questionnaire survey. These were administered through e-mail and were received through the same means. A total of twenty-three (23) questionnaires representing 88.8% response rate achievement recorded on questionnaire administration. Differential statistics statistical tool was used for data analysis. The sample frame consist of Project Managers, Civil Engineers, Planning Engineers, Quantity Surveyors, Servitude Administrators and Environmental Project Manager. Regarding the qualification of respondents, bachelors' degrees predominate. Most of the respondents have working experience above six years and have handled large project more than six. Based on these, the data obtained can be deem reliable. The Cronbach alpha values were >.70, which is satisfactory.

3.1 Data Presentation and Analysis

This section presents the data obtained of the study and the analysis.

Table 1 presents the respondents' rating of the influence of project delivery challenges encountered during the Project Life Cycle at Umgeni Water. It is notable that all factors in the category have MSs $> 2.60 \leq 3.50$, which indicates that the factors have between a near minor to moderate/moderate influence on project delivery at Umgeni Water.

Table 1. Project delivery challenges encountered during the Project Life Cycle at Umgeni Water

S/No	Factors	Mean score	Standard deviation	Rank
1	Insufficient planning and designing done during the project	3.5	1.2	1
2	Inadequate attributes and capabilities endorsed by the project manager that hinder the implementation of a good project management approach	3.3	1.3	2
3	Unsatisfactory cost estimating resulting in errors and oversights	3.2	1.0	3
4	Inadequate participation and contribution by the client during the project	3.1	1.3	4
5	Communication problems within the project	3.1	1.2	5
6	Oversight of critical quality factors during the project	2.8	1.3	6
7	Project risks that are poorly identified, assessed, mitigated and controlled during the project	2.8	1.4	7

The factor with the most significant influence is insufficient planning and designing done during design stage of the project (MS = 3.5). It is evident that when a project is not properly thought out, planned and designed, these short-falls are felt during the construction phase noticeably affecting time, cost and quality.

Following this factor is inadequate attributes and capabilities endorsed by the project manager that hinder the implementation of a good project management approach (MS = 3.3). A project manager is an individual who is appointed to steer the project team that is accountable for accomplishing the project purposes and has the necessary abilities to apply understanding, capabilities and tools to project activities in meeting project objectives. It could be catastrophic to a project, should a project manager be incompetent, inexperienced or lack leadership skills.

The least significant factor is project risks that are poorly identified, assessed, mitigated and controlled during the project (MS = 2.8). Project risk is a specialised field of expertise and requires knowledge with experience to execute successfully. It is normally assessed and controlled by consultants and contractors, therefore this may be the reason why the factor is the least influential.

Impact on Project Delivery

Table 2 presents the respondents rating regarding influencing factors that impact on project delivery at Umgeni Water. All factors in this category have MSs > 2.80 3.10, which indicates that these factors have between a near minor to moderate/moderate influence on infrastructure project delivery at Umgeni Water.

The most significant influencing factor impacting on project delivery at Umgeni Water is planning and designing (MS = 3.1). The probable reason for this is that respondents are fully aware of time constraints given to properly plan and design including not putting enough emphasis into front-end loading at the beginning of a project.

Table 2. Influencing factors impacting on project delivery at Umgeni Water

S/No	Impact	Mean score	Standard deviation	Rank
1	Planning and designing	3.1	1.1	1
2	Influence cost estimating	3.0	1.3	2
3	Implementation of a good project management approach	3.0	1.2	3
4	Level of risk management	2.9	1.2	4
5	Level of participation and contribution by the client	2.8	1.3	5
6	On quality of product	2.8	1.2	6
7	Quality of project communication	2.8	1.2	7

The second most significant influencing factor is cost estimating (MS = 3.0). The most likely reason for this is that respondents realise at Umgeni Water estimates are generally calculated based on incomplete project scope definition, designs, specifications and information.

The third most significant influencing factor is implementation of a good project management approach (MS = 3.0). The most possible reason for this is that respondents are conscious that Umgeni Water employs young inexperienced project managers to undertake complex projects. These project managers tend to have weak leadership skills and lack coordination capabilities.

The least significant factor in this category is project communication (MS = 2.8). The most probable reason for this could be that Umgeni Water has effective communication procedures in place that addresses matters around project communication.

3.2 Discussion

Findings from the data analysed through mean score ranking indicates that insufficient planning and designing done during the project is the most influential project delivery challenge at Umgeni Water, which is consistent with [10], conclusion that the lack of effort invested at the beginning of the project is the most significant problem that hinders project delivery. The second most influential project delivery challenge is inadequate attributes and capabilities endorsed by the project manager that hinder the implementation of a good project management approach. This is supported by [20], who argue that employing a weak project manager to steer a project to its successful completion is very unlikely to occur. [20] further explain that the most substantial challenge that obstructs project delivery is incompetent and inexperienced project managers. Unsatisfactory cost estimating resulting in errors and oversights is the third ranked influence, which is in alignment with [24], deduction that commencing and continuing with incorrect cost estimates can be the single most crippling factor to a project.

The findings do not back the view of prior researchers, who found, inadequate participation and contribution by the client during the project, communication problems within the project, oversight of critical quality factors during the project and project risk management as very significant problems that have a major influence on project delivery. [5, 16, 25], and [11]. These results may be due to the fact that Umgeni Water has good client relations, established project communication systems, and adequate quality and risk management procedures in place.

According to [26], water infrastructure is still one of South Africa's most important requirements and 20% of South Africans live without clean drinkable water. Regrettably [27] reported that in South Africa the delivery of large scale water infrastructure projects decreased by 30% from 2015 to 2016. It is becoming increasingly apparent that currently and in the future, water infrastructure project implementation will be at the forefront of the construction industry, which will require, the need for sound project implementation strategies and models to achieve successful project delivery.

4 Conclusions and Recommendations

4.1 Conclusion

Based on the analysis of data, the following conclusions are reached:

Insufficient planning and designing done during the project and inadequate attributes and capabilities endorsed by the project manager that hinder the implementation of a good project management approach are the most significant project delivery challenge at Umgeni Water. Unsatisfactory cost estimating resulting in errors and oversights is observed as challenge in the delivery of Umgeni water.

4.2 Recommendations

The following are recommendation to mitigate these project delivery challenges at Umgeni Water:

The project team should endorse a front-end loading approach when implementing projects and sufficient time must be given to the planning and designing team to execute their work successfully in order to mitigate insufficient planning and designing issues experience on the project.

Regular checks and updates must be done to determine if the cost estimates and time periods are still valid and applicable. Umgeni Waters estimating there is adequate project scope defined to start the estimating process to guide against adverse influence on cost estimating.

The organisations project stakeholders must hold regular meetings to assist the project manager. Relative to the evaluation of the progress and settling issues of concern to ensure smooth progression of the project. This will positively influencing the implementation of a good project management approach.

References

1. Wideman, M.R.: Modelling Project Management, pp. 1–20. AEW Services, Vancouver (2003)
2. Project Management Institute (PMI): A Guide to the Project Management Body of Knowledge. PMI, Pennsylvania (2013)
3. Elbeltagi, E.: Construction Project Management (Lecture Notes). Construction Management. Department of Construction Management and Structural Engineering (2009). http://www.dphu.org/uploads/attachements/books/books_2569_0.pdf
4. McDowell, R.: The Benefits of Implementing a Project Management Methodology. (Presentation), Russona Consulting. APMG International, New York (2013)
5. Folifac, A.F.: National Water Policies and Water Services at the extremes: What Challenges must be faced in bridging the gap? Learning from the South Africa Experience (2007). http://www.infrastructuredialogue.co.za/wp-content/uploads/2012/10/National-Water-Policies-and-Water-Services-at-the-extremes.pdf
6. Budget Review-Public-Sector Infrastructure Update. Published by Department of National Treasury, Research conducted by National Treasury (2016)
7. Kulatunga, K., Amaratunga, D., Haigh, R.: Researching "Construction Client and Innovation": pilot study and analysis (2008). http://usir.salford.ac.uk/9816/1/construction_client.pdf
8. Chan, C.A., Scott, D., Chan, L.A.: Factors affecting the success of a construction project. J. Constr. Eng. Manage. **130**, 153–155 (2004)
9. Trigunarsyah, B., Al Solaiman, S.: The impact of client involvement on project performance: case of the kingdom of Saudi Arabia. In: Proceedings of the CIB World Building Congress Conference, Tampere, Finland, pp. 258–269 (2016)
10. van der Weijde, A.G.: Front-End Loading in the Oil and Gas Industry. Delft University of Technology. M.Sc., Delft University of Technology (2008)
11. Smith, J.N., Merna, T., Jobling, P.: Managing Risks in Construction Projects. Blackwell Publishing, Oxford (2006)
12. Al-Shibly, H.H., Louzi, B.M., Hiassat, M.A.: The impact of risk management on construction projects success from the employees perspective. Interdisc. J. Contemp. Res. Bus. **5**(4), 12–43 (2013)
13. Odendaal, M.M., Vermaak, F.N.S., du Toit, E.: Cost estimation and management over the life cycle of metallurgical research projects. South. Afr. Bus. Rev. Spec. Ed. **2015**, 137–156 (2015)
14. Ojo, K.G., Odediran, J.S.: Significance of construction cost estimating risks in Nigeria. Int. J. Civ. Eng. Constr. Sci. **2**(1), 1–8 (2015)
15. Renault, Y.B., Agumba, N.J.: The issue of communication in the construction industry: a case of South Africa. In: Joint International Conference (JIC) on 21st Century Human Habitat: Issues, Sustainability and Development, 21–24 March 2016, Akure, Nigeria, pp. 1729–1738 (2016)
16. Gamil, Y., Rahman, A.I.: Identification of causes and effects of poor communication in construction industry: a theoretical review. Emerg. Sci. J. **1**(4), 239–247 (2018)
17. Gwahula, R., Augustine, W.P.: An assessment of critical factors affecting quality performance of government financed construction projects: evidence from Tanzania: business management and strategy. Bus. Manage. Strategy **7**(2), 82–101 (2016)
18. Delgado-Hernandez, J.D., Aspinwall, E.: A framework for building quality into construction projects. Total Qual. Manage. **19**(10), 1013–1028 (2008)

19. Andi, A., Santoso, W.K., Tali, C.R.: Basic, actual and effectiveness of leadership styles of construction project managers in Surabaya. Civ. Eng. Dimension **6**(1), 32–39 (2004). http://puslit.petra.ac.id/journals/civil
20. Otara, A.: Perception: a guide for managers and leaders. J. Manage. Strategy **2**(3), 21–24 (2011)
21. Hyvari, I.: Success of projects in different organizational conditions. Proj. Manage. J. **37**(4), 31–41 (2006)
22. Olawale, Y., Sun, M.: Cost and time control of construction projects: inhibiting factors and mitigating measures in practice. Constr. Manage. Econ. **28**(5), 509–526 (2010)
23. Banaitiene, N., Banaitis, A.: Risk Management in Construction Projects (2012). http://dx.doi.org/10.5772/51460
24. Torp, O., Klakegg, J.O.: Challenges in cost estimation under uncertainty—a case study of the decommissioning of Barsebäck nuclear power plant. J. Adm. Sci. **6**(4), 1–21 (2016)
25. Saeed, N.M.N., Hasan, S.A.: The effect of total quality management on construction project performance. J. Sci. Technol. **17**(2), 11–30 (2012)
26. Petterson, D.: Water infrastructure – unlocking SA's economy. Infrastructure News and Service Delivery (2016). www.infrastructurene.ws/2016/05/27/water-infrastructure-unlocking-sas-economy
27. Naidoo, R.: Water construction projects take a dive in 2016. Infrastructure news and Service Delivery (2016). www.infrastructurene.ws/2016/12/…/water-construction-projects-take-a-dive-in-2016.pdf

Exploring the Quality Management Methods Adopted by Contractors in Fast-Track Construction Projects in Eastern Cape

Fidelis Emuze and Michael Oladokun(✉)

Department of Built Environment, Faculty of Engineering, Built Environment and Information Technology, Central University of Technology, Bloemfontein, Free State, Republic of South Africa
{femuze,moladokun}@cut.ac.za

Abstract. The aim of this research is to explore the quality management methods adopted by contractors in fast-track construction projects by identifying the challenges experienced by the contractors and how these challenges are mitigated. To achieve the study objectives, a qualitative case study research design was applied using the narratives, in form of interviews, extracted from the contractors, clerk of works, quantity surveyors and project managers. The case studies were from projects in East London, Queenstown and Aliwal North in the Eastern Cape Province of South Africa. The findings from the study suggest that majority of those interviewed have an understanding of the concept of fast-track construction. The study also discovered that the most essential quality management methods adopted by contractors in fast-track construction were checklist, design-build, schedule compression technique, quality control plan, and Plan-Do-Act-Check methods. The study concludes that the fast-track construction approach can aid construction projects to be completed on time without jeopardising the quality standard. However, the contractors need to be meticulous in the application of the fast-track construction approach considering the challenges inherent therein.

Keywords: Construction industry · Eastern Cape · Fast-track construction · Quality management · South Africa

1 Introduction

The construction industry has been an epicentre of criticism over the years considering its performance and productivity when compared to other industries [1]. As such, there is the need to challenge the management approaches being used by the construction companies. Clients demand improved service quality, faster building and innovations in technology. This is why the construction industry has borrowed the concepts such as Total Quality Management (TQM) and Lean from the manufacturing industry [2]. The concept of fast-track construction is also gaining momentum.

Fast-track construction refers to the method of shortening the normal time it takes to carry out different phases of construction projects so that the total time taken for the project is less than the normal time it takes to finish up such projects [3]. That is,

C. Aigbavboa and W. Thwala (Eds.): CIDB 2019, *The Construction Industry in the Fourth Industrial Revolution*, pp. 477–484, 2020.
https://doi.org/10.1007/978-3-030-26528-1_48

fast-tracking compresses the project schedule by progressive design and construction phases going on simultaneously as well as site preparation works and procurement of major long-term delivery are done in parallel [4]. It is worthy of note that in traditional construction on the one hand, projects follow a predetermined sequence, concept, schematic design, design development, construction documents and construction administration [4]. On the other hand, fast-track construction overlaps these stages to allow for construction to begin earlier and proceed at a faster pace [4]. The main challenge in the construction sector is to create certainty about delivery on time, on budget and to the targeted quality and safety. Fast-track construction strategy is therefore being used to achieve shorter project duration [1].

Many contractors that participate in fast-track construction use many fast-track techniques, such as temperature-controlled concrete, the adoption of high-strength and high-performance concrete, and the use of ready mix and pumped concrete. Slip forms and moving forms are other techniques that speed up construction, which in turn demand a detailed planned parallel working on reinforcement. Other techniques include the use of pre-fabricated structures, adoption of modular construction techniques, the use of left-in shutter reinforcement and the adoption of fly ash concrete and self-compacting concrete. The selection of technology, methodology, materials along with trained man power, proper planning and co-ordination are essential to a successful fast-track construction [4].

Quality management in fast-track construction, unlike in standard construction projects has a unique and detailed procedure for approval. The contractor gets informed at the tender stage about the quality requirements and is required to have quality management plans and a full-time quality assurance manager. Also, the sub-contractors are expected to fill out inspection test plans. With each inspection the relevant parties sign documents that indicate that there is approval of the quality before work can continue to the next stage [5].

However, it is of note that within the realm of construction project management research, most studies of quality management have been in line with standard construction, especially in South Africa. Thus this study explored the quality management processes adopted by contractors in fast-track construction projects in the Eastern Cape Province of South Africa.

2 Methodology

For the objectives of the study to be fulfilled, the research firmed up a research approach capable of answering the research questions. As such, the case study research approach was adopted. Considering the fact that the study aimed at exploring the quality management processes of fast-track construction in the Eastern Cape Province of South Africa and given that the perspectives of contractors in South Africa cannot be generalized without potential errors in the process, the study opted to gain access to contractors' experiences and knowledge engaged in fast-track construction qualitatively. The study used interviews to generate data from four case study projects in the Eastern Cape Province of South Africa. The design and the data collection methods follow the guidelines in Yin [6].

The questions that were included in the interview protocol were first piloted with the experienced academics within the Department of Built of Environment at the Central University of Technology, Free State. This is mainly to ensure the validity and reliability of the instrument of data collection as well as to remove any form of ambiguity in the questions posed. The interviews were basically focused on the project participants that were directly involved with the engagement of quality methods in fast-track construction in a bid to reflect the actual situation on fast-track construction sites. This enabled the researchers to make proper and accurate findings. Participants on the study included the contractors, clerk of works, quantity surveyors and project managers on the case study projects. The interviews took place on-site in their site offices. The interviews were tape recorded before being transcribed. The data were analysed in a thematic manner based on the interview protocol.

It is equality important to briefly report on the four case study projects. The first case study project is a construction of an office block with the client being the Department of Agriculture located in East London. The implementing agent for the project is the Department of Public Works (DPW). The main contractor for the project is S Builders with SDL as the principal agent. The contract is for a duration of 30 months with a contract sum of R 124 000 000. The second case study project is the construction of an early childhood development centre by the Department of Education with the implementing agent being the DPW and project is located in Queenstown. T construction is the main contractor whereas HA is the principal agent for the project. The duration for the project is eight months with a contract sum of R 12 000 000. The third case study project located at Aliwal North of the Eastern Cape Province is the construction of 314 housing units at Sterkspruit with the Department of Human Settlements being the client and implementing agent as the same time. SC contractors is both the main contractor and the principal agent for the project. The duration of the project is for two years with a contract sum of R 52 116 762.30. The fourth and the last case study project is the construction of 539 housing units at Sterkspruit as well belonging to the Department of Human Settlements as both the client and implementing agent. KT is both the main contractor and the principal agent for the project. The duration of the project is for two years at R 72 000 000 as contract sum.

3 The Results and Discussion

This section of the paper thematically presents the findings emanating from the interviews conducted regarding the case study projects. The findings discussed in this section of the paper are based on the understanding of fast-track construction by the interviewees on the case study project, issues relating to quality management on fast-track construction method, and identification of the challenges associated with quality management in fast-track construction technique.

3.1 Theme 1: Understanding of Fast-Track Construction

For this theme, there are quite a number of questions that were posed to the interviewees. Among those questions are questions relating to their understanding of fast-track construction concept, question relating to the methods used in fast-track construction projects, effectiveness of those methods, amongst others.

Regarding the understanding of the concept of fast-track construction, the interviewee 1 for the Case study 1 project identified fast-track construction as a labour-intensive project involving so many labour and tradesmen working concurrently on the project. According to interviewee 2 working as well on Case Study 1 project, based on his experience on the project, he admitted that the fast-track construction method leads to an increased labour on site. Interviewee 3 for the Case study 2 concisely defined fast-track construction as a kind of construction method in which so many activities are done concurrently. The understanding of interviewee 4 working on Case study 3 project is very apt as he put the fast-track construction approach as "the project delivery approach to start construction before the design is complete. It is the overlapping of tasks to reduce the time needed to complete a project. The contractor is selected early enough before the commencement of other processes in such a way that the contractor supplies its input to the design development and bids for the project before the design is completed. In this type of project, a guaranteed maximum cost is provided by the constructor. The purpose is to shorten the time to completion." In the opinion of interviewee 5 working on the Case Study 4 project, the fast-track construction method is "a method that can be used to reduce the overall period of a project." From the responses recorded from the interviewees on the case study projects, it is clear that majority of them have a little bit understanding what the concept of fast-track construction is all about.

For the question relating to the methods used for fast-track construction, interviewee 1 on the Case Study 1 project explained that the Plan-Do-Act-Check approach is being used on the site. He further indicated that they had to use labour scheduling approach in order to know those that will work overtime and the kind of activities that will commence simultaneously without jeopardising the quality standard of the work and safety of the workmen. Apart from that he also mentioned that material scheduling is taking seriously as a way of ensuring fast-track construction project go as scheduled by ordering materials in advance for fast-tracking construction on site. The interviewee 2 in the Case Study 1 project also mentioned that another method adopted includes the use of latest technology and making use of knowledgeable skilled workers to fast-track work on site. The interviewee 3 in the Case study 2 project, based on his experience on the project, was of the opinion that the management and supervision of labourers and the use of machinery are some of the methods they use to fast-track construction on site. He reiterated that these methods are effective when managed correctly. In Case study 3 project, the interviewee 4 mentioned that the method used to fast-track construction is the form of the procurement method used for the delivery of the project, which is Design-Build method. He argued that it is effective in achieving fast-track construction projects when the designer/engineer and the contractor are working as a single-unit on the project. As such, both the design and construction phase can overlap. In this way, construction documents need not to be completely finished prior to the commencement of construction. According to him, this does not only enhance the flexibility for changes during the project itself, but radically shortens the construction timelines, which is the main tenet of fast-track construction approach as enunciated Wearne [7] and Bogus et al. [5].

About the effectiveness of the methods of fast-track construction, the interviewee 4 working on the Case Study 3 project indicated that it is extremely difficult to maintain quality in fast-track construction as the construction phase starts prior to the completion of the design phase. The drawings are not yet complete and the specifications are not in place. The construction methodology needs to be first decided on. He further reiterated that communication and collaboration are crucial for success of the fast-track project. The interviewee 4 on the Case Study 4 project explained that setting and maintaining high quality standards in fast-track construction projects can be a very big challenge. In his words "there are so many moving parts on a daily and weekly basis that it can become quite difficult to oversee. The project team needs to make sure that top quality work from subcontractors still gets achieved without sacrificing the schedule through the use of a formal quality program to be able to communicate to the team members so that they can enforce the execution there of." In short, it can be deduced from the opinions expressed by the interviewees that the effectiveness of the fast-track construction methods still cast shadows in their minds.

3.2 Theme 2: Quality Management Methods Available to Contractors in Fast-Track Construction

Kanji and Wong [8] argued that there is an upsurge in the adoption of quality management as strategy by construction companies to resolve the quality problems and meet the needs of the customers. The quality management methods need to be put in place, utilized and monitored. As such, this sub-section discusses in-depth the insights from the interviewees regarding the quality management methods available to contractors in fast-track construction projects. Majorly, they were asked questions relating the methods in place for achieving quality in fast-track construction projects, regulations governing how the quality management methods are implemented for fast-track constructions, amongst others.

Interviewee 1 in Case study 1 claimed that the methods available differ from existing and new building. He states that "a person with knowledge in that trade carries out quality checks, persons such as an on-site engineer, having different supervisors in each trade to oversee work done and having a knowledgeable foreman." The interviewee 2 working as well on Case Study 1 agreed that "having a good experienced team to overlook each trade on site is one of the most important quality management method in fast-track construction". The interviewee 4 on Case Study 3 declared that to achieve quality during fast-track construction they submit quality management plan to the Department of Human Settlements (DoHS), being the client, before the design stage. This plan defined the roles and responsibilities of the project team members, management and reporting structures and identified the procedures to be followed to ensure satisfactory delivery of all services on site. Further, they also mentioned that the plan covered the necessary documents for the performance of the project's stakeholders for the stages of the project, which includes project inception, preliminary design, detailed design, documentation and procurement, contract administration and inspections and project close-out. They also argued that the quality management plan for the project included a checklist tool of all the information required in each stage to ensure that the activities are not overlooked during the stages. According to interviewee 4, the quality control measures used on site were the use of good quality workmanship in all activities, site staff meetings and testing of materials. Interviewee 5 working on Case Study 4 project stated that every construction project is unique and the approach to setting quality standards must reflect that. Further, he mentioned that it is very important that the construction manager has a quality control plan as they do have in the Case Study 5 project. Accordingly, they developed a pre-plan checklist, a proactive assessment tool that identifies every potential quality related issues that might be encountered on each job.

Regarding the question about the regulations governing the implementation of the quality management methods, Interviewee 1 on the Case Study 1 project stated that all professionals working on the project are registered with their respective Councils. However, the submission of interviewee 4 working on the Case Study 3 project is more apt by explaining that the quality management plan (QMP) is designed and guided by the National Housing Code containing the minimum National technical norms and standards for the creation of houses to be constructed through the application of the National Housing Programme. According to him, it is also guided by the National Home Builders Registration Council (NHBRC). He gave further information that this is in the Act No. 95 of 1998 of the Housing Consumers Protections Measures, which empowers all home builders to register with the NHBRC. The NHBRC has also published the Home Building Manuals and the National Building Regulations, which consists of performance standards for buildings dealing with conventional construction only. Conversely, interviewee 5 working on the Case Study 4 project argued that there are regulations that govern fast-track construction but there are no restrictions from adding where necessary. They usually provide clients with a copy of the Quality Management methods for approval.

On the question relating to workers' adherence to quality management methods on site, interviewee 4 on the Case Study 3 explained that in order to ensure that workers on site adhere to these quality management methods they train the foremen and team leaders to guarantee that they understand the standard of works expected, provide adequate good tools and monitor the works at intervals. Both interviewees 1 and 2 on the Case Study 1 project agreed that constant training and supervision are conducted. Based on the experience of interviewee 5 on Case Study 4 project, he indicated that a quality control plan (QCP) is provided to sign for each project and have weekly meetings to see if the QCP is still in order and if there are any adjustments to be made. This method is used to ensure that workers on site adhere to quality management methods.

A question was also posed to the interviewees on whether or not the way quality is managed in the fast-track construction and convectional/traditional construction projects are the same. Interviewee 4 on Case Study 3 project stated that quality management is highly required in fast track projects as the speed of works is high and the design phase may not be complete in order to know the specifications. The response of interviewee 5 on Case Study 4 project aligned with that of interviewee 4 and further reiterated that the quality management methods are not managed the same way as in conventional and fast-track construction. Conventional projects have a lot more room for errors that fast track construction eliminates. The interviewees 1 and 2 on Case Study 1 project agreed that the management methods are not managed the same because the fast-track construction projects require more supervision than the conventional construction projects.

3.3 Theme 3: Identifying the Challenges Associated with the Quality Management Methods in Fast-Track Construction

It is important to understand that quality management methods are implemented by construction companies to ensure good quality standards and to achieve targeted timeframes; however, if these are not done correctly it creates challenges to fast-track construction. To this end, this sub-section reports the findings regarding the identification of challenges associated with the quality management methods in fast track construction and how to overcome those challenges.

According to interviewee 1 on Case study 1 project, inclement weather conditions, plant and equipment break down, availability of material and delays in availability of detailed drawings are the big challenges associated with the quality management methods in fast-track construction. The response of the interviewee 2 on Case study 2 project mentioned that the attitude from labourers and inclement weather conditions are part of the challenges they do encounter in fast-track construction projects. Interviewee 5 on Case Study 4 project indicated that every project is unique and faces challenges of its own that is why they have a general quality plan. Tin response to the question, the interviewee 4 working on Case Study 3 project argued that the challenges that are faced when applying quality management in fast-track projects is the involvement of subcontractors and suppliers. He reiterated that quality management methods work against speed and that speed is a requirement in fast-track construction projects. As such, this requires skilled and experienced workforce on site.

In a bid to overcome the challenges, the interviewees were asked about what their respective companies are doing based on the experience gained so far on the fast-track construction projects. Interviewee 3 on Case Study 2 project stated that they organise workshops on site purposely to train workers on communication and working together. Interviewee 1 on Case Study 1, suggested that there must be a maintenance management plan for plants and equipment should any of them breakdown.

4 Conclusions

The paper explored the quality management methods used by contractors in fast-track construction projects. The findings from the study indicates that stakeholders interviewed have the requisite understanding of fast-track construction concept. Though the understanding differs a little bit with big contractors with more experience being more knowledgeable about fast-track construction as compared to small contractors that do not have as much experience.

The study also discovered that the most essential quality management methods adopted by contractors in fast-track construction were checklist, design-build, schedule compression technique, quality control plan, and Plan-Do-Act-Check methods. It was also confirmed that the quality management methods in fast-track construction are very effective when done correctly by a trained work force, though this cast shadows in the minds of some of them. When these quality methods in fast-track construction are done correctly, many fast-track projects can be completed timeously. Implementing these quality management methods effectively can assist is improving working performance. The study discovered that inclement weather conditions, plant and equipment break down, availability of material and delays in availability of detailed drawings are the big challenges associated with the quality management methods in fast-track construction.

In conclusion, the fast-track construction approach can aid construction projects to be completed on time without jeopardising the quality standard. However, the contractors need to be careful in the application of the fast-track construction approach considering the challenges inherent therein. As such, there is the need for the contractors to implement a proper resource management. This may mean that the construction companies will have to offer continuous training to their on-site workers.

References

1. Das, K.A.: Strategies of fast-track mode of construction. Int. Res. J. Eng. Technol. **5**(4), 4995–4999 (2018)
2. Hoonakkera, P., Carayona, P., Loushine, T.: Barriers and benefits of quality management in the construction industry. Total Qual. Manage. **21**(9), 953–969 (2010)
3. Heigers, A.S.: Fast Track Construction in S.A. – A Critical Investigation. Thesis submitted in partial fulfilment of the requirements for the degree of BSc in Construction Management (1988)
4. Kalirajan, L., Sivagnana, T.: Fast-track construction – the need of the hour. In: International Conference on Advances in Materials and Techniques, January 2010
5. Bogus, S., Molenaar, K.R., Diekmann, J.E.: Strategies for overlapping dependent design activities. Constr. Manage. Econ. **24**(8), 829–837 (2006)
6. Yin, R.K.: Case Study Research: Design and Methods, 5th edn. SAGE, California (2013)
7. Wearne, S.H.: Fast-track project duration. Int. J. Project Manage. **2**(4), 240–241 (1984)
8. Kanji, G., Wong, A.: Business excellence model for supply chain management. Total Qual. Manage. **10**(8), 1147–1168 (1998)

Using Case Study Design to Investigate the Delivery of Building Energy Retrofit Projects

Chikezirim Okorafor[1(✉)], Fidelis Emuze[1], and Dillip Kumar Das[2]

[1] Department of Built Environment, Faculty of Engineering, Built Environment and Information Technology, Central University of Technology, Bloemfontein, Free State, South Africa
kezirim042@yahoo.com, femuze@cut.ac.za

[2] Department of Civil Engineering, Faculty of Engineering, Built Environment and Information Technology, Central University of Technology, Bloemfontein, Free State, South Africa
ddas@cut.ac.za

Abstract. The potential for reductions in carbon emissions in the built environment is significant. However, unlocking the potential presents a problem. To better comprehend the problem and advance solutions, a desktop case-study research design on retrofits was adopted to generate textual data. The textual data were analysed. These reveal different innovative and proactive tools required to attain efficiency in the delivery of building energy retrofit projects (BERP). The study unravels the retrofit features of the project, recorded challenges, and lessons learnt from each of the project. Subsequently, a literal replication of the study were carried out and the result shows that no single BER projects are the same in terms of delivery due to many factors such as age, use, materials, thermal mass, location, orientation and occupancy as the study suggested. The research concludes with a motivation for socio-technical approach to retrofitting exercise.

Keywords: Building · Energy · Projects · Retrofit · South Africa

1 Introduction

The built environment consumes a lot of energy. Buildings use about one-third of the world's energy [1]. In the world today, the buildings sector accounts for 40% of the primary energy use, compared to 32% for the industrial sector and 28% for the transportation sector [1]. The use of electric power and heat in the buildings sector also accounts for about 40% of world carbon emissions. However, advances in energy-efficient technologies hold the promise in reducing energy consumption in buildings [1, 2]. In most cases, energy retrofit technologies perform well in theory based upon analytical building models and simulations. However, once the project is completed and occupants move in, the actual performance could be problematic [3, 4]. In furtherance to this assertion, the paper seeks to investigate how projects were delivered internationally with the intention to bridge the performance gap.

C. Aigbavboa and W. Thwala (Eds.): CIDB 2019, *The Construction Industry in the Fourth Industrial Revolution*, pp. 485–496, 2020.
https://doi.org/10.1007/978-3-030-26528-1_49

2 Literature Review

The terms retrofitting have a multifaceted definition based on context. For the purposes of this research, the term retrofitting is defined as the installation of individual or multiple energy efficiency measures to an existing building [5]. An energy efficiency measure is any technology that improves the energy performance of the building, such as loft insulation, advanced heating controls, and renewable energy generation technologies [5, 6]. While some high-performance buildings achieve their environmental performance goals, many do not [3, 4]. Due to global warming and depletion of natural resources, the world has recognised the impact of CO_2 emission and its attendant consequences. The perception is that the built environment is major contributor of this CO_2 emission. In support of the above view, authors such as Lombard assert that existing buildings are major consumer of energy and as such responsible for CO_2 emission, with the majority of the energy being used to power heating, ventilation and air conditioning (HVAC) systems, lighting and office equipment [7]. Building energy retrofit project will bring about economic gains, while enhancing social wellbeing and engendering sustainable development. However, energy retrofit projects encompass additional considerations and requirements in terms of process, material, expertise, and technology. All these factors constitute difficulties in the delivery of the project. According to Miller and Buys, retrofitting existing buildings is more challenging, than designing a new sustainable building from scratch [8]. Sustainable retrofits are viewed as expensive disruptive process. In view of the disruptive process, building occupants show resistance to change [8, 9]. Retrofit processes often include addition of existing walls, opening walling elements, removal and installation of Heating, Ventilation and Air Conditioning (HVAC) elements, and strengthening of frames [10]. This leads to costly consequences such as demolishing, lengthy construction time and occupant relocation. Such disruptions' associated with these processes usually deters building owners from retrofitting their buildings [11, 12]. Despite these impediments, the potentials of energy retrofit of existing buildings remain promising.

3 Research Method

According to Gray and Sutrisna, research is an organised way of exploring a problem to get the best solution [13, 14]. Researchers' chosen methodology provides the foundation for the utilisation of a particular philosophy, approach, strategy and choice. Because of this, the study is situated in an inductive paradigm; a case study research design was adopted. Inductive research entails the search for pattern from observation and the development of explanations–theories-for those patterns through series of hypotheses [15, 16]. This approach aims to generate meanings from the data set collected to identify patterns and relationships to build a theory. Since, Saunders et al and Neuman, has confirmed inductive reasoning as been based on learning from experience, patterns, resemblances and regularities in experience are observed in order to reach conclusions (or to generate theory) [17, 18]. The author has used "bottom-up" approach to describe the phenomenon that is being studied [19]. A desktop case study was chosen to gather data on how energy retrofit of an existing building was carried out in the international arena. Three case studies were conducted on retrofit projects. The projects have been completed, and their document analysis is available in the public domain. The case studies provide greater understanding of how energy retrofits has been done across the globe. The data extracted from the desktop studies included key retrofit features, challenges, and lessons learnt. The projects was selected because of their uniqueness, the public attention they have received, the features of retrofitting that they demonstrate, and the availability of data from numerous sources about the same project.

4 Findings

For clarity, the Table 1 below provides an overview of the desktop cases.

Table 1. Overview of three desktop case studies

Number	Country	Project name
1	Australia	Szencorp Building in Melbourne, Australia
2	Singapore	Zero Energy Building in Singapore
3	China	MGM Macau Resort in Sé, Macau, China

Source: Researchers (2019).

4.1 Desktop Case Study 1 - Australia: Szencorp Building

Key Retrofit Features of the Project: The building were installed with an occupier integrated controls system of sensors, so that services such as air-conditioning and lighting are only provided if the area is occupied; on-site power generation from different sources, including multiple solar panel arrays and a ceramic fuel cell; lift controls and the lift car were completely modernised, for smoother, safer operation and reduced energy consumption; using rainwater capture and grey water recycling for flushing, waterless urinals, dual-flush toilets, low-flow taps, and cut-off sensors on basin faucets; natural ventilation through automated opening windows. The lighting of the building utilises new-generation triphosphor and T5 lamps, and dimmable DSI ballasts controlled via an intelligent occupancy-based system, achieving 1.4 watts per 100 lx. An integrated sensor and management system for occupancy lighting, HVAC and security control were installed. A ceramic fuel cell was installed, which generates low-emission, off-grid energy, with the potential to provide for >30% of the building's energy requirements on-site. Two solar photovoltaic (PV) grids (one amorphous) were also installed, which generate 5.5 kW. With the ceramic fuel cell, these grids will potentially ensure zero grid energy consumption in future. The ceiling height was increased (reclaimed from the old building plenum), allowing for use of thermal mass for improved energy efficiency. Use of the Drykor dehumidification unit, which removes 94% of all micro-organisms and 77% of particles larger than 5 microns from the airspace, which helps to overcome "sick building syndrome", was also introduced thereby increasing the productivity of the occupants. A refrigerant leak-detection and -monitoring system was installed to ease operation.

Recorded Challenges: The challenges include selection of the right retrofit components for the whole building, developing a performance-based design, engineering capacity, and developing financial modelling for the project.

Lessons Learnt From the Project: The Szencorp Group looked at life cycle costing when undertaking this retrofit, as they benefit directly from the building's improved efficiency and performance. The lesson is that the strategy is paying off, with energy savings of over 70% and water savings of 94% less than the industry average in the second year of occupation after project completion. The company is also benefiting from a perceived overall productivity increase of 13% because of improved internal environment quality. As a demonstration and partly experimental project, Szencorp believes "the money and effort expended would be difficult to justify for a single building". However, they have proven that "the investment has more than paid for itself in terms of the learning's, the profile the project has received and the ability of the owner to develop a new level of business services in the rapidly growing market of leading edge green buildings".

4.2 Desktop Case Study 2 - Singapore: Zero Energy Building

Key Retrofit Features of the Project: The building was cooled by natural ventilation. Building occupants in Singapore appreciate some air movement, as it reduces the effective temperature, and the new HVAC and natural ventilation systems provide increased indoor air movement. A solar chimney system was chosen for natural ventilation for the building. Four chimneys on the roof, which are the end of a series of partially hidden ducts along the building envelope, are the most visible part of the system. The system starts with exposed vertical ducts along the west façade, which then bend to follow the curved roof and eventually connect with the prominent central chimneys. When exposed to sunlight, the vertical ducts heat up, creating internal hot air, which expands, becoming lighter, and rises (the buoyancy effect) and, in turn, 'sucks' warm indoor air through various inlets, drawing ambient air through the façade into the interior. In the building, air movement of up to 394 fpm (2 m/s) has been measured and has changed the thermal acceptability from unacceptable to acceptable. This improved thermal comfort was determined through predicted mean vote (PMV) and predicted percentage dissatisfied (PPD) and was reconfirmed through an occupant survey. The cooling system is designed specifically for the tropics. Energy efficiency is achieved by cooling fresh and re-circulated air separately and by having separate fan controls with variable speed to match localised demand. In tackling heat gains, the original building envelope with exposed concrete walls and metal roofs that had little shading would get heated up during the day and would re-radiate the heat into the interior, due to the absence of insulation. The overall strategy was to add a cooling skin to the building envelope. Sunshades and vertical green walls were added on the western side, and the roof received a layer of photovoltaic (PV) modules. The PV roof was elevated about 1 ft (300 mm) off the metal roof and had horizontal gaps between the modules to ensure ventilation and cool the PV modules and the metal roof below. The cooling skins served additional purposes beyond shading. Some sunshades on the façade had PV modules on the upper parts, generating additional electricity. Others had reflective films, doubling as light shelves, redirecting daylight deeper into the building. The green walls and the roof system support the study of their shading and evaporative cooling effect on reducing heat transfer and resulting cooling energy use.

In addressing day lighting, an innovative design concept was to direct the windows towards the sun, or rather to collect the zenith light from the roof and the façade and redirect it to where it is needed. Several advanced day lighting systems were installed and tested for providing daylight for some selected zones, including vertical and horizontal hollow light guides and ducts, external light shelves and customised double-glazing with integrated adjustable blinds, electro-chromic films, and semi-transparent PV modules. In conclusion, the concept of collecting bright zenith light on roofs and façades and directing it into deep building zones was found to be an effective and innovative alternative or supplement to electric lighting, and it provided excellent colour neutrality. However, this solution required more space and planning compared to electric lighting, and it slightly increased the mean radiant temperature, by 1 °F (0.5 °C). While integrating the photovoltaic. The energy target for the building was to be net-zero, i.e., to produce as much electricity as the building consumes over the course of one year. As there is no heating required, all energy was electric for air-conditioning, ventilation, lighting and plug loads, which was estimated to be about 706,300 kBtu (207 MWh), or 14.6 kBtu/ft^2 (55.3 kWh/m^2) per year. To produce an equivalent amount of electricity with PV modules, it became clear that the building roof would need to be completely reserved for PV modules. After a few iterations to define the benefits of electricity generation with PV modules versus energy savings through solar chimneys, roof greening, or reflective coatings, a PV system of 190-kWp capacity, covering some 16,577 ft^2 (1,540 m^2), was chosen. A large grid-connected system designed to produce a maximum electricity yield was installed on the roof. Therefore, a performance-based invitation to bid was launched. The supplier had to guarantee a certain amount of electricity production, which provided the motivation to install as well as operate and maintain the PV system efficiently. PV systems were also installed on the façades, designed here to demonstrate the variety of PV technologies and their multi-functionality, such as serving as sunshades, railings, opaque and semi-transparent walls, and windows. These smaller systems were off-grid, meaning that their DC electricity was consumed on the spot by a cell phone charger. Both grid-connected and off-grid systems are owned and operated by the BCA, following the requirements for electrical power systems set by Singapore's Energy Market Authority (EMA) and the design guidelines on conservation and development control formulated by the Urban Redevelopment Authority (URA).

Recorded Challenges: A further challenge is that space use may change over time. Here, some of the classroom spaces planned for natural ventilation was converted to air-conditioned spaces with a different use. Responding to increased energy use is another key take-away point. This requires constant monitoring to identify areas for further energy savings. For example, the initial lighting was using T5 lamps, but after replacing them with LED lamps, the energy consumption was reduced by about 40%, partially absorbing the increased energy consumption for the enlarged air-conditioned space. There were some difficulties, mainly due to the lack of experience and artisanship in installing green building technologies properly on-site. This was especially true if it was the first of its kind, such as the solar chimney system and the PV façades. Most of the extra work could fortunately be supported by the accompanying research projects, which also brought in foreign experts and their experience. What also turned out to be essential was the call for a performance-based arrangement for the building-integrated PV system, unlike the usual capacity-based arrangement. For the performance-based arrangement, the supplier had to ensure that the specified annual electricity generation is achieved.

Lessons Learnt From the Project: Many lessons were gathered regarding the generation of accurate energy models, enabling of monitoring and verification, designing for maintenance, and responding to increasing energy use. The integrated design process with all stakeholders at the early stage of the project was beneficial in setting the stage and identifying best practices. The design-build-operate approach was also beneficial, as it considered the operational costs, too, which are usually ignored in the standard design-build-sell approach. Simulations on energy savings and yield were instrumental in sizing different energy systems. However, building accurate and integrated energy models with occupancy schedules and dynamically responsive systems was challenging. Occupancy schedules are very difficult to predict, but their resulting energy loads have a strong impact on the predicted energy consumption of a building. Actual and predicted occupancy schedules usually differ, especially if the prediction is outdated. The planning of the project included some reserves, e.g. for extension of air-conditioned spaces. In fact, the energy use intensity (EUI) of the building has increased by 15% over the first two years, and it keeps increasing, due to converting more of the naturally ventilated spaces into air-conditioned zones. Nevertheless, with additional energy efficiency measures, it has remained a net-zero-energy building over the first five years.

Building information modelling (BIM) was used to create and communicate design aspects. However, not all of the green building systems were a part of the standard building products library, and they had to be created and added first. Multifunctional objects, such as electricity-producing semi-transparent PV windows, are difficult to represent in BIM. Traditionally, windows only have thermal and optical properties, and not electrical, and PV modules have only electrical properties, and even if they are integrated in the building envelope, they remain mono-functional energy generators. Therefore, BIM was used for integration and communication purposes, but not as a front end for energy performance simulation. Energy simulations were performed independently from BIM, with the locally prevailing tools for PV system sizing, building energy performance, and HVAC and day-lighting systems.

4.3 Desktop Case Study 3 - China: MGM Macau Resort

Key Retrofit Features of the Project: The chilled water network was designed as a de-coupler system, with two de-coupler bypasses. The chilled water design supply and return temperatures were 7 °C and 12 °C (45 °F and 54 °F), respectively. The team disabled the de-coupler bypasses and converted the system into a full variable primary system. The result was a reduction of 876,196 kWh (41%) of pumping power in the first year post-modification. The secondary sides of the –7 °C (19 °F) heat exchangers' differential bypass were replaced with a variable-speed pumping system, using variable-speed drives. The on-off-type control valves on the primary sides of the heat exchangers were replaced with modulating valves. This has reduced the pumping power on the secondary sides of the plate heat exchanger and has improved the distribution transformer on both sides. The oversized cooling water pump impellers were trimmed, with all the springs inside the constant-flow valve serving each chiller removed.

The cooling water flow was balanced using manual butterfly valves. Variable-speed drives were installed on all cooling tower fans. During spring and autumn, there were 2 × n numbers of cooling fans in operation, with n number of operating chillers. Chilled water supply temperature was reset linearly and inversely against outdoor air temperature, to minimise chiller power use, by reducing chiller lift between evaporators and condensers. The boiler plant initially consisted of two liquefied petroleum gas-fired (LPG-fired) steam-tube boilers, with an output of 7,800 kg/h at 10 bar (17,196 lb/h at 4,014 in. w.c.). The transferred energy was provided to facilities in the hotel and casino areas for pool heating, laundry, kitchens, space heating, and domestic hot water. After a feasibility study, the two larger boilers were replaced with two 1,500 kg/h (3,301 lb/h) smaller boilers as the primary source of steam for kitchens, laundry, and the pool, with three new 880 kW (250 ton) heat pumps at N + 1 configuration. The heat pumps were installed to serve as the heat source for space heating, space dehumidification reheat, and domestic hot-water heating. One larger boiler was kept and not removed, to cater for extremely low-temperature weather, e.g. below 8 °C (46 °F) outdoor temperature. There are 1,000 fan coil units (FCUs) in the nearly 600 guest rooms, suites and villas. All existing capacitance motors/fans were removed from the FCUs and replaced with electronic commutated motors. All FCUs are connected via a wireless mesh network through digital thermostats to the hotel management system, which can determine the room temperature and the on/off status of the FCUs, depending on occupancy.

Other retrofit measures such as variable-speed drives were installed for exhaust fans in busy kitchens, so the chefs can select cooking mode (normal speed) or standby mode (reduced speed) of fan operations. Air handling unit (AHU) economiser mode was introduced to enable free cooling of indoor spaces when the outdoor temperature is low. Enthalpy wheels were also installed in AHUs to facilitate pre-cooling/pre-heating of outdoor air. Programmable thermostats are used to control runtimes, fan speeds, and minimum adjustable temperatures of the FCUs in the back of house office areas. Over 95% of the traditional halogen, filament, fluorescent, and cold cathode lighting at MGM Macau was replaced with LED lighting, resulting in an annual electricity reduction of 4,807,462 kWh. In addition, timer switches are used in plant rooms, and daylight sensors are used to control the operations of aesthetic façade lighting on the building perimeter. Introduction of the primary variable system has helped to significantly reduce the maintenance cost of the chiller plant, through the removal of the 10 primary chilled-water pumps and the related pipe-work. Energy-monitoring dashboards were installed to enhance staff buy-in and visibility. MGM Facilities Management introduced a plant-monitoring energy dashboard, which tracks the energy performance of both the chillers and the heat pumps. Regular reporting is done on the chillers and the heat pumps, and month-to-date electricity consumption of various components is measured. Monthly utility summaries are created automatically and are reviewed regularly, to ensure that the chiller plant and the heat pumps are running in optimal condition. Energy consumption data is stored in 15-minute intervals, which allows International Performance Measurement and Verification Protocol (IPMVP)-compliant energy-saving projects, analysis of plant operating data through the monitoring-based commissioning (MBCx) platform, and record-keeping for the ISO50001 system, in which MGM Macau is certified. By deploying the energy-performance dashboards in addition to the Building Management Systems (BMS) functional dashboards, the contractors have strengthened data visibility and have made energy management part of the BMS control operation. This allows feedback on initiatives and operational issues, where lessons are learnt and mistakes are not repeated.

Recorded Challenges: The MGM Grand Hotel project is very complex, influenced by a wide range of HVAC systems. Several challenges were experienced such as ability to quantify and compare the relative cost and performance attributes of a proposed design in a realistic manner and even integration of experts in the project and optimal installation of technical requirements, as envisaged during the start-up of the project.

Lessons Learnt From the Project: Building energy retrofit is better served where building energy simulation is properly undertaken and tasks are properly assigned and integrated. This will help to engender sustainability in the delivery process.

5 Discussion of the Findings from the Cross-Case Analysis

The cross-case case analysis draws upon the findings of each case study and sets out to find literal replication across each case. This analysis provides significant evidence of literal replication. It is summarised in Table 2.

Table 2. Literal replication from the cross-case analysis

Theme	Szencorp Building, Australia	Zero Energy Building, Singapore	MGM Macau Resort, China
Retrofit features	Occupier integrated controls system of sensors, on-site power generation from different sources, including multiple solar arrays and ceramic fuel cell, and natural ventilation throughout. Automated opening windows were also installed. The lighting utilises new-generation triphosphor and T5 lamps, dimmable DSI ballasts controlled via an intelligent occupancy-based system, achieving 1.4 watts per 100 lx. An integrated sensor and management system for occupancy lighting. HVAC and security control were also installed. The ceiling height was increased, allowing for use of thermal mass for improved energy efficiency	An integrated approach was adopted in engendering deep retrofit in the building. For example, solar chimney systems were chosen for natural ventilation for the building. In tackling thermal gains, the strategy was to add a cooling skin to the building envelope. The cooling skin served additional purposes beyond shading. Some sunshades on the façade had PV modules on the upper parts, generating additional electricity. Others had reflective films, doubling as light shelves, redirecting daylight deeper into the building. The green walls and the roof system support the study of their shading and evaporative cooling effect on reducing heat transfer and resulting cooling energy use. In addressing day lighting, an innovative design concept was to direct the windows towards the sun, or rather to collect the zenith light from the roof and the façade and redirect it to where it is needed	• The chilled water network was designed as a de-coupler system, with two de-coupler bypasses. The team disabled the de-coupler bypasses and converted the system into a full variable primary system. Chilled water supply temperature was reset linearly and inversely against outdoor air temperature, to minimise chiller power use, by reducing chiller lift between evaporators and condensers. There are 1,000 fan coil units (FCUs) in the nearly 600 guest rooms, suites and villas. All existing capacitance motors/fans were removed from the FCUs and replaced with electronic commutated motors. All FCUs are connected via a wireless mesh network through digital thermostats to the hotel management system, which can determine the room temperature and the on/off status of the FCUs, depending on occupancy. *Energy-monitoring dashboards were installed to* enhance staff buy-in and visibility. MGM Facilities Management introduced a plant-monitoring energy dashboard, which tracks the energy performance of both the chillers and the heat pumps.

(*continued*)

Table 2. *(continued)*

Theme	Szencorp Building, Australia	Zero Energy Building, Singapore	MGM Macau Resort, China
			Regular reporting is done on the chillers and the heat pumps, and month-to-date electricity consumption of various components is measured. Monthly utility summaries are created automatically and are reviewed regularly, to ensure that the chiller plant and the heat pumps are running in optimal condition
Recorded challenges	Selection of the right retrofit components for the building, developing a performance-based design, engineering capacity, and developing financial modelling for the project	The use of building space use may change over time, considering its nature. There were some difficulties, mainly due to the lack of experience and artisanship in installing green building technologies properly on-site. Most of the extra work was supported by the accompanying research projects, which also brought in foreign experts and their experience	The project is very complex, influenced by a wide range of HVAC systems. Several challenges were experienced such as ability to quantify and compare the relative cost and performance attributes of a proposed design in a realistic manner and even integration of experts in the project and optimal installation of technical requirements, as envisaged during the start-up of the project
Lessons learnt	The Szencorp Group looked at life cycle costing when undertaking this retrofit	The integrated design process with all stakeholders at the early stage of the project was beneficial in setting the stage and identifying best practices. The design-build-operate approach was also beneficial, as it considered the operational costs, too, which are usually ignored in the standard design-build-sell approach	Building energy retrofit is better served where building energy simulation is properly undertaken and tasks are properly assigned and integrated

Source: Researchers (2019).

6 Conclusions and Way Forward

From the case studies reviewed, it can be deduced that no energy retrofit project deliveries are the same. The retrofitted case study buildings show that each building has its own unique retrofit features. This can be attributed to different factors, such as the building orientation, the building assessment, the detailed energy survey, the technical analysis, the retrofit technologies used, the cost-benefit analysis, the nature of the building, the implementation plan, and the geographical location, as identified in the study. This observation resonates with the suggestion that no single intervention is capable of delivering substantial reductions in energy usage, and, instead, a series of measures is uniquely required as per each project [20]. This is shown in the reviewed case studies. While there are numerous technologies that have the potential to address carbon emissions in existing buildings, there remain problems associated with their deployment, which is clearly identified in the study. This entire consideration makes each building energy retrofit project optimisation a problem. However, in addressing this problem, an integrated approach should be adopted and tasks should be properly assigned and integrated [20, 21].

In sum, this paper provides a detailed technical description of three cases. The reviewed projects show innovative ways to deliver energy retrofit in existing buildings. Techniques available are not limited to using green technologies to improve the productivity of the workforce, indoor air movement and lighting. Nevertheless, various challenges highlighted in the study need to be addressed. Such impediments drag the concept of energy retrofit backwards in curbing CO2 emission. The action should be targeted at making the best use of a situation or resource in implementing energy efficiency (EE) drives in any energy retrofit project delivery. The guiding principle should be taking each project life cycle as a point of departure in arriving at right retrofit components for the whole building, developing a performance-based design, scouting for engineering capacity, and developing financial modelling for the delivery of the project. The limitations of the desktop cases require future inquiries that embrace pragmatism as a philosophical approach so that both breadth and depth of knowledge on energy retrofit can be developed.

References

1. Wolfe, A.K., Malone, E.L., Heerwagen, J., Dion, J.: Behavioral Change and Building Performance: Strategies for Significant, Persistent, and Measurable Institutional Change: Prepared for the U.S. Department of Energy under Contract DE-AC05-76RL01830 Pacific Northwest National Laboratory Richland, Washington 99352 (2014)
2. Abrahamse, W., Steg, L., Vlek, C., Rottengatter, T.: A review of intervention studies aimed at household energy conservation. J. Environ. Psychol. **25**, 273–291 (2005)
3. Turner, C., Frankel, M.: Energy Performance of LEED for New Construction Buildings. New Buildings Institute, Vancouver, WA (2008). www.usgbc.org/ShowFile.aspx?DocumentID=3930
4. Cole, R.J.: Motivating stakeholders to deliver environmental change. Build. Res. Inf. **39**, 431–435 (2011)

5. Winkler, H.: Long-term mitigation scenarios: technical appendix. Prepared by the Energy Research Centre for the Department of Environmental Affairs and Tourism, Pretoria, October 2007. http://www.erc.uct.ac.za/Research/publications/07-Winkler-LTMSTechnical percentage20Report.pdf. Accessed 15 Aug 2018
6. Unruh, G.C.: Understanding carbon lock-in. Energy Policy **28**(12), 817–830 (2010)
7. Lombard, L.J.: Empirical study into the benefits of the retrofitting of existing commercial buildings to improve energy efficiency. MSc research report. University of the Witwatersrand, Johannesburg (2012)
8. Miller, E., Buys, L.: Retrofitting commercial office buildings for sustainability: tenants expectations and experiences. In: Management and Innovation for a Sustainable Built Environment, vol. 3, pp. 33–50 (2011)
9. Ma, Z., Cooper, P., Daly, D., Ledo, L.: Existing building retrofits: methodology and state-of-the-art. Energy Build. **55**, 889–902 (2012)
10. Wilkinson, S.: The sustainable retrofit potential in lower quality office stock in the central business district. In: Management and Innovation for a Sustainable Built Environment. Amsterdam (2011)
11. Cheung, M., Foo, S., Granadino, J.: Seismic retrofit of existing buildings: innovative alternatives. Public works and government services, Canada, pp. 1–10 (2000)
12. Wilkinson, S.: Analysing sustainable retrofit potential in premium office buildings. Struct. Surv. **30**, 398–410 (2012)
13. Gray, D.E.: Doing Research in the Real World, 3rd edn. SAGE publishing Ltd, London (2014)
14. Sutrisna, M.: Research methodology in doctoral research: understanding the meaning of conducting qualitative research. Working paper presented in ARCOM Doctorial Workshop. John Moores University, Liverpool, 12 May 2009
15. Goddard, W., Melville, S.: Research Methodology: An Introduction, 2nd edn. Blackwell Publishing, Lansdowne (2004)
16. Bernard, H.R.: Research Methods in Anthropology, 5th edn, p. 7. Alta Mira Press, Lanham (2011)
17. Saunders, M., Lewis, P., Thornhill, A.: Research Methods for Business Students, 6th edn. Pearson Education Limited, Harlow (2012)
18. Neuman, W.L.: Social Research Methods: Qualitative and Quantitative Approaches. Allyn and Bacon, Boston (2003)
19. Lodico, M.G., Spaulding, D.T., Voegtle, K.H.: Methods in Educational Research: From Theory to Practice, p. 10. Wiley, San Francisco (2010)
20. Hewitt, N.J.: Retrofit of existing buildings – a major technical, economic and social challenge. Int. J. Ambient Energy **33**(1), 1 (2012). https://doi.org/10.1080/01430750.2012.666893
21. Pielke, R.A.: The carbon economy and climate mitigation: an editorial essay. WIREs Clim. Change **1**, 770–772 (2010)

Developing a Framework for Deploying Unmanned Aerial Vehicles (UAVs) to Improve Construction Safety Management

Lesiba George Mollo[1(✉)], Fidelis Emuze[1], and John Smallwood[2]

[1] Department of Built Environment, Central University of Technology, Free State, Bloemfontein, South Africa
lmollo@cut.ac.za

[2] Department of Construction Management, Nelson Mandela University, Port Elizabeth, South Africa

Abstract. The use of unmanned aerial vehicles (UAVs) is gaining traction in the construction industry. UAVs are used to address productivity, inspection and safety matters. This paper presents a conceptual framework that would help to deploy UAVs on construction sites for safety inspection purposes. The paper was compiled from the data obtained in the literature. The keywords unmanned aerial vehicles and safety in construction were used to search the relevant database. The findings reveal that the use of UAVs would help the construction professionals improve safety on construction sites through the capturing of visual images and video clips on the site project from a bird's eye view. The collected data (images or video clips) could be analyzed to identify risks and hazards that might cause accidents on construction sites. There is major scope for using UAVs to improve construction safety when appropriate guidelines are implemented.

Keywords: Construction · Safety · Unmanned aerial vehicle

1 Introduction

The nature of construction work is complex and unique when compared to other industries. Construction industry embraces different types of risks and hazards that are faced by workers daily. Thus, there is a need to reduce risks and hazards in construction and this need encouraged researchers and practitioners to search for innovative methods, technics, and technologies. Despite innovative methods of construction or the adoption of technologies, safety at work is a complex phenomenon, and often the adopted method of construction works exposes the workers to hazards that might result with accidents causing either injury or fatality [1]. To solve this reported problem, most of the construction organizations have opted to design safe work procedures that have the primary purpose of managing safety on sites [2]. In addition, safety at work is concerned with the workers, and most of the workers have an informal and oral culture of risks, in which safety is rarely openly expressed in the workplace [3].

C. Aigbavboa and W. Thwala (Eds.): CIDB 2019, *The Construction Industry in the Fourth Industrial Revolution*, pp. 497–504, 2020.
https://doi.org/10.1007/978-3-030-26528-1_50

There are several factors contributing to the unsafe working environment in construction. For instance, the worker's safety-related behaviors and their compliance with safety rules are influenced by the site supervisors' decisions or actions. Site supervisors decide how to control certain risks on construction sites, and this makes it necessary for the workers to behave in certain ways [4]. The unsafe behavior of the workers on construction sites is recognized as one of the factors causing accident causation [5]. This is because the knowledge and attitudes of the workers towards safety often differ from individual to individual. Also, safety practices, norms and attitudes are continuously negotiated between the workers and site supervisors [3]. Additionally, it is reported in the HSE [4] report that it is better to eliminate the risks and hazards in a way which are not reliant on human behavior.

To ameliorate reported safety problems causing accidents in construction, several tools are deployed on project sites. In the context of this paper, a tool called unmanned aerial vehicles (UAVs) is chosen to address safety problems in construction. UAVs has been used in construction for a variety of purposes for more than a decade [6]. It is reported that in the United Emirates of Arab (UAE), Australia and the United State of America (USA), UAVs are used to perform site inspections and violations detections on construction sites [7–9].

However, in South Africa, the construction industry is not utilizing emerging technology to improve safety when compared to other developed countries. In essence, there is little utilization of such technology in South African construction, despite UVAs been in existence for more than a decade. Thus, the aim of this paper is to propose how to deploy a UAVs for on sites in South Africa. The deployment framework emphasizes the effectiveness of UAV as a tool for enhanced safety regarding construction site objectives.

2 Research Method

In this research, the authors applied systematic literature review to answer the questions which asked, how will the use of UAVs help the safety manager to improve safety on construction sites? The literature review included papers that investigated UAVs and safety topics in the construction industry. The adopted literature search is completed by reviewing the titles and abstracts of the published papers between 2010 and 2019. Firstly, the author would review the title, followed by an abstract, and a full paper would be screened. The adopted flowchart of this systematic literature study is highlighted in Fig. 1. The databases used to search for the published papers included a Science Direct database (https://www.sciencedirect.com/), International Council for Research and Innovation in Building and Construction (CIB) database (http://www.irbnet.de/daten/iconda/CIB_DC31504.pdf), and Google Scholar databases (https://scholar.google.co.za/). To identify the relevant papers for this study, the keywords such as unmanned aerial vehicles (UAVs), safety management systems and construction industry, were used to search the databases. In this study, 26 authors were cited from the reviewed papers.

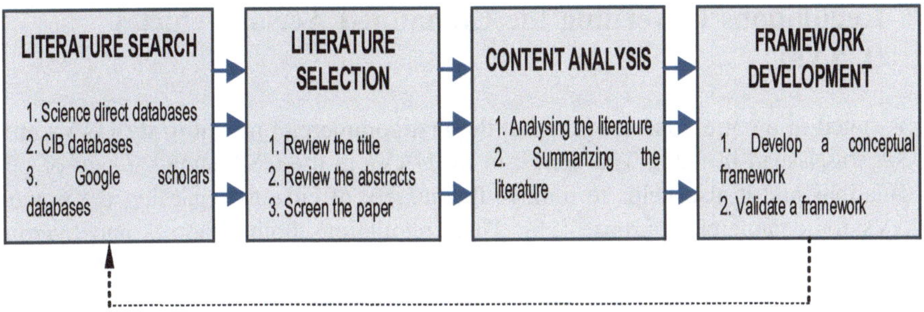

Fig. 1. The flowcharts of the study

3 Background of the Unmanned Aerial Vehicles (UAVs)

This section outlines the history or background of unmanned aerial vehicles (UAVs). So, UAVs, which is commonly known as drones, are aircrafts systems that can fly without a pilot and passengers on board [10]. The aircraft system includes the UAVs or drones (remotely controlled aircraft), the control system, satellite-based equipment, communication links and an operator (pilot) needed to operate or fly the aircraft effectively and safely [11]. The operations of the UAVs depend on the participation of a pilot being a person. Historically, UAVs were designed for the purpose of military services [8]. In the year 2006, the Federal Aviation Administration (FAA) in the USA issued a first commercial UAVs permit to grant permission for its use for business purpose [12].

According to the report by Dronethusiast [12], the first commercial UAVs in the USA were used by the government agencies for disaster relief, border surveillance and wildfire fighting, and while the Agricultural Industry started using UAVs to inspect pipelines and spray pesticides of farms. In the industrial sectors, the agricultural industry dominated the use of UAVs for improving agricultural production, and other industries such the real estate has taken advantages of this technology to conduct aerial survey and mapping of planned developments [11].

UAVs in the construction industry has been adopted to carry out site inspections and design site layout plans [13]. It is also used to monitor the working conditions of people, and to identify the potential risks and hazards on sites [14]. According to Hubbard *et al.* [15], UAVs provides a platform to improve safety management by providing real-time visual information to monitor the work on sites. The high definition camera which is installed on UAVs provides visual assets (images and video clips), which can be analyzed to improve the safety management system on construction projects [14].

4 Regulations Governing the Unmanned Aerial Vehicles (UAVs)

It is stated in the previous section, that the first commercial permit to fly UAVs in the USA was issued in 2006. The permit to fly UAVs in the USA shows a sign of good regulations and it also helps to remove the anxiety of citizens regarding the usage of UAVs for commercial purpose [16]. Thus, regulations both endorses and overturns innovations. Good regulations should be able to apprehend social values and objectives and is effective in promoting economic activities by reducing direct and indirect costs [16]. UAVs in the USA is regulated by the Federal Aviation Industry (FAA), which is the main agency for managing civil aviation [6]. The FAA regulates UAVs by dividing the UAV uses into two categories namely, to fly for hobby purposes and fly for commercial use [6].

While, in South Africa, the use of UAVs is regulated by the South African Civil Aviation Authority (SACAA). In terms of Part 101 of the Civil Aviation Regulations, 2011, the pilot operating a UAVs shall be appropriately trained on the UAVs and qualified for the area and type of operation. In addition, the SACAA acknowledges that numerous businesspeople interested in obtaining a UAVs operator certificates for business purpose are required to provide aerial work service South African Civil Aviation Authority [17]. The operator or pilot of a UAVs is compelled to observe all statutory requirements relating to liability, privacy and any other laws enforceable by any other authorities when flying a UAVs.

5 Unmanned Aerial Vehicles (UAVs) on Construction Sites

This section outlines the impact of UAVs in the construction industry. As elaborated in the introduction section that this study focuses on the deployment of UAVs for enhanced safety regarding construction site objectives. Safety in the construction industry is a serious concern and the industry is regarded as one of the most hazardous industry contributing to high accidents rate worldwide [2, 3, 18]. The causes of accidents in the construction industry are rooted in the working conditions of people. Notably, construction sites could be overcrowded with workers who are exposed to high risks duties such as operating at height and outdoors and with heavy machinery and equipment [19]. There are numerous factors contributing to the worker's safety through the interactions among the workers and the work environment [20]. Thus, it is important to investigate the working environment of the workers to improve safety in the construction industry. The use of modern technologies on construction sites is one of the most recent interventions used to improve safety management system on construction sites.

As explained by Tatum and Liu [11] UAVs are used to improve the working conditions on construction sites. For example, UAVs could be used to investigate the worker's safety-related behaviors through site inspections. However, it is critical for a UAVs operator to investigate the site project which UAVs would be deployed to and to determine the area which would be suitable to capture the images and videos of a site project [21]. It should be acknowledged that safety inspections are the responsibility of a safety manager on a site project [8]. The safety inspection process is carried out by regularly walking and taking direct observations of the site activities or tasks by a safety manager.

The competency of a safety manager on construction sites can be increased significantly with the use of UAVs [11]. The use of UAVs in the construction industry is continually changing the safety managers task of direct observation on site works and interacts with the workers [15]. UAVs improves site inspection by providing real-time inspection and surveillance from the site project in the form of a high-definition (HD) images and videos [9]. Figure 2 highlights the images that were taken on a Brazilian construction site using a UAVs for a safety inspection. It can also be noted that the images captured using a UAVs on a site project helps the safety manager to be able to identify the risk and hazards which might cause accidents. Additionally, the images and video clips captured using a UAVs provide reliable data suitable to improve the safety inspection as indicated in Fig. 2 [21].

It should be acknowledged, that safety inspection process in the construction industry is based on the three main aspects, which includes, being regular, implementing direct observation and leading interaction with workers [22]. Therefore, it is critical for the safety manager to align safety inspection process with the safety system, especially when evaluating the working conditions of a site project [23]. It is further reported that safety inspections through visual aspects (images and video clips) also provide data to improve the working conditions at a site project [24].

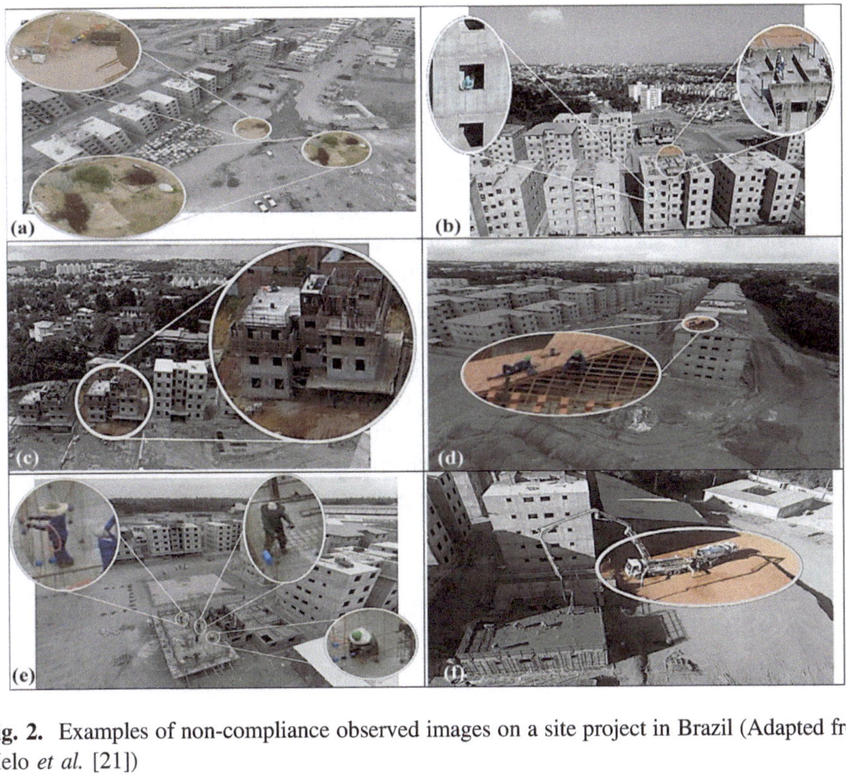

Fig. 2. Examples of non-compliance observed images on a site project in Brazil (Adapted from Melo *et al.* [21])

In addition, Fig. 2 shows the example of non-compliance observed images on a project in Brazil. Plate (a) demonstrate waste unprotected from rain; (b) workers without hard-hats and personal fall arrest systems; (c) safety platforms not installed on the entire perimeter of the building; (d) workers on the roof unprotected from falling; (e) inappropriate use of hard-hats; (f) there is no isolation of areas for loading and unloading of materials.

6 Discussion and Way Forward

Based on the reported literature regarding the construction challenges relating to safety, it can be concluded that indeed safety at work is a complex phenomenon as emphasized by Rae and Provan [2]. Accident experienced in construction is described as unavoidable, happens unexpectedly and the construction industry is high risk when compared to other industries [25]. Hence, the causes of accidents are also rooted in the working conditions on construction sites. To solve this reported safety problem, a proposed conceptual UAVs framework sets a standard ground on how to improve safety in construction using visual safety inspections. According to Mendes *et al.* [24] the use of visual technologies can contribute positively to the entire process of safety inspections and improving safety practice on construction sites. Additionally, it is reported by several authors such as Melo *et al.* [21], Hubbard *et al.* [15], Tatum and Liu [11], Irizarry *et al.* [8], and Moud *et al.* [6] that the application of UAVs for safety inspections is used to identify hazards and improve the unsafe working conditions on construction sites.

The application of UAVs for safety inspections is indicated in Fig. 2. While Fig. 3 presents a conceptual framework guideline for deployment of UAVs in the construction industry. The following procedures must be followed when deploying UAVs for safety inspections. Firstly, an operator that might be a safety manager must set a UAV. Thereafter, an operator must fly a UAV to start with the site surveillance. During the surveillance, an operator will capture the images or video clips of the site under surveillance (Fig. 2 shows the images captured using UAVs on a Brazilian site project). The images or videos clips will be saved either on a smartphone or iPad, which was used as a signal transmitter. Thereafter, a database will be created, and the images or video clips will be saved on a computer. The images or video clips would be analyzed with the determination of identifying the risks and hazards, which might cause accidents, and the working conditions on sites. Thereafter, the data would be processed to help the safety management team to improve the safety management system based on the analyzed data of the images or video clips (Fig. 2 shows the images which were used to identify hazards on a Brazilian site project).

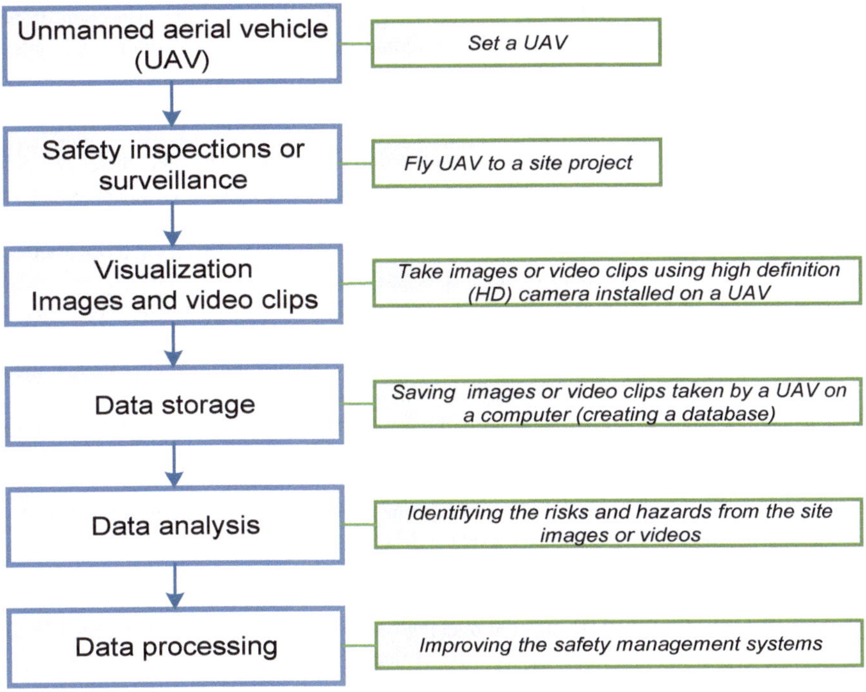

Fig. 3. A conceptual UAVs framework to improve safety in construction

Despite the formulation of a UAV deployment UAVs to improve safety in construction, there is still a need to validate the UAVs framework on an actual research project. To test this deployment framework, new studies are under development in order to evaluate the impact of safety inspection using UAVs on a construction site.

References

1. Yanar, B., Lay, M., Smith, P.M.: The interplay between supervisor safety support and occupational health and safety vulnerability on work injury. Saf. Health Work **10**(2), 1–8 (2018)
2. Rae, A., Provan, D.: Safety work versus the safety of work. Saf. Sci. **111**, 119–127 (2019)
3. Kines, P., Andersen, L.P.S., Spangenberg, S., Mikkelsen, K.L., Dyreborg, J.: Improving construction site safety through leader-based verbal safety communication. J. Saf. Res. **41**, 399–406 (2010)
4. Health and Safety Executive (HSE). Reducing error and influencing behaviour. London: Health and Safety Executive (2009)
5. Wang, D., Wang, X., Xia, N.: How safety-related stress affects workers' safety behavior: the moderating role of psychological capital. Saf. Sci. **103**, 247–259 (2018)
6. Moud, H.I., Shojaei, A., Flood, I., Zhang, X., Hatami, M.: Qualitative and quantitative risk analysis of unmanned aerial vehicle flights over construction job sites. In: Eighth International Conference on Advanced Communications and Computation, INFOCOMP, Barcelona, Spain (2018)

7. Ashour, R., Taha, T., Mohamed, F., Hableel, E., Abu Kheil, Y.: Site inspection drone: a solution for inspecting and. In: IEEE 59th International Midwest Symposium on Circuits and Systems. Abu Dhabi, UAE (2016)
8. Irizarry, J., Gheisari, M., Walker, B.N.: Usability assessment of drone technology as safety inspection tools. J. Inform. Technol. Constr. **17**, 194–212 (2012)
9. Schriener, J., Doherty, P.: Drones show potential to aid jobsite safety and efficiency, 25 July 2013. http://constructionexec.com/article/drones-show-potential-to-aid-jobsite-safety-and-efficiency. Accessed 15 Mar 2019
10. Kardasz, P., Doskocz, J., Hejduk, M., Wiejkut, P., Zarzycki, H.: Drones and possibilities of their using. J. Civil Environ. Eng. **6**(3), 1–7 (2016)
11. Tatum, M.C., Liu, J.: Unmanned aircraft system applications in construction. Procedia Eng. **196**, 167–175 (2017)
12. Dronethusiast. The History of Drones (Drone History Timeline From 1849 To 2019) (2014). https://www.dronethusiast.com/history-of-drones/. Accessed 11 Mar 2019
13. Canis, B.: Unmanned Aircraft Systems (UAS): Commercial Outlook for a New Industry. Congressional Research Service (CRS), Washington DC (2015)
14. Melo, R.R., Costa, D.B.: Contributions of resilience engineering and visual technology to safety planning and control process. In: Saurin, T.A., Costa, D.B., Behm, M., Emuze, F. (eds.) Proceedings of the Joint CIB W099 and TG59 International Safety, Health, and People in Construction Conference, Salvador, Brazil, pp. 13–22 (2018)
15. Hubbard, B., Wang, H., Leasure, M., Ropp, T., Lofton, T., Hubbard, S., Lin, S.: Feasibility study of UAV use for RFID material tracking on construction sites. In: Associated Schools of Construction Proceedings of the 51st Annual International Conference. ASC, Texas (2015)
16. Nakamura, H., Kajikawa, Y.: Regulation and innovation: How should small unmanned aerial vehicles be regulated? Technol. Forecast. Soc. Chang. **128**, 262–274 (2018)
17. South African Civil Aviation Authority (SACAA). Remote pilot licence: training, examination and application for RPL requirements. SACAA, Johannesburg (2015)
18. Li, H., Lu, M., Hsu, S., Gray, M., Huang, T.: Proactive behavior-based safety management for construction safety improvement. Saf. Sci. **75**, 107–117 (2015)
19. Cheng, E.W.L., Ryan, N., Kelly, S.: Exploring the perceived influence of safety management practices on project performance in the construction industry. Saf. Sci. **50**, 363–369 (2012)
20. Zhang, P., Li, N., Jiang, Z., Fang, D., Anumbad, C.J.: An agent-based modeling approach for understanding the effect of worker management interactions on construction workers' safety-related behaviors. Autom. Constr. **97**, 29–43 (2019)
21. Melo, R.R., Costa, D.B., Álvares, J.S., Irizarry, J.: Applicability of unmanned aerial system (UAS) for safety inspection on construction sites. Saf. Sci. **98**, 174–185 (2017)
22. Toole, T.M.: Construction site safety roles. J. Constr. Eng. Manage. **128**(3), 203–210 (2002)
23. Saurin, T.A.: Safety inspections in construction sites: a systems thinking perspective. Accid. Anal. Prev. **93**, 240–250 (2016)
24. Mendes, C.M., Silveira, B.F., Costa, D.B., Melo, R.S.: Evaluating USA-image pattern recognition system application for safety guardrails inspection. In: Saurin, T.A., Costa, D.B., Behm, M., Emuze, F. (eds.) Proceedings of the Joint CIB W099 and TG59 Conference Coping with the Complexity of Safety, Health, and Wellbeing in Construction. Salvador, Brazil, pp. 23–32 (2018)
25. Asanka, W.A., Ranasinghe, M.: Study on the impact of accidents on construction projects. In: 6th International Conference on Structural Engineering and Construction Management, pp. 58–67. CSECM, Kandy (2015)

Communication Channel at Pre-contract Phase and Construction Project Delivery in South-West, Nigeria

E. Aladeloba Adesina[1(✉)], I. Idoro Godwin[1], and Osunsanmi Temidayo O[2]

[1] Department of Building, University of Lagos, Akoka, Yaba, Lagos State, Nigeria
aladelobasina@yahoo.com, gidoro@unilag.edu.ng
[2] SARChI in Sustainable Construction Management and Leadership in the Built Environment, Faculty of Engineering and Built Environment, University of Johannesburg, Johannesburg, South Africa
osunsanmidayo@gmail.com

Abstract. The advent of the fourth industrial revolution has disrupted the channel of communication within the construction industry especially at the pre-contract phase of a construction project. Thus, this study investigated the communication channel adopted at the pre-contract phase of a construction project in Nigeria. The study adopted random sampling from collecting data from construction professionals in South West of Nigeria. A total of 480 questionnaires was distributed to the respondents and used for the analysis. The questionnaire was analysed using SPSS V 20 using the mean item score. The findings from the analysis show that the communication channel adopted by the construction professionals consists of print, face to face and electronic media. The study discovered that the fourth industrial revolution has not disrupted the communication practice in Nigeria as the print media remains the predominant channel of communication.

Keywords: Communication · Communication channel · Construction project · Information

1 Introduction

The construction project contributes significantly to the development of any nation [1]. The contribution includes the workforce employment of which the industry is responsible for about 20% of the total workforce [2]. Ansah et al. [3] provided another means in which the construction industry contributes to the economy. They affirmed that the infrastructure provided by the industry serves as a backbone for the economy. In support of this statement Koolwijk et al. [4] concluded that the world depends on the construction industry to solve its infrastructural problems. Hence it is reasonable to postulate that the industry contributes significantly to the growth of the economy through the provision of infrastructure.

C. Aigbavboa and W. Thwala (Eds.): CIDB 2019, *The Construction Industry in the Fourth Industrial Revolution*, pp. 505–511, 2020.
https://doi.org/10.1007/978-3-030-26528-1_51

Unfortunately, Meng [5] stipulated that the industry has not been able to effectively contribute to the economic growth of the country. The ineffectiveness is attributed to numerous factors major amongst them which is the unclear communication process adopted by construction professionals [6]. Olaniran [7] related the poor communication process among construction professionals to the assemblage of different profession under one flag which typifies the display of different principles, and skills formulated in a manner. This, no doubt, explains the reason for non-cooperation among distinct professionals and passage of incomplete information. Dainty [8] corroborated this claim when he observed that the communication process tends to be characterized by an unacquainted group of people coming together for a brief period.

It, therefore, becomes imperative that appropriate communication channel is used to resolve design and construction problems which, according to [9] are inevitable projects and undesirably disturb cost, schedule and safety performance. Research studies have also discovered that the communication process used in project delivery is far from being effective. For instance, Olaniran [7] observed that the use of the traditional system of documentation or communication which is dominant in the delivery of construction project, will not aid collaboration and coordination of all players and encourage cross-discipline communication between distinct professionals. It will also constitute a major challenge as document transfer in construction projects and between members are rarely incompatible and complete format. Wu et al. [10] also identified conflicting design information, delaying issuance of revised drawings, dimensional inaccuracies as critical factors emanating from incompetent design and resulting in delays in project delivery time. This is because most distinct professionals have conflicting priorities and different objectives. This, therefore, creates the need for examining the communication channel adopted within the Nigeria construction industry.

In Nigeria, previous study by Ikediashi and Ogwueleka [11] and Olaniran [7] focused on communication systems and their effect on construction project delivery. The study revealed that some organizations make use of communication channel with varying formality while some others focus on both formal and informal communication routes. In the cases where the communication process was studied, the focus was on just linear communication instead of taking into account, the project team communication in Nigeria.

Although these studies investigated several aspects of communication in construction, studies on "project characteristics, Communication channel at pre-contract stage and project performance in south-west Nigeria" appear to be relatively scarce. Hence the need to appraise communication channel adopted at pre-contract phase of construction project delivery.

2 Construction Industry and the Communication Process

The construction industry is such an influential economic sector, which utilizes very heavy human, material and financial resources characterized by different professionals working together to achieve a common goal [5]. Olanrewaju and Abdul-Aziz [12] posited that a construction project is termed to be successful when it is completed at the stipulated cost and time satisfying the client quality expectation. In other to achieve project success, construction professionals engage in relationships among themselves that are contractually driven [9]. Brewer et al. [13] asserted that the relationship is built on effective communication among the construction project team members. In support of this assertion Adeleke et al. [14] submitted that excellent communication among the construction professionals facilitates an effective relationship among them. Wu et al. [10] opined that poor communication among the construction professionals is responsible for conflicts and lack of mutual respect and trust that in return leads to design errors that negatively affect cost. Ofori [15] and Ajayi and Osunsanmi [16] reported that some construction defects occur because of poor communication. According to Ikediashi and Ogwueleka [11], a delay in the supply of necessary information during construction may result in 21–30% delay of the total project.

"This implied that communication is essential for construction project success. Brewer et al. [13] affirmed that proper communication planning for all the project stakeholders is crucial as it ensures the success of a construction project. Also, Dainty et al. [9] argued that passage of information is significant to the running of any construction project and must be properly managed. It can be inferred from the above that improvements in communication will result in an increase in the quality of the building and a reduction in the level of defect occurrence."

2.1 Communication Channel in the Construction Industry

"Dainty et al. [9] categorizes construction communication channels like print, electronic or face-to-face (interpersonal). Acar et al. [17] asserted that the most used channel among construction professionals is the print channel, which consumes about half of communication time due to its low level of effectiveness. Although, Olaniran [7] on the other hand observed that project team members are exposed to various communication channels, but selecting the most appropriate medium or media becomes a critical issue for professionals. Dainty et al. [9] reported that the media choice should match the ambiguity or complexity of the information being passed and the agenda of the information attempting to achieve."

Presently the construction industry has become a global arena with increased complexity and competitiveness due to the influx of global participation in projects [18]. Fundamental changes resulting from the worldwide growth of the construction industry is evident in the adoption of technological and scientific innovations [19]. Modern construction is complex and has physical realities, which point to a need for more technology such as information and Communication Technology (ICT) [20] as the growth of ICT presents more opportunities for its integration in construction.

3 Research Methodology

The appraisal of the communication channel is a significant factor towards ensuring the effective flow of construction information. The methodology adopted in this study to appraise communication channel adopted by construction professionals entails two primary steps. In the first step literature relating to communication channels among construction professionals and the construction industry were reviewed. The second step entails designing questionnaire based on the reviewed literature and self-administered to the construction professionals within South-West Nigeria. The construction professionals comprise of the architect, builders, civil engineers, structural engineers, quantity surveyors, mechanical engineers and Electrical Engineers.

"A convenience sampling was used, while the questionnaire was distributed online using google forms and through personal administration after ethical approval was received from their respective firms. The questionnaire was divided into two (2) sections covering the communication channel and the respondent personal information. A total of four hundred and eighty questionnaires (480) was received and used for the analysis after scrutinizing for errors out of four hundred and eighty-two (482) that was distributed to the respondents. The response to the questionnaire was then inputted into SPSS (Statistical Package for Social Science) version 23. The study adopted a quantitative research design study, using mean item score."

4 Discussion and Implication of Findings

4.1 Adoption of Communication Channel at Pre-contract Phase of Project Delivery

Communication channel consists of three communication practices namely: print, face-to-face and electronics [9]. Print communication channel consists of seven communication practices namely: daily reports, annual reports, memos, policy, manuals, posters, brochures and newsletters. "Face-to-face communication practice consists of six practices namely: team meetings, speeches, focus groups, brown bag lunches, social events and gatherings and management by wandering around". The electronic channel consists of ten practices namely: "email and voicemail, instant messaging systems, intranet, chat rooms, video conferencing, blogs, podcasts, electronic town-hall meetings, business TV and wikis". The level of adoption of each communication channel was measured using a 5-point Likert scale namely: nil, rare, moderate, often and always. Respondents were requested to indicate the scale that best represented their perception of the level of adoption of each communication channel in the projects selected. The level of adoption of the three communication practices and their sub-practices were analysed using Mean Score (MS).

The results in Table 1 revealed that the respondents perceived two communication practices in print communication medium namely: reports and annual reports to be often adopted. "The findings of this study coincide with the study conducted by [11] which indicated that report writing is the predominant means of communicating in Nigeria construction industry". Whereas the remaining five practices namely: memos, policy manuals, posters, brochures and newsletters are moderately adopted at the pre-contract phase of project delivery.

On electronic medium, Table 1 showed that Email with a mean score of 3.90 is often used by the construction professionals for communicating. Whereas instant messaging and intranet are moderately used by the respondents. Electronic means of communicating like chat rooms, video conferencing, blogs, podcast and business TV are rarely used by the construction professionals.

Table 1. Mean scores of the level of adoption of practices in each communication channel at the pre-contract phase.

Practices of the communication channel	Mean	Rank	Adoption
Print media	**3.08**		
Daily reports	3.73	1	Often
Annual reports	3.54	2	often
Memos	3.23	3	Moderate
Policy manuals	3.19	4	Moderate
Posters	2.96	5	Moderate
Brochures	2.54	6	Moderate
Newsletters	2.54	7	Rare
Electronic media	**2.26**		
Email	3.90	1	Often
Instant messaging systems	3.17	2	Moderate
Intranet	2.67	3	Moderate
Chat rooms	2.31	4	Rare
Video conferencing	2.21	5	Rare
Blogs	1.81	6	Rare
Podcasts	1.71	7	Rare
Electronic town-hall meetings	1.71	8	Rare
Business TV	1.65	9	Rare
Wikis	1.48	10	Nil
Face to face Media	**3.02**		
Team meetings	3.98	1	Often
Speeches	3.71	2	Often
Focus groups	2.90	3	Moderate
Brown bag lunches	2.69	4	Moderate
Social events and gatherings	2.67	5	Moderate
Management by wandering around	2.15	6	Rare

1 = Nil, 2 = Rare, 3 = Moderate, 4 = Often, 5 = Always, N = Number of respondents, TS = Total score, MS = Mean score, Rk = Rank. 1–1.49 = Nil, 1.50–2.49 = Rare, 2.50–3.49 = Moderate, 3.50–4.49 = often, 4.50–5.00 = always

Concerning face to face channel for communication respondents perceived the adoption of one communication channel namely: Team meetings and Speeches to be often adopted while three communication channel in face to face medium namely: "Focus groups, Brown bag lunches and Social events and gatherings" were perceived to be moderately adopted. The last communication practice in face to face namely: management by wandering around was perceived to be rarely adopted.

Table 1 also shows a general mean score for all the communication channels; print medium, electronic medium and face to face media. The table showed that the print channel has the highest mean score of 3.08, therefore, implying that traditional form of passing communication is the predominant communication channel in the Nigeria construction industry. Whereas the electronic communication channel was perceived to be moderately used and the second most adopted channel. Face to face communication Channel was perceived to be rarely used and the least adopted communication channel.

5 Conclusion

Construction projects as activities within the construction industry are based on information flow which is an indication of the involvement of a diverse number of people participating in a network of communication at different levels. Communication plays a vital role in all stages of a construction project either at the pre-contract or post-contract phase of the projects. Any shortfall therefore or insufficient communication may contribute to projects failure. Therefore, it is imperative that an appropriate communication channel should be used for construction management.

The communication channel was broken down into three namely; print, face-to-face and electronic media. The study discovered that the print channel is the predominant communication channel in the Nigeria construction industry. Whereas the electronic communication channel was perceived to be moderately used and the second most adopted channel. Face to face communication Channel was perceived to be rarely used and the least adopted communication channel. It can be deduced that the fourth industrial revolution has not disrupted the channel of communication within the Nigeria construction industry. The study recommends that construction firms should adopt a more electronic channel of communication due to its effectiveness for passing information.

Acknowledgements. The authors like to appreciate everybody that has contributed to this article directly or indirectly.

References

1. Daniel, E., Oshineye, O., Olalekan, O.: Barriers to sustainable construction practice in Nigeria. Presented at the Proceeding of the 34th Annual ARCOM Conference, Belfast, UK, 3–5 September 2018
2. Olanrewaju, A., Anavhe, P.: Perceived claim sources in the Nigerian construction industry. Built Environ. Project Asset Manage. **4**, 281–295 (2014)
3. Ansah, R.H., Sorooshian, S., Mustafa, S.B.: The 4Ps: a framework for evaluating construction projects delays, vol. 13 (2018)
4. Koolwijk, J.S.J., van Oel, C.J., Wamelink, J.W.F., Vrijhoef, R.: Collaboration and Integration in project-based supply chains in the construction industry. J. Manage. Eng. **34**, 04018001 (2018)
5. Meng, X.: The effect of relationship management on project performance in construction. Int. J. Project Manage. **30**, 188–198 (2012)
6. Cheung, S.O., Yiu, T.W., Lam, M.C.: Interweaving trust and communication with project performance. J. Constr. Eng. Manage. **139**, 941–950 (2013)
7. Olaniran, H.: On the role of communication in construction projects in Nigeria. J. Environ. Sci. Technol. **2**, 048–054 (2015)

8. Dainty, A.: Methodological pluralism in construction management research. Adv. Res. Methods Built Environ. **1**, 1–13 (2008)
9. Dainty, A., Moore, D., Murray, M.: Communication in Construction: Theory and Practice. Routledge, London (2007)
10. Wu, G., Liu, C., Zhao, X., Zuo, J.: Investigating the relationship between communication-conflict interaction and project success among construction project teams. Int. J. Project Manage. **35**, 1466–1482 (2017)
11. Ikediashi, D.I., Ogwueleka, A.C.: Assessing the use of ICT systems and their impact on construction project performance in the Nigerian construction industry. J. Eng. Des. Technol. **14**, 252–276 (2016)
12. Olanrewaju, A.L., Abdul-Aziz, A.-R.: Building Maintenance Processes and Practices: The Case of a Fast Developing Country. Springer, Singapore (2014)
13. Brewer, G., Gajendran, T., Chen, S.: The use of ICT in the construction industry: critical success factors and strategic relationship in temporary project organizations. In: CIBW78 Information Communication Technology in Construction, pp. 19–21 (2005)
14. Adeleke, A., Windapo, A.O., Khan, M.W.A., Bamgbade, J., Salimon, M.G., Nawanir, G.: Validating the influence of effective communication, team competency and skills, active leadership on construction risk management practices of Nigerian construction companies. J. Soc. Sci. Res. 460–465: 6, 2018
15. Ofori, G.: Nature of the construction industry, its needs and its development: a review of four decades of research. J. Constr. Developing Countries **20**, 115 (2015)
16. Ajayi, O., Osunsanmi, T.: Constraints and challenges in the implementation of total quality management (TQM) in contracting organisations. J. Constr. Project Manage. Innov. **8**, 1753–1766 (2018)
17. Acar, E., Kocak, I., Sey, Y., Arditi, D.: Use of information and communication technologies by small and medium-sized enterprises (SMEs) in building construction. Constr. Manage. Econ. **23**, 713–722 (2005)
18. Ankrah, N., Proverbs, D., Debrah, Y.: Factors influencing the culture of a construction project organisation: an empirical investigation. Eng. Constr. Architect. Manage. **16**, 26–47 (2009)
19. Doloi, H., Sawhney, A., Iyer, K., Rentala, S.: Analysing factors affecting delays in Indian construction projects. Int. J. Project Manage. **30**, 479–489 (2012)
20. Afolabi, A., Owolabi, J., Ojelabi, R.A., Oyeyipo, O., Aina, D.: Development of a web-based tendering protocol for procurement of construction works in a tertiary institution. J. Theor. Appl. Inform. Technol. **95**, 1595–1606 (2017)

The Implications of Not Utilizing a Quantity Surveyor as Principal Agent for Residential Construction Projects

Tascha Bremer(✉), Henry Deacon, Mart-Mari Els, and IG Mare

Department of Quantity Surveying and Construction Management, University of the Free State, Bloemfontein 9300, South Africa
bremert@ufs.ac.za

Abstract. Clients who plan to build their own homes are not fully aware of all the roles played by professional consultants during construction projects. The problem exists that many negative consequences with regard to cost, time and quality performance arise on self-build residential construction projects where a principal agent is not appointed by the client.

The purpose of this study is to determine the necessary roles of a quantity surveyor serving as principal agent on residential construction projects; and subsequently establishing the possible implications of only using a building contractor to execute and manage such projects.

The research design is based on a case study that involved three residential construction projects in Bloemfontein, Free State. The data was collected by means of semi-structured interviews with various stakeholders involved in these projects, with a purposive sampling method applied.

The findings indicate that the public is uninformed of the important services that a quantity surveyor, serving as principal agent, fulfill with regard to achieving project success within time, cost and quality constraints. The primary reasons for not utilizing quantity surveyors as principal agents on self-build residential construction projects are that clients have a lack of knowledge of the services provided by construction consultants along with a perception that consultants' fees are expensive. Clients further have the personal belief that they can successfully manage the building contractor, as well as monitor and control their projects, on their own.

The study concludes that future self-build homeowners (clients) should appoint a quantity surveyor as principal agent on their projects, which would assist with the management of time, cost and quality parameters.

This study would be valuable to the residential construction industry by increasing the awareness of the roles and duties fulfilled by professional construction consultants, especially those of the quantity surveyor serving the dual role of principal agent.

Keywords: Construction industry · Principal agent · Quantity surveyor · Residential property

C. Aigbavboa and W. Thwala (Eds.): CIDB 2019, *The Construction Industry in the Fourth Industrial Revolution*, pp. 512–520, 2020.
https://doi.org/10.1007/978-3-030-26528-1_52

1 Introduction

Quantity surveyors are the "cost economists" of the construction industry and play an important role in the cost management of a project to safeguard the interests of the client [9]. Their level of expertise are used to draft and interpret contract documents, therefore enabling them to forecast final costs, execute regular and fair valuations, prevent disputes, and monitor the progress of a project [3].

The core services offered by a quantity surveyor is to control the construction cost and advise the client on VAT implications, insurance valuations and dispute resolution [17]. The quantity surveyor will also consult with the client to ensure that the best contractor is appointed when there is a need for additional specialists and subcontractors on a project [10]. However [12] stated that quantity surveyors have expanded their "normal duties" and are offering services beyond their scope of works.

The South African Council for the Quantity Surveying Profession (SACQSP) lists all the crucial services that a quantity surveyor provides during the life-cycle of a project. This life-cycle consists of six stages namely, inception (stage 1); concept and viability (stage 2); design development (stage 3); documentation and procurement (stage 4); construction (stage 5) and close-out (stage 6). These services are crucial to a project's success and there is no better professional to implement these services than the Quantity Surveyor [17].

[3] further more lists numerous services which covers a wide range of activities implemented by quantity surveyors such as tendering, claims- and value management, dispute resolution, life-cycle costing, value engineering, cost benefit analysis, value for money assessments, research consultancy, valuation, management contracting, feasibility studies, construction dispute resolution, cost planning, change control, cost estimation and risk analysis. [13] also found that quantity surveying appears to be relevant in the banking, law, manufacturing (especially oil and gas), project- and asset management, insurance and taxation.

Recommended fee scales of quantity surveyors are posted by ASAQS and SACQSP, where the fee scales are only to be used as a basis for fee calculation. It appears that quantity surveyors provide considerable amounts of discount on professional fees, which indicates a concrete level of competitiveness among practicing firms. Hence, the quantity surveying industry proves to offer affordable fees while still generating substantial profits [18].

[16] proclaims that the principal agent plays an important role on behalf of the client especially in today's complex and dynamic multi-disciplinary environment in which a building agreement is undertaken. The principal agent is regarded as an agent who acts on behalf of the client [8]. Any construction professional (architect, quantity surveyor, project manager, and engineer) can be appointed as the principal agent [7]. In the built environment, the principal agent will play a dual role, thereby acting as a consultant in the profession he/she is contracted, and as an agent representing the client in the discharge of the legal relations of the respective parties as they relate to that service in terms of the building contracts [16].

The services of a principal agent are also detailed according to the six stages of a construction project, which concentrates on procurement, contracts, quality management, inspection and program management.

[16] states that there are different approaches by the professions in calculating the principal agent's fees. [3] published that the principal agency fee is calculated at 45% of the total quantity surveying fee for a project.

[22] determined that time, cost and quality are interlinked with one another. Therefore an impact on one of the factors would had have an effect on the other. The complexity of project management with regards to time, cost and quality should not be underestimated and that minor mistakes can lead to costly expenses. A successful project is one that is completed within the required performance and that all time, cost and quality goals are achieved [15] (Fig. 1).

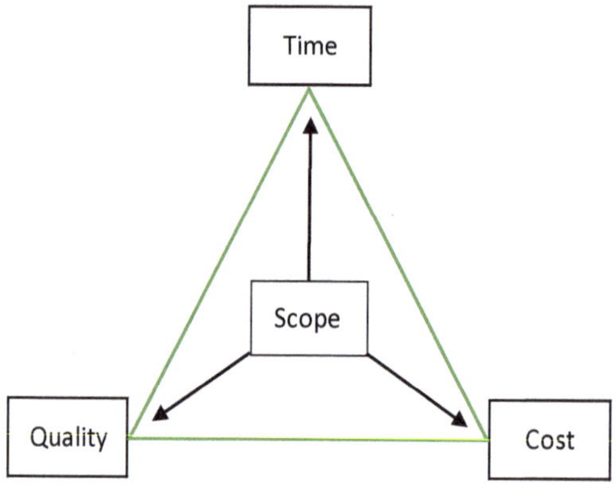

Fig. 1. Critical success factors of a project (The Triple Constraint) Source [22]

Secondary data has shown that there are numerous factors which could lead to projects delays. These factors could either emanate from the client or the contractor, or from neither [6]. The effect of time delays could also have an effect on project cost and sometimes cancellation of a contract [22]. The JBCC PBA (2014), clause 23.0 deals with the revision of the date for practical completion and extension of time and this could add substantial value to the time management of a project and also benefit the parties in terms of extension of time claims.

[19] also refers to a cost overrun as "*a cost increase*" or "*budget overrun*" and it involves unforeseen costs incurred which are not accounted for in the estimate. [11] states that a project cost overrun is measured as the difference between the initial estimated project cost and the actual construction cost after the project has been completed.

[14] showed that there are various factors which may cause cost overruns and that some the factors pose more risk than others. [15] discovered that a lot of projects experience cost overruns varying between 20.4% and 44.7%, depending on the type of project. Project cost management is found to be a complex process.

Quality management forms an integral part of improving project performance and preventing cost and quality defects. Quality control during construction therefore entails that the necessary inspections and measures are followed to ensure that all components of a building adheres to the correct standards and requirements [2].

[5] found that there are numerous quality defects which may arise during construction, with some being more severe than others. The JBCC Principal Building Agreement Edition 6.1 – March 2014 is recommended to clients, for it allows the principal agent to regularly inspect the quality of work done by a contractor and issue defects lists that ultimately enforces a contractor to rectify defects within a specific time period [8]. A suitable procurement process is essential to obtain the best contractor to complete the works and who is concurrently managed by means of a proper contract. Clients who are more inexperienced with procurement and contracts in construction should obtain professional advice to assist them with the process [20]. A quantity surveyor acting as Principal Agent is the ideal professional to manage the cost of a project, reduce the risk of cost overruns and ensure quality by regularly inspecting the works [7].

2 Research Methodology

The research method applied in this study is a qualitative approach that includes a case study of three residential projects in the city of Bloemfontein, which is located in the Free State province of South Africa. In this study both primary and secondary sources are utilized and reference is made for both these sources. A purposive sampling method was applied as the selected respondents are in the position to provide the best information to achieve the goals of this research.

Furthermore, primary data was collected by means of semi-structured interviews with various stakeholders involved in the selected projects. Secondary sources include estimates, cost reports as well as lists of delays and defects from the homeowners. The primary data was analyzed and themed, and secondary data was analyzed, summarized and compared. Triangulation of the data will give a better understanding of the problem.

3 Research Results and Discussion

3.1 Case Studies Reponses

Scope:	House 1: Residential 2-storey building built on a reinforced strip foundation, that includes 5 bedrooms (3 with own balconies), 5½ bathrooms, kitchen,	House 2: Residential 2-storey building built on a reinforced strip foundation, that includes 3 bedrooms, 2½ bathrooms, kitchen, scullery, dining area,	House3: Residential 2-storey building built on a steel structure, that includes 3 bedrooms, 3½ bathrooms, kitchen, scullery, dining area, TV and living area,

(*continued*)

(*continued*)

	scullery, living area, dining area, TV room, study room, entertainment room, entrance hall, covered patio, store room, double garage and a single garage. A swimming pool, and outdoor "boma"	TV and living area, covered patio including a built in braai, double garage and a single garage. Apartment linked with house: 1 Bedroom, 1 bathroom and small kitchen area. Swimming pool	covered patio and two double garages. No swimming pool
Finishes:	Roof covering: Concrete roof tiles. Walls: Plaster and paint. Windows and doors: Aluminium. Floors: Ceramic tiles. Paving: Prefabricated paving blocks. Internally: Walls: Ceramic tiles (bathrooms) and plaster and paint finishes. Floors: Ceramic tiles, Laminated flooring and vinyl tiles. Doors: Aluminium, semi solid and solid timber doors. Ceilings: Nailed-up ceilings of gypsum plasterboard	Roof covering: Concrete roof tiles Walls: Plaster and paint and stone cladding. Windows and doors: Aluminium. Floors: Ceramic tiles. Paving: Prefabricated paving blocks. Internally: Walls: Ceramic tiles (bathrooms), plaster and paint finishes and wall paper. Floors: Ceramic and vinyl tiles. Doors: Aluminium, semi solid and solid timber doors. Ceilings: Nailed-up ceilings of gypsum plaster board	Roof covering: Concrete roof tiles Walls: Plaster and paint and face brick features. Windows and doors: Aluminium. Floors: Ceramic tiles. Paving: Prefabricated paving blocks. Internally: Walls: Ceramic tiles (bathrooms) and plaster and paint finishes. Floors: Ceramic and vinyl tiles. Doors: Aluminium, semi solid and solid timber doors. Ceilings: Nailed-up ceilings of gypsum plaster board
Size:	1000 m^2	590 m^2	270 m^2
Estimate value	R 3 400 000.00	R 2 360 000.00	R 1 080 000.00
Final cost:	R 4 200 000.00	R 2 006 000.00	R 1 480 000.00
Cost overrun:	R 800 000.00	−R 354 000.00	R 400 000.00
Planned constr period:	9 Months	6 Months	9 Months
Actual constr period:	12 Months	9 Months	12 Months
Late completion:	3 Months	3 Months	3 Months

(*continued*)

(continued)

Penalties applicable:	No	Yes, for 1 month only	Yes, for three months
Quality defects:	Yes	Yes	Minor
1 Building contractor	Yes	Yes	Yes
Quantity Surveyor	No	No	No
Principal Agent	No	No	No

The case studies have shown that no quantity surveyor, quantity surveyor as principal agent, or any other professional as principal agent had been appointed on the projects. Two out of three projects had incurred cost overruns, three out of three time delays, and two out of the three experienced quality defects when the projects were completed. The researcher therefore calculated the project related quantity surveyor's value of professional fees (from the SACQSP fee scale and a market related fee) and compared the value of fees with the cost overruns incurred in the projects. The fees of a principal agent were also added to the value of the quantity surveying fees to determine if it would still have been financially viable to appoint a quantity surveyor as principal agent (Fig. 2).

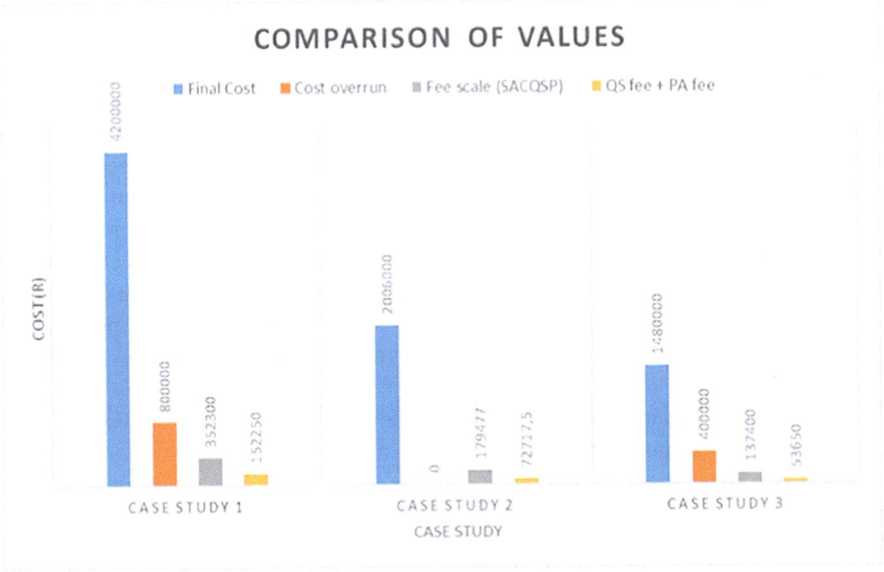

Fig. 2. Comparison of cost overruns with professional fees (QS + PA)

It can be observed that the professional fees are lower compared to the cost overruns. The researcher calculated that the professional fees only forms 3.63% of the total contract value compared to the average percentage of 32.55% of the estimate value that was cost overruns. [10] stated that people have the perception if they employ a quantity surveyor, it will only increase the project cost, but a good quantity surveyor in return will actually save the client money because of cost control.

The services of a principal agent would certainly have reduced the risk of time delays, disputes between parties, unrectified defects and poor contract administration. [16] supported this by stating that the principal agent plays an important role on behalf of the client and eases perplexing obligations, especially in today's complex and dynamic multi-disciplinary environment. This proves that it would still have been financially viable to employ a quantity surveyor also acting as principal agent.

3.2 Interview Responses and Findings

The researcher had found that the clients had a relative perception of the term "quantity surveyor", but few actually grasped the term "principal agent". Responses showed that clients are not fully informed of all the roles and duties of a quantity surveyor and principal agent. Some clients believed that most of the services are only rendered during the initial stages of a project which contrasted with literature as [4]'s statement and [17]'s list of a quantity surveyor's roles and duties indicates that a quantity surveyor is engaged throughout the duration of a project. Clients did not have a relative idea of the average fee quantity surveyors ask, especially for residential construction projects. The fees of quantity surveyors are also believed to be very expensive. [21]'s opinion aligned with this, as he remarked that a quantity surveyor's services are mostly associated with large scale projects and luxury components.

It appears that the quality of estimates done for the projects were not to the desired standards and two out of the three projects incurred cost overruns as a result. [1] stated that the purpose of effective cost control is to ensure that the final contract value does not widely exceed the initial estimate value accepted by the client. The responses also link with one of the main problems identified that no sufficient contracts forms and procurement processes are used to appoint the best contractor which may lead to several consequences. [2] clarified the important services of a principal agent concerning contract management including procurement strategies. [20] embarked that it is essential to obtain professional advice regarding the best selection of procurement methods and to ensure that the client's needs, requirements, project goals and objectives are clear.

The majority of the projects showed quality defects when the projects were completed and many disputes arose regarding the rectification of defects. Literature has shown that with a proper contract form like the JBCC and the services of a principal agent, the risk of defects and concurrent disputes would certainly have been reduced. There was a question raised on whether the clients considered rendering the services of a quantity surveyor and principal agent. The majority of the respondents did however not consider the need for the services, but in the end observed that these services are inevitable.

Results have shown that a quantity surveyor also acting as principal agent would definitely improve the efficiency of a project in terms of time, cost and quality. A complete study done by [7], had shown that the quantity surveyor is highly recommended to act as the principal agent, also adding substantial value to a project. The responses have shown the appointment of a quantity surveyor is considered an inevitability in the current construction industry.

3.3 Hypothesis Testing

Hypothesis 1:
The non-employment of a quantity surveyor on residential construction projects, could lead to poor cost management, which results in cost overruns.

The results proved that without the services of a quantity surveyor, the degree of cost management will be reduced which leads to cost overruns. The employment of a quantity surveyor will certainly reduce the risk of cost overruns in a residential construction project.

Hypothesis 2:
The non-employment of a quantity surveyor on residential construction projects, could lead to poor time management, which results delayed completions.

The results proved that without the services of a quantity surveyor, the degree of time management will be reduced which leads to delayed completions. The employment of a quantity surveyor will certainly reduce the risk of time delays in a residential construction project.

Hypothesis 3:
The non-employment of a quantity surveyor on residential construction projects, could lead to poor quality management, which results in quality defects.

The results proved that without the services of a quantity surveyor, the degree of quality management will not always be reduced and lead to quality defects. The employment of a quantity surveyor will however, reduce the risk of quality defects in a residential construction project.

4 Conclusion and Recommendation

It can clearly be seen that clients are not fully informed of the important services a quantity surveyor provides and to how important the services are to project success, especially with regards to time, cost and quality. The non-employment of quantity surveyors on residential construction projects are mainly because of the lack of knowledge, perception of very expensive professional fees and the personal belief of self-control of the project.

The researcher had achieved to clarify the importance of a quantity surveyor and why it is recommended to utilize his/her services on residential projects. All the research questions have been answered, which supports the significance of the study.

It is recommended that future homebuilders and clients appoint a quantity surveyor as principal agent on their projects. Even the professional advice from a quantity surveyor will help clients to reconsider the self-management of time, cost and quality of a project.

References

1. Ashworth, A., Perera, S.: Cost Studies of Buildings, 6th edn. Routledge, London (2015)
2. Amoah, C.: Project Procurement Management - Chapter 11 & 9. University of the Free State, Bloemfontein (2018)
3. ASAQS (2018)
4. Cartlidge, D.: Quantity Surveyor's Pocket Book, 3rd edn. Routledge, London (2017)
5. Hasan, M.I.: Minimizing defects in building construction project. J. Teknol. **78**, 79–84 (2015)
6. Islam, M., Trigunarsyah, B., Hassanian, M., Assaf, S.: Causes of delay in construction projects. In: ICCEPM, Busan, Korea (2015)
7. Janse van Rensburg, J.A.: Which consultant as principal agent can contribute the most to the cost effectiveness of a building project. University of Pretoria, Pretoria (2009)
8. JBCC: Principal Building Agreement Edition 6.1 (2014)
9. Mataa, S.: Project Costing and Financial Feasibilities (n.d.)
10. McCoy, I., Joyce, S.: Middelton Advice: Do I Need a Quantity Surveyor, p. 1 (2016)
11. Niazi, G.A., Painting, N.: Significant factors causing cost overruns in the construction industry in Afghanistan. Procedia Eng. **182**, 510–517 (2017)
12. Olanrewaju, A., Anahve, P.J.: Duties and responsibilities of quantity surveyors in the procurement of building services engineering. Procedia Eng. **123**(SI), 352–360 (2015)
13. Oke, A., Aigbavboa, C., Wellington, T.: Personnal management practices in quantity surveying firms. J. Construct. Proj. Manage. Innov. **6**, 1633–1651 (2016)
14. Ramabodu, M.S., Verster, J.J.P.: Factors contributing to cost overruns of construction projects. In: 5th Built Environment Conference, Association of Schools of Construction of Southern Africa, Durban, South Africa, pp. 131–143, 18–20 July 2010
15. Ramabodu, M.S.: Procurement guidelines for project success in cost planning of construction projects. University of the Free State, Bloemfontein (2014)
16. Richards, P.E.: The role of the principal agent across the built environment professions, February 2017
17. SACQSP: Quantity Surveying Profession Act 2000 (Act No 49 of 2000). s.l.: SACQSP (2015)
18. Snyman, E., Cruywagen, H.: Affordability of quantity surveying services on construction projects in South Africa. In: COBRA Research Conference, Brisbane, Australia (2005)
19. Shanmugapriya, S., Subramanian, S.K.: Investigation of significant factors affecting time and cost overrun in indian construction projects. Int. J. Emerg. Technol. Adv. Eng. **3**, 734–740 (2013)
20. Thwala, W., Mathonsi, M.: Selection of procurement systems in the South African construction industry: an exploratory study. Acta Commercii **12**, 13–26 (2012)
21. Visser, I.G.: The benefits of utilizing a quantity surveyor. University of Pretoria, Pretoria (2009)
22. Wilson, R.: Mastering Project Time Management, Cost Control, and Quality Management, 1st edn. FT Press, Upper Saddle River (2015)

A Review of Critical Project Management Techniques to Enhance Construction SMEs Project Performance

Chijioke Emere(✉), Innocent Musonda, and Chioma Okoro

Department of Construction Management and Quantity Surveying, University of Johannesburg, Johannesburg 2000, South Africa
cemere@uj.ac.za

Abstract. Performance is linked with implementation of project management techniques. Literature suggests that problems such as inefficiency, poor budgetary issues, improper planning, scheduling and control in projects execution are associated with inadequate implementation of the essential techniques for meeting project deliverables. These shortcomings are prevalent among Small and Medium Enterprises (SMEs) in the developing countries including South Africa. Hence, the primary objective of this paper is to review the critical project management techniques for the improvement of construction project performance especially among SMEs in South Africa. The study is based on previous literature on construction SMEs project performance as well as project management techniques implementation. The literature review centered on both international and South African context. The study revealed that critical path method, work breakdown structure, and earned value management analysis were the most occurring project management techniques from the sampled literature. Findings from this study are envisaged to be beneficial to construction stakeholders in developing relevant project management techniques to improve the performance of SMEs.

Keywords: Construction SMEs · Project performance · Project management techniques

1 Introduction

Adequate implementation of project management techniques is a concern in the construction industry especially in this 21st century. Conventional operational procedures are no longer sufficient for settling the issues surrounding quite a large and complex environment such as construction [1]. Hence, construction projects now require a highly specialized approach such as the discipline of project management and the implementation of its techniques in achieving the desired deliverables [1]. Projects are termed successful particularly when they are finished within the planned time, budget,

This work is supported by the University of Johannesburg through its National Research Fund Scholarship.

C. Aigbavboa and W. Thwala (Eds.): CIDB 2019, *The Construction Industry in the Fourth Industrial Revolution*, pp. 521–528, 2020.
https://doi.org/10.1007/978-3-030-26528-1_53

and quality of which are the popular constraints of project management in executing a project [2]. As organizations continue to move toward project-based management to get more done with fewer resources, the need for project management and its techniques became more and more important [3, 4].

However, literature indicates that construction SMEs are fraught with many project management related problems such as inefficiency, poor budgetary issues, improper planning, scheduling and control in their projects execution, partly because of inadequate implementation of project management techniques [5–7]. It appears that many construction SMEs are negligent or not maximising the usage of formal, recognised techniques and very little attention is being paid to improving the status quo. Hence, schedule delays, cost overrun, poor quality, client dissatisfaction, loss of contract and unsustainable business growth [5, 8–10]. Improving project performance therefore requires implementation of effective project management techniques to improve performance. Therefore, continuous research on this topic is relevant.

Project management techniques are systematic procedures or practices that project managers use to achieve project deliverables [11]. They are important to the performance of construction SMEs, which account for about 78% of companies in the South African construction industry (CI) which contributes about 10% to the GDP [12, 13]. Moreover, SMEs in the construction industry are major platforms for job creation to solve the prolonged unemployment rate of about 27.7% among South Africans and to foster economic growth [10, 14]. Therefore, the performance of SMEs is of paramount concern. However, SMEs are characterized by high failure rate and are below par in their project performance efficiency and effectiveness especially among the lower grades of the CIDB [5]. Consequently, most construction SMEs have a short life span of approximately 6 months due to liquidation arising from inefficiency and poor cash flow management [15]. Projects among emerging contractors in South Africa suffer project management deficiency and quality is compromised due to negligence and shortcuts in project execution [16]. Statistics South Africa [17] and Monks [18] declared that the rate of project failure among SMEs in South Africa is about 75%. CIDB [5] found that 5 in 10 buildings deteriorate few months after erection, which is related to poor quality - a project management problem. Therefore, research on critical project management techniques for the success of SMEs is important.

However, previous related literature seemed to use the terms, “tools” and “techniques” synonymously [1, 19]. Other studies, Flemming and Koppelman [20], as well as Burke [21] discussed the use of Earn Value Management in aiding the project management plan, and control the project in comparable units, so that it can be accurately tracked and controlled. Other studies, like Milosevic [11], Simons and Lucarelli [22], and Kliem and Ludin [23], identified Work Breakdown Structure (WBS) as a key technique in managing scope. PESTEL (Political, Environment, Social, Technological, and Legal factors) Analysis, deliverable breakdown structure, PERT, Make or Buy analysis etc. are among the vital techniques discussed by the Project management method guide [24]. Appropriate evaluation and implementation of project management techniques towards the attainment of project objectives is invaluable.

The objective of this study is to identify the techniques that are relevant for construction project performance improvement. The succeeding sections of the paper present an overview of the method used in conducting the study. Subsequently, the findings from the desktop study are presented and conclusions drawn from the findings are presented.

2 Research Method

The study sought to establish the critical techniques for the improvement of project performance. The paper is part of a wider study being conducted on a Master's degree project. Consequently, the findings from a review of literature on the techniques are presented. Various sources were consulted including journals, conference proceedings, dissertations and theses, to meet the objective of the study. Databases including Science Direct, Ebscohost, Google and Google Scholar were consulted. The materials were selected based on their possession of relevant keywords including project management, performance, SMEs, success, construction industry and techniques. The identified techniques were ranked according to the frequency of occurrence among the sampled studies. The findings from the review are presented hereunder.

3 Findings on Project Management Techniques for SMEs Performance Improvement

Some project management techniques have been identified by many authors as reliable for improving various aspects of project performance, irrespective of the chosen methodology [11, 25–28]. For instance, for cost management, Earn Value Management is vital [11, 23, 29, 30]. For time management, techniques such as CPM, PERT, milestone analysis, are essential [21, 30–32]. For project risk management, SWOT analysis and Delphi are the notable project management techniques [11, 30]. For project human resource management and communication, conflict management, stakeholder analysis/management and responsibility matrix are some of the essential techniques to be used [11, 19] and 'bidder's conference' and 'Make or Buy' analysis are useful for procurement [1, 11].

Findings from this study portrayed important techniques, which influence project performance in both efficiency and effectiveness. Table 1 presents a summary of the identified techniques from the sampled literature. As shown in the table, the most cited techniques were Critical Path Method, Work Breakdown Structure, Earn Value Management, and Milestone Analysis. Some of these techniques are discussed hereunder.

3.1 Critical Path Method (CPM)

This technique helps the PM determine the start and finish dates for all the activities and identifies the sequence of activities that form the critical path [21]. Burke [21] affirms that the challenge of the PM is *to develop a network diagram and perform the CPM calculation to produce a logical project timeline that will serve as the project plan framework.* According to PMBOK [2], the CPM is necessary even after the use of other techniques such as the Work Breakdown Structure (WBS), to represent work activities for effective time management.

3.2 Work Breakdown Structure (WBS)

According to PMBOK [2], WBS is the process of subdividing project deliverables and project work into smaller, more manageable components. This technique "enables the project manager define the scope of work" [21]. Hence, the purpose behind the WBS is to subdivide the scope of work into a sensible lump that are less demanding to gauge, plan, and relegate to a dependable individual or department for consummation. Similar work should be grouped together in the breakdown to enhance efficiency of production [21].

3.3 Earned Value Management Analysis

Earn Value Management is regarded as a vital technique for project cost management [11, 23, 29, 30]. According to Burke [21], this is one of the special techniques within the project integration knowledge area that aids the PM plan and controls the project in comparable units, so that it can be accurately tracked and controlled". It can also be used for project cost management and as well as the assessment of project risk [11, 29, 30].

3.4 Quality Benchmarking

Benchmarking is a notable technique for project quality management [1, 19]. Quality benchmarking, also referred to as "best-practice benchmarking", or "process benchmarking", is used in strategic management, to evaluate various aspects of organizations' work-tasks and processes in relation to best-practice. This is usually within a peer-group, which is defined for the purposes of comparison [33]. Organizations are enabled to develop plans on how to make improvements or adapt specific best practices, with the aim of increasing some aspect of performance.

3.5 SWOT Analysis

This is a risk management technique that can be used to examine the strength, weakness, opportunity and threat of project criteria and deliverables for effective contingency planning [24]. This can also be essential for generating courses of action should problems occur.

3.6 Make or Buy Analysis

A Make or Buy Analysis helps the organization to take an informed decision about what to outsource and what not to outsource [24]. "Portfolio managers and project sponsors are often faced with the dilemma to make or buy, considering the availability and skills of resources at hand. The various factors to be considered for a Make or Buy Analysis include cost comparison, technology and business processes, supplier related information, and support systems" [24].

Table 1. Project management techniques

Technique	Source															Total
	Abbassi and Al-Mharmah (2000)	Burke (2013)	Cetindama et al. (2012)	Chou and Yang (2012)	Fleming and Koppelman (2002)	Golini et al. (2012)	Kerzner and Kerzner (2017)	Micheal et al. (2014)	Milosovic (2003)	Newell (2002)	Patanakul et al. (2010)	Stojcetovic et al. (2013)	PM^2 Guide (2016)	PMBOK (2012)	White and Fortune (2002)	
Critical Path Method	x	x		x		x	x	x	x		x		x	x	x	11
Work Breakdown Structure (WBS)	x	x		x		x	x	x	x		x			x	x	10
Earn value Management		x		x	x	x	x		x		x		x	x		9
Milestone Analysis						x		x		x	x	x	x	x	x	8
Program Evaluation and Review Technique (PERT)	x						x	x			x		x		x	6
Stakeholder Management (Interest/Influence Matrix)				x		x					x		x	x		5
SWOT Analysis			x								x			x	x	4
Lesson Learned											x		x	x	x	4
Make or Buy Analysis				x							x		x	x		4
Analogous Estimating				x							x			x		3
Parametric Estimating				x							x			x		3
Quality review technique				x							x			x		3
Quality Benchmarking				x							x			x		3
Bidders conference				x							x			x		3
Conflict Management				x										x		2
Delphi Technique											x			x		2
Brainstorming								x				x				2
Change control											x			x		2
Meetings													x			1
PESTEL Analysis													x			1
Deliverable Breakdown Structure (DBS)													x			1

3.7 Stakeholder Analysis (Interest/Influence Matrix (SIIM))

Here, interest and influence of each stakeholder in the project are analyzed, facilitated and documented [24]. It is of utmost importance to know the stakeholders and their relevance for the project to identify project champions and possible detractors.

3.8 Lessons Learned

Capturing Lessons Learned is a way of identifying development/improvement areas within a project for the aim of helping similar projects avoid certain pitfalls in the future [24]. Information that can be captured includes lessons learned from management of risks, quality issues, and outsourcing or contractor issues, change requests and so on.

In summary, the above-discussed techniques are not exhaustive. However, it is notable that the techniques are usable to improve various aspects of project performance and are therefore essential for SMEs in the construction industry, especially CPM, WBS and Earned Value Management Analysis.

4 Conclusion

The paper set out to establish project management techniques mostly identified in existing literature. The study found that the critical path method, work breakdown structure and earned value management analysis were the most frequently occurring project management techniques among the sampled literature. These techniques, if judiciously applied by Construction SMEs, will go a long way in ensuring that targeted objectives in a project are achieved. Project objectives when achieved increase the overall productivity of the organization, which in turn affects the economic abundance and business sustainability of the company, as well as the gross domestic product of any economy. Good project management practices such as the implementation of techniques in project execution should be given priority among construction SMEs for achieving efficiency, effectiveness, and meeting of the project objective of time, cost and quality.

Being a literature review presentation, the findings of the study are not exhaustive and therefore further research is necessary, even with the use of primary data, to determine critical techniques used among construction SMEs to improve project performance.

References

1. Chou, J.S., Yang, J.G.: Project management knowledge and effects on construction project outcomes: an empirical study. Proj. Manage. J. **43**(5), 47–67 (2012)
2. PMBoK, A.: A guide to the project management body of knowledge (PMBOK guide). Project Management Institute, Inc. (2013)
3. Horine, G.M.: Project Management: Absolute Beginners Guide, 3rd edn. Que Publishing, Indianapolis (2013)

4. Patah, L.A.: Avaliação da relação do uso de métodos e treinamentos em gerenciamento de projetos no sucesso dos projetos através de uma perspectiva contingencial: uma análise quantitativa. [Evaluation of the relationship between the use of methods and training in project management and the success of projects through a contingency perspective: a quantitative analysis]. Ph.D. in Production Engineering (Dissertation) – Universidade de São Paulo, São Paulo (2010)
5. CIDB. SMEs Development. www.info.gov.za/View/DownloadFileAction?id=68205. Accessed 27 Apr 2018
6. Ihesiene, U.C.: A survey-based study of project management problems in small and medium scale enterprises (SME) in Nigeria. Eur. Sci. J. **10**(25) (2014). 1857 7881
7. Moilwa, S.: Factors constraining the development of professional project managers in small and medium size construction enterprises in South Africa. (Unpublished Master's thesis). University of Witswatersrand, Johannesburg (2013)
8. CIDB. Construction quality in South Africa: a client's perspective. CIDB, Pretoria, South Africa (2011)
9. PwC. Trends, challenges and future outlook: capital projects and infrastructure in East Africa, Southern Africa, and West Africa. SA construction. www.pwc.co.za/infrastructure. Accessed 26 June 2018
10. Statistic South Africa. Gross Domestic Product: Fourth Quarter 2017 (2017). www.statssa.gov.za
11. Milosevic, D.Z.: Project Management Toolbox. Wiley, New York (2003)
12. Stats, S.A.: Statistics South Africa. Formal census (2011)
13. Statistics South Africa. Gross domestic product: Second quarter. Statistical Release PO441. Statistics South Africa, Pretoria (2014)
14. Shakantu, W.M.: Contractor development. In: Ofori, G. (ed.) New Perspectives on Construction in Developing Countries, 1st edn. Spon Press, New York (2012)
15. Gillam, S.: SA builder/bouer, awards for excellence. J. Build. Ind. Fed. S. Afr. **961**, 28–29 (2004)
16. Mavetera, N., Sekhabisa, K., Mavetera, C., Choga, I.: Factors influencing success of construction projects by emerging contractors in South Africa: a case of mahikeng area. J. Corp. Ownersh. Control **13**(1), 1–9 (2015)
17. Statistics South Africa. Quarterly labour force survey. Statistics South Africa (2010)
18. Monks, P.: Sustainable Growth of SMEs. Unpublished Master's research report, Nelson Mandela Metropolitan University (2010)
19. Patanakul, P., Iewwongcharoen, B., Milosevic, D.: An empirical study on the use of project management tools and techniques across project life-cycle and their impact on project success. J. Gen. Manage. **35**(3), 41–66 (2010)
20. Fleming, Q.W., Koppelman, J.M.: Earned value management. Cost Eng. **44**, 32–36 (2002)
21. Burke, R.: Project management: planning and control techniques. New Jersey, USA (2013)
22. Simons, G.R., Lucarelli, C.M.: Work breakdown structures. In: Pinto, J.K. (ed.) The Project Management Handbook. Jossey-Bass Inc., San Francisco (1998)
23. Kliem, R.L., Ludin, I.S.: Tools and Tips for Today's Project Manager. Project Management Institute, Newtown Square (1999)
24. PMM. Project Management Methodology Guide. European Union. http://europa.eu/pm2. Accessed 21 Nov. 2018
25. Ghanim, A., Munassar, F., Dahlan, A.R.A.: Project and change management success factors from Malaysian government departments and agencies perspective. IOSR J. Bus. Manage. **11**(2), 36–45 (2013)
26. GMC PMP: Communications skills for project managers. AMACOM Div American Mgmt Assn (2009)

27. SE Project. Software Project Management: Methodologies and Techniques. Technische Universiteit Eindhoven (2004)
28. Špundak, M.: Mixed agile/traditional project management methodology–reality or illusion? Procedia Soc. Behav. Sci. **119**, 939–948 (2014)
29. Fleming, Q.W., Koppelman, J.M.: Earned value project management. Project Management Institute, Newton Square (2000)
30. Newell, M.W.: Preparing for the Project Management Professional (PM) Certification Exam. AMACOM, New York (2002)
31. Kerzner, H., Kerzner, H.R.: Project Management: A Systems Approach to Planning, Scheduling, and Controlling. Wiley, New Jersey (2017)
32. Michael, J., Deepak, T., Venishri, P., Siow, Y.T.: Ranking the factors that influence the construction project management success-Malaysian perspective. Civ. Environ. Res. **6**(1), 80–88 (2014)
33. Chen, H.L.: Benchmarking and quality improvement: a quality benchmarking deployment approach. Int. J. Qual. Reliab. Manage. **19**(6), 757–773 (2002)
34. Abbassi, G.Y., Al-Mharmah, H.: Project management practice by the public sector in a developing country. Int. J. Project Manage. **18**(2), 105–109 (2000)
35. Misic, M., Sarkocevic, Z.: Quality tools in project management. Centre for Quality (2013)
36. Centindamar, D., Wasti, N.S., Beyhan, B.: Technology management tools and techniques: factors affecting their usage and their impact on performance. Int. J. Innov. Technol. Manage. **9**(5), 1250036 (2012)
37. White, D., Fortune, J.: Current practice in project management - an empirical study. Int. J. Project Manage. **20**(1), 1–11 (2002)

An Investigation of Causes of Deterioration for Health Physical Infrastructure Transitioned from Donor to National Management

Bwalya Helena Mutale[1], Erastus Mwanaumo[1(✉)], Nathan Musonda[2], and Sampa Chisumbe[1]

[1] Department of Civil and Environmental Engineering, School of Engineering, University of Zambia, Great East Road, Lusaka, Zambia
erastus.mwanaumo@unza.zm
[2] School of Post Graduate Studies, University of Lusaka, Lusaka, Zambia

Abstract. Maintenance of physical infrastructure transitioned from donor management to national ownership has been a challenge in the health sector leading to a considerable number of physical infrastructure dilapidating to a deplorable state. This study investigated the causes of deterioration for Health Physical Infrastructure Transitioned from Donor to National Management in Zambia. A mixed method approach was adopted for the study. Qualitative and quantitative data was collected through interviews and questionnaires respectively. The findings revealed that there was lack of onsite maintenance technicians in the health centres. Further, that though maintenance plans were in place in most urban facilities as compared to rural health facilities adherence to such plans was hindered due to lack of funding. The study therefore, recommends improved budget allocation for maintenance activities to the Ministry of Health, assigning of qualified maintenance technicians in health centres. Further, that Government should consider having a maintenance policy to achieve sustainability for built assets transferred from donor to national management.

Keywords: Transition · National Management · Maintenance · Physical Infrastructure · Health centres · Donors

1 Introduction

Infrastructure is the basic physical and organisational structures needed for the operation of a society, this include; industries, buildings, roads, bridges, health services, governance and many others [1]. Zambia has in the past years since independence in1964 benefitted from numerous donor aid and support in various infrastructure developments projects, more especially in the construction of structural infrastructure in the health care sector. Several hospitals, clinics and health posts in urban and rural parts of the country have been built through the support of donor aid from various western donors. Management of these health centres by the donors has resulted in consistent and timely maintenance works being carried on these facilities to ensure that

C. Aigbavboa and W. Thwala (Eds.): CIDB 2019, *The Construction Industry in the Fourth Industrial Revolution*, pp. 529–537, 2020.
https://doi.org/10.1007/978-3-030-26528-1_54

unnecessary destruction and decay to these structures does not hinder the service provision being offered to the communities [2].

Maintenance of physical infrastructure in all sectors be it health, roads, ICTs etc. is a critical component of the economy in that it is known to improve outcomes of people using and accessing services from these sectors. Physical structures in general, overtime if not taken care of exhibits signs of dilapidation and this has been known to affect work in terms of poor productivity, efficiency and time management from the health care workers. The issue of lack of maintenance to infrastructure has engaged practitioners and researchers for a long time to find best possible ways to encourage communities and countries to take up maintenance of various infrastructures. According to Doman [3] "public infrastructure in most developing countries, is often left to decay due to lack of maintenance. The article further indicates that lack of maintenance strategies and funding are the common causes of non-functional maintenance departments."

In Zambia's health sector, the major setback in lack of maintenance of physical structures can be attributed to inadequate funding and non-adherence to maintenance schedules [4]. Maintenance of physical structures can be improved to yield better results if issues of maintenance are taken more seriously. To continuously provide quality health care to the citizens and boost the morale of health care workers, maintenance activities are of utmost importance [5]. The focus for most donors is moving away from physical structural construction and management to encouraging local ownership and management. More so, to foster sustainability of project or programs through transitioning their projects from donor management to national ownership through the government by fully engaging the relevant ministry from project initiation, completion and implementation [6]. Therefore, the aim of the study was to investigate the causes of deterioration of health physical infrastructure transitioned from donor support to national ownership.

2 Maintenance

All built infrastructures are subject to aging, wear and tear in the performance of their functions and deterioration by exposure to outside operating environment. Hence, left to themselves, built infrastructures will eventually become inefficient, unreliable and fail [7]. Maintenance is defined as the act of keeping something in good condition by checking or repairing it regularly [8]. Maintenance works are conducted to infrastructure to maintain, revive and prolong the life of the structure. Maintenance work is defined as the combination of technical and administrative actions taken to preserve or protect a structure, system or equipment to function properly [9]. To increase the quality of maintenance works, maintenance culture has been recognized as an important aspect to this [9].

Maintenance culture defines the way of thinking, behaviour, perceptions, values and underlying assumptions of a person or group or society that considers maintenance as a matter that is of priority and practices it in their life [9]. Florence [10] further asserts that maintenance culture is not universal in nature and is usually derived or learned through a person making maintenance a natural daily practice that can be followed and emulated by others. Therefore, it has been observed through literature that to conduct effective maintenance on infrastructure requires on set maintenance culture from the relevant authorities mandated to conduct maintenance works.

3 Maintenance of Physical Infrastructure in the Health Sector

Infrastructure serves a central delivery mechanism in achieving sustainable economic development and in the generation of quality social-economic development in a country [11]. Physical infrastructure consists of telecommunications, sewerage and water systems, roads, railways, airports, airline, education and health. In this study, the focus is on social infrastructure in the health sector. Health infrastructure is critical to the development of any economy and society. In the health sector, government efforts and resources have been supplemented by donor finances in further developing the hospitals, clinics and other rural health centres. The reality is that all the elements and components that make up an engineering infrastructure unavoidably deteriorates with time due to inherent defect in design and construction and affects user activities and this overtime requires maintaining [1]. Maintenance-free or self-sustaining infrastructure is highly desirable but not feasible. The inadequacy of maintenance of infrastructure in developing country's cities has profound consequences for economic and social development. The ability of these cities to support productive public economic activities is severely hampered by inadequate service delivery, deteriorating infrastructure often caused by poor maintenance [12]. The main aim of infrastructure maintenance is to preserve a building in its initial effective state and in most developing countries maintenance of building infrastructure in rural and urban settlements is inadequate and usually non-existent [13].

Most governments in developing countries including Zambia have poor infrastructure maintenance systems due to, lack of maintenance policies, insufficient funding and therefore rely mostly on donor support for such services [14]. Donors in most cases have funded the construction of new building infrastructure projects instead of maintaining or refurbishing already existing infrastructure because this has been seen as a more tangible asset that is mostly appreciated and highly favoured by most donor aid recipients [3].

Preference usually decided by the government has been more on the construction of new structures as opposed to refurbishing existing health centres due to the excessive cost of maintenance, limited financial resources and other pressing expenditure needs which extend to both rural and urban settings. It is argued that one major weak link in managing health infrastructure in Africa and Zambia is that of maintenance. Sustainability of any infrastructure and equipment is dependent on maintenance; therefore, donors seeking to assist in developing good maintenance systems should put measures in place such as; developing a culture of maintenance, which requires having the right artisans, technicians, technologist and engineers to provide the necessary skills. A major problem in the health sector is the dilapidated status of most facilities, which has affected safety of users and patients [15].

3.1 Health Infrastructure Maintenance Policy in Zambia

In general, the health care infrastructure in Zambia is in a desperate state of refurbishment and development needs [15]. With regards to the National Health Strategic plan 2011–2015 under the Infrastructure and equipment component, the core elements that were monitored included; Capital Investment Plan (CIP) developed and under implementation; with significant support from Ministry of Health/Government of the Republic of Zambia and partners; Construction of new health facilities and staff houses continued; Construction of maternity units and mothers' waiting shelters continued as well as Engagement of provincial equipment engineers for maintenance of medical equipment.

Unfortunately, no considerations or plans were incorporated to factor in physical infrastructure maintenance towards the already existing buildings in the strategic plan. Maintenance policies are critical to the conduct of maintenance works [16]. However, currently the country has no maintenance policies formulated and in use by either MoH or MWS.

3.2 Challenges Faced in Managing Maintenance in Transitioned Physical Infrastructure

Several authors have cited many reasons as contributing factors to challenges being experienced during the transitioning of infrastructure management from NGO management to national ownership such as;

- Lack of maintenance funds to support maintenance works from the responsible ministries and departments. Maintenance is often regarded as a necessary expense that belongs to the operating budget and should be balanced with the annual work schedule that has been decided [12].
- During project implementation, the government as major stakeholders is not fully involved and engaged in the implementation of activities and are more focused on actual structures being completed on time there by leaving much of the responsibilities of the works being done to be managed by thee donor.
- Lack of integrating transition plans into the project implementation stage with emphasis on capacity building local personnel to foster continuation and reduce any negative impact.
- Scarcity of trained man power caused by experienced staffs that have migrated in search of greener pastures within the diaspora therefore leading to over dependency on expatriates [17].
- The poor maintenance culture and inadequate resource allocation in the past have been identified as the main inadequacies in the development, management sustainability of national infrastructure in Zambia [18].

4 Research Method

The study was carried out using a mixed method approach that involved the collection of data in both quantitative and qualitative approaches. The quantitative approach involved the distribution of thirty (30) questionnaires in thirty (30) purposefully selected health centres of which fifteen (15) were urban health centres while the other fifteen (15) were rural health centres in Eastern, Lusaka and Western provinces of Zambia. The respondents for the questionnaires included administrators of the sampled health centres. Twenty-four (24) questionnaires (representing 80% response rate) were collected and analysed. A questionnaire was used in order to ensure anonymity of the respondents, which resulted in the respondents being more comfortable and free to express their view. Furthermore, interviews were also conducted with respondents' in-charge of maintenance from MoH, Ministry of Works and Supply (MWS), and Centre for Infectious Diseases Research in Zambia (CIDRZ) issues of inquiry included maintenance scheduling, maintenance policy, skills knowledge, availability of man-power, maintenance financing, transition management and recommendations on improvements that can be made to foster continuous maintenance.

5 Findings

5.1 Maintenance Strategies in the Health Facilities

The findings revealed that there was lack of onsite maintenance technicians in the health centres. More so, that the health centres relied mostly on MoH to manage most of the facility maintenance requirements. Further, an assessment was carried out to determine whether the health centres had maintenance schedules and plans. This was ranked first with SD of 1.313 as shown in Table 1. Similarly, an assessment on whether the respondents understood the written sentences and paragraphs in the maintenance schedule or plan was ranked second with SD of 1.213 as indicated in Table 1. This suggest that most of the respondents understood contents of the maintenance schedule or plan.

Table 1. Factors assessing the level of usage of maintenance schedules and plans.

Factors	Weighting	Mean	SD	Rank
Does the facility have a maintenance schedule or plan?	24	3.625	1.313	1
Understanding the written sentences and paragraphs in the maintenance schedule	24	3.417	1.213	2
Are you able to prepare budgets and reports for all anticipated maintenance works on time as indicated per schedule?	24	3.250	1.073	3
Do you conduct random tests and inspections of infrastructure to ensure they are in working order?	24	3.208	1.179	4
Are you able to install, replace and maintain infrastructure as specified in the maintenance schedule?	24	2.375	0.924	5

Source: Own computations. Mean = 3.175, Standard Deviation = 1.14026024.

5.2 Factors that Can Lead to Improvement in Maintenance

Respondents were asked to rank the key factors that would assist to improve maintenance of physical infrastructure transitioned from donor support and managed by Ministry of Health. All factors were ranked according to the standard deviation to adduce the most crucial factors that can help improve maintenance of physical infrastructure in the health centres. Conducting regular inspections and audits on maintenance works was ranked first with SD of 0.487. This shows that the respondents saw this conduct of regular inspections and audits as a critical factor that assist in timely identification of maintenance works. Having clear policy guidelines on maintenance distributed and displayed in the health centre was ranked second with SD of 0.504. This result indicates that the respondents thought this was a very important factor to assist in conducting maintenance activities efficiently. Strict management practices by MoH to nurture maintenance activities was ranked third with SD of 0.659. Good communication and feedback by all stakeholders on maintenance works undertaken was ranked fourth with SD of 0.590 as can be seen in Table 2.

More so, Provision of smart technologies to alert maintenance officers when to conduct scheduled maintenance activities was ranked fifth with SD of 1.021. Assigning permanent maintenance officers at the health centres was ranked sixth with SD of 0.728, continuously building capacity in maintenance officers was ranked seventh with SD of 0.690. Awarding incentives to health centres adhering to maintenance schedules was ranked eighth with SD of 0.637 and lastly, ranked in ninth position is the factor to remove zonal maintenance officers with SD of 1.050 as shown in Table 2.

Table 2. Factors assessing improvement of maintenance of physical infrastructure in the health centers.

Factors	Weighting	Mean	SD	Rank
Conducting regular inspections and audits on maintenance works in all health centres by MoH	23	1.348	0.487	1
Clear policy guidelines on maintenance distributed and displayed in the health centre	24	1.417	0.504	2
Strict management practices by MoH to nurture maintenance activities and schedules	24	1.500	0.659	3
Good communication and feedback by all stakeholders on maintenance works undertaken	24	1.500	0.590	4
Provision of smart technologies to alert maintenance officers when to conduct scheduled maintenance activities	24	1.542	1.021	5
Assigning permanent maintenance officer at the health centre	23	1.565	0.728	6
Continuously build capacity in maintenance officers	24	1.708	0.690	7
Award incentives to health centres adhering to maintenance schedules	24	1.833	0.637	8
Removal of zonal maintenance officers	24	3.333	1.050	9

Source: Own computations. Mean = 1.749597333, Standard Deviation = 0.707211789

5.3 Availability of Maintenance Policy

All the interviewees acknowledged that both MoH and MWS had no maintenance policy. However, that MoH had started the process of developing a maintenance policy and went as far as production of a draft maintenance policy. Higher authorities unfortunately did not approve this instead that draft policy was converted to maintenance guidelines for MoH for use in all health centres and hospitals.

Interviews on various maintenance challenges affecting the built infrastructure due to the absence of the policy revealed that; District Health Office (DHO) had zonal maintenance technicians that respond to minor facility repairs. While the health centres do not have maintenance technicians on site, zonal maintenance technicians are responsible to attending to maintenance requests in their assigned zones. Further, that there are no skills transfer on maintenance works conducted by the donors to the district or MoH members of staff. More so, that the major contributing factor to lack of maintenance of infrastructure in the health centres is the non-existence of financial resources for maintenance works.

However, it was revealed that the MWS began the process of drafting the overall maintenance policy in 2016 and currently the draft copy has been completed and circulated to all ministries and other major government stakeholders such as spending agencies and the provinces for comments before the final adoption and approval of the maintenance policy. The interviewee affirmed that once the maintenance policy is approved, this will be the only maintenance policy to be used by all ministries and other government stakeholders.

6 Discussion

The study established that there was lack of onsite maintenance technicians in the health centres. More so, that the health centres relied mostly on MoH to manage most of the facility maintenance requirements. On the level of usage of maintenance schedules and plans. The study revealed the usage of maintenance schedules and plans to be more prevalent in urban health centres as compared to the rural health centres. Additionally, that the lack of implementation of actual maintenance works was hampered by inadequate financial resources.

On viable options that can assist, improve the maintenance of physical infrastructure in the health centres. Respondents were asked to rate the measures to help improve maintenance in the health centres. Nine (9) measures were identified. Conducting regular inspections and audits on maintenance works was ranked first with SD of 0.487. This factor was also recognised by the O&M Best Practices guide 2010 which strongly recommends implementing reactive, preventative and corrective maintenance activities to prevent infrastructure from ruin and dilapidation. Having clear policy guidelines on maintenance distributed and displayed in the health centre was ranked second with SD of 0.504. These results agree with the parliamentary report on health [14] which indicated that there is no policy on maintenance of infrastructure in the health sector and there in need to formulate a maintenance policy. Dhillion [16] further asserts that having a maintenance policy is one of the most crucial elements of effective maintenance management.

Furthermore, that strict management practices by MoH to nurture maintenance activities was ranked third with SD of 0.659. This factor was also recognized by Sania et al. [9] who noted that maintenance culture is not universal in nature and is usually derived or learned through a person or organisation making maintenance a natural daily practice that can be followed and emulated by others. While good communication and feedback by all stakeholders on maintenance works undertaken was ranked fourth with SD of 0.590 as can be seen in Table 3. This crucial factor agrees with Dhillion [16] who identified eight elements of effective maintenance management. Under equipment records that is among the elements identified, he mentions that records or documents are useful when procuring latest items or equipment to determine operating performance trends, troubleshooting breakdowns, making replacements or modifications and identifying areas of concerns and investigating incidents.

7 Conclusion and Recommendations

This study investigated the causes of deterioration for Health Physical Infrastructure Transitioned from Donor to National Management. It was established that though maintenance plans were in place in most urban facilities as compared to rural health facilities adherence to such plans was hindered due to lack of funding for maintenance activities. The results suggest that an aspect of health management that is overlooked is Built infrastructure maintenance. This issue concerns all levels of the health system ranging from the clinic to the hospital levels. The study therefore, recommends improved budget allocation for maintenance activities to the Ministry of Health, assigning of qualified maintenance technicians in health centres to be in-charge of maintenance works at facility level. Further, that Government should consider having a separate maintenance policy for Ministry of Health in order to sustain built infrastructure assets transferred from donor to national management.

References

1. Ondiege, P., Moyo, J.P., Verdier-Chouchane, A.: Developing Africa's Infrastructure for Enhanced Competitiveness (2012). https://www.fig.net/resources/proceedings/fig_proceedings/fig2012/papers/ts01c/TS01C_oyedele_6119.pdf. Accessed 30 May 2017
2. Ministry of Health: Joint Annual Review 2011, Health Sector Performance Review. Government of the Republic of Zambia (2012). http://www.nationalplanningcycles.org/sites/default/files/planning_cycle_repository/zambia/jar_report_final_submission.pdf. Accessed 12 Feb 2017
3. Dornan, M.: Aid and the maintenance of infrastructure in the pacific (2012). http://devpolicy.org/aid-and-the-maintenance-of-infrastructure-in-the-pacific20120329/. Accessed 24 Apr 2017
4. Government of the Republic of Zambia: National Health Policy. Government Printers, Zambia (2012)
5. Crye, L.: Transition of Management and Leadership of HIV Care and Treatment Programs to Local Partners: Critical Elements and Lessons Learned. Arlington, VA: USAID's AIDS Support and Technical Assistance Resources, AIDSTAR-One, Task Order 1 (2011)

6. USAID Zambia Country Development Cooperation Strategy 2011–2019. https://www.usaid.gov/sites/default/files/documents/1860/CDCS_Zambia_December_20192.pdf
7. Lekan M.A, Anosike. M.N., Oginde, A.: Sustainability Strategies in Engineering Infrastructure-Maintenance in Developing Countries Selected State in South Western Nigeria (2011). https://www.iiste.org/Journals/index.php/CER/article/view/10849. Accessed 5 May 2019
8. Oxford Advanced Learner's Dictionary, 7th edn. Oxford University Press (2006). ISBN 13:978-0-19-4316613
9. Sania, S.I.A., Mohammeda, A.H., Misnanb, M.S., Awang, M.: Determinant Factors in Development of Maintenance Culture in Managing Public Asset and Facilities. International Congress on Interdisciplinary Business and Social Science (2012). https://pdfs.semanticscholar.org/9c34/e8f92511ff3cea70541a21415665fc3353c0.pdf
10. Florence, T.: An empirical analysis of asset replacement decisions and maintenance culture in some government organizations located in Ogbomoso and Ilorin metropolis as case study. J. Manage. Soc. **1**(3), 01–09 (2011)
11. Zambia Development Agency: Infrastructure Sector Profile (2014). http://www.zda.org.zm/?q=content/infrastructure. Accessed 30 Mar 2017
12. Zakaria, H., Ahmad, K. A.S., Aiyub, K.: Financial factor affecting maintenance management in safety and health practices (2012). www.ukessays.com
13. Zawawi, E., Kamaruzzaman, S., Ali, A.: Retail Leisure Property **9**, 349 (2010). https://doi.org/10.1057/rlp.2010.16
14. Zambian Parliamentary Committee Report on Health (2014). http://www.parliament.gov.zm/sites/default/files/documents/committee_reports/REPORT%20OF%20THE%20COMMITTEE%20ON%20HEALTH%202014.pdf
15. Report on the Health Care Sector and Business Opportunities in Zambia (2013). https://www.swecare.se/Portals/swecare/Documents/Report-on-the-Health-Care-Sector-and-Business-Opportunities-in-Zambia.pdf
16. Dhillion, B.S.: Engineering Maintenance: A Modern Approach. CRC Press (2002)
17. Aulick, K.: Learning Agenda mini- case 10, Tanzania, Localization; Three partner models (2011). www.developmentiscapacity.org/sites/default/files/10-localisation_ThreePartnerTransitionmodels-TanzaniaLAMinicase.pdf. Accessed 13 Feb 2017
18. Muya, M., Lusambo, E., Suzzane, R., Banda, I., Mubemba, C., Hibaajene, S., Mang'elele, I., Mukelabai, G., Chisanga, K.: Infrastructure Watch Culture: Zambia's Infrastructure Report Card (2017). http://www.hrpub.org/download/20170130/CEA2-14808276.pdf. Accessed 30 Mar 2017

Success Strategies for Competitive Advantage in the Ghanaian Construction Industry: A Delphi Study

Matthew Kwaw Somiah[1,2](✉), Clinton Ohis Aigbavboa[3], and Wellington Didibhuku Thwala[3]

[1] University of Johannesburg, APK Auckland-Kingsway, Johannesburg, South Africa
brokwaw@yahoo.com
[2] Takoradi Technical University, Takoradi, Ghana
[3] Sustainable Human Settlement and Construction Research Centre, University of Johannesburg, APK Auckland-Kingsway, Johannesburg, South Africa

Abstract. Though strategy appears to be the most mentioned attribute for firms' competitive advantage, the success strategies for competitive advantage of firms in the Ghanaian construction industry in not known. This Delphi study identified the success strategies for competitive advantage in the Ghanaian construction industry and determined the relative impact of each of the success strategies. It used structured questionnaire in soliciting views from 15 experts. Data was analysed using the median, mean, and standard deviation. In all 14 success strategies were identified to impact competitive advantage with service differentiation strategy and being technologically independent ranking 1st among the strategies. It is recommended that firms should highly prioritize service differentiation strategy, and be technologically independent since these strategies recorded a very high impact on competitive advantage. The findings of this study will form the basis for future competitive advantage studies and inform policy direction in the Ghanaian construction industry.

Keywords: Competitive advantage · Construction industry · Ghana · Strategies · Success

1 Introduction

Firms' competitive advantage is derived from set of attributes that aid firms to perform better than their rivals in a competitive industry [1]. Firms with competitive advantage enjoy cost advantage, larger market share, profit advantage, technology advantage, and optimizes stakeholders' satisfaction benchmarked against best performing competitors in the industry [1–5]. However, as to what actually constitute the set of attributes that will give firms' competitive advantage, researchers' views have varied significantly over the years. It is attributable to the subjective nature of the competitive advantage concept and the varying political, social, cultural and economic settings among nations [4, 6]. Some known attributes for competitive advantage of firms include firms'

C. Aigbavboa and W. Thwala (Eds.): CIDB 2019, *The Construction Industry in the Fourth Industrial Revolution*, pp. 538–546, 2020.
https://doi.org/10.1007/978-3-030-26528-1_55

structure, factor conditions, strategy, resources and competencies, and demand conditions in a particular industry in previous studies (see [1, 3, 5, 7]). Though strategy appears to be the most recognized and mentioned among the set of attribute for firms' competitive advantage in previous studies (see [1, 3, 5, 7]) the industry and national specific success strategies have varied among the researchers. In relation to the construction industry in Ghana, there is a dearth of studies that identified the set of success strategies for competitive advantage for firms. It is against this backdrop that this current research seeks to identify the success strategies for competitive advantage in the Ghanaian construction industry and to determine the relative impact of each of the success strategies. The specific objectives that governed this current study were:

- to identify the success strategies for competitive advantage in the Ghanaian construction industry; and
- to determine the relative impact of each of the success strategies in the Ghanaian construction industry.

The findings of this study will form the basis for future studies on strategies for competitive advantage. Empirically, it informs firms and stakeholders in the Ghanaian construction industry of the set of success strategies for competitive advantage. It unravels the success strategies for policy formation and direction. Though the findings of this research may largely represent what is prevailing in the Ghanaian construction industry, it could also be of help to other developing nations whose construction industry has similar characteristics with Ghana. Again, the findings of this research were limited to the views of the experts that constituted the Delphi panel and that further study to validate this qualitative study is recommendable. Construction firms are registered building and road contractors who undertake civil engineering works in Ghana. They include both foreign and local contractors. The subsequent sections of this current study is structured under literature review, methodology and results and discussions.

2 Literature Review

2.1 Strategies for Competitive Advantage in Previous Studies

Strategies are practical plan that firms use to achieve their mission and objectives in a competitive industry [8]. Strategies impact structure and competitive advantage of firms [2, 3, 8]. However, the set of strategies that impact competitive advantage have varied among researchers. For example, [7] identified focus, growth, differentiation, and cost leadership strategies to have impacted competitive advantage of firms in the Hong Kong. Whiles, adopting lean construction, core values as well as firms' mission and objectives in industry were identified by [5] to have constituted the strategies for competitive advantage in the UK, Finland, and Sweden. Subsequently [6] informed that unique technology, and client relations strategies impact competitive advantage of firms in China. Thus, informed by previous studies on strategies for competitive advantage, in this current study, strategies are the set of practical plans that impact competitive advantage of construction firms; and it includes: lowest price strategy, growth strategy, among others as informed by the literature reviewed. The next section of the study will theoretical explain strategies for competitive advantage.

2.2 Theories of Firms' Competitive Strategy

In explaining the strategies for competitive advantage, the competitive strategy theory by [7] and the three generic competitive strategies theory by [2, 3] were drawn upon.

2.2.1 Competitive Strategy Model

The model posits that competitive advantage is derived from a combination of resource, strategy and performance in the Hong Kong. Further, it informs that strategy informs the resources required and again influences the firm's performance in an industry [7]. The theory informs that strategies for competitive advantage include: focus, differentiation, growth and cost leadership strategies [7].

2.2.2 The Three Generic Competitive Strategies Model

According to [2], three generic strategies give firms' competitive advantage in a competitive industry. The generic strategies were: the differentiation strategy, the overall cost leadership strategy, and the focus strategy. In relation to the overall cost leadership strategy, [3, 9] posited that, firms will achieve cost advantage if the aggregate cost of performing their value activities is lower than that of their rivals in the same industry; and sustainability of cost advantage in the market is achieved when products of firms are unique and cannot be imitated [3, 9]. This assertion is not different in the construction industry as firms that achieve overall cost advantage are very competitive [7]. Furthermore, depending on the nature of competition in the industry, firms that assume differentiation strategy position have competitive edge over their rivals if, clients or buyers in the industry find their products or services to be of higher value, and of more importance to them [2, 3, 9]. This position can be associated with technology, brand or design image, customer service, features or dealer network [6, 9]. The third generic strategy for competitive advantage is the focus strategy. This could be, focusing on a segment of the product line, focusing on a particular buyer group, or focusing on a particular geographic market ([2, 3, 9]. It must be emphasized that, the three generic strategies are alternative approaches to neutralizing the competitive forces by rivals [2], in a particular industry, so as to have competitive edge or advantage [2, 3, 6].

Nonetheless, Porter's generic competitive strategies theory considered only one firm at a time thus, neglected the issue of having competitive advantage through the strategy of synergy or collaboration between firms. According to [5], firms' competitive advantage is also determined by ability to collaborate with other firms to get tasks executed. Thus, the need to consider synergy, which could be the fourth strategy an entity may adopt for competitive advantage in a given industry. However, this fourth strategy could be adopted in conjunction with any of the three generic strategies at the same time. This view is affirmed by [10] when the researchers contended that, 'construction firms' unique capabilities are not rooted solely within the company but are spread across networks of relational ties'; suggesting that firms could obtain competitive advantage in the construction industry through the ability to collaborate, and make use of the capabilities of other firms, at times, in project delivery.

Thus, the literature reviewed informs that none of the theories is exhaustive in explaining the success strategies for competitive advantage. Hence, a synthesis of the literature reviewed will give a more comprehensive framework which will guide this current study to identify the success strategies for competitive advantage in the Ghanaian construction industry, as well as determining the relative impact of each of the strategies. As a result, the proposed framework governing the study was made up of the following strategies: the appropriateness of the pricing strategy of the firm, having the appropriate growth strategy, having the appropriate focus strategy in relation to firm's size, having a brand identity in the industry, having a brand image in the industry, having a brand communication in the industry, sustaining clients' loyalty in the industry, adopting service differentiation strategy, adopting the appropriate lean construction strategy, adopting the appropriate financial synergies, corporate management synergies, being technological independent, adopting the appropriate operative synergies, maintaining cordial firm-client relations.

3 Methodology

This study adopted a two-stage approach to research. The first stage was a literature review which aided in identifying the strategies for competitive advantage. The second stage was the use of the Delphi method to seek experts' view on the success strategies for competitive advantage in the Ghanaian construction industry and the relative impact of each of the strategies. In all fifteen (15) experts completed three rounds of the Delphi process. The Delphi method was used based on the principle that, forecasts (or decisions) from a structured group of persons are more accurate than forecasts from unstructured groups (see [11, 12]). More so, it is suitable for studies that seek to solicit opinions/views from a group of experts in an attempt to build consensus in experts' opinions, and predict the timing and likelihood of a future event [13]. [14] accentuated that the Delphi technique could be used for both qualitative and quantitative study with the aim to establish consensus in experts' opinions through rounds of survey [14, 15]. It uses structured questionnaire in soliciting data from experts [16]. Thus, ideal for this qualitative study. The experts were drawn from industry and academia. The common criteria for selecting them was that, an expert should have had both theoretical and practical knowledge on strategies for competitive advanatge. Again, the panel of experts, were carefully selected, to represent a broad spectrum of opinion on the issue being examined (see [16–18]). After each round of the Delphi survey, a statistical estimate of the experts' opinions was calculated and analyzed using the median, mean, standard deviation and interquartile deviation. The adopted scale for measuring consensus was:

1. Strong consensus - median 9–10, mean 8–10, interquartile deviation (IQD) ≤ 1 and $\geq 80\%$ (8–10);
2. Good consensus - median 7–8.99, mean 6–7.99, IQD $\geq 1.1 \leq 2$ and $\geq 60\%$ 79% (6–7.99); and
3. Weak consensus - median ≤ 6.99, mean ≤ 5.99 and IQD $\geq 2.1 \leq 3$ and 59% (5.99).

The adopted scale was based on a 10-point impact scale where 1 and 2 represent no impact; 3 and 4 represent low impact; 5 and 6 represent medium impact; 7 and 8 represent high impact; 9 and 10 represent very high impact (see [16]). Ethically, the credentials of the experts were kept confidential throughout the study.

Further, informed by the aim, the research paradigm as well as the major philosophical considerations or positions of this research, this research adopted a qualitative methodology, and followed a subjectivity epistemology stance (see [19]). Phenomenal about subjectivity epistemology is that, meanings to reality are constructed (constructivism) and interpreted (interpretivism) by the researcher, thus aligned to the concept of constructivism and interpretivism [20]. Though reliability and generalization of findings are undermined as mainly data generated in qualitative studies are heavily impacted by the values and views of the researcher [21, 22]. Nonetheless, some advocates of qualitative research accentuate that, the weakness of non-generalization and reliability of qualitative research is overcome when a qualitative research is based on sound theoretical reasoning [23]. Thus, this study was theoretical underpinned by the competitive strategy theory by [7] and, the three generic competitive strategies theory by [2, 3].

4 Results and Discussions

Objective 1: to identify the success strategies for competitive advantage in the Ghanaian construction industry;
In relation to objective one, fourteen (14) success strategies for competitive advantage in the Ghanaian construction industry were identified. The success strategies consisted of: the appropriateness of the pricing strategy of the firm, having the appropriate growth strategy, having the appropriate focus strategy in relation to firm's size, having a brand identity in the industry, having a brand image in the industry, having a brand communication in the industry, sustaining clients' loyalty in the industry, adopting service differentiation strategy, adopting the appropriate lean construction strategy, adopting the appropriate financial synergies, corporate management synergies, being technological independent, adopting the appropriate operative synergies, and maintaining cordial firm-client relations. Each of the strategies had a median value fallen within the range of very high impact or a high impact. The IQD score suggest strong consensus in the view of the expert panelists.

Objective 2: to determine the relative impact of each of the success strategies for competitive advantage in the Ghanaian construction industry
The outcome of the analysis as shown in Table 1 informs that, out of the 14 success strategies for competitive advantage in the Ghanaian construction industry, three (3) of the strategies had very high impact (VHI: 9.00–10.00) on competitive advantage in the Ghanaian construction industry, whiles (11) had high impact (HI: 7.00–8.99) base on the 10 point impact scale this study employed. None of the strategies was found not to have had an impact on competitive advantage in the Ghanaian construction industry. More so, the respective IQD scores of the strategies revealed that strong consensus was achieved for all the (14) strategies as they obtained scores ranging from 0.00 to 1.00.

Furthermore, their respective standard deviation (σx) revealed consistency in the responses of the experts as their respective (σx) values was at most (1).

Table 1. Success strategies for competitive advantage in the Ghanaian construction industry

Strategies	Median (M)	Mean ($\bar{x}$)	Standard deviation (σx)	Interquartile deviation (IQD)	Mean ranking
The appropriateness of the pricing strategy of the firm	8	8.13	0.64	0.50	6th
Having the appropriate growth strategy	8	8.07	0.70	0.50	8th
Having the appropriate focus strategy in relation to firm's size	9	8.73	0.80	0.00	1st
Having a brand identity in the industry	8	8.13	0.74	1.00	6th
Having a brand image in the industry	8	8.07	0.70	0.50	8th
Having a brand communication in the industry	8	8.07	0.59	0.00	8th
Sustaining clients' loyalty in the industry	8	8.00	1.00	1.00	12th
Adopting service differentiation strategy	9	8.73	0.80	0.00	1st
Adopting the appropriate lean construction strategy	8	8.07	0.59	0.00	8th
Adopting the appropriate financial synergies	8	8.00	1.00	1.00	12th
Corporate management synergies	8	8.27	0.80	1.00	4th
Being technological independent	9	8.73	0.80	0.00	1st
Adopting the appropriate operative synergies	8	8.00	0.65	0.00	12th
Maintaining cordial firm-client relations	8	8.21	0.70	1.00	5th

Specifically, having the appropriate pricing strategy had a mean value of (8.13) indicating that it had a high impact on competitive advantage. This is consistent with the opinion of [3] that, firms that have price advantage in competitive industry have competitive advantage. More so, having the appropriate growth strategy obtained a mean value of (8.07) indicating that it had a high impact on competitive advantage. This supports the findings of [7] that appropriateness of a growth strategy in competitive industry impacts on competitive advantage. In addition, having the appropriate focus strategy in relation to firm's size had a mean score of (8.73) indicating that it had a high impact on competitive advantage. This also supports the revelation by [3] that, firms whose strategy is to focus on a particular industry, location or a segment of the production chain tends to have competitive advantage over their rivals in the industry. More so, having a brand identity in the industry recorded a mean value of (8.13); whereas having a brand image in the industry also recorded a mean value of (8.07). Having a brand communication in the industry recorded a mean score of (8.07); and sustaining clients' loyalty in the industry also had a mean score of (8.00). Having brand identity, brand image, brand communication and sustaining clients' loyalty by their respective mean values had a high impact on competitive advantage. Thus, supporting the findings of [24] that brand identity, brand image, brand communication and brand loyalty significantly impact on competitive advantage [24]. Further, adopting service differentiation strategy had a mean value of (8.73) indicating that it had high impact on competitive advantage. This confirms to the study by [3] that, service differentiation impact on competitive advantage in a competitive industry. In addition, adopting appropriate corporate management synergies had a mean value of (8.27); adopting the appropriate financial synergies recorded a mean value of (8.00); adopting the appropriate operative synergies had (8.00) mean value. Accordingly, the respective mean value indicates a high impact on competitive advantage (see Table 1). This supports the studies by [5, 10] that having the appropriate synergies in business impact on competitive advantage. More so, adopting the appropriate lean construction strategy had a mean score of (8.07); being technological independent had a mean of (8.73) whiles maintaining cordial firm-client relations had a mean value of (8.21).

Thus, it is evident from the discussions that, the findings of this study are largely consistent with the findings of previous studies on firms' competitive advantage, it is the relative impact values that each strategy recorded that differed. Therefore, it is empirically evident from the views expressed by the experts that the strategies for competitive advantage in other industries do impact on competitive advantage in the Ghanaian construction industry. Specifically, in the Ghanaian construction industry, having the appropriate focus strategy in relation to firm's size; adopting service differentiation strategy; and being technological independent jointly emerged 1st among the various strategies for competitive advantage; whiles strategies such as adopting the appropriate operative synergies; adopting the appropriate financial synergies; and sustaining clients' loyalty in the industry, jointly emerged 12th among the success strategies for competitive advantage.

5 Conclusions

In relation to objective one of this study, this research concludes that fourteen (14) strategies constitute the success strategies for competitive advantage in the Ghanaian construction industry. The fourteen success strategies for competitive advantage include having the appropriate focus strategy in relation to firm's size, adopting service differentiation strategy; and being technological independent. In relation to objective two, this study concludes that, relatively, in the Ghanaian construction industry, having the appropriate focus strategy in relation to firm's size; adopting service differentiation strategy; and being technological independent jointly emerged 1st among the success strategies for competitive advantage in the Ghanaian construction industry; whiles success strategies such as adopting the appropriate operative synergies; adopting the appropriate financial synergies; and sustaining clients' loyalty in the industry, jointly emerged 12th among the success strategies for competitive advantage. It is therefore recommended that owners and managers of firms should highly prioritize focus strategy, service differentiation strategy, and being technologically independent since these success strategies for competitive advantage had very high impact on competitive advantage in the Ghanaian construction industry. The findings of this study will add on to existing literature on strategies for competitive advantage; could form the basis for future competitive advantage studies; and inform decision making by owners and managers of construction firm in Ghana. A future study that quantitatively validates these qualitative findings will be a novelty.

References

1. Wang, H.: Theories for competitive advantage. In: Hasan, H. (ed.) Being Practical with Theory: A Window into Business Research, pp. 33–43. Theori Research Group, Wollongong (2014)
2. Porter, M.E.: Competitive Strategy: Techniques for Analyzing Industries and Competitors, Collier Macmillan, New York, Free Press, London (1980)
3. Porter, M.E.: Competitive Advantage: Creating and Sustaining Superior Performance, New York, Collier Macmillan, Free Press, London (1985)
4. Porter, M.E.: Location, Competition and Economic Development: Local Clusters in the Global Economy. Econ. Dev. Q. **14**, 15–31 (2000)
5. Flanagan, R., Jewell, C., Ericsson, S., Henrics, P.: Measuring Construction Competitiveness in Selected Countries, the University of Reading, Reading: School of Construction Management and Engineering, the University of Reading (2005)
6. Lu, W.S.: A system for assessing and communicating contractors' competitiveness, Doctoral Thesis, The Hong Kong Polytechnic University (2006)
7. Yongtoa, T.: Contractors competitiveness and competitive strategy in Hong Kong, Doctoral dissertation, The Department of Building and Real Estate (2008)
8. Ingram, D.: Strategy and structure of an organization. http://smallbusiness.chron.com/strategy-structure-organization-2765.html. Accessed 23 Jan 2018
9. Mekic, E., Mekic, E.: Supports and critiques on Porter's competitive strategy and competitive advantage (2014)
10. Kao, C.C., Green, S.D., Larsen, G.D.: Emergent discourses of construction competitiveness: localized learning and embeddedness. Constr. Manage. Eco. **27**(10), 1005–1017 (2009)
11. Rowe, G., Wright, G.: Expert opinions in forecasting. role of the Delphi technique. In: Amstrong, J. (ed.) Principles of Forecasting: A Handbook of Researchers and Practitioners. Kluwer Academic Publishers, Boston (2001)
12. Ameyaw, E.E., Hu, Y., Shan, M., Chan, A.P., Le, Y.: Application of Delphi method in construction engineering and management research: a quantitative perspective, J. Civil Eng. Manage. 1–10 (2016)
13. Lang, T.: An overview of four futures methodologies (Delphi, environmental scanning, issues management and emerging issue analysis. Manoa J. Fried and Half-fried Ideas (about the future), 7 (Occasional paper number seven) (1995)
14. Hanson, W.E., Creswell, J.W., Clark, V.P., Petska, K.S., Creswell, J.D.: Mixed methods research designs in counseling psychology. J. Couns. Psychol. **52**(2), 224–235 (2005)

15. Miller, M.M.: Enhancing regional analysis with the Delphi method. Rev. Reg. Stud. **23**(2), 191–212 (1993)
16. Aigbavboa, C.O.: An integrated beneficiary centered satisfaction model for publicly funded housing schemes in South Africa, Doctorate Thesis, University of Johannesburg (2014)
17. Loo, R.: The delphi method: a powerful tool for strategic management. Policing, Int. J. Police Strat. Manage. **25**(4), 762–769 (2002)
18. Tilakasiri, K.K.: Development of new frameworks, standards and principles via Delphi data collection method. Int. J. Sci. Res. (IJSR) **4**(9), 1189–1194 (2015)
19. Jean, K.: Livelihood Strategies Among Farm Youth in Rwanda. Michigan State University, Michigan (1992)
20. Creswell, J.W.: Research Design: Qualitative, Quantitative and Mixed Methods Approaches. Sage, Thousand Oaks (1994)
21. Bryman, A.: Social Research Methods. Oxford University Press, Oxford (2001)
22. Sarantakos, S.: Social Research, 3rd edn. Palgrave Macmillan, Basingstoke (2005)
23. Mitchell, J.C.: Case and situation analysis. Sociol. Rev. **31**(2), 187–211 (1983)
24. Ghodeswar, B.M.: Building brand identity in competitive markets: a conceptual model. J. Prod. Brand Manage. **17**(1), 4–12 (2008)

Accuracy Influencing Factors for Pre-tender Cost Estimates for the Roads Sector in Zambia

Chipulu Chipulu[1], Erastus Mwanaumo[1(✉)], Balimu Mwiya[1], Lubinda Haabazoka[2], and Sampa Chisumbe[1]

[1] Department of Civil and Environmental Engineering, School of Engineering, University of Zambia, Great East Road, Lusaka, Zambia
erastus.mwanaumo@unza.zm
[2] Graduate School of Business, University of Zambia, Lusaka, Zambia

Abstract. Inaccurate pre-tender cost estimates do not only contribute to cost overruns but also results in failure to implement viable projects. This study investigated factors that influence the accuracy of pre-tender cost estimates in the Zambian roads sector. The methodology adopted was a mixed method which included the use of both qualitative and quantitative approaches. The study established that the most severe accuracy-influencing factors for pre-tender cost estimates were completeness of the project information, which had SI of 87%. Clarity and details of the drawing as well as the specification having SI 85% was ranked second. The study therefore, recommends that, more time should be allocated to the design team to comprehensive prepare drawings and coordinate various players such as mechanical and civil consultants before the pre-tender estimate can be sought.

Keywords: Cost estimate · Accuracy factors · Pre-tender · Roads sector · Zambia

1 Introduction

For low-income countries infrastructure investment providing access to energy, clean water and basic transport may mean the difference between life and death [1]. The Zambian government recognises that infrastructure development is important in achieving economic growth and development [2]. However, decisions concerning financial resources to be set aside for infrastructure development are usually made based on pre-tender cost estimates of construction project cost [3]. This therefore, places a demand on the accuracy of pre-tender cost estimates used on construction projects.

Evans, Lanham and Marsh [4] defines cost as the expenditure necessary for the attainment of a goal and cost estimation as the prediction of the cost prior to undertaking the activity. Enshassi, Mohamed and Abdel-Hadi [5] defines pre-tender cost estimating as the final costing of the work carried out by a consultant on behalf of a client before tenders are received. On the other hand, Aibinu, Dassanayake and Chau [6] defines pre-tender cost estimation as the forecasting of the cost of a project during the planning and design stage. From these two definitions, one can note that pre-tender estimates are prepared before inviting bids. Further, consultants on behalf of the clients

C. Aigbavboa and W. Thwala (Eds.): CIDB 2019, *The Construction Industry in the Fourth Industrial Revolution*, pp. 547–555, 2020.
https://doi.org/10.1007/978-3-030-26528-1_56

prepare pre-tender cost estimates during project planning and design. Pasco and Aibinu [7] indicate that a pre-tender estimate is an important source of information during project planning and design. This means that pre-tender cost estimates helps the client to make decisions on whether to make financial commitment to a project or not. Further, Oladokun, Oladokun and Odesola [8], contend that a pre-tender cost estimate is one of the important aspect of the clients' expenditure control. This means it used to help the client get value for money.

Though clients rely upon pre-tender estimates during budgeting, project selection and bid evaluation, pre-tender estimates are susceptible to inaccuracies because they are often prepared within limited timeframe and with limited information about project scope [6]. Underestimated pre-tender cost estimates can result in a non-viable project being implemented while overestimated pre-tender cost estimate could lead to viable projects being abandoned [6]. Therefore, accuracy of pre-tender cost estimates is critical to construction clients' decisions to undertake a project. In addition, inaccurate early or pre-tender estimates can lead to lost opportunities, wasted development effort, and lower than expected returns [8].

Aibinu et al., [6], contend that inaccurate estimation of uncertainties is the underlying cause of project cost overruns in construction. However, Ali & Kamaruzzaman [9], states that the biggest factor that contributes to overruns of budget is inaccurate estimation of original cost of a project. Kaliba [11] also added to this school of thought by contending that insufficient initial cost analysis is one of the major causes for cost overruns. These three authors all agree that cost estimation significantly contribute to overruns on construction projects. Cantarelli, Flyvbjerg, Molin and van Wee [12], further states that inaccurate estimates which includes pre-tender estimates makes it particularly difficult to manage large projects and often lead to cost overruns that further increases the burden on the country's GDP. In Zambia, it is not uncommon for final construction projects costs to be more than the pre-tender cost [11]. This imply that cost overruns are common on most construction projects in Zambia. The presence of overruns suggests the existence of inaccurate pre-tender cost estimates on most construction projects undertaken in Zambia.

1.1 Study Aim

The aim of this study was to establish the factors that affects pre-tender cost estimates for roads sector in Zambia.

2 Literature Review

Cong, Mbachu and Domingo [13], in there study of the factors influencing the accuracy of pre-contract stage estimation of final contract price in New Zealand, established thirty-seven factors which could influence the final contract price. The three most influential factors being; poor tender documentation, complexity of design & construction, and completeness of project information. Other factors relating to project, client and contractor characteristics, design consultants and tendering conditions, estimating practice and external factors were reported. Concordance analysis indicated high level of agreement among survey participants in the rank ordering of the relative importance of the identified factors.

Conversely Nor, *et al.*, [14] by investigating the factors affecting construction cost of industrialised building systems (IBS) projects ranked cost factors regarded as pertinent by the IBS stakeholders. The key identified factors affecting cost being those associated with the main project characteristics, contract procedures and procurement methods, contractors' and consultants' attributes and design parameters besides external market conditions. However, factors grouped under the consultants and design parameters, external factors and government/authority requirements indicate low RII index while project/IBS characteristics, contractor attributes, external market conditions denote high RII index. They further assent that the failure by the cost estimator to consider these factors during cost estimation could result in inaccuracy of the resulting cost estimate.

Ramabodu [15] uncovered twelve factors that affects the accuracy of a cost estimate. These include building function, type of contract, conditions of contract, contract sum, price intensity, contract period, number of bidders, good/bad years, procurement basis, project sector (public, private or joint) and number of price items of drawing.

By administering questionnaires to the clients, consultants and contractors of Kaduna state of Nigeria, Alumbugu, *et al.* [16], assessed the factors that affect the accuracy of pre-tender cost estimate in Kaduna state of Nigeria. This team of researchers analysed the data collected in this survey using the relative importance index and correlation analysis. They found that a statistically significant relationship existed between the clients, consultants, and contractor on the essential factors affecting accuracy of pre-tender cost estimate. The study concluded that: Experience and skill level of the consultants, project teams experience on the construction type, clear and detail drawings and specification, completeness of cost information, accuracy and reliability of cost information are the most significant factors affecting accuracy of pre-tender cost estimate. It was recommended that clients should always engage the services of skilled professionals to undertake the consultancy services in cost estimating process.

In his study, Ardianto [17] "analysis of the factors affecting the accuracy of early cost in construction projects" also identified certain factors that affected the quality and accuracy of cost estimates. Ardianto [17] stated that the quality of cost estimates associated with the accuracy and completeness of the supporting elements depends on; nature of the project, the level of data and information available, techniques and methods used ability of estimator, and other factor considered while preparing the estimate. These factors were developed into fifty-four (54) variables as the basic elements of the preparation of early cost estimates. Through the use of multiple linear regression analysis, significant factors (<0.05) which affects the level of accuracy of the early cost estimates were obtained, as follows.: Geographical location/Site location; quality level of the estimator/estimator Expertise; Availability of project information/Documents used in preparing the estimate; the availability of scope of the project; the availability of project documents (preliminary and Size of Project/Scope of the project.

Pasco & Aibinu [7] in their study concluded that project procurement method, location of project and project size are the three most important project related factors driving the level of bias and inconsistencies in early stage estimates when compared with the accepted tender sum.

Other authors like Enshassi, *et al.* [5], using a survey questionnaire in the Gaza strip identified five critical factors that affect the accuracy of a pre-tender cost estimates. These are; (1), materials (prices/availability/supply/quality/imports), (2) closure and blockade of borders, (3) project team's experience in the construction type, (4) the experience and skill level of the consultant and (5) clear and detailed drawings and specifications.

Review of the above pieces of research seems to suggest that different factors are having impacts in certain region and not the other. For instance closure of the boarders is a critical factor that affect the accuracy of pre-tender estimates in the Gaza strip Enshassi et al. [5], but that is not the case for New Zealand were procurements method are very critical as far as the accuracy of pre-tender cost estimate is concerned (Cong et al., [13]. While this is true for some factors, it is not the case for others. For instance, the majority of these scholars agree that the level of skill and competency of the estimator is a factor that cannot be overlooked when discussing the accuracy of pre-tender cost estimates.

There are limited studies conducted on the accuracy influencing factors for the pre-tender cost estimates in Zambia roads sector. This study endeavoured to cover this gap by investigating the factors that affect the accuracy of pre-tender cost estimates for the roads sector in Zambia.

3 Methodology

The methodology adopted for this study was a mixed method which included the use of both qualitative and quantitative approaches. Literature review was used to collect secondary data while primary data was collected using a self-administered semi-structured questionnaire. The respondents were purposively sampled, and these included; road construction clients, consultants and contractors and financiers. This group of respondents was requested to rate the pre-tender cost estimates unit cost factors on a five point ordinal Likert scale in which "important" and "very important" were assigned weights of one and five respectively. The target sample size was 60. This sample size was purposely selected in order to achieve the aim of the study within the time of the study. Analysis of collected data through the questionnaire was quantitatively undertaken through calculating the severity indices Kaliba [11] using Eq. 1.

$$\mathbf{SI} = \frac{\sum_{K=1}^{n} \mathrm{S_K P_K}}{5 * \sum_{K=1}^{n} \mathrm{P_K}} * 100 \tag{1}$$

Where

SI = is the severity index

S_k = is the assigned severity weight assigned to option k

P_k = is the number of participants who responded in favour of an option k

n = is the Total number of respondents

Internal consistency of the data was measured employing Cronbach's alpha calculated in SPSS. Hair, Anderson, Tatham, and Black [18], as cited by Peter and Peter [19] contends that the Cronbach's alpha as a measure of internal consistency of items overcomes the potential problem of half splits. According to Jonsson and Svingby [20], a Cronbach's alpha greater than 0.7 indicates a higher degree of internal consistency. The cronbach's alpha is usually computed using the Eq. 2 [21].

$$\text{Alpha} = \frac{NC}{\mathrm{V} + (N-1) * \mathrm{C}} \tag{2}$$

Where

Alpha = Cronbach's alpha
N = the number of items
V = the average variance
C = the average inter-item covariance

4 Findings

The questionnaire survey was conducted between June 2016 and August 2016. The questionnaire was distributed to Sixty (60) respondents made up of consultants, clients/client representatives, contractors and financiers. The response rate was 66.7%. Baruch & Holtom, [22], observed that the average and response rate from surveys that utilised data from individuals was 52.7% and 35.7% for surveys that utilised data from organisations. Therefore, a response rate of 66.7 is acceptable. Further similar studies like the ones conducted by Ali & Kamaruzzaman [10], Jayesena (2005) and Eshofonie [23] had sample sizes of 30, 62 and 70 respectively. A sample size of 60 can therefore be relied.

Table 1. Internal consistency reliability test for pre-tender cost estimate accuracy influencing factors (cronbach's' alpha)

Cronbach's alpha	Cronbach's alpha based on standardized items	No of items
0.856	0.853	26

The Cronbach's alpha calculated was 0.856 as can be noted from Table 1. According to Jonsson & Svingby [20], a Cronbach's alpha greater than 0.7 indicates a higher degree of internal consistency. This therefore implies that the data collected concerning the factors that affect pre-tender cost estimates unit rates factors was reliable and had a higher internal consistency reliability [19].

The most severe accuracy-influencing factors for pre-tender cost estimates as opined by the respondents is completeness of the project information, which had SI of 87%. Clarity and details of the drawing as well as the specification having SI 85% was ranked second while project size with SI 71% was the least ranked major accuracy-influencing factor above the seventy percent threshold. The other major factors that affects the accuracy of pre-tender estimates as opined by the respondents are shown in Table 2.

Table 2. Factors affecting the accuracy of pre-tender cost estimates in the roads sector in Zambia

No.	Roads pre-tender cost estimate accuracy influencing factors	Severity index (SI) %	Rank
1	Completeness of project information	87	1
2	Clarity and details of the drawing and specifications	85	2
3	Experience and skill level of the cost estimator	84	3
4	Complexity of design and construction	84	3
5	Accuracy and reliability of cost information	83	4
6	Experience and skill level of consultants	78	5
7	Geographical location/site location	75	6
8	Estimating practice	74	7
9	Materials (prices/availability/supply/quality/imports)	74	7
10	The time given to developing an estimate	74	7
11	Project team experience on the construction type	73	8
12	Project characteristics (Nature of Project)	72	9
13	Cost estimating techniques and methods used	72	9
14	Insufficient understanding of site conditions	72	9
15	Contract period	71	10
16	Project size	71	10

From the results shown in Table 2, the respondents regarded 16 factors as major. Comparing these findings with Cong, *et al.*, [13], it can be observed that in both studies the complexity of design and construction and the completeness of project information are very essential factors to pre-tender cost estimates' accuracy. This is so because incomplete project information may cause the estimator to leave out certain essential factors as they do not have the whole scope of the project. In addition, for an estimate to be considered as good or accurate, it must be based on the description of the project that has a well defined content and which also has identifiable risks areas [24]. The findings of this study also agree with Alumbugu, *et al.*, [16] in that, both studies established that the experience and skill level of the consultants engaged in the design of the structures, Clarity and details of the drawing and specifications, and accuracy as well as reliability of cost information are critical factors to the accuracy of pre-tender cost estimates. The failure to consider these factors will result in a poor and inaccurate estimate because inexperienced consultants are likely to neglect factors which are seemingly not critical during design but a very essential during construction. Further, as inexperienced consultants may not be able to understand which information should be included on the design drawings and specifications, the clarity and details of the design drawings and specifications would be affected. One can not leave out the quality and expertise of the cost estimators in the quest for an accurate pre-tender cost estimate. The US Government Accountability Office [24] and Ardianto [17] both agree with the findings of this study that the experience and skill level of the cost estimator is a pertinent consideration in cost estimation. They also agree that the availability and completeness of project information also affects the quality of the established estimate, which agrees with the findings of this study.

However, the respondents opined that the type of risk analysis conducted has low influence on the quality of pre-tender cost estimate as can be noticed from the low factor importance index. This finding contradicts with the US Government Accountability Office [24] recommendations. According to the US Government Accountability Office [24], risk and uncertainty analysis is very essential to the accuracy of an estimate. Further Mislick & Nussbaum [25] contends that an estimate can not be accurate if there has been insufficient analysis and subsequent addressing of risks and uncertainities. This is so because, the risk identified determine the amount of contigency sums to be reserved in order to mitigate the forseen risks and uncertainities. According to Mulcahy [26] contigency reserves are used to mitigate the known risks which are specifically identified risks and for unidentified risks, management reserves are used to cover them. Insufficient risk and uncertainty analysis results in padding of activity estimates which results in unrealistic contingency reserves. Therefore the failure of any estimator to conduct a thorough risk analysis may result in a pre-tender cost estimate which is inaccurate. From the findings of this study, risk analysis is not thoroughly considered in pre-tender cost estimate development. However the failure to undertake a comprehensive risk analysis would lead to inaccuracy in the pre-tender estimates produced.

5 Conclusion

This study aimed to establish the factors that influence or affect the accuracy of pre-tender cost estimates in the Zambia roads sector. The study established that the most severe factor that affect the accuracy and quality of the estimate was completeness of project information. Clarity and details of the drawing including specification, experience and skill level of the consultant and the complexity of design were regarded as the second, third and fourth important factors affecting the accuracy of pre-tender cost estimates respectively. The study therefore, recommends that, more time should be allocated to the design team to comprehensive prepare drawings and coordinate various players such as mechanical and civil consultants before the pre-tender estimate can be sought.

The study was limited to the investigation of factors affecting the accuracy of pre-tender estimates in the Zambian roads sectors. However, the results obtained can be relied on when considering the accuracy of pre-tender estimates in other sectors of the construction industry of Zambia.

References

1. Henckel, T., McKibbin, W.J.: The economics of infrastructure in a globalized world: issues, lessons and future challenges. J. Infrastruct. Policy Dev. **1**(2), 254–271 (2017). https://doi.org/10.24294/jipd.v1i2.55. https://systems.enpresspublisher.com/index.php/jipd/article/view/55/68
2. Goverment of the Republic of Zambia: Zambia's National Budget. Ministry of Finance, Lusaka (2015)
3. AbouRizk, S., Babey, G., Karumanasseri, G.: Estimating the cost of capital projects: an empirical study of accuracy levels for municipal government projects (2008). http://www.researchgate.net/profile/Gordon_Babey/publication/237188753_Estimating_the_cost_of_capital_projects_An_empirical_study_of_accuracy_levels_for_municipal_government_projects/links/54419d770cf2a6a049a5be01.pdf?inViewer = 1&pdfJsDownload = 1&origin = pub. Accessed 7 Dec 2015
4. Evans, D.K, Lanham. J.D., Marsh, R.: Cost estimation method selection: matching user requirements and knowledge availability to methods (2012). https://pdfs.semanticscholar.org/835d/0ddf6eb55d08afc05d3aeb8f6b6f1826d419.pdf. http://diiconference.org/wp-content/uploads/2016/10/DII-2016-PROCEEDINGS-1.pdf
5. Enshassi, A., Mohamed, S., Abdel-Hadi, M.: Factors affecting the accuracy of pre-tender cost estimates in the gaza strip. J. Constr. Developing Countries, **18**(1), pp. 73–94 (2013). http://diiconference.org/wp-content/uploads/2016/10/DII-2016-PROCEEDINGS-1.pdf
6. Aibinu, A.A., Dassanayake, D., Chau, T.V.: Use of artificial intelligence to predict the accuracy of pre-tender building cost estimate. In: Management and Innovation for a Sustainable Built Environment Conference, Amsterdam, 20–23 June 2011. http://diiconference.org/wp-content/uploads/2016/10/DII-2016-PROCEEDINGS-1.pdf
7. Pasco, T., Aibinu, A.A.: Project Factors Influencing the Accuracy of Early Stage Estimates. Heritance Kandalama, Sri-Lanka, University of Stanford, UK (2008). http://diiconference.org/wp-content/uploads/2016/10/DII-2016-PROCEEDINGS-1.pdf
8. Oladokun, G.M., Oladokun, A.A., Odesola I,A.: Undated. www.irbnet.de. http://www.irbnet.de/daten/iconda/CIB15880.pdf. Accessed 18 Aug 2016
9. Oberlender, G.D., Trost, S.M.: Predicting accuracy of early cost estimates based on estimate quality. J. Constr. Eng. Manage. **127**(3), 173–182 (2001)
10. Ali, A., Kamaruzzaman, S.: Cost performance for building construction projects in Klang Valley. J. Build. Perfom. **1**(1), 110–118 (2010)
11. Kaliba, C.: Cost Escalation, Schedule overruns and Quality Shortfalls on Construction Projects, A Thesis submitted in partial fulfilment of the requirements of the University of Zambia for a Degree of Master of Engineering. Lusaka: University of Zambia, Lusaka. Unpublished (2010)
12. Cantarelli, C.C., Flyvbjerg, B., Molin, E.J.E., van Wee, B.: Cost overruns in large-scale transportation infrastructure projects: explanations and their theoretical embeddedness. Eur. J. Transp. Infrastruct. Res. **10**(1), 5–18 (2010)
13. Cong, J., Mbachu, J., Domingo, N.: Factors influencing the accuracy of precontract stage estimation of final contract price in New Zealand. Int. J. Constr. **4**(2), 51–64 (2014)
14. Nor, A.A.B., et al.: Factors Influencing the Construction Cost of Industrialised. Elsevier, Malaysia (2012)
15. Ramabodu, M.S.: Procurement Guidelines for Project Success in Cost Planning of Construction project, Bloemfontein, Free State, University of the Free State: A thesis Submitted in partial fulfilment of the requirements of University of the Free State for the Degree of Doctor of Philosophy (2014)

16. Alumbugu, P.: Assessment of the factors affecting accuracy of pre-tender cost estimate in Kaduna state Nigeria. IOSR J. Environ. Sci,. Toxicol. Food Technol. **8**(5), 19–27 (2014)
17. Ardianto, D.: Analysis of factors Affecting the Accuracy of Early Cost Estimates in Construction Projects, Johor Bahru, Johor, Malaysia: A thesis submitted in fulfilment of the requirements of Universiti Teknologi Malaysia for the degree master of science construction Management (2011)
18. Hair, J.F., Anderson, R.E., Tatham, R.L., Black, W.C.: Multivariate Data Analysis, 5th edn. Prentice-Hall International Inc., New Jersey (1998)
19. Peter, R., Peter, V.: An assessment of the internal consistency of measures of constructs used to revise the innovation decision framework. In: Conference Proceedings Academy of World Business, Marketing & Management Development, vol. 3, no. 1, July 2008
20. Jonsson, A., Svingby, G.: The use of scoring ubrics: reliability, validity and educational consequencies. ScienceDirect **2**, 131–141 (2007)
21. Alinaitwe, H., Ruth, A., Tindiwensi, D.: Investigation into the causes of delays and cost overruns in Uganda's public sector construction projects. J. Constr. Develop. Countries, Issue **18**(2), 33–47 (2013)
22. Baruch, Y., Holtom, B.C.: Survey response rate levels and trends in organizational research (2009). https://www.researchgate.net/publication/228079609Survey_Response_Rate_Levels_and_Trends_in_Organizational_Research
23. Eshofonie, F.P.: Factors Affecting Cost Construction in Nigeria, a thesis submitted in Patial fulfilment of the requirements of University of Lagos Akoka for the degree of master of Construction Management. University of Lagos, Akoka, Lagos (2008)
24. US Goverment Accountability Office (2009). GAO COST ESTIMATING AND ASSESSMENT GUIDE, Best Practices for Developing and Managing Capital Program cost, Washington DC: US Goverment Accountability Office
25. Mislick, G.K., Nussbaum, D.A.: Cost Estimation Methods and Tools. Wiley & Sons Inc, New Jersey (2015)
26. Mulcahy. R.: PMP Exam Prep By Rita Mulcahy, 2013 Eighth Edition, Rita's Course in a Book for Passing the PMP Exam Paperback – 2013. http://www.parliament.gov.zm/sites/default/files/images/publication_docs/2015%20Budget%20Speech_0.pdf

An Investigation into the Critical Barriers to the Practice of Design for Construction Safety in Nigeria

Adekansola Labo-Popoola[1(✉)], Abdul-Majeed Mahamadue[1], Patrick Manu[2], Clinton Aigbavboa[3], and Krzysztof Dziekonski[1]

[1] Department of Architecture and the Built Environment, University of the West of England, Bristol BS16 1QY, UK
Adekansola2.Labo-Popoola@live.uwe.ac.uk
[2] School of Mechanical, Aerospace and Civil Engineering, University of Manchester, Manchester M13 9PL, UK
[3] Department of Construction Management and Quantity Surveying, University of Johannesburg, Johannesburg, South Africa

Abstract. The construction industry is one of the most dangerous industries due to its high accident rates. In Nigeria, higher than average rates of accidents have been recorded. Several practices have been applied to improve this situation albeit there remains an appreciably high rate of occurrence of accidents. Research has shown that the early consideration of the safety risk inherent in designs can help prevent their occurrence later during the construction of the facilities. This concept is called Designing for Construction Safety (DfCS). Despite the advantages of DfCS, there remains barriers that prevent designers from applying the concept. This study therefore aimed to investigate the barriers to the practice DfCS in Nigeria focusing on Civil engineers. The study used a quantitative approach through survey questionnaires for data collection. Results showed that while there exists a very high level of awareness of DfCS, the extent of practise is very low. Results further showed that this is attributed to: designers' exposure to liability, increased cost and time, designers' lack of safety expertise among others. The study made some recommendations that could help reduce or eliminate these barriers. In general, this study has provided a pathway for mainstreaming the practice of DfCS in Nigeria.

Keywords: Construction site safety · Design · Design for construction safety · Nigeria

1 Introduction

Accident prevention and control continues to be a challenge in the construction industry [1]. This is because the industry remains one of the world's most dangerous sectors [2]. Although, several initiatives to improve safety have been identified, statistically, the construction industry still has one of the worst safety records compared to other industries such as mining and transportation [1]. All across the world [1–4] it is reported that construction workers are more likely to be killed at work than those in

C. Aigbavboa and W. Thwala (Eds.): CIDB 2019, *The Construction Industry in the Fourth Industrial Revolution*, pp. 556–565, 2020.
https://doi.org/10.1007/978-3-030-26528-1_57

other industries. Due to this rate, construction site safety (CSS) has become a major concern in the industry, including developing countries like Nigeria [4]. The situation in Nigeria is worse than what exists in developed countries due to "*lack of concern, accurate records and statutory regulations on health and safety*" [4]. In 2005, more than 20 workers died within 24 h while constructing a four-story building in Port Harcourt and with similar incidents being a regular occurrence across other states in Nigeria [5].

Due to this continuous occurrence of safety problems, several ideas to mitigate against it have been proposed. One of the advanced initiatives is to involve designers in considering safety during design [6]. According to Behm [7] the concept of design for construction safety (DfCS) is "*the consideration of CSS in the design of a project*". It was further defined by Toole et al., [8] as "*the deliberate consideration of CSS in the design phase of a construction project, with the goal of reducing inherent risk to construction workers*".

Usually, CSS is the responsibility of the contractor [9], however, evidence has shown that the early decisions made by designers play a major role in the safety of construction workers on site [10]. In the UK, research has shown that out of 100 construction accidents, 47% of possible accidents would have been reduced if DfCS was considered [11]. Similarly, in Nigeria, 50% of site accidents have been attributed to design errors [12]. Therefore, CSS can be improved if designers consider the impact of construction methods [13].

In developed countries such as the UK, the significance of DFCS is recognized and provided for in the Construction (Design and Management) Regulations (CDM), putting the safety responsibilities on all parties of the project [9]. Though the concept has been identified as a means of improvement, there are still some barriers to its implementation. In a pilot study to check the viability of DfCS, Gambetese et al., [14] highlighted some of the barriers to implement DfCS including; construction contract structure, designers' knowledge, and safety experience.

DfCS research has been limited to developed countries such as the UK, USA, Australia, etc. In developing countries like Nigeria, where the state of OSH is more tragic [15], there is a dearth of research addressing the need for designers to factor safety of construction workers in their designs. This represents a gap in knowledge about the practice of DfCS in Nigeria. Hence, proactive measures are needed to reduce and control CSS issues [4]. DfCS can be a proactive measure, however, it could be impeded by some barriers [7]. Therefore, it is important to investigate these barriers to implementing DfCS in Nigeria. This will facilitate proactive improvement within the industry by promoting and enhancing the implementation of DfCS, hence, improving CSS. The aim of this study is to investigate the barriers to DfCS in Nigeria among Civil Engineers. Civil engineers were chosen because despite their contribution to design most DfCS studies have often focussed on Architects [11, 14] hence leaving a gap. To address this study (1) investigated the level of awareness of DfCS among Civil Engineers and (2) the critical barriers for the practice of DfCS in Nigeria.

2 Designing for Construction Safety

Traditionally, CSS has been the responsibility of the contractor [16]. However, research has shown that there is a link between the design and CSS [7] and [9]. Many safety issues have aroused due to lack of design consideration [16]. The European Foundation for the Improvement of Living and Working Conditions in 1991 found that lack of consideration of CSS decisions at design stage resulted in about 60% of fatal accidents on construction sites [7]. Therefore, the most important stage in the construction process is the design stage, because, it is where the conceptual ideas are turned into constructible realities [17]. The link between design decisions and CSS has been well established in several literatures. While reviewing 224 fatality investigation reports, [7] discovered that 42% of the reports were linked to the design and the accidents could have been reduced or eliminated during the project design phase. [18] also discovered that approximately 22% of 226 injuries that occurred, demonstrated a link between designs and accidents. Similarly, [11] reviewed 100 construction accidents and discovered that the likelihood of accidents in 47% of these cases would have reduced if there were changes in the permanent design. In South Africa, [19] revealed that of all the factors affecting safety, design ranked highest. Therefore, [7] emphasized that designers have a role to play in the safe delivery of construction projects. Historically, the concept of DfCS was first conceived in 1985 by the International Labour Office [20]. The office recognized that there is a need for design professionals to consider CSS. It was therefore recommended that the safety of workers who will be involved in the erection of a facility and other civil engineering works, be factored into the design process [7]. DfCS is a process where design professionals categorically consider CSS during the design phase of a facility [21] towards the realisation of a safe construction site [8]. Szymberski, [22] explained that ideally, construction worker safety must be the priority in the conceptual design phase. He proposed a model of time/safety influence which indicates that the ability to influence CSS reduces when it is left until the construction phase. This means that the earlier the potential hazards are identified, the safer the construction activities will be.

The construction industry in various countries has recognised the effectiveness of DfCS, to enhance occupational safety [20]. This has led to the release of CDM regulation in the United Kingdom [7], [17] and [23], and other regulations in other developed countries [24].

While developed countries like the United Kingdom have been able to put in place methods to improve CSS, developing countries like Nigeria should embrace this idea of DfCS. [4] suggested that using proactive methods is a better way to deal with site hazards and accident. Such proactive method is using the approach of DfCS.

2.1 Barriers to Designing for Construction Safety

It is obvious that the concept of DfCS has its merits if implemented. However, there remain barriers to the implementation of the concept. Some of these barriers have resulted in a low level of implementation of the concept [14]. These barriers usually originated from the designer's misperceptions and concerns. [25] maintained that the most critical barriers are designers' concern about increased liability, lack of safety expertise and increased cost. Gambetese et al., [14] further noted that other barriers include; designers' lack of interest, reduced design creativity and schedules problems.

These barriers are usually attributed to the impacts of the implementation of the concept [14]. Bello [27] explained that designers are not motivated and willing to implement DFCS due to the absence of interest. Also, designers' lack of safety expertise relates to the absence of the designers' knowledge of the concept, DfCS education and training, difficulty in implementing the concept and inadequacy of DfCS implementation tools and resources. Finally, clashing project cost, time and design targets could suffer due to reduced design innovations and increased schedule, cost of DFCS [27]. These are general barriers mostly reported in developed country contexts or architects and it is worth investigating their role in the uptake and understanding of DfCS practice in developing countries such as Nigeria and for the civil engineering context.

3 Research Methodology

To achieve the research aim, this study employed a quantitative approach. To fully utilize this approach, this study gathered the indicators of awareness and barriers from literature and reports. The barriers were defined for the respondents to indicate the extent to which they impede the practice of DfCS using a 5-point Likert scale, ranging from 1 (strongly disagree) to 5 (strongly agree). The questions were initially pilot-tested to determine the appropriateness and clarity and duly amended. The questions also included respondents' demographic information, safety education and access to DfCS tools.

The respondents were selected using purposive sampling. This involves identifying the needed samples and approaching them for eligibility [28]. Previous studies such as [29] have been focused on just architect. With civil engineers responsible for 51% of design errors in Nigeria [12], their assessment is much important. Hence, civil engineers have been sampled. Over 500 invitations were sent to potential participants. However, 268 confirmed interest in the study. Therefore, a total of 268 questionnaires were administered to Civil Engineers. The respondents cut across different Nigerian engineering professional directories. These include the Nigerian Society of Engineers (NSE), Nigerian Institution of Civil Engineers (NICE) and Nigerian Institute of structural engineers (NiStructE). A total of 106 questionnaires were returned achieving a 39.5% response rate. The response rate for norm survey questionnaire described by Takim and Adnan (2008) is usually 20–30%. Therefore, a response rate of 39.5% is adequate for analysis in this study. However, 3 questionnaire responses were rendered invalid due to incompletion. Hence, 103 questionnaires were used for data analysis.

4 Data Analysis

This presents the analysis of the survey carried out among the designers, to determine the extent of safety design practices, the barriers to the practice of DfCS. The analysis involved two steps: organising the data for analysis and describing the data

(Descriptive Statistics). The analysis was structured into two main segments: (1) Data organisation was supported by the Statistical Package for Social Sciences (SPSSv23) software for the preparation, transformation and storage of the data; and (2) The descriptive analysis involved describing the basic features of the respondents by summarising and aggregating the results. The analysis used tables including (frequencies, percentages and means), simple graphs and charts.

5 Results and Discussions

5.1 Demographic Information

The result shows that all the respondents are civil engineers. Over 57% and 52% of the respondents have at least 5 years' experience in their role and at least 11 years' experience in the construction industry respectively. Over 80% of the respondents are members of a professional body. In terms of education, most of the respondents have at least a BSc. Overall, the demographic information illustrates an experienced group of respondents and suitable to provide adequate responses.

5.2 DfCS Awareness

From the survey, most designers (92.2%) are aware of DfCS (Table 1). However, 70% identified that they practice at a very low level. This justifies the study of [7] that designers do not practice safety in design. It is, therefore, no surprise that 51% of construction site hazards in Nigeria [12] have been attributed to design issues. From the data analysis, the barriers responsible for this low extent have been identified and they are discussed in sections below.

Table 1. Awareness and Practice of DfCS

DfCS awareness			The extent of DfCS practice		
Awareness of concept	Frequency	Percentage (%)	Extent of practice	Frequency	Percentage (%)
Yes	95	92.2	Very low	70	68
No	7	6.8	Very high	33	32
Total	102	99.0			
Unspecified	1	1.0			
Total	103	100.0	Total	103	100.0

5.3 Barriers to the Practice of DfCS in Nigeria

The top-rated barriers to DfCS in Nigeria are discussed in this section. As shown in Table 2, the mean values reveal that 9 of the 10 barriers have a mean value of approximately 3. Therefore, there are 9 barriers to DfCS in Nigeria. These barriers are discussed below.

Table 2. Barriers to the practice of DfCS

Nos	Barriers	Mean	Ranking
1	Designers' exposure to liability	3.28	1
2	Project time requirement	3.25	2
3	Project cost requirement	3.24	3
4	Limited designer influence in the selection of a contractor.	3.15	4
5	Project environmental sustainability requirement	3.11	5
6	Your knowledge and understanding of construction site operations and methods.	3.09	6
7	Possibility of designing for safety to interfere with the contractor's methods of construction.	3.06	7
8	Your knowledge and understanding of construction health and safety issues.	3.03	8
9	Project aesthetic requirement (Innovative Designs)	2.92	9
10	Little or no motivation	2.45	10

Designers' Exposure to Liability. The findings show that the top-rated barrier to DfCS in Nigeria is the fear of designers' exposure to liability. This is in line with the study of [14], that designers think that DfCS will increase their liability. There are liability implications in situations where a designer specifies a DfCS measure as an approach or procedure for the contractor. This is because if a measure is specified by a designer, in the event of a safety incident, the designer will bear the cause of the incidents. Therefore, designers believe that it will go against the clause in the contract by interfering with the contractor's means and methods.

Project Time Requirements: Another top-rated barrier to DFCS is the likelihood of increasing the project schedule. The increased project time and schedule may be because of the time needed to fix additional protective features included in the design [30]. The time is however dependent on the approach of DfCS used. In cases where a feature is eradicated, there may be reduced project time, but where there is increased time, it becomes a barrier [27].

Also, because time will be allocated for safety criteria, there may be an increased project design time [26]. Furthermore, because of liability concerns, designers will have to excessively review designs before submission [27].

With the implementation of DFCS, there is a likelihood of increased project delays. 45% of the respondents believe that DfCS will act as a barrier. This is like the finding of [14] which found out that about 50% of surveyed designers believed that the DfCS will lead to delays in schedules.

Project Cost Requirements: Implementing DfCS may lead to a rise in direct and indirect costs [14, 27]. Checked for the impact of designing for safety and 74% of surveyed designers mentioned increased project costs as an impact. Therefore, it is no surprise as the cost requirement is rated third barrier in this study. The increased project costs may be because of additional protective features included in the design [30]. This cost is however dependent on the approach of DfCS used. In cases where a feature is eradicated, there may be reduced project costs, but where there is increased cost, it becomes a barrier [27].

[31] explained the project might generate some savings by removing the need for temporary protection during construction. The designer would not have to bear the cost of construction in a traditional procurement system [32]. However, for a design and build project, the net project cost savings may increase their project boundary. This is because, in a design and build project, only a firm is approached for both design and construction [32].

Increase in cost from implementing DFCS may also arise through the training of staff. This is true as design firms will have to train all their designers about safety practices in design [26].

Also, because designers become responsible for CSS, they will opt for insurance and thus resulting in an increased cost for the design firms [27]. This is because regardless of the circumstance, plaintiff lawyers will claim that the designers are responsible for construction safety [26].

All these forms of additional costs may drive designers to charge more additional fees [27]. This will make the design firms less competitive against firms who do not implement the concept [26].

Designers' Lack of Safety Expertise: It is crucial that designers possess a certain level of expertise in awareness of construction processes to effectively contribute to CSS [26], [27] and [30]. In Nigeria, this study has shown that designers' lack of safety expertise and understanding of CSS issues are barriers to practice DfCS. This is like the study of [14], where it was identified that designers' lack of safety expertise and construction experience are the most critical barrier to implement DfCS.

To improve DfCS, designers need to understand the concept and process. Also, it is important that they have adequate awareness of construction tasks and their coordination [27]. Designers' implementation will also be eased by an understanding of safety standards and regulations, which deal with safety design skills and not designers' obligations [30].

Another related issue of safety expertise is the limited amount of academic knowledge of CSS management [26]. Researchers have found that the rate of designers with academic knowledge of CSS is low. American research in 2002 found out that only 20% of 75 design firms showed that 50% of their employees had received safety training, during which 70% showed that less than 25% of their employees had received safety training [26] and [30]. In Nigeria, this study found out that 75% of the respondents do not have formal training on DFCS while 41% have received continuous development training on DfCS. With a low extent of design practices (18%), it is evident that their level of knowledge is a major barrier but not different from the context of other designers such as architects in Nigeria [29].

Reduced Project Quality and Innovative Designs: DfCS may lead to reduced project quality and innovative designs. Because aesthetic functional features add value to projects, removing elaborate features to favour construction safety can result in diminished creativity in design and possibly project quality [27]. In a study by [14], several surveyed designers said that DfCS would lead to decreased project quality. Reduction in design quality and project complexity were also identified as barriers to the concept [14].

6 Conclusions

The barriers to the practice DfCS in Nigeria have been investigated. This has been done by analysing and discussing the findings from the data. The highlight barriers will facilitate proactive improvement within the industry by promoting and enhancing the implementation of DfCS. As discussed earlier, the research sample is limited to civil engineers in Nigeria, hence these conclusions will be attributed mainly to this sample. From the findings of this study, it can be concluded that:

The level of awareness of DfCS is very high at 92%. However, the extent of the practice is very low at only 68% similar to Architects [29] in Nigeria. All indicated barriers attributed to Liability, Schedule Problems, Cost, Expertise, design creativity and motivation were rated high thus making them major barriers to DfCS among Civil engineers in Nigeria.

It is recommended that legislation similar to CDM, UK be enacted to incentivise mainstreaming of DfCS in Nigeria.

References

1. Hu, K., Hazhir, R., Tonya, S., Woodrow, W.: Factors influencing the risk of falls in the construction industry: a review of the evidence. Constr. Manage. Econ. **29**(4), 397–416 (2011)
2. Carter, G., Smith, S.D.: Safety hazard identification on construction projects. J. Constr. Eng. Manage. **132**(2), pp. 197–205 (2006)
3. HSE: An introduction to health and safety: Health and safety in small businesses. Suffolk: HSE Books (2003)
4. Idoro, G.I.: Health and safety management efforts as correlates of performance in the Nigerian construction industry. J. Civ. Eng. Manage. **14**(4), 277–285 (2008)
5. Agwu, M.O., Olele, H.E.: Fatalities in the Nigerian construction industry: a case of poor safety culture. Br. J. Eco., Manage. Trade **4**(3), 431–452 (2014)
6. Korman, R.: Wanted: new ideas. Panel ponders ways to end accidents and health hazards. Engineering News-Record **247**(27), 26–29 (2001)
7. Behm, M.: Linking construction fatalities to the design for construction safety concept. Saf. Sci. **43**(8), 589–611 (2005)
8. Toole, M., Hervol, M., Hallowell, M.: Designing for construction safety. Mod. Steel Constr. **46**(6), 55–59 (2006)
9. Gambatese, J.A., Behm, M., Rajendran, S.: Design's role in construction accident causality and prevention: perspectives from an expert panel. Saf. Sci. **46**(4), 675–691 (2008)
10. Nashwan, D., Mohamad, K.: Proceedings of the 13th International Conference on Construction Applications of Virtual Reality, CONVR 2013, London. www.convr2013.com, 30–31 October 2013

11. Gibb, A., Haslam, R., Hide, S., Gyi, D.: The role of design in accident causality. In: Designing for Safety and Health in Construction: Proceedings, Research and Practice Symposium. UO Press, Eugene, Ore (2004)
12. Ayininuola, G., Olalusi, O.: Assessment of building failures in Nigeria: Lagos and Ibadan case study. African J. Sci. Technol. **5**(1), 73–78 (2004)
13. Rafael, S., Jennifer, W., Dana, S., Gabriel, R., Wei, Z., Aviad, S.: Safety by design: dialogues between designers and builders using virtual reality. Constr. Manage. Econ. **33**(1), 55–72 (2015)
14. Gambatese, J.A., Behm, M., Hinze, J.W.: Viability of designing for construction worker safety. J. Constr. Eng. Manage. **131**(9), 1029–1036 (2005)
15. Okoye, P.U., Okolie, K.C.: Exploratory study of the cost of health and safety performance of building contractors in South-east Nigeria. Br. J. Environ. Sci. **2**(1), 21–33 (2014)
16. Lingard, H.C., Rowlinson, S.M.: Occupational Health and Safety in Construction Project Management. Spon Press, London (2005)
17. Thorpe, B.: Health and Safety in Construction Design. Gower Publishing Ltd., Aldershot (2005)
18. Behm, M.: Establishing the link between construction fatalities and disabling injuries and the design for construction safety concept. Ph.D., Oregon State University, Corvallis, Oregon (2004)
19. Smallwood, J.: The influence of designers on occupational safety and health. In: Proceedings 1st International Conference of CIB Working Commission W99, pp. 203–213. Implementation of Safety and Health on Construction Sites, Lisbon, Portugal (1996)
20. Bhattacharjee, S., Ghosh S., Young-Corbett, D.: Safety improvement approaches in construction industry: a review and future directions. In: Tulio, S. (ed) 47th ASC Annual International Conference, 6–9 April 2011. Associated School of Construction, United States (2011)
21. Toole, T.M., Gambatese, J.: The trajectories of prevention through design in construction. J. Saf. Res. **39**(2), 225–230 (2008)
22. Szymberski, R.: Construction project safety planning. TAPPI J. USA **80**(11), 69–74 (1997)
23. Horne, K., Gibb, A., Pavitt, T., Haslam, R.D.: Influences on designing for health: results from interviews and questionnaires. Constr. Manage. Econ. **1**(2), 101–110 (2003)
24. Bluff, L.: Regulating safe design and planning of construction works: a review of strategies for regulating OHS in the design and planning of buildings, structures, and other construction projects (Working Paper 19). The Australian National University, Canberra (2003)
25. Toole, T.: An overview of prevention through design in construction. http://www.designforconstructionsafety.org/Documents/An%20Overview%20of%20Prevention%20through%20Design%20in%20Construction.docx (2007)
26. Toole, T.M.: Increasing engineers' role in construction safety: opportunities and barriers. J. Prof. Issues Eng. Educ. Pract. **131**(3), 199–207 (2005)
27. Bello, A.M.: Minimizing impediments to design for construction safety (DFCS) implementation on capital projects. Ph.D., Carnegie Mellon University (2012)
28. Easterby-Smith, M., Thorpe, R., Jackson, P.R.: Management Research, 3rd edn. SAGE, London (2008)
29. Manu, P., Poghosyan, A., Mshelia, I.M., Iwo, I.T., Mahamadu, A-M., Dziekonski, K.: Design for occupational safety and health of workers in construction in developing countries: a study of architects in Nigeria. Int. J. Occup. Saf. Ergonomics (2018)

30. Yu, X., Li, C.Q., Zhang, G.: Impediments to implementing design for construction safety. In: Proceedings of the 19th International Symposium on Advancement of Construction Management and Real Estate, pp. 313–326. Springer, Heidelberg (2015)
31. Gambatese, J., Hinze, J., Haas, C.: Tool to design for construction worker safety. J. Architectural Eng. **3**(1), pp. 32–41 (1997)
32. Ashworth, A., Hogg, K., Higgs, C., Willis, C.J.: Willis's Practice and Procedure for the Quantity Surveyor. Wiley-Blackwell, Chichester (2013)

Determinants of the Petroleum Fuel Supply Chain Performance in Zimbabwe: A Case of the National Oil Infrastructure Company

Lovemore Tendedziso[1], Erastus Mwanaumo[1(✉)], Nathan Musonda[2], Mwewa Mambwe[1], and Lubinda Haabazoka[3]

[1] Department of Civil and Environmental Engineering, School of Engineering, University of Zambia, Great East Road, Lusaka, Zambia
erastus.mwanaumo@unza.zm

[2] School of Post Graduate Studies, University of Lusaka, Lusaka, Zambia

[3] Graduate School of Business, University of Zambia, Lusaka, Zambia

Abstract. The petroleum fuel supply chain in Zimbabwe has been faced with multiplicity of challenges from perennial erratic fuel supplies to poor product lead times. Despite introduction of direct importation scheme for private players, erratic fuel supplies have remained persistent. The study aimed at identifying factors affecting petroleum fuel supply chain and propose strategies to National Oil Infrastructure Company for the improvement of the petroleum fuel supply chain performance in Zimbabwe. The study exploited a descriptive case study through a mixed method approach in which semi-structured interviews and self-administered structured questionnaire were used to collect primary data from 110 respondents constituting 100% response rate. Findings indicated that petroleum fuel sector is highly fragmented with sub-optimal use of the pipeline in the transportation hence negatively affecting supply. The study concluded that poor performance of the petroleum fuel supply chain had significant downstream effects curtailing the productive capacity of other sectors.

Keywords: Performance · Petroleum fuel · Supply chain · Strategies

1 Introduction

Globally, petroleum fuel supply chains (PFSCs) have exhibited unique characteristics given the criticality of the commodities they are associated with and have continued to grow as important energy consumption source world over. The significant role of a PFSC performance in value chain models have received wider recognition across various sectors and the world over. Crandall et al., [1] opines that supply chain performance seeks to ensure improved inventory availability, reduce supply lead-times, as well as exhibit effective and efficient consistent supply of petroleum fuel products to meet the end-user's demands. However, the World Bank [2] noted that there exist a number of bottlenecks with regards to petroleum fuel energy flows particularly in developing countries that depend on petroleum energy sources.

Over the years, petroleum fuel industry in Zimbabwe has seen greater turbulence and volatility [3]. The Zimbabwean PFSC has witnessed a number of challenges in the

C. Aigbavboa and W. Thwala (Eds.): CIDB 2019, *The Construction Industry in the Fourth Industrial Revolution*, pp. 566–576, 2020.
https://doi.org/10.1007/978-3-030-26528-1_58

past decades, with the country faced with growing erratic petroleum fuel supplies, poor lead times and market shortages, emergence of parallel markets, product adulteration, thus, making the PFSC a source of great ambivalence amongst a number of stakeholders. Hence, the need for the study which sought to identify strategies that can be adopted by National Oil Infrastructure Company (NOIC), as a case study, and other stakeholders in order to improve the petroleum fuel supply chain performance in Zimbabwe. While previous studies have mainly focused on the generic challenges facing the petroleum sector in Zimbabwe, this study managed to tackle the matter from a supply chain perspective thus tracking the petroleum value chain.

2 Literature Review

The significant role of the supply chain performance in the conceptualization of the value chain models has received wider recognition across various sectors. The central objective of a supply chain is to deliver superior customer value at less cost to the supply chain as a whole [3]. According to Kojima and Sexsmith [4] when supply chain is used the objective of its use is to improve performance by augmenting speed, dependability, quality, cost reflective and flexible delivery of goods and services, making reliable, flexible and responsive value chain. Nonetheless, that has not been so globally, because it has not been used to embrace its sustainability concept as a much wider determinate of performance [5]. Increase in delivery performance is possible if suitable channel policies are utilised that support the sustainability of a PFSC and further underlined that fundamental goals for oil companies was to deliver fuel and its products safely, at the right time with the best price. Nonetheless, Tunio [6] argues that this is not so, due to the challenges facing the oil and gas industry of untimely delivery of the petroleum product to the consumer that creates shortages.

World demand for petroleum fuels has been on the increase, affecting delivery as was reported by the United States Energy Information Administration in its 2011 International Energy Outlook, which further projected that the world's energy consumption will increases by 53% by the year 2035 as reported by Cohen [7]. Cousins et al. [8] revealed that petroleum products are widely used in most countries and have far-reaching micro-economic and macro-economic effects due to the ever growing demand even though recently they have without supply chains that are value and sustainable oriented. According to Iovion [9] when complexity arises in a supply chain, the downstream suffer negative supply chain dynamics resulting in severe or wide downstream product shortages. This mainly is as a result of poor coordination in the supply chain that may result in both demand and supply volatility.

Furthermore, lack of adequate equipment or infrastructure in terms of transportation capacity significantly affect smooth flow of the petroleum fuel prompting panic in the supply chain [10]. In the Sub-Saharan Africa, Kimani [11] and Mbara and Bryceson [12] argued that some of the major drivers of illicit supply chains are to do with failure to meet the demand which causes an increase in prices, hence cause supply shortages of essential goods like drugs, fuel, food and other consumables.

From various studies reviewed such as studies by Grover and Malhotra [16], Osoro [13] and Crandall et al. [1] asserted that certain factors that play an important role in determining the success or failure of PFSC performance and include transportation facilities, infrastructure situational factors such as storage facilities, structure or design of the supply such as planning, management factors, environmental uncertainty, and the regulatory framework. Other studies reviewed also indicated information vagueness, conflicting objectives and decision variables [14].

According to the Zimbabwe Energy Council (ZEC) [15] the petroleum energy sector in Zimbabwe is faced with supply chain bottlenecks from erratic fuel supplies to poor product lead time causing market shortages, while Zimbabwe Revenue Authority (ZIMRA) [16] also attested that petroleum imports increased by 47% from 207.6 million litres in 2017. Critical supply chain management issues have since emerged around fuel pipeline utilization in Zimbabwe as can be seen in the surge of road transport in fuel importation which is currently estimated at 52% compared to rail at 12% and pipeline at 34% usage [17]. Another significant challenge has been that of intense product adulteration and illicit fuel supply chains that continue to threaten the sustainability of the formal petroleum fuel supply chain in Zimbabwe. According to ZIMRA [12] the country losses close to US$1 billion annually as a result of illicit petroleum fuel supply chains due to shortages.

Additionally, the fragmented supply of petroleum fuel in the sector has been triggered by poor capacity utilization of the pipeline as small players fail to constitute economies of scale for its optimal use. Further, the absence of a viable supply chain framework affects the petroleum supply chain which is under the control of National Oil Infrastructure Company (NOIC).

3 Research Method

The research methodology used in the study was a combination of both qualitative and quantitative approach and adopted a descriptive case study design. A similar research method was applied in exploratory studies by Matveeva [18] who looked at the issues in the supply chain management in the oil industry and Cousins et al. [8] whose research was on performance measurement in strategic buyer-supplier relationships as a mediating role of socialization. The target population of the study comprised of operational and management staff of NOIC. Other key respondents to this study were drawn from fuel retailers and government regulatory bodies, that is Zimbabwe Energy Regulatory Authority (ZERA) and ZIMRA. Purposive sampling was applied in the choice of retail companies to provide clarity, insight and understanding about the study issues while the sampling used for operational and management staff at the NOIC, was accidental as well as availability sampling methods respectively. To the information rich segments such ZERA and ZIMRA, expert and maximum variation sampling was applied.

Data was collected from both management and operational staff at NOIC using a self-administered structured questionnaire designed using a 5-point Likert scale. This was to provide freedom from bias and for the respondents to remain anonymous, thus helping to obtain accurate information as much as possible [19]. Studies by Grover and Malhotra [20] in operations and supply chain management theory and management, qualified the use of questionnaire to reduce bias, while the Likert scale provide quality, increases response rate and reduces frustration levels in answering [19].

Semi-structured interview questions were developed and utilized to gather the perspectives and experiences of the fuel retailers, customers and other key stakeholders on the performance of the PFSC. According to Kimani [11], interviews reveal more than can be detected or reliably assumed from observing a situation, of which qualitative data was backed by prose interpretations. The study identified variables. The independent variables which are the supply chain factors include supply chain infrastructure situational factors, institutional situational factors and network strategic factors; while improved product flows, product supplies and demand responsive supply chain were dependent variables for supply chain performance. These were modified from the study by Tunio [6]. Statistical Packages for Social Sciences (SPSS) version 22.0 was used to analyse quantitative data to show the distribution of respondents, while qualitative data was analysed using content analysis and cross tabulation [19].

4 Results

The following are the findings of the study. All response rate on interviews and questionnaire survey gave 100%. Ten (10) respondents who were interviewed managed to provide essential data used by the researcher. Hence is credible for further analysis as the response rate was over 70% [19]. The following are findings to the study on assessing the determinants of the PFSC in Zimbabwe, with NOIC as a case study.

4.1 Reliability Test of Results

A reliability of at least 0.60 at $\alpha = 0.05$ significance level of confidence was accepted using Cronbach's Alpha Coefficient which is a measure of internal coefficient according to Kimani [11]. This reliability threshold was adopted based on similar studies by Kojima et al. [4] and Cousins et al. [8] (see in Table 1).

4.2 Petroleum Fuel Supply Chain Profile in Zimbabwe

To gather respondent's opinion on the level of the PFSC performance in Zimbabwe, most of the respondents, 45.5% considered the PFSC as being poor, while 23.6% viewed as very poor. Only 12.7% deemed the petroleum supply chain excellent as indicated in Table 2.

Further, the study sought to understand the extent to which supply chain practices used influence the petroleum supply chain performance (See Table 3). The higher the mean (3,77), the greater the extent of agreement while the greater the standard deviation (1.142), the greater the level of variation in the responses. The results show that there is limited application and poor implementation of supply chain metrics which does not meet international standards.

Table 1. Cronbach's alpha reliability coefficients results for the questionnaire factors.

	Mean	Standard deviation	Cronbach's alpha reliability coefficients	Evaluation based on Iovion [2]
What is the current petroleum fuel supply chain profile in Zimbabwe?	8.6669	2.58068	0.715055	Acceptable
Factors affecting the supply chain performance in Zimbabwe	8.99646	3.02830	0.757562	Acceptable
Downstream effects of the petroleum supply chain in Zimbabwe	8.4966	2.51733	0.688859	Acceptable
Strategies of improving the petroleum supply chain performance in Zimbabwe	7.99674	2.007166	0.620156	Acceptable
Overall reliability			0.695408	Acceptable

Table 2. Current petroleum fuel supply chain performance rating.

		Frequency	Percent	Valid percent	Cumulative percent
Valid	Excellent	14	12.7	12.7	12.7
	Good	12	10.9	10.9	23.6
	Neutral	8	7.3	7.3	30.9
	Poor	50	45.5	45.5	76.4
	Very poor	26	23.6	23.6	100.0
	Total	110	100.0	100.0	

Table 3. Supply chain performance metrics at NOIC.

Rating by respondents	Mean	Std. Dev.
Inventory availability in PFSC meet expected levels?	3.77	1.142
Order cycle times are within meet expected standards?	3.60	1.030
Delivery frequency is within expected standards?	3.68	0.238
Delivery reliability meet expected standards	3.61	1.087
Order fill rate meet expected standards	3.75	0.231
Deliveries are made on time and in full (OTIF)	3.46	0.303
There is high information visibility in the PFSC	3.19	0.415
Average	3.58	0.635

4.3 Factors Affecting the Petroleum Fuel Supply Chain Performance

Information and Communications Technology (ICT) Deployment in the PFSC Industry. About 45.45% affirmatively said that there was need for ICT deployment in the industry while only 4.55% gave a negative response. This indicated that there is limited use of ICTs in the PFSC at NOIC (See Fig. 1).

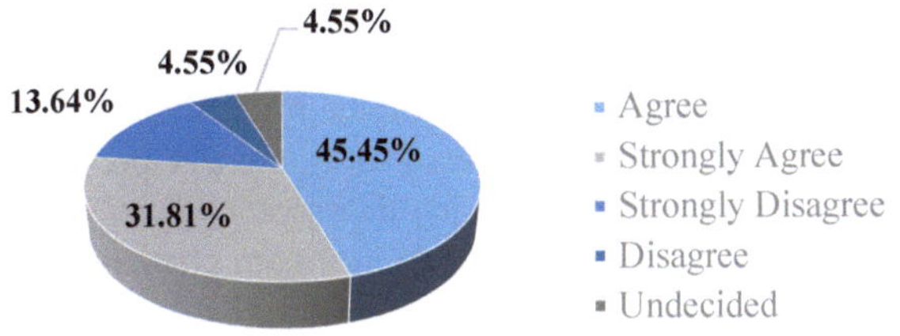

Fig. 1. ICT resources as vital factors in PFSC performance

Synchronized Processes as Factors in Improving PFSC Performance. Respondents views on the process alignment or value added processes help in the improved performance of PFSC at NOIC indicated that 31.8% strongly agreed to the notion while 27.3% strongly disagreed and 36.4% agreed. The above information shows that the majority understood the importance of supply chain process alignment in the PFSC.

Environmental Uncertainty Contributes to PFSC Performance. The results showed that the relationship between environmental factors and the PFSC performance is positive and significant as can be seen in the number of respondents that agreed to this notion of 35.50% while only 10.90% strongly disagreed to this notion.

Transportation Permits Efficient Product Flow and is Fully Optimised. About 45.5% of the results were non-affirmative that the transport infrastructure permits efficient product flow and is fully optimised and only 9.10% agreed to this. The modal share in the PFSC and their level of utilisation was established, of which majority pointed to the use of pipeline = 39%, road = 48%, Rail = 12% and Air = 1% as portrayed in Fig. 2.

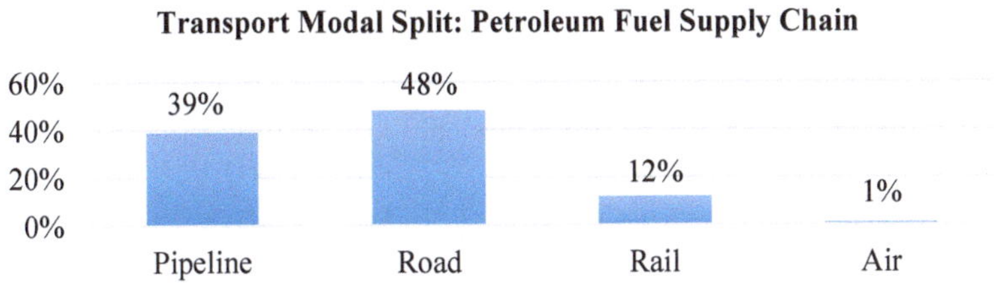

Fig. 2. Modes of transport used in the petroleum fuel supply chain in Zimbabwe

Regulatory Framework and Policy Factors. Majority of the respondents, 36.36% strongly agreed to the notion that the regulatory framework and policies impact on the PFSC activities at NOIC. While, 9.09% strongly disagreed to this view. The results obtained therefore suggest that there is a strong relationship between regulatory framework and the overall PFSC performance.

4.4 Downstream Effects of the Petroleum Supply Chain Performance

Some of the perceived downstream challenges of the PFSC by the respondents showed that lack of the supply chain resulted in widespread fuel shortages (25.45%), high fuel prices, (22.73%), thriving parallel markets (20%), impact on industrial productivity (11.82%) as well as performance decline in other sectors (6.36%). The results showed that the petroleum industry has significant downstream challenges.

4.5 Barriers of the Petroleum Fuel Supply Improvement Optimisation at NOIC

Some of the notable barriers to PFSC performance in Zimbabwe as obtained from the results include lack of management support (30%), policy barriers (25%), poor supply chain strategy (22%), poor resource capability (18%) and resistance to change (12%). Those citing lack of management support and policy barriers pointed corruption and unethical practices as major challenges undermining the petroleum supply chain performance.

4.6 Strategies of Improving the Petroleum Fuel Supply Chain Performance

The strategic options considered as most effective and viable were controllable variables/factors which stakeholders in the petroleum can effectively pursue as indicated in Table 4. The higher the mean, the greater the extent of agreement while the greater the standard deviation, the greater the level of variation in the responses.

Table 4. Strategic options for the petroleum fuel supply chain improvement

Statement	Mean	Standard deviation
Supply chain design/configuration	3.88	0.446
Transportation optimisation	3.73	1.041
Demand management	3.73	1.116
Sector taxation regimes and regulations	4.04	0.038
Investment in alternatives	3.01	2.783
Average	3.68	1.085

With the application of supply chain strategy, managers can make a wise decision to satisfy customers' needs at affordable costs and maintain a responsive supply chain.

5 Discussion

One of the objectives of the study was to establish the current PFSC performance profile in Zimbabwe. From the results, it can be deduced that the PFSC has been failing to meet the aggregate demand and this has resulted in massive stock outs in the sampled fuel retail outlets. The PFSC chain is prone to supply chain breakdowns and is notably characterised by erratic fuel supplies. Christopher [3] equally acknowledged the importance of effective distribution strategy as a basis for supply chain competitiveness. The likeness of the results could be attributed to the existing reservations on performance of the current PFSC especially in terms of demand fulfilment, product pricing and stock or inventory availability. Sound distribution strategies are premised on high delivery flexibility and reliability competitiveness through logistics as opposed by Misund et al. [17].

According Harrison and Van Hoek [10] on the importance of competing through logistics, they highlighted the importance of sound distribution strategies premised on high delivery flexibility and reliability. While Mbara [12] acknowledged the importance of an effective distribution strategy as a basis for supply chain competitiveness. It can be said that supply chain optimisation is a product of value stream mapping as it affects the system-wide performance. Additionally, managing product lead-times is critical in handling major supply chain disturbances which may culminate into intense supply chain distortions. The central objective of a supply chain is to deliver superior customer value at less cost to the supply chain.

The study also revealed that the PFSC had been failing to meet the aggregate demand resulting in massive stock outs in most retail outlets. These results found credence in Crandall et al. [1] on effective demand management as well as Tunio [6] who concurred that managing product lead-times is critical in handling major supply chain disturbances which may culminate into intense supply chain distortions.

The decline in the performance profile of the PFSC was driven by low capacity utilisation in the fuel pipeline from Beira, Mozambique to Zimbabwe. Lack of certainty of supplies is likely to destabilize the entire supply chain. While the use of guaranteed off-take agreements, forward purchases, options and swaps are some of the critical measures that can be explored to ensure stability of supplies. In a study conducted by Osoro [13] it concluded that a well-established and optimised supply chain should foster growth as well as sustainable operations. The PFSC in Zimbabwe is affected by a number of supply chain bottlenecks with seamless flow of the petroleum fuel into the country. To this end, the study has managed to unravel lack of viable framework in the current PFSC to counter supply chain bottlenecks. Apparently, a number of authorities such as Kojima et al. [4] Osoro [13] have demonstrated that viable supply chain framework is critical for an efficient flow and the sustainability of the supply chain.

Ramakrishnan and Ma [5] proposed intelligent ways of supply chain optimisation by using dynamic simulation and modelling ICT tools that are critical in a supply chain as they determine the extent to which significant and proprietary information is communicated to. The relationship between environmental factors and the PFSC performance was found to be a positive and significant, flexible and reliable planning and control system of the supply chain that reduces uncertainties. Christopher [3] who stated increasingly ICT capabilities are being used for supply chain optimisation, while

Matveeva [18] opines that ICT tools are critical in a supply chain as it determines the extent of critical and proprietary information is communicated. Therefore, with a level of complexity in the environment, supply chain optimisation has a potential to make a significant contribution to resolve the challenges.

Furthermore, transportation of fuel is not fully optimised and this affects performance of the fuel supply chain flow into Zimbabwe. Optimisation also covers good storage facilities along the supply chain as they are the heart of improved product quality in oil and gas supply chains. Pipeline was found to be the most dominant form of transport in oil and gas supply chains followed by bulk rail transport and road transport. This was also opined by Grover and Malhotra [20] and further proposed for various digital methods to be used to ensure smooth flow of petroleum fuel. Commodities such as oil, gas, and petrochemicals require specific modes of transportation such as pipelines, vessels or tankers, and railroads.

Some government regulations and policies negatively affect the PFSC. For instance, the study revealed that the delay in the disbursement of foreign currency by central bank, Reserve Bank of Zimbabwe results in delays in procurement and delivery of fuel into the country by fuel importers. Christopher [3] insists that the external dynamics are equally crucial for PFSC performance and that the impact of unplanned and unforeseen events can have severe financial effects across the network as a whole. Over the years, the oil and gas industry has continued to face growing challenges, from stricter government regulation, political risks, competition, emergent new comers and political hostilities, which has affected price hike and shortages.

The study sought to establish the downstream effects of the petroleum as result of a lack in supply chain improvement and managed to ascertain that lack of PFSC negatively affect a number of sectors and stakeholders. The results showed that the petroleum industry has significant downstream challenges. Majority of the respondents identified lack of management support, policy barriers and poor supply chain strategy as major setbacks to the petroleum fuel supply chain improvement in Zimbabwe. Kimani [11] avers that lack of knowledge as well as financial challenges were some of the major challenges confronting supply reform efforts in emerging firms.

6 Conclusion and Recommendations

The purpose of the study was to identify the strategies that can be adopted by NOIC in order to improve the PFSC performance in Zimbabwe. There is a general poor performance in the PFSC performance by NOIC caused by factors internal to the supply chain and influenced by external regulations and policies. Nonetheless, the study deduced strategies that can be adopted by NOIC to improve performance based on the results and the literature reviewed. These include carefully configuring, planning and implementing the PFSC according to best practice; redesigning or configuration of the supply chain; transportation optimisation and maintenance; alignment and streamlining of the petroleum supply chain processes; management and sound distribution resource capability; and improved ICT management.

The study recommends that management at NOIC and the government of Zimbabwe consider transport optimization and dynamic simulation through the use of pipeline for the movement of petroleum fuel and storage capacity expansion in line with growing aggregate demand. There is need for a more robust infrastructure management and maintenance of bulk terminals. The study recommends that the government through the Ministry of Power and Energy, invest in more sustainable fuel alternatives thus substituting the petroleum based energy whose consumption has nearly doubled in the last three years.

References

1. Crandall, R., Crandall, W.R., Chen, C.: Principles of Supply Chain Management, 2nd edn. CRC Press, Taylor & Francis Group, New York (2015)
2. World Bank.: Petroleum Product Markets in Sub-Saharan Africa: Comperative Efficiency Analysis of 12 Countries, Washington D.C.: World Bank (2014)
3. Christopher, M.: Logistics and Supply Chain Management, 4th edn. Pearson Education, Edinburgh (2012)
4. Kojima, M., Mathews, W., Sexsmith, F.: Petroleum Markets in Sub-Saharan Africa: Analysis and Assessment of 12 Countries. World Bank, Washington D.C. (2010)
5. Ramakrishnan, P., Ma, Y.: Adaptative Supply Chain Systems: IoT Based Conceptual Framework, pp. 62–68 (2018)
6. Tunio, W.A.: Supply Chain Practices and Optimisation in Petroleum Industry. Supply Chain Practices in Oil and Gas Industry, pp. 4–15 (2013)
7. Cohen, B.: Statistical review of the world energy full report. Energy Econ. (2016)
8. Cousins, P.D., Lawson, B., Squire, B.: Performance measurement in strategic buyer-supplier relationships: the mediating role of socialization mechanism. Int. J. Oper. Prod. Manag. **28**(3), 238–258 (2013)
9. Iovion, F.: Relationship marketing by energy companies. Rev. Int. Comperative Manage. **16**(5), 558–572 (2015)
10. Harrison, A., van Hoek, R.: Logistics Management and Strategy: Competing through the supply chain, 4th edn. Pretince Hall, Harlow (2012)
11. Kimani, C.W.: Supply chain management challenges in Kenya petroleum industry: case of national oil corporation of Kenya. Int. J. Soc. Sci. Entrepreneurship **1**(3), 231–246 (2013)
12. Mbara, T., Bryceson, D.F.: Petrol pumps and economic slumps: rural-urban linkages in Sub-Saharan Africa's globalisation process. J. Econ. Soc. Geogr. **94**(3), 335–349 (2008)
13. Osoro, A.: Challenges affecting performance of supply chain systems in the petroleum industries in Kenya. J. Energy Technol. Policy, 82–92 (2015)
14. Wilson, H., Daniel, E., McDonald, M.: Factors of success in customer relationship management. J. Mark. Manage. **18**(1), 193–219 (2011)
15. Zimbabwe Energy Council.: Petroleum Fuel Update by Energy Council. Zimbabwe Situation, Harare (2017)
16. Zimbabwe Revenue Authority.: Revenue Performance Report for the First Half of 2016. ZIMRA, Harare (2016)
17. Misund, B., Osmundsen, P., McMillan, D.: Valuation of proved vs. probable oil and gas reserves. Cogent Econ. Finance **5**(1) (2017)

18. Matveeva, K.: Exploring Current Issues in the Supply Chain Management in the Oil and Gas Industry (2015)
19. Tashakkori, A., Creswell, J.W.: The era of mixed methods. J. Mix. Methods **1**(3), 2–7 (2007)
20. Grover, V., Malhotra, M.K.: Transaction cost framework in operations and supply chain management research: theory and measurement. J. Oper. Manage. **21**, 457–473 (2011)

Ethics and Stakeholder Engagement for Industry/Construction 4.0: A Systematic Review

Kamil Okedara[1(✉)], Obuks Ejohwomu[1], and Paul Chan[2]

[1] School of Mechanical, Aerospace and Civil Engineering (MACE), University of Manchester, Manchester, UK
{Tunde.okedara, Obuks.ejohwomu}@manchester.ac.uk

[2] Department of Management, Design and Construction Management, Deft University of Technology, Delft, Netherlands
p.w.c.chan@tudelft.nl

Abstract. Industry 4.0 is an offshoot of the first, second and third industrial revolutions. Which is an indication of its globalization. In the same vein, construction 4.0 which takes its context after the industry 4.0 is expected to lead to the fourth construction revolution? Ethics and stakeholder engagement have been individually studies and written about. Both industry and construction 4.0 require an ethical lens and an engagement pattern towards a rapid realization of their objectives. This paper quantitatively x-rays researchers' interests in the fields, and the subject areas so far covered in their studies. However, little or no studies have applied the ethical lens and engagement pattern into the revolutionary concepts. We applied systematic review method (the first in the field), to discover the linear and the polynomial nature of researches in the fields. This discoveries is hoped to open up further researches towards the rapid development of the fourth I&C revolutions through the application of ethics and stakeholder theories.

Keywords: Industry 4.0 · Construction 4.0 · Ethics · Stakeholder engagement · Organizational culture · Project · Programme & Portfolio (PPP)

1 Introduction

Every endeavour requires a way to achieving success. This is what we refer to as the ethics needed for the success. This ethical approach requires an engagement pattern in achieving the successful outcomes (Okedara et al. [1]). The endeavour could be a project, programme or portfolio (PPP), it could be a concept such as "construction 4.0" or "industry 4.0". This paper, therefore researches into the extent of application of ethics and stakeholder concepts in both industry 4.0 and construction 4.0 (I&C 4.0) in order to ensure success of the revolutions. To understand this approach, we will look at the perspectives of success, engagement and ethics.

Successful business, organizational or project management strategies have been variously discussed and described. Strategic management success may be measured through the use of scorecard which supports the alignment and management of

C. Aigbavboa and W. Thwala (Eds.): CIDB 2019, *The Construction Industry in the Fourth Industrial Revolution*, pp. 577–587, 2020.
https://doi.org/10.1007/978-3-030-26528-1_59

corporate activities to their strategic relevance [2], while Figge et al. [2] highlighted the importance of strategic management aligning with corporate activities, others have also come up with various criteria that can cumulatively be necessary for envisaged success. For instance, it is averred that organizational success depends not only on how the organization makes the most of human competences, but also how it stimulates commitment to the organization [3]. The turbulent organizational environment is not left out of the discourse, this has been attributed to why organizations seek competitive advantage through organizational learning [4], in line with Namada [4], this learning, as part of knowledge management orientations, is said to comprise of 'the personal knowledge orientation' and 'organizational knowledge orientation' [5]. To further buttress the significance of business and organizational management success, process automation involving modelling processes and workflow [6], a fourth industrial revolution phase has been introduced. The term "industry 4.0", according to [7] was established ex-ante for a planned "4th industrial revolution", the term being a reminiscence of software versioning Methodology. **Industry 4.0** which is also commonly referred to as the fourth **industrial** revolution is a name given to the current trend of automation and data exchange in manufacturing technologies. It includes cyber-physical systems, the Internet of things, cloud computing and cognitive computing [8]. In the same vein, similar revolutions are being introduced to construction with "the fourth construction revolution tagged "construction 4.0". Though industry 4.0 describes the trend towards digitisation and automation of the manufacturing environment with potentials to benefits in terms of improvement in productivity and quality [9], the concept has not gained much attention in the construction industry [9] hence the need to not only ensure awareness but for researchers to dig deep into the area.

To attempt a broader understanding, ethics and stakeholder engagement will be presented in perspective of fourth industrial and construction revolutions, also known respectively as industry 4.0 and construction 4.0. Our definition for the purpose of this paper will also be stated. Stakeholder engagement and ethics have variously undergone definitions and perspectives over the years. Ethics could be rightness or wrongness of behaviour [10]. "The definition of ethics is said to be at abstract level [10] as it is difficult to have a consensus on what is morally right or wrong, good or bad, ethical or unethical". Despite the growth of ethics in business, business ethics researchers are yet to agree upon ethical perspective for the discipline [11]. [12], in their paper on human research ethics while quoting OED definition define ethics as "The codes of conduct or moral principles recognized in a particular profession, sphere of activity, relationship, or other context or aspect of human life. Attempting a definition for the purpose of this research, which I hope may transcend industry, project and business. Ethics can be a way of doing things or as a process culmination or path to an engagement outcome. This process culmination or path, when tested over time, can become norm in industry, business or project, it can also translate to culture or governance. When ethics is measured, sometimes in projects, to have given expected outcome or surpass expectation, the approach becomes adopted and becomes an ethical approach for such projects or business. This sometimes can become organizational norms or best practices, deviation from or adherence to those approaches at that point in time, may become an indication of good or bad practices etc.

Stakeholder engagement on the other hand, has gone through different phases of definitions, perspectives, and development of topics and terms. Scholars have debated what it means to be a stakeholder, what obligations, if any, firms owe to stakeholders, and whether or not shareholders ought to be granted precedence over other stakeholders [13]. In addition, 'Many accounts of stakeholder activities focus on the attributes of the organizations (or industry) or the attributes of the stakeholders rather than on the attributes of the relationship between organizations (or Industry) and stakeholders'. In the last few years, however, greater attention has been given to thinking about what it means to engage stakeholders (bracket mine) (Johnson- Cramer et al. 2003). As stakeholder ethics grows in recognition and in being a veritable tool for project and engagement success, thought about which actions firms or industry must or must not perform in order to meet philosophical, social, legal or environmental moral standards (this could be the fourth industrial or construction revolutions), greater attention is now being paid to relationships between firms and its stakeholders. [13]. This has made approaches to stakeholder engagement to progress through some stages (Reed, 2008). Theories have therefore moved from awareness of the 1960s [14] to adding local angle to the debate in the 1970s (Pretty, 1995), from strategies of engagement in the 1980s to increasing community participation and democratizing engagement in stakeholder governance [15].

With the above background on ethics, stakeholder engagement, industry 4.0 and construction 4.0, there is the need to examine the essence of this review and the research question, the growing trend in industry 4.0 and the consequential increased interest in construction 4.0. Both industry and construction (I&C) 4.0 can be subjected to process ethics which will not only enhance their evolution, but also has the tendency to catalyse growth, replicability and development in the fields. The twin area of ethics and stakeholder engagement have the capacities to establish proper record of developments and a path that can be followed and replicated for the realisation of the objectives of fourth I&C revolutions.

To this end, the following research question, as a preliminary or foundation setting for further research in the field of ethics, stakeholder engagement, and industry 4.0 and construction 4.0 are set for this systematic review objective.

RQ1: How has researchers' interest in the field been growing?
RQ2: How has this growth impacted other subject area and how indeed it has affected the globalization of I & C 4.0s

Before we venture into answering the research questions, let me quickly add that this is the first research linking ethics, stakeholder engagement, industry 4.0 and construction 4.0 and it is also the first systematic review in the fields.

2 Methodology

To answer the research questions in the introduction above, the literatures on ethics, stakeholder engagement, industry 4.0 and construction 4.0 were systematically reviewed and synthesised. Other literatures were also studied in other to give an impactful analysis of the 28 papers under review. Systematic reviews, which originated from the medical and health sciences, are a thorough and transparent way of mapping and assessing the evidence in a particular topic area [16]. The adoption was after a thorough review of other methods such as bibliometric analysis [17] and analysis of meta – narratives in the literature [18]. These other methods have been stated to rely "on quantitative methods to identify keyword frequencies and the changing trends of keyword associations" [19]. To further justify our choice of review, aside from systematic review being known for its transparency and being an evidence-based approaches that identify key scientific contributions to a field and differ from narrative reviews by adopting a replicable, scientific and transparent process (Tranfield et al. [16]), it also allow researchers to study the strength of the published evidence while still attempting to remain as natural and unbiased as possible. This is possible because of the inherent evidences in the review analysis and synthesis. When facts are presented, scientific conclusions becomes easier. Though the process may be cumbersome and detailed, systematic review, according to Tranfield et al. [16], is still acknowledged as the most efficient for identifying and evaluating literatures. In addition, systematic reviews, aside from not only being important for advancing the field of study, it is also a veritable tool for advancing and informing management practice [1, 20].

2.1 Systematic Literature Review Approach

In the process of reviewing the empirical evidences in the existing literature, systematic review was considered as best and comprehensive option for identifying the gaps in ethic in stakeholder engagement, this was also buttressed by Tranfield et al. and Booth et al. [16, 21]. This review was done with the protocol in Fig. 1.

Stage I–Planning the review

Phase 0 - Identification for the need for a review

Phase 1 - Preparation of a proposal for a review

Phase 2 - Development of a review protocol

Stage II–Conducting a review

Phase 3 - Identification of research

Phase 4 - Selection of studies

Phase 5 - Study quality assessment

Phase 6 - Data extraction and monitoring progress

Phase 7 - Data synthesis

Stage III–Reporting and dissemination

Phase 8 - The report and recommendations

Phase 9 - Getting evidence into practice

Fig. 1. Stages of a systematic review [Source: Tranfield, Denyer and Smart [16]]

For the purpose of this paper, Fig. 1 has been collapsed to the five steps below, which the review is subjected to:

- Identification of research;
- Selection of studies;
- Study quality assessment;
- Data extraction and monitoring progress; and
- Data synthesis

The process of arriving at the 18 document for final review is analytically captured in the flowchart.

Figure 2: Flow chart showing search and analytical *process*. This makes the process transparent and it confirms the rigour and reliability of the systematic review [16].

The coding structure for the systematic review follows the Tranfield protocols. The basis for the coding, amongst others, is to ensure that all vital information are well captured in the coding system and hence in the analysis. Microsoft excel was utilized for the coding step of the systematic review.

The 18 papers were subjected to the coding structure (as in Fig. 2) utilization of Microsoft excel made the coding easy as well as the analysis. The coding uses the bibliometric details such as year of publication, details of authors, journal and article title. In addition, further categorization in terms of the study sector, author's nationality

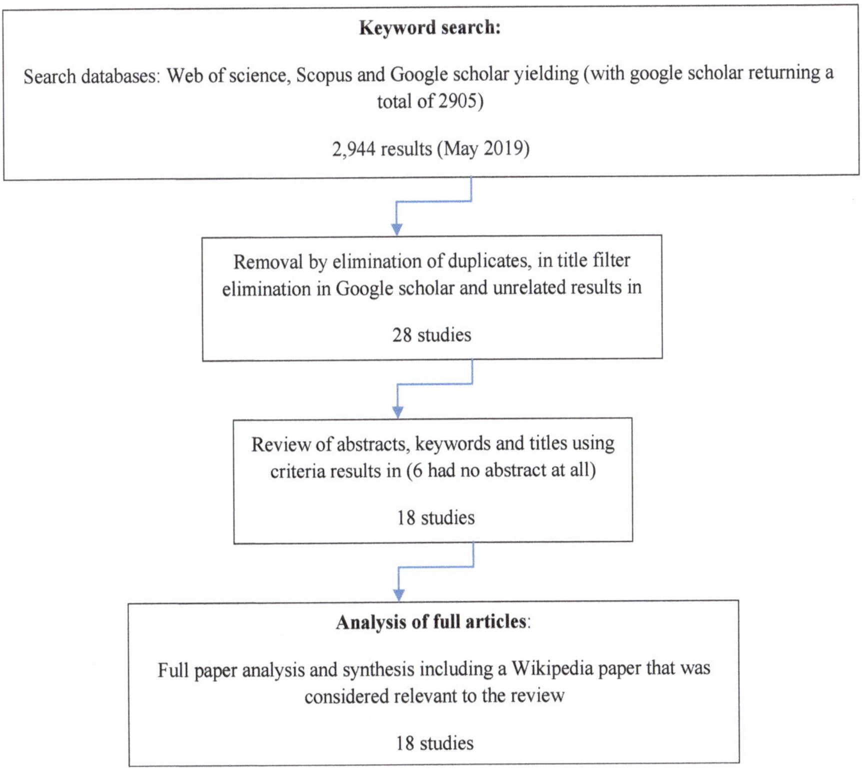

Fig. 2. Flow chart showing search and analytical process

and in some cases, sample size of the research. This sometimes makes it easy to observe saturation trend across research articles. These 28 studies that scaled through to full analysis and synthesis also scaled through the inclusion and exclusion, as shown in criteria when subjected it.

The review commenced with identification of research. In order to increase the credibility of the research, we did not impose any time limitation or be limited to a time in the past. The result is all the databases could return. To also increase our review horizon and identify literatures, we did not limit our types of publications. The types returned were what the databases could return back during each search. Our search included reviews, journal articles, book chapters, conference papers. This is a standard practice in review studies, since these sources are considered 'certified knowledge' and enhance the results' reliability [22]. Our conclusion that this is the first study and review in ethics, stakeholder engagement, industry 4.0 and construction 4.0 was enhanced by the outcome of our search results. This further created the need to keep varying our search keywords (as most databases were returning zero results). In all the databases, the search strings "ethics" and "stakeholder engagement" and "industry 4.0" on one hand yielded zero result except for google scholar that brought 75 results but eventually gave zero when the "in title" option was used and the search strings "ethics" and "stakeholder engagement" and "construction 4.0" on the other, also yielded zero results from all the databases. To enhance our review quality, the following combination of search string was finally adopted "ethics" AND "industry 4.0" on one hand and "construction 4.0" were separately searched for the review. We applied the search string to the titles, keywords and abstracts of publications in academic databases. In particular, we used Scopus, Web of Science and Google scholar databases to search for all publications whose topics cover at least one keyword from the selected sets of search string. These databases were chosen for their known broad and multidisciplinary scope, their popularity in academia and are reputed for being the most commonly used sources for review studies [23, 24].

2.2 Inclusion and Exclusion Criteria

To arrive at the 18 studies for review, the following inclusion and exclusion criteria were used. These are captured in table below (Table 1):

Table 1. Table showing inclusion and exclusion criteria

Exclusion	Inclusion
No abstract	Articles
Unrelated publications	Book chapters
	Conference paper
	Review paper
	Workshop output
	Wikipedia

3 Results, Findings and Discussion

The review, as previously suggested, brought out series of analytical outputs that would not just determine the future direction of research in the field of ethics, industry 4.0 and construction 4.0 (ESEI&C 4.0), it also brings to the fore, relational and deductive researches that will extensively contribute to both practice and academics and may lead to rapid development in the field. To quantitatively confirm that development and growth in both industry 4.0 and construction 4.0 has not been studied in relation to the needed ethics and stakeholder engagement approaches, we would begin with a frequency analysis of our researches in the fields.

3.1 Frequency Analysis

The frequency analysis is a veritable tool to projecting, descriptively and graphically, the pattern of the review over the past years. This can be an indication of either growing or declining interest in the discourse of ethics in stakeholder engagement, industry 4.0 and construction 4.0.

The search results from the databases brought the earliest time studies were conducted in the field to be 2015, which is also a confirmation that both industry 4.0 and construction 4.0 are recent additions into research. From Fig. 3: *Frequency Analysis – Linear* below, we can also see that it is an emerging field. The r-square value of 0.6563 is an indication of its still low emergence.

Fig. 3. Frequency analysis – linear

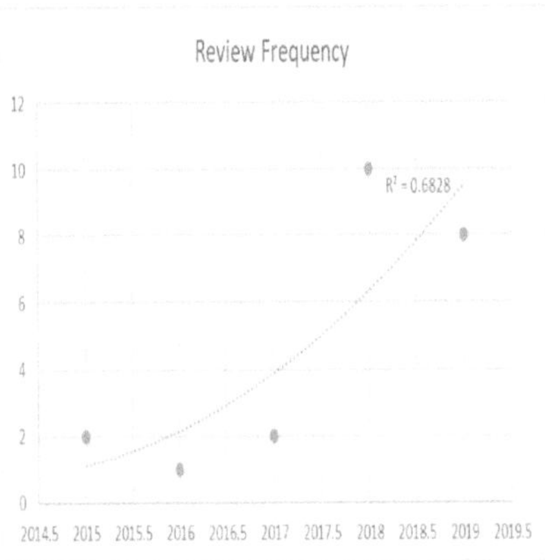

Fig. 4. Frequency analysis – polynomial

Above notwithstanding, if we attempt an analysis of Fig. 4: *Frequency Analysis – Polynomial*, we would observe that the r-square is higher than Fig. 3: *Frequency Analysis – Linear*. This is an indication that the frequency is more polynomial than linear. This can be attributed to the inconsistency in the interest researcher have shown in the field. The highest study of 10 studies was in 2018 which was a leap jump from 2 in 2017. Though 2019 has not ended we can't predict what the 2019 pattern will end up with (the authors expect an increase because there are already 8 studies in less than half of the year). The pattern that gave rise to a higher polynomic r-square vale can be seen from studies from 2015 to 2018 (Fig. 5).

The authors also applied the moving average as shown in Fig. 5 to indicate that though the field is emerging, the interest researchers are showing in the fields is still low - with the lowest average being only one study in 2016 and highest of nine studies in 2019. (Kindly note that these studies are predominantly in education and are not in ethics field). This makes the subject area analysis necessary to strengthen our conclusions.

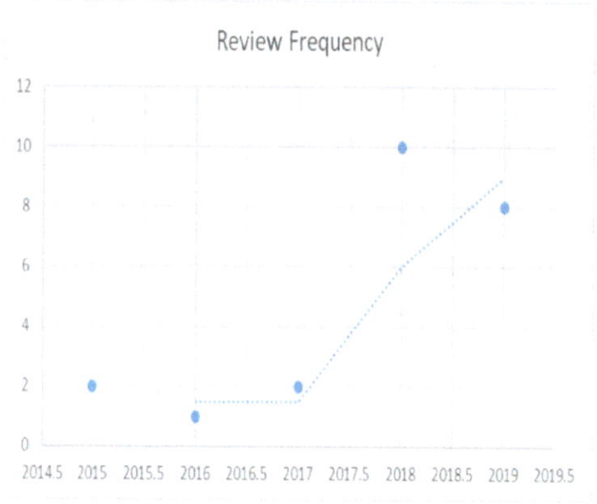

Fig. 5. Frequency analysis – moving average

Table 2. Review analysis of subject areas

Review subject area	Frequency
Advanced process control	1
Automation	1
Construction 4.0	1
Cybersecurity	1
Construction (dam & steel)	2
Earthmoving machinery	2
Education	3
Hunan-computer interaction	1
Innovation	1
Robotics, AI etc.	1
Sustainable development	1
Not defined subject area	8
Total	**23**

3.2 Subject Area Analysis

With the frequency analysis in 2 above, there is the need to analyse further and see the subject areas researchers have been studying or expanding the frontiers of knowledge in relation to how ethics and stakeholder engagement are being applied in the field of industry 4.0 and construction 4.0. This will further show how the field is developing and how researchers are spreading the application or implementation of industry 4.0 and construction 4.0 to various fields of study and by implication its level of globalization.

Table 2*: Review Analysis of Subject Areas* shows the breakdown of subject areas that both industry 4.0 and construction 4.0 are being studied within the context of the review. The pictorial analysis of this table can be seen in the figure below

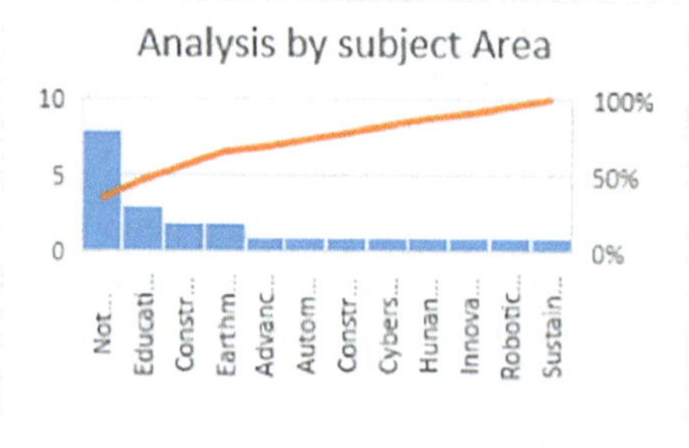

Fig. 6. Subject areas analysis

Both Table 2*: Review Analysis of Subject Areas* and Fig. 6*: Subject Areas Analysis* give a clearer pattern of emerging subject fields being studied. The highest study goes to an area the authors could not readily attribute to a subject concern hence the tag "undefined". Let's quickly add that the publications without abstracts are included in this categorization. The second in the series is the academic or education field. This predominantly focus on amending curriculum and or preparing academics with the probable challenges envisaged by the movement to fourth industrial revolution. This can be seen in the following included discus: Learning Framework in the Industrial Age 4.0 in Higher Education [25], Vocational Education and Training of construction machinery operators in Spain (Corral, [26]), On a Frame Work of Curriculum for Engineering Education 4.0 [27], Fourth Industrial Revolution and the future of Engineering: Could Robots Replace Human Jobs? How Ethical Recommendations can Help Engineers Rule on Artificial Intelligence [28]. These titles clearly show the focus of the study in the educational field. This brings us to the second conclusion that industry 4.0 and construction 4.0 study requires expansion of fields and subject areas. So far, within the limits of this analysis, industry and construction 4.0 studies have been restricted to only eleven clearly defined fields. It may interest us that none of this studies is ethics or stakeholder engagement related.

4 Conclusion

From the forgoing, the following conclusions can be derived from the systematic review study:

1. That researchers have not yet shown interest in ethics and stakeholder engagement in both industry 4.0 and construction 4.0. This therefore calls for concerted efforts at establishing an ethical approach to both fourth industrial and construction revolutions. Further studies in this field may reveal the path to rapid growth and replication into other sectors, lesson learnt can be studied as an ethical approach and developmental tool [29].
2. From the coverage of the subject areas, it can be seen that the highest defined interest is in the education filed and has more to do with curricular and framework setting. This is also a clarion call for industrial and construction researchers and practitioners to look deeper into studies that will enhance the rapid development of industry and construction 4.0 by bringing out the ethics of engagement in core industries and construction with a view to enhancing rapid development and expansion of the fourth industrial and construction revolutions.

References

1. Okedara Kamil, E.O., Chan, P., Collinge, B.: Ethics in stakeholder engagement for successful organizational and Project Management: A systematic Review, British Academy of Management (2019)
2. Figge, F., Hahn, T., Schaltegger, S., Wagner, M.: The sustainability balanced scorecard - linking sustainability management to business strategy. Bus. Strateg. Environ. **11**(5), 269–284 (2002)
3. Tripathi, K., Agrawal, M.: Competency based management in organizational context. Glob. J. Financ. Manag. **6**(4), 349–356 (2014)
4. Namada, J.M.: Organizational learning and competitive advantage. In: Handbook of Research on Knowledge Management for Contemporary Business Environments, p. 19 (2018)

5. Sanchez, R.: Knowledge Management and Concepts for Theory and Practice, pp. 29–30 (2006)
6. Zur Muehlen, M.: Organizational management in workflow applications – issues and perspectives. Inf. Technol. Manag. **5**(3/4), 271–291 (2004)
7. Lele, A., Wikipedia: Industry 4.0. Wikipedia (2019). https://en.wikipedia.org/wiki/Industry_4.0. Accessed 19 May 2019
8. Wikipedia: Industry 4.0. Wikipedia (2019). https://en.wikipedia.org/wiki/Industry_4.0. Accessed 19 May 2019
9. Oesterreich, T.D., Teuteberg, F.: Understanding the implications of digitisation and automation in the context of industry 4.0: a triangulation approach and elements of a research agenda for the construction industry. Comput. Ind. **83**, 121–139 (2016)
10. Lewis, P.V.: Defining 'business ethics': like nailing jello to a wall. J. Bus. Ethics **4**(5), 377–383 (1985)
11. Robin, D.: Toward an applied meaning for ethics in business. J. Bus. Ethics **89**(1), 139–150 (2009)
12. Aminian, E., Stewart, I., Fenn, F.: Human research ethics – is construction management research concerned? Constr. Manag. Econ. **35**(11), 665–675 (2017)
13. Noland, J., Phillips, R.: Stakeholder engagement, discourse ethics and strategic management. Int. J. Manag. Rev. **12**, 39–49 (2010)
14. Arnstein, S.R.: A ladder of citizen participation. J. Am. Plan. Assoc. **35**(4), 216–224 (1969)
15. Ortas, E., Álvarez, I., Garayar, A.: The environmental, social, governance, and financial performance effects on companies that adopt the United Nations Global Compact. Sustainability **7**(2), 1932–1956 (2015)
16. Tranfield, D., Denyer, D., Smart, P.: Towards a methodology for developing evidence-informed management knowledge by means of systematic review. Br. J. Manag. **14**(3), 207–222 (2003)
17. Pollack, J., Adler, D.: Emergent trends and passing fads in project management research: a scientometric analysis of changes in the field. Int. J. Proj. Manag. **33**(1), 236–248 (2015)
18. Padalkar, M., Gopinath, S.: Six decades of project management research: thematic trends and future opportunities. Int. J. Proj. Manag. **34**(7), 1305–1321 (2016)
19. Chan, P.W., Ejohwomu, O.: How Does Project Management Relate to Productivity? A Systematic Reviewed of Published Evidence, p. 40, July 2018
20. Cummings, S., Daellenbach, U.: A guide to the future of strategy? Long Range Plann. **42**(2), 234–263 (2009)
21. Booth, A., Papaioannou, D., Sutton, A.: Systematic Approaches to a Successful Literature Review. SAGE Publications Ltd, London (2013)
22. Saggese, S., Sarto, F., Cuccurullo, C.: Evolution of the debate on control enhancing mechanisms: a systematic review and bibliometric analysis. Int. J. Manag. Rev. **18**(4), 417–439 (2016)
23. Tachizawa, E.M., Wong, C.Y.: Towards a theory of multi-tier sustainable supply chains: a systematic literature review. Supply Chain Manag. **19**, 643–653 (2014)
24. Beske-Janssen, P., Johnson, M.P., Schaltegger, S.: 20 years of performance measurement in sustainable supply chain management – what has been achieved? Supply Chain Manag. **20**(6), 664–680 (2015)
25. Winanti, Gaol, F.L., Napitupulu, T.A., Soeparno, H., Trisetyarso, A.: Learning framework in the industrial age 4.0 in higher education. In: 2018 Indonesian Association for Pattern Recognition International Conference (INAPR), pp. 227–232 (2018)
26. Corral Álvarez, E.: Vocational education and training of construction machinery operators in Spain. Carreteras **4**(222), 68–72 (2018)

27. Jeganathan, L., Khan, A.N., Raju, J.K., Narayanasamy, S.: On a frame work of curriculum for engineering education 4.0. In: 2018 World Engineering Education Forum - Global Engineering Deans Council (WEEF-GEDC) (2018)
28. Hoeschl, M.B., Bueno, T.C.D., Hoeschl, H.C.: Fourth industrial revolution and the future of engineering: could robots replace human jobs? How ethical recommendations can help engineers rule on artificial intelligence. In: 2017 7th World Engineering Education Forum (WEEF), pp. 21–26 (2017)
29. Ching, J.T.W.: To assess the feasibility of constructing "Developmental Relationships" as a relevant mentoring model for the diploma in business & social enterprise (2011)

An Assessment of Risks Associated with the Use of Second Hand Tyres in Zambia from the Perspective of Enforcement Authorities

Bridget Kamanga[1], Erastus Mwanaumo[1(✉)], Nathan Sefuka Musonda[2], and Mwewa Mambwe[1]

[1] Department of Civil and Environmental Engineering, School of Engineering, University of Zambia, Great East Road, Lusaka, Zambia
erastus.mwanaumo@unza.zm

[2] School of Post Graduate Studies, University of Lusaka, Lusaka, Zambia

Abstract. The study evaluated the effects of driving second hand degraded tyres and their contribution to accidents in Zambia. A triangulation method was adopted which consisted of case studies on the dangers of using second hand tyres and the use of secondary data from Zambia Police. Structured interviews were applied and a questionnaire was distributed randomly to a sampled population of 46 respondents from distributed 66. Data from Zambia Police and Road Transport and Safety Authority was used to analyze recorded accidents that occurred in the period 2006–2015. Results revealed that users and second hand tyre dealers are ignorant of the tyre quality expectations and that they had never been inspected by Zambia Bureau of Standards which did not have standards on the use of second hand tyres at the time of the study. The study suggested the need for authorities to improve safety for all road users by ensuring that imported second hand tyres meet the manufacturer's minimum recommendation.

Keywords: Accidents · Pressure · Prevention · Tyre · Wear

1 Introduction

Road traffic accidents have been identified among the leading cause of death and disability in the world. The World Health Organization (WHO) described traffic accidents as 'hidden epidemics' with a forecast to be the 5th leading cause of death worldwide and the second leading cause of disability in many developing countries by 2030 [1]. In the Africa region, ten out of over 50 countries account for roughly 70% of all road traffic deaths [2]. These include Nigeria, Ghana, Tanzania, Kenya, Uganda, South Africa, Republic of Congo, Madagascar, Ethiopia, and Mozambique. Consequently, the WHO declared the period between 2011 and 2020 as a 'Decade of Action for Road Safety' [2]. In Zambia, the epidemic is growing. World Health Organisation data shows that death from traffic accidents increased by 20% between 1990 and 2002 [3].

Motor Vehicle Crashes are the leading causes of injuries and disabilities in Zambia today [4]. The age group laying between 16 to 35 years had become highly susceptible to

C. Aigbavboa and W. Thwala (Eds.): CIDB 2019, *The Construction Industry in the Fourth Industrial Revolution*, pp. 588–598, 2020.
https://doi.org/10.1007/978-3-030-26528-1_60

crashes. The road traffic accidents are estimated to be costing the nation millions of kwacha as was seen from the reports obtained from Zambia Police and Road Transport and Safety Agency [5]. The report further stated that some of these accidents were caused by reckless driving such as over-speeding, driving extremely slowly, drunken drivers, inexperienced drivers, tyre bursts, and ignorance of drivers. Most of these causes of accidents could be avoided through intervention of government measures and discipline among road users. The study sought to understand the extent to which the number of accidents attributed to the use of second hand or defective tyres, and if the state agencies inspect the condition of tyres both at vehicle inspection centres and at accident scenes.

2 Literature Review

According to Black Circles [6] on the composition of tyres and their identification parameters, tyres use natural and synthetic rubber as their primary raw material. Other materials include Sulphur, Zinc oxide, and carbon black which are just a few to mention [6]. At a glance, all tyres fitted on vehicles have a serial identification number. This is used to determine the age of the tyre [7]. The date of manufacture of the tyre is determined by the last four numbers to the nearest week. For example, a tyre with serial identification of 4809 was manufactured in the 48th week of 2009.

All tyres being manufactured are made with a code that tells the maximum load capacity of the tyre. Nonetheless, Osueke and Uguru-Okorie [7] and Murray and Lopez [1] on incorrect load and speed indices, stated that there is a tendency of the illegal use of the wrong tyres with insufficient load or speed indices. One of the causes of accidents has been ignorance of drivers on the load capacity of the vehicles originating from the tyres [7]. The speed indices indicated on the tyres tells a lot on how much speed the vehicle can be subjected to. It also helps the driver to determine what type of a tyre to use for the vehicle so that it matches with the speed of the vehicle.

Further, the maximum load for vehicles should not be greater than the tyres maximum load [1] which is indicated on the sidewalls of the tyre. The maximum load capacity of the tyre is assumed to be operating at the proper tyre inflation. Hence, it is not recommended to mix tyres on a car with different speed ratings as postulated by the Road Traffic and Safety Agency in the Lusaka Times [4].

According to Manas et al. [8] bad inflation pressure of tyres either under-inflation or over-inflation causes tyres to wear easily and increases poor vehicle handling by drivers. Over or under-inflated tyres traditionally manifested classic wear patterns. When a person pumps his/her tyre with low pressure, it causes the tyre to collapse slightly which causes the weight of the vehicle to be carried on the tyre sidewalls than on the center of the tread [6]. This results in the tread wear to be greater towards the sides than in the center. Hence, it is important to make sure that one maintains the correct inflation pressure as indicated on the tyres. This helps to keep vehicle handling and braking at its best, improves fuel efficiency and tyre life.

In developed countries like Japan, the law stipulates that automobiles must pass an automobile safety inspection and are allowed to run on Japanese public roads only when the tread depth of the tyres is 1.6 mm or larger [9]. For tyres seeming to be brand new, those above 6 years from the date of manufacture should not be comfortably used because of the ageing effect and the high risk of causing accidents. More or so, in Norway and most European Union countries, an extensive programme of periodic motor vehicle inspection including the state of the tyres which are measured and tested to verify their roadworthiness [10]. While in the United Arab Emirates, it was found that 5%–9% of traffic accidents are caused by mechanical defects, with the state of tyres, brakes and lights often being the major factors [11]. A set of inspection surveys help to collect information on the environmental parameters that could affect the efficiency of tyres depending on the geographical weather conditions and how they affect the tyres.

A study conducted in South Africa, by Vogel and Bester [12] indicated that from an average of 404 accident reports analysed 75.4% were due to human errors, 14.5% were risk factors due to environmental causes and 10.2% were due to vehicle factors. The vehicle factors comprised of faulty brakes and tyres. They further indicated that the road traffic burden was becoming a worsening global disaster.

In Zambia, there is no known legislation which restricts the importation of second hand tyres. Only 28 countries, representing 416 million people (7% of the world's population) have adequate laws that address the main five risk factors which are speed, drink-driving, helmets, seat-belts and child restraints [3]. Tyres are seen lying alongside the roads in direct sunlight. These tyres have different quality standards and brands. They range from a low grade to an average high grade [13].

Moreover, there is also an increase in the use of second hand tyres which easily wear without close monitoring. According to the report by Zambia Bureau of Standards [14] of the 4, 322 motor vehicles inspected in Lusaka 1,895 (44%) vehicles had to pay penalties at the point of entry into the country due to their non-compliance to the standards of roadworthiness including tyres. The second hand tyres have also increased because of retreading which involves grinding down the surface or casing of a worn tyre until it is smooth and then gluing a new veneer of tread onto it [15]. There are a number of dangerous chemicals in tyres amongst them is latex and synthetic rubber. This means that they cannot be burnt safely and cannot easily be disposed of. Additionally, Tyre separation is the most common cause for tyre blowouts. This is as a result of insufficient bonding of the tread in the tyres. Because of this, it causes the tyre to slowly separate. Some other conditions under which a vehicle tyre can cause an accident include: inability to grip road in weather; over inflation or under inflation; poor traction or tread design; cheap road construction having potholes; used or the prolonged use of tyres; tread separation; weak outer tyre structure and blowouts according to Osueke and Uguru-Okorie [7].

3 Research Methodology

A triangulation method was adopted which consisted of case studies on the dangers of using second hand tyres, the use of secondary data from Zambia Police, and use of interviews, observations and questionnaire survey. This approach was used to validate the data obtained by means of secondary sources from Zambia Police on accident reports and the structured interviews with Road Transport and Safety Agency officers and Police Traffic Officers. The triangulation approach was also adopted by Bun [16] and Ponboon [17] in which they both assessed the factors of road traffic accidents.

Triangulation approach was adopted in order to increase credibility and validity of the research [18]. A self-administered questionnaire was distributed to some selected government departments. The questionnaire was self-administered in order to reduce bias, and allow anonymity and honesty in answering the questions. The target group being the sampled population were divided into four groups including: road users and bus drivers; tyre resellers and tyre menders; Road Transport and Safety Association and Zambia Police Traffic officers; and customs officers and Zambia Bureau of Standards Inspectors. In depth discussions were also carried out with Zambia Revenue Authority custom officers at Chanida border and Zambia Bureau of Standards concerning the risks involved in driving passenger vehicles fitted with second hand tyres.

A total of 10 Traffic officers from Zambia Police headquarters participated in the research through interviews and answering the questionnaires in a rather more detailed approach. Only 66 questionnaires were distributed of which 46 responses were collected from government departments such as Police Traffic officers, Road Transport and Safety Agency agents and motor vehicle users. Only 3 inspecting officers from Zambia Bureau of Standards and 6 Road Transport and Safety Agency traffic officers and 3 from research and statistics office at Road Transport and Safety Agency were subjected to an in-depth interview. Secondary data was also obtained using data from Zambia Police and Road Transport and Safety Agency on the accidents caused by tyres that occurred in the pass 4–5 years. The study area was in Lusaka province as it recorded the maximum import of second hand vehicles and tyres.

4 Findings

From secondary data obtained and reviewed, the statistical survey on the number of second hand motor vehicle importation, it was found that Zambia was amongst the top third importer of second hand vehicles in Southern Africa in the years 2010 and 2011. The results obtained indicate that there has been a rapid increase in importation of second hand used cars, which are normally fitted with second hand tyres (See Table 1). The data also give a general overview of the number of imported second hand tyres imported into the country.

Table 1. Vehicle importation statistics in Southern Africa between 2010 and 2011

Name	Total	
	2010	2011
Angola	8	1
Botswana	3151	10256
Congo (DRC)	1227	2613
Lesotho	67	372
Malawi	1002	1656
Mozambique	3524	7890
Namibia	298	591
South Africa	0	41727
Swaziland	808	521
Zambia	**3554**	**9815**
Zimbabwe	1167	1725

4.1 Number of Accidents and Fatalities

According to the Road Traffic and Safety Agency on accident statistics obtained during interviews, accidents and fatalities are linked to the total number of registered motor vehicles. An estimate of every 10,000 registered motor vehicle indicators shows the ratio of accidents and fatalities to the number of registered motor vehicles which in turn establishes the extent to which road crashes contribute towards public health problems. Table 2 shows that an increase in the number of accidents per 10,000 motor vehicles and number of fatalities per 10,000 motor vehicles from 135 to 136 and 9 to 10 respectively. As at September 30th, 2015 the cumulative number of registered motor vehicles stood at 636,378. By looking at these accident statistics, one can tell that there has been a shady increase in the number of vehicles entering into the country.

Table 2. Number of accidents and fatalities per 10,000 vehicles.

Year	Number of vehicles registered	Number of accidents	Number of fatalities	Number of accidents per 10,000 vehicles	Number of fatalities per 10,000 vehicles
2015 (1st quarter)	620,758	7,782	465	125	7
2015 (2nd quarter)	636,378	8,617	563	135	9
2015 (3rd quarter)	651,280	8,839	641	136	10
Jan to Sept 2015	651,280	25,238	1,669	388	26

Source: Road Transport and Safety Agency [5]

Additionally, during the interview at Road Traffic and Safety Agency, road traffic accident factors were categorised according to human error; pedestrians; mechanical failure; road defects; and weather conditions. From this categorisation, contributory factors caused by mechanical failure highlighted that tyres were ranked highest in causing road traffic crashes. In the first quarter of 2015 tyres accounted for 66 crashes out of 132. However, by the third quarter, tyres contributed 62 crashes out of 122 from the motor vehicle mechanical failure related cases. The second in ranking was motor vehicle brake failure accounting for 45 cases of 122. This is illustrated in Table 3.

Table 3. Number of road traffic crashes due to motor vehicle mechanical failures.

Contributory factors	3rd quarter	1st quarter	2nd quarter	3rd quarter
Mechanical failure	2014	2015	2015	2015
Brakes	28	37	33	45
Tyres	**76**	**66**	**49**	**62**
Steering	10	11	5	2

(*continued*)

Table 3. (*continued*)

Contributory factors	3rd quarter	1st quarter	2nd quarter	3rd quarter
Mechanical failure	2014	2015	2015	2015
Springs	9	4	2	3
No front lights	7	1	2	0
No rear lights/reflection	1	0	1	0
Unattended vehicle running away	4	10	7	4
Smashed windscreen	2	1	0	0
Vehicle overloaded	4	2	3	6
Total	141	132	102	122

Source: Road Transport and Safety Agency [5]

4.2 Knowledge on Dangers of Using Second Hand Tyres

From the administered questionnaire to understand whether or not users are aware of the dangers of using second hand tyres, it was discovered that 64.30% of road users including bus drivers indicated that they did not know the dangers of importing or driving vehicles with second hand tyres. Meaning more than 50% of the sampled population are at risk of buying these second hand tyres (See Fig. 1).

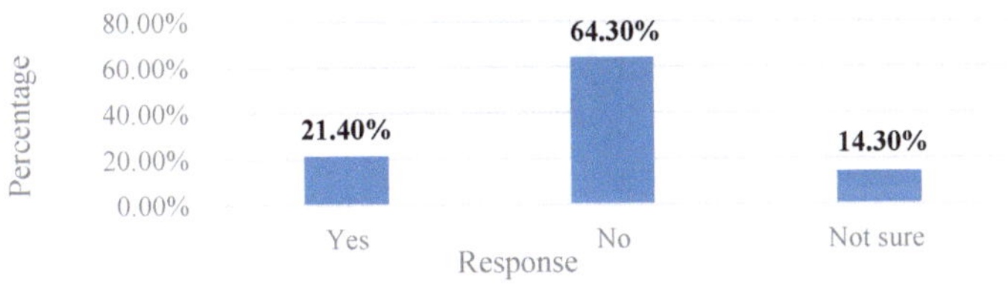

Fig. 1. Percentage of user with knowledge on the dangers of importing of using second hand tyres

Main Causes of Tyre Bursts. Road users including bus drivers interviewed mentioned that overloading was the main causes of tyre bursts. Other reasons for tyre bursts discovered from the interviews are inappropriate tyre inflation, usage of worn out tyres, extreme temperatures and exceeding tyre limit.

4.3 Understanding the Writing on the Tyres

About the level of awareness on the understanding the meaning of the writings on the tyres, 72% of the respondents admitted that did not understand. Only 14% had an understanding of the meaning of the writings (See Fig. 2). On the legal limit of 1.6 mm tread wear, only 7% indicated understanding of the limit as shown in Fig. 3.

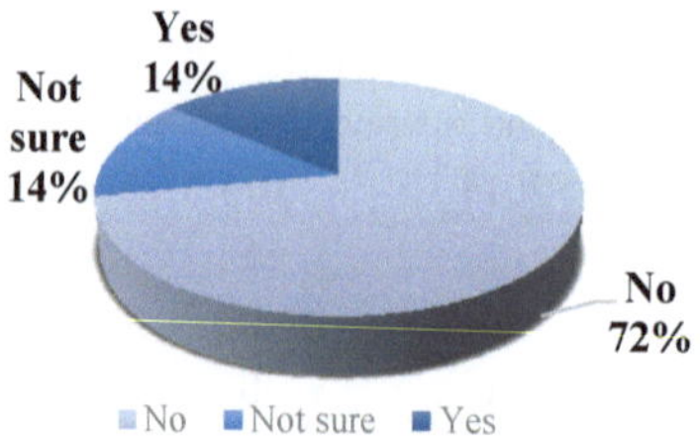

Fig. 2. Percentage of respondents who understand the writing on the tyres

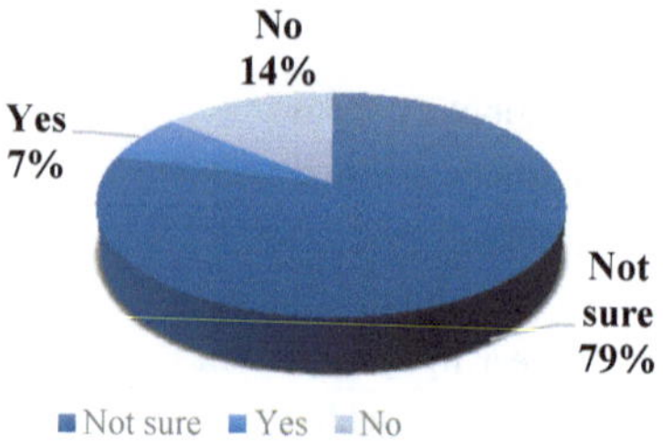

Fig. 3. Percentage of understanding the legal limits of tyres

4.4 Provision of Professional Advice When Selling Second Tyres

On the provision of advice when selling second hand tyres, 61% of the 46 respondents mentioned that tyres dealers did not provide professional advice when selling the tyres. This is because they too, did not understand the markings on the tyre. They would only ask for which tyre size the user is interested in (See Fig. 4).

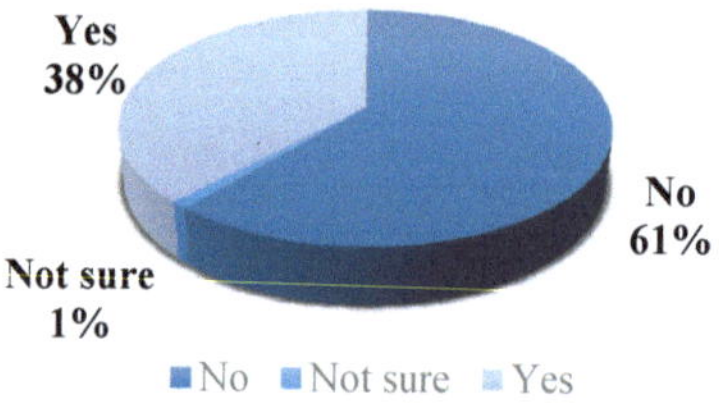

Fig. 4. Percentage values of whether car dealers provide professional advice when selling tyres

4.5 Possible Risks Associated with Second Hand Tyres

The possible risks associated with the second hand tyres were identified through in-depth interviews with the sampled population from Road Transport and Safety Agency. It was highlighted that the following risks could highly contribute to motor vehicle accidents due to the use of second tyres:

Inability to Grip on the Road Due to Wet Weather Conditions. Treads easily finish making the surfaces smooth and easy for the vehicle to skid off the road at high speed.

Tread Separation. If the tyre was retreaded and bonding was not properly done, the tyre easily separates at high motion.

Over Inflation or Under Inflation Causing a Tyre to Rapture. Over inflated tyres bulge, forming cracks in the middle causing burst at high temperatures. Similarly, when under-inflated, the rim casing rests on the tyre and easily wears causing outward cracks.

Ignoring the Expiring Date of Tyres. It was discovered that the majority of the tyres which were being sold had the date of manufacture beyond the recommended 4 years [8] which is the maximum performance threshold.

Chemical Reactions as a Result of the Tyres Exposure to the Atmosphere or Direct Sunlight. It was discovered that tyre resellers are not knowledgeable of the proper storage of tyres or their direct exposure to sunlight. Resellers left their tyres in open direct sunlight which had great effect on the tyre life.

Further, according to the statistics shared during the interviews with the Road Traffic and Safety Agency, at least 275 recorded accidents between the years 2003 to 2014 were related to tyres as a major contributing factor to road traffic accidents.

4.6 Interview Summary with Zambia Police and Road Transport Safety Agency

In order to understand the extent, the law enforcement record data on accidents caused by the prolonged use of second hand tyres, the law enforcers indicated that there was no record to this effect. This was because there are very few causes of accidents reported pointing to the prolonged use of tyres, hence they did not check the usage apart from accident causations.

To understand the extent to which the enforcement agents check for tyre fitness, it was discovered that the fitness was determined at both vehicle inspection centres and at an accident scene. Additionally, they mentioned that the guidelines provided in the Roads and Road Traffic Act Cap 464 of the Laws of Zambia on the physical condition of the tyre, was followed.

5 Discussion

In this study on the assessment of factors associated with the use of second hand tyres on light vehicles in Zambia, the critical factor was the link between accidents and second hand tyres having the minimum recommended tread depth and expired tyres. Generally, tyres determine the stability and safety of vehicles on the road according to Murray and Lopez [1]. Since tyres are either designed for cold regions or for hot regions. In Zambia, the weather is different from other regions and temperature range between 25 and 35° all year round [4]. From the findings, it was discovered that some road users do not check for whether the tyre is meant for the hot or cold season. Bun [16] affirms the notion that when a person uses tyres meant for the cold season in hot weather, the tyre may rapture. Should the tyre rapture when the vehicle is in high speed, it may lead to a traffic accident. Results also revealed that authorities need to improve safety of all road users by ensuring that imported second hand tyres meet the manufacturer's minimum recommendation by checking if the treads are clearly marked and checking their expiring dates.

In Zambia, most vehicles imported are from Japan, with a small volume being imported from Singapore, United Arab Emirates (Dubai), England (U.K.), and U.S.A [14]. These vehicles come with second hand tyres which are manufactured according to the patterns they are made from. However, the study revealed that quality of the tyres is not checked at point of entry and only checked for certificate before issuing a clearance certification, the valuation of the vehicle, model and year of make as it determines the value of what has to be paid to the receiver of taxes. However, Emuze and Smallwood [19] postulated that inspections of imported tyres from the point of entry into the country need to be checked for quality and usability before they are certified for use in the countries.

Additionally, the Zambia Bureau of Standards reveled that they only inspected brand new tyres to check for the standards by measuring their tensile strength before they could be allowed to be sold or used on the roads. The standard which is used to test for tyres is called Pneumatic tyres specification ZF-437 [9]. However, they further confirmed that they did not check second hand tyres because the country had no standards for them at the time of the study. Standards on the use of second hand tyres are significant as they help the users know the acceptable years and life span before disposal, which gives the remaining tensile strength before the tyre becomes obsolete and ready for disposal. This notion was affirmed by Christensen and Rune [10] in which they said that when the standards are known, it is easy to enforce the requirements at the point of importation. Also, the local resellers of second hand tyres would be informed and advise the user on how long a tyre can be used before it becomes a risk factor to cause an accident.

Lack of enforcement and control on the importation of second hand tyres increases the chances of importing substandard second hand tyres. It is important that goods being brought into any country are controlled whenever they are imported. It is relevant if such legislation is put in place for controlling the importation of goods. This was affirmed by the Zambia Bureau of Standards and Road Transport and Safety Agency [5]. Further, the Zambia Police equally agreed to this view as was suggested by Osueke and Uguru-Okirie [7]. Zambia Bureau of Standards and Road Transport and Safety Agency officers admitted that they pay little or no attention in checking for quality of tyres at the border or during fitness tests.

The treads and the markings should be visible and the tread depth should not be less than 1.6 mm Yamashita and Yamashita [9] and Zambia Bureau of Standards [14]. When checking for wear and tear. The study discovered that this was not followed in most cases as most people and second hand tyre dealers were not aware. Other causes that were highlighted which may lead to an accident include under inflation of tyres, over inflation of tyres and little to no thread depths on tyres, poor tread design and tread separation, and prolonged of tyres. These factors were equally affirmed by studies conducted by Ponboon et al. [17], Schatz [13] and World Health Organisation [3]. Overall results revealed that motorists did not know the dangers of driving second hand tyres.

Regular checks on tyre dealers should be carried out and improved by the responsible authorities such as Road Transport and Safety Agency, Traffic Police and Zambia Bureau of Standards at various locations in the country forming part of the guideline for the selling of second hand tyres in the country.

6 Conclusion and Recommendations

The study concluded that the causes of road traffic accidents depend on a list of factors which can be broadly divided into: vehicle operator or driver factors; vehicle condition factors; road pavement condition factors and environmental factors. It can be concluded that there are several risks associated with driving vehicles fitted with second hand tyres as overall results revealed that users are ignorant of minimum tyre quality expectation. They do not understand the markings on the tyre, especially on how to check for the expiry date. In the end, they tend to buy expired tyres which are not road worthy. Results revealed that authorities need to improve safety of all road users by ensuring that imported second hand tyres meet the manufacturer's minimum recommendation by checking if the treads are clearly marked and checking their expiring dates.

The study recommends that the need for authorities to improve safety for all road users by ensuring that imported second hand tyres meet the manufacturer's minimum recommendations and hence restrict the importation of second hand tyres. Further recommendations from the study indicates that for quality purposes, there is need to introduce licenses for selling second tyres and conduct awareness on the markings and period of usage of the second tyres.

References

1. Murray, C.J., Lopez, A.T.: A Comprehensive Assessment of Mortality and Disability From Diseases, Injuries and Risk Factors in 1990 and Projected to 2020. Harvard University Press, Cambridge (1996)
2. World Health Organisation: Global Status Report on Road Safety. http://www.who.int/violenceinjuryprevention/roadsafetystatus/2009/
3. World Health Organisation: WHO Road Traffic Injuries: Fact sheet No. 358 (2013)
4. Lusaka Times.: Road Transport and Safety Agency Identifies major causes of accidents (2013). http://www.lusakatimes.com
5. Road Transport and Safety Agency: Third quarter accidents report for 2015, Statistics and Research Unit, RTSA, Lusaka, Zambia (2015)
6. Black Circles: Black Circles. http://www.blackcircles.com/general/history
7. Osueke, E.C., Uguru-Okirie, D.C.: The role of tyre in car crash, its causes, and prevention. Int. J. Emerg. Technol. Adv. **12**, 54 (2012)
8. Manas, M., Stanek, M., Pata, V., Manas, D.: Wear of tyre threads. J. Achiev. Mater. Manuf. Eng. **37**(2), 538–543 (2009)
9. Yamashita, M., Yamashita, S.: Dust resulting from tyre wear and the risk of health hazards. J. Environ. Prot. **4**, 509–515 (2013)
10. Christensen, P., Elvik, R.: Effects on accidents of periodic motor vehicle inspection in Norway. Accid. Anal. Prev. **39**(1), 47–52 (2007). Science Direct
11. Selim, M.Y.E., Maraqa, M.A., Hawas, Y.E., Mohamed, A.M.O.: Assessment of vehicle inspection and emission standards in the United Arab Emirates. J. Transp. Res. **16D**(4), 332–334 (2011)
12. Vogel, L., Bester, C.L.: A relationship between accidents types and causes, South Africa (2003)
13. Schatz, J.: On the road in Zambia. Lancet **372**(1637), 435–436 (2008)

14. Zambia Bureau of Standards (ZABS): ZABS increases inspections on second-hand car. Lusaka Times, Lusaka (2015)
15. Thomas, P.: The Ecologist. http://www.theecologist.org/archive.html/2007
16. Bun, E.: Road traffic accidents in Nigeria. Afr. Med. J. **3**(2), 34–36 (2012)
17. Ponboon, S., Bontob, N., Tanaboriboon, Y., Islam, M., Kanipong, K.: Contributing factors of road crashes Thailand: evidence from the accidents in-depth study. J. East. Asia Soc. Transp. Stud. **8**, 1–13 (2010)
18. Perumal, T.: Research Methods in Competitive Intelligence. Open University Malaysia (2010)
19. Emuze, F., Smallwood, J.: Construction motor vehicles accidents in South Africa. In: Proceedings of CIB W099 International Conference, 2012: Modelling and Building Health and Safety, Singapore (2012)

Professional Construction Health and Safety Agent (CHSA) Practice Issues in the South African Construction Sector

Claire Deacon(✉)

Nelson Mandela University, Port Elizabeth, South Africa
Claire.deacon@mandela.ac.za

Abstract. The introduction of professional construction H&S agents (CHSAs') in 2013 to the built environment professional (BEP) team are now practicing in the construction H&S space. The CHSAs' have specific contributions and actions that dovetail into the construction life cycle. The CHSA activities on projects are managed by clients, but more so by BEPs on projects. Despite the statutory requirements that govern the role of the CHSAs', a number of issues relate to the practice of the CHSA at professional and candidate levels on construction projects.

Research conducted among CHSAs' indicate BEPs' lack of H&S knowledge; tenders are given to non-professional CHS categories of registered persons. Collectively the findings indicate very high risks to projects, clients, and the public at large. The CHSAs' are further deemed to be at risk from the lack of support by the DoL and clients and in their professional capacity.

Keywords: Built environment professionals · Construction · Health and safety · Health and safety agents

1 Introduction

In 2013 the SACPCMP required persons practicing as the construction H&S clients 'agent' to register with the South African Council for the Project and Construction Management Professionals (SACPCMP) in a specified category of 'Professional Construction H&S Agent' (PrCHSA), or candidate CHSA (CanCHSA). The two categories will be referred to as the CHSA for ease of reference.

The roles and responsibilities of the CHSA are legislated in the Construction Regulations (CRs), and the SACPCMP [1, 3, 6]. The requirement was further reinforced by an exemption notice issued by the Department of Labour (DoL) in 2014. As such, each construction project requires such an appointment as part of the client team to provide expert H&S input to the professional team and the client. Such appointment is required to be registered with the SACPCMP [5].

However, there are currently fewer than 240 registered CHSAs', including candidates that have been registered with the SACPCMP. Given the low number of professionals available, one would expect that the professionals would be oversubscribed with work, as each project over the value of R40 million requires a CHSA to be

C. Aigbavboa and W. Thwala (Eds.): CIDB 2019, *The Construction Industry in the Fourth Industrial Revolution*, pp. 599–605, 2020.
https://doi.org/10.1007/978-3-030-26528-1_61

appointed on the project. As the group of CHSAs professionals are such a small community who interact together fairly often, a number of aspects were being continually raised by the CHSAs', that affect their practice. A number of themes emerged, that relate to the lack of projects and insufficient work to sustain practice, among other.

The concept and framework of the study was provided by the practice issues against the prescribed requirements to practice, among other.

2 Review of the Literature

2.1 Statutory Requirements

The practice of construction H&S has two sets of legislation that guides the sector. The SACPCMP has published through Act 48 of 2000 the registration criteria for those practicing H&S in the construction sector, at one level. The Construction Regulations (CRs) 2014 prescribe the requirements of the sector to only use those registered persons on projects at another level. The requirements are relatively prescriptive in the CRs, relating to the roles from the client, designer and contractors, and highly prescriptive by the SACPCMP [4, 6].

2.2 Registration Criteria for the Practice of Health and Safety with the South African Construction Project Management Professions Council

South Africa is one of a number of countries in Africa where professional registration of H&S professionals is required to practice. While such countries are not necessarily limited to the construction sector, no other countries in the world have such registration requirements [1].

The SACPCMP is one of six statutory councils that require BEPs to register in order to practice in their discipline. Each council regulates the registration requirements and monitors disciplinary aspects where there is risk to the public through registered persons not adhering to their scope of practice. The PrCHSA practices in all six stages of the construction life cycle with the client team [6, 7].

In order to meet the registration criteria, a CHSA is required to have a mixture of formal training, with a minimum of a diploma in H&S and a range of short courses. Project experience is required which is commensurate with the level of such formal training. A range of documentation is required to be submitted with each application. Two reports are required to be submitted that discuss successes and perceived failures relative to the following nine knowledge areas:

- Procurement Management;
- Cost management;
- Hazard Identification Management;
- Risk Management;
- Accident or Incident Investigation Management;
- Legislation and Regulations Management;
- Communication Management, and
- Emergency Preparedness Management [8].

Each application is reviewed by three assessors for compliance. Successful applicants are invited to an interview to verify their knowledge across the six stages by a group of peers [6].

There are two other H&S categories that are required to be registered, however these are generally appointed by the contractors, the criteria are similar, except the applicants write exams and are not considered professionals, but 'registered persons'. The titles for the groups are the construction H&S Manager (CHSM), and Officer (CHSO). The CHSM and CHSO work from stages four to six of the construction life cycle [8].

There are six construction stages that are most often used in the construction sector that have tasks undertaken by the various BEPs throughout a project, dependent upon their role. The stages, while not entrenched in law, can be related to in a broad sense, when reviewing the requirements of the CRs. Table 1 indicates the stages of the construction sector, as well as the three categories of H&S registration and the practice areas of each category.

Table 1. Project stages and practice areas per SACPCMP registered category of H&S.

Project stage		Category practice areas	
		CHSA	CHSO/CHSM
Stage 1:	Project initiation and briefing	✓	N/A
Stage 2:	Concept and feasibility	✓	N/A
Stage 3:	Design and development	✓	N/A
Stage 4:	Tender documentation and procurement	✓	✓
Stage 5:	Construction documentation and management	✓	✓
Stage 6:	Project close-out	✓	✓

2.3 The Construction Regulations

Any person, or organisation that employs persons, or does 'work', is required to comply with the Occupational Health and Safety Act (OHSA), No. 85 of 1993, and its regulations (RSA, 1993). The CR's were introduced originally through statutory promulgation in 2003 and amended in 2014. The 2014 amendments increased, inter alia, the requirements for the level of competence and accountability among construction stakeholders, which includes the client and the BEPs [1, 3].

An agent is defined by the CRs as a person who represents a client [2] and is responsible for guiding and ensuring that the scope of work relating to the project. The CRs imply that the CHSA is to be appointed by the client prior to the BEPs who would be managing the project.

The DoL published a guideline to the CRs where it states that the client may appoint a CHSA or a CHSM as their agent based on the scope and risk profile of the project [2].

3 Procurement

Procurement practices in South Africa in the public sector are set out by the National Treasury, against a number of finance management laws. Treasury supplies standard forms of contract for government for public sector, that are not only restricted to the construction sector. National Treasury closely monitors the procurement processes at all levels of government.

However, in a study at provincial and district levels, Deacon [1] reports that stages 1 and 2 do not result in the inclusion of the CHSA, as the activities really just are costing exercises, with BEPs reportedly having to 'thumb suck crazily' about project costs. The result in terms of stages 1 and 2 are blurred at best, and mostly an administrative exercise [1]. Smallwood and Deacon [8], indicate in their study that 27% of CHSAs are appointed at stage 1, none at stage 2; 18% at stage 3 and 27% at stages 4 and 5. Figure 1 indicates the stages at which each of the categories of construction H&S are appointed on projects. In most cases the CHSA is appointed from stage 4 to 6; but what is notable is the number of CHSOs that state they are appointed from stage 2 [1].

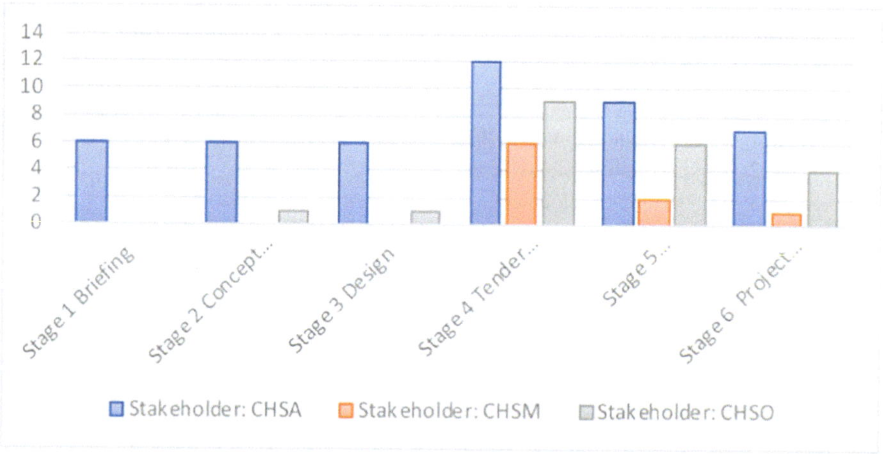

Fig. 1. Appointment of construction H&S categories across the 6 stages of work.

4 Competencies in the Built Environment

Smallwood and Deacon [8] cite Smallwood and Haupt who investigated the importance of 79 knowledge areas and 50 skills relating to the CHSA. The study resulted in 7 composite areas ranked in the following order: H&S project administration, design, law, management and management of parameters. Construction technology and technology, planning, and financial management, among other. Given the wide range of composite areas and statutory requirements, Deacon [1] cites findings where various stakeholders in the BEP and H&S categories were asked about their levels of confidence (LoC) about H&S and the stages of work. The CHSAs indicated a LoC, as did the CHSMs and CHSOs relative to their areas of work. The BEPs all indicated low levels of H&S LoC across the 6 stages. Figure 2 indicates the range of BEPs who participated in Deacon's study and their levels of confidence as reported [1].

4.1 Training

There is no specific construction H&S undergraduate training in South Africa for H&S practitioners in the main, and many who practice in the construction H&S have 'migrated' from other industries. As there is a requirement for experience in the sector as part of the application criteria, the CHSAs generally have many years of experience. As the criteria require the knowledge of the 6 stages of work, the CHSA is expected to have a specific level of competence in the area, as reported under heading 3 [8].

5 Research Methodology

The aspects explored relate to the transferring of the statutory framework into the general practices of H&S. Such practices relate to how the services of the PrCHSA are procured, and when, and the issues that are encountered in relationship to the BEPs who are managing projects.

A total of 104 invitations to complete an online questionnaire using Google Forms© were sent to a convenience sample that was determined by their CHSA status at the time of the survey. The sample included candidate CHSAs registered with The Association of Construction Health and Safety Management (ACHASM). The questionnaire consisted of 17 questions, 3 of which were close-ended and the balance open ended. The close-ended questions were 6-point Likert scale type questions. The response rate was 43%.

6 Research Findings and Discussion

A total of 80% of the Respondents were PrCHSAs and 20% were categorized as CanCHSAs.

Respondents indicated an average of 8 years of experience in the sector, which in most cases exclude the number of years prior to registration. In most cases the CHSAs would have between 3 and 7 years of experience in order to meet the minimum registration criteria.

Respondents indicated that they had worked in other industries prior to construction, namely: telecommunications, occupational health and medicine, emergency services and H&S auditing, environmental health, the South African Defense Force, mining, chemical, and forestry, among other. The finding supports the literature in terms of most CHSAs migrating into the sector.

Qualifications of Respondents ranged from Diplomas' in Safety Management (NADSAM), Degree in Medicine, Level 5 Paramedic, Bachelor of Technology in Safety Management and Environmental Health, Master of Science, National Diploma of Electrical Engineering, Project Management, and a PhD.

CHSAs mainly source their work from both the public and private sector, but Respondents indicated that they needed to obtain work from wherever they could.

Figure 1 indicates the Respondents indication of which stage they are appointed across the 6 stages. A total of 57% agree/strongly agree that they are appointed from stage 1, with 38% not in agreement of being appointed during the early stages. The outcome partially supports the findings by Deacon [1].

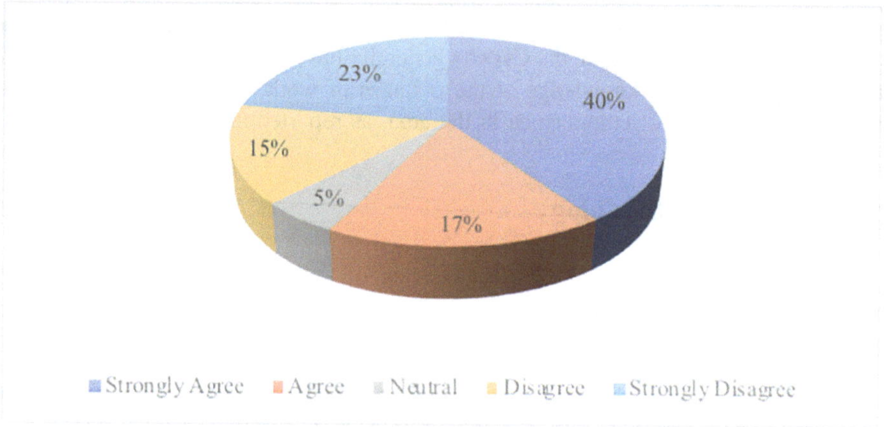

Fig. 2. The appointment of the CHSA is from stage 1.

A total of 85% of Respondents believe that clients expect the CHSA to accept responsibility for all stages, even when not appointed for the previous stages. The National Treasury tender documents are standardized and issued as such with minimal changes by client bodies.

A total of 85% of Respondents believe that the reason they don't have enough work is that projects are awarded to the lower levels of registered persons. As the DoL has issued a Guideline to the CRs (2014), CHSMs are able to apply for projects, which indicates that there will be fewer projects being awarded to the CHSA category.

A total of 87% of respondents indicating that the LoC of BEPs is inadequate for projects. The study by Deacon [1] indicated that the BEPs themselves indicated that their knowledge was very limited in terms of H&S. The findings, while not from BEPs themselves, is further confirmed by the H&S categories of the BEPs.

A total of 84% of respondents agree that only the CHSA categories should be accepted to represent clients on projects. The scope of work and level of liability carried by the CHSA categories requires competence and experience, which the lower levels of registered persons do not have. Project risk is increased when inexperienced persons are placed on projects.

7 Conclusions and Recommendations

The CRs require the CHSA to be appointed by the client, as the liability of a project rests on the shoulders of the CHSA. The requirements of the CRs and the SACPCMP scope of work, the PrCHSA should be appointed at stage 1, however, most only get appointed at stage 5, which increases project risk and liability for the client.

In most public sector tender documents the service level agreement will expect the CHSA to accept responsibility for stages 1–6, even though all the preliminary work has been completed with no H&S input.

Clients are noted as lacking the required knowledge relating to H&S and therefore prefer to divest themselves of the responsibility of the H&S issues on a project.

It is strongly recommended that ongoing training and development is undertaken for the BEPs and clients. Further awareness is required to change the procurement processes and ensure the responsibilities are aligned to the stage of appointment, and project specific. Ongoing awareness among clients, the DoL and statutory bodies of the practices occurring, and addressing risk issues to limit liability and project risk. Further research among the CHSM and CHSO on their perceptions and aspects relating to issues is needed to identify if similar issues are prevalent.

References

1. Deacon, C.H.: The effect of the integration of design, procurement, and the effect of the integration of design, procurement, and construction relative to health and safety. Ph.D. thesis, Nelson Mandela Metropolitan University (2016). https://doi.org/10.13140/RG.2.2.21197.74726
2. Republic of South Africa: Construction Regulations, 2014: Guidelines. Government Printer, Pretoria, South Africa (2017)
3. Republic of South Africa: Construction Regulations. Government Printer (2014)
4. Republic of South Africa: CR_2014_Exemption_2_aug 14 re CHSA CHSO reg.pdf
5. Republic of South Africa: Notice regarding Application for Construction Work Permit of Construction Work in terms of the Construction Regulations 2014. Government Printer, Pretoria (2015)
6. SACPCMP: Registration Rules for Construction Health and Safety Agent (PrCHSA) (2013)
7. Smallwood, J., Deacon, C.: Construction ergonomics: construction health and safety agents' (CHSAs') perceptions and practices. In: Advances in Intelligent Systems and Computing, pp. 477–485 (2019). https://doi.org/10.1007/978-3-319-96068-5_54
8. Smallwood, J., Deacon, C.: The performance of construction health and safety agents. In: ISEC 2017 - 9th International Structural Engineering and Construction Conference: Resilient Structures and Sustainable Construction (2017)

Author Index

C. Aigbavboa and W. Thwala (Eds.): CIDB 2019, *The Construction Industry in the Fourth Industrial Revolution*, pp. 607–608, 2020.
https://doi.org/10.1007/978-3-030-26528-1